SOLID STATE PHYSICS LITERATURE GUIDES
Volume 2

# SEMICONDUCTORS
## Preparation, Crystal Growth, and Selected Properties

# Solid State Physics Literature Guides

Prepared under the auspices of the Research Materials Information Center,
Oak Ridge National Laboratory

## General Editor: T. F. Connolly

*Solid State Division*
*Oak Ridge National Laboratory\**
*Oak Ridge, Tennessee*

Volume 1: Ferroelectric Materials and Ferroelectricity—1970

Volume 2: Semiconductors—Preparation, Crystal Growth, and Selected Properties—1972

Volume 3: Groups IV, V, and VI Transition Metals and Compounds—Preparation and Properties—1972

Volume 4: Electrical Properties of Solids—Surface Preparation and Methods of Measurement—1972

\*Oak Ridge National Laboratory is operated by Union Carbide Corporation for the U.S. Atomic Energy Commission.

# SOLID STATE PHYSICS LITERATURE GUIDES
## Volume 2

# SEMICONDUCTORS
## Preparation, Crystal Growth, and Selected Properties

### Edited by
### T. F. Connolly

*Research Materials Information Center*
*Solid State Division*
*Oak Ridge National Laboratory*
*Oak Ridge, Tennessee*

**IFI/PLENUM • NEW YORK-WASHINGTON-LONDON • 1972**

Library of Congress Catalog Card Number 74-133269

ISBN 978-1-4684-6203-6        ISBN 978-1-4684-6201-2 (eBook)

DOI 10.1007/978-1-4684-6201-2

© 1972 IFI/Plenum Data Corporation

Softcover reprint of the hardcover 1st edition 1972

A Subsidiary of Plenum Publishing Corporation

227 West 17th Street, New York, N.Y. 10011

United Kingdom edition published by Plenum Press, London

A Division of Plenum Publishing Company, Ltd.

Davis House (4th Floor), 8 Scrubs Lane, Harlesden, London, NW10 6SE, England

# Introduction

The material in this collection is based mainly on papers actually received by the Research Materials Information Center, although some references are included on specific recommendations. While this might exclude a few relevant papers, it also excludes a much larger number of nonpertinent references that might be chosen on the basis of deceptive titles or inadequate abstracts.

For any collection of this sort the question of what should or should not be included often involves individual bias or differences of opinion of the meaning of terms, and the criticisms of two experts in the same field are often contradictory. For this reason most compilations in this series, in addition to organization under appropriate subject headings, contain one or more sections entitled "Reviews, Bibliographies, and Compilations," in which references to peripheral related subjects are deliberately included. In all other sections the effort is to be as specific as possible, with borderline references kept to a minimum.

At this writing, there are over 70,000 searchable references in the RMIC collection on solid-state inorganic materials science, and the coverage of the field is good back to 1960, although many earlier references are included. Still, there will be omissions and errors in compilations drawn from the collection, and any pointed out to us will be corrected in future editions. (Such corrections should be sent to the Center and not to the publisher.)

The timeliness of these compilations, as well as our ability to answer daily inquiries, depends very largely on the continued receipt by the Center of all papers, reprints, reports, and preprints within our scope. These should be mailed to

T. F. Connolly
Research Materials Information Center
Oak Ridge National Laboratory
P. O. Box X
Oak Ridge, Tennessee 37830

# Introduction

# Preface

Since 1963 the Research Materials Information Center (RMIC) has been answering inquiries on the availability, preparation, and properties of ultrapure inorganic research specimens. It has been possible to do this with reasonable efficiency by searching an automated, coded microfilm collection of the report and open literature and of data sheets and questionnaires provided by commercial and research producers of pure materials.

With the growth of the collection to over 70,000 documents and the increase in the demand for more general background information, it has been necessary to compile bibliographies on an increasing variety of subjects. These have been used as indexes to the microfilmed documents for more efficient searching, and in the past distributed in response to individual requests. However, their size and number will no longer permit so casual and uneconomic a method of distribution. The "ORNL Solid State Physics Literature Guides" is a practical alternative.

This second in the series, a bibliography on the preparation and crystal growth of semiconductors, is arranged, as the Table of Contents indicates, according to the chemical groups of the compounds. In those cases where the class represents a new or revived interest, or where the literature is sparse, references to properties of the compounds are indexed.

Two recently developed areas of special interest are covered separately (in Sections 17 and 18) from the standard listing by compounds. These are amorphous semiconductors and ion-implantation doping of semiconductors. Since the RMIC does not cover organic compounds in any detail, the section on organic semiconductors is far from comprehensive, but the books and other reviews listed are well referenced and should be an adequate introduction to the literature.

Materials whose semiconducting properties are of less immediate practical interest, such as the transition-metal oxides, are not included (they will be covered in a bibliography on the Group IV-V-VI transition metals and selected compounds, to be published as a volume in this series).

The "General" subsections include, in addition to bibliographies, compilations, and surveys, those papers that could not be placed in the more precisely defined categories — usually because they refer to more than one of the materials covered in the section.

**Availability of Documents**

U. S. Government contractor reports, usually identified by an alpha-numeric report number, can be purchased from

National Technical Information Service
U. S. Department of Commerce
Springfield, Virginia 22151

and often on request from the issuing installation.

USAEC reports are also available from

> International Atomic Energy Agency
> Kaerntnerring A 1010
> Vienna, Austria

> National Lending Library
> Boston Spa, England

Monographs and reports of the National Bureau of Standards are for sale by

> Superintendent of Documents
> U. S. Government Printing Office
> Washington, D. C. 20402

Theses, listed as Dissertation Abstracts + number, are available in North and South America from

> University Microfilms
> Dissertation Copies
> P. O. Box 1764
> Ann Arbor, Michigan 48106

and elsewhere from

> University Microfilms, Ltd.
> St. John's Road
> Tylers Green
> Penn, Buckinghamshire
> England

# Contents

# Contents

# Addendum

The following papers were received too late to be included in the main body of the bibliography.

## Preparation, Crystal Growth

Progress in the science and technology of monocrystals
W. C. Ellis
Advances in Materials Research, Vol. 5, John Wiley and Sons, Inc., New York (1971), pp. 315–376
Crystal growth, semiconductors, vapor deposition, zone melting

Review: Mechanisms of crystal growth from fluxed melts
D. Elwell and B. W. Neate
J. Mater. Sci. 6:1499–1519 (Dec. 1971)
Basic mechanisms and origin of defects; 80 refs.

Review of the preparation of single crystals by fused melt electrolysis and some general properties
W. Kunnmann
Preparation and Properties of Solid State Materials, Vol. 1: Aspects of Crystal Growth (R. A. Lefever, ed.) Marcel Dekker, Inc., New York (1971), pp. 1–36

Physical processes in epitaxial growth
B. Lewis
Thin Solid Films 7:179–217 (1971)

Role of mass transfer in crystallization processes
W. R. Wilcox
Preparation and Properties of Solid State Materials, Vol. 1: Aspects of Crystal Growth (R. A. Lefever, ed.) Marcel Dekker, Inc., New York (1971), pp. 37–182

Chemical Vapor Deposition
John M. Blocher, Jr. and James C. Withers, eds.
Electrochemical Society, Inc., New York 10017, 861 pp.

Crystal growth mechanisms: Energetics, kinetics and transport
R. L. Parker
Solid State Physics, Vol. 25 (H. Ehrenreich, F. Seitz, and D. Turnbull, eds.), Academic Press, New York and London (1970)

Vapor-phase growth of several III-V compound semiconductors
J. J. Tietjen, R. E. Enstrom, and D. Richman
RCA Rev. 31:635–646 (1970)
Review; 46 refs.

Epitaxic films of lead chalcogenides and related compounds
J. N. Zemel
Solid State Surface Sci. 1:291–403 (1969)
167 refs.

## General, Reviews

The Physics of Semimetals and Narrow Gap Semiconductors
D. L. Carter and R. T. Bate, eds.
Proceedings of 1970 Conference held in Dallas, Texas, Pergamon (1971), 594 pp.

Defect Electronics in Semiconductors
Herbert F. Matare
Wiley-Interscience, New York (1971), 640 pp.

Handbook of Electronic Materials, Vol. 2: III-V Semiconducting Compounds
M. Neuberger, ed.
IFI/Plenum, New York (1971), 115 pp.
Extensive tabulation of properties with references for each data point

Handbook of Electronic Materials, Vol. 5: Group IV Semiconducting Materials
M. Neuberger, ed.
IFI/Plenum, New York (1971), 67 pp.

Crystal Chemistry and Semiconduction in Transition Metal Binary Compounds
J. P. Suchet
Academic Press, New York (1971)

Semiconductors
Helmut F. Wolf
Wiley-Interscience, New York (1971), 544 pp.
Summarizes the theoretical aspects of semiconductor behavior. Includes a selection of experimental and other data for the design and operation of semiconductor devices

Diffusion in semiconductors (excepting oxides)
Diffusion Data 4:360–369 (1970)
(Data compilation)

Ion Implantation
Proceedings of the European Conference, Reading, England,
September 7–9, 1970
CONF-700945; Peter Peregrinus Ltd., Stevenage, England
(1970), 241 pp.

First International Topical Conference on Ion
Implantation in Semiconductors, Thousand Oaks,
California, May 4–7, 1970
Lewis T. Chadderton, ed.
CONF-700544
Radiation Effects 6:1–318 (1970)

Non-radiative transitions in semiconductors
P. T. Landsberg
Phys. Stat. Sol. 41(2):457–489 (1970)
Critical review of theory

Handbook of Thin Film Technology
L. I. Maissel and Reinhard Glang
McGraw-Hill, New York (1970), 1200 pp.

Effects of uniaxial stress on the optical properties of semiconductors
F. H. Pollak
Proceedings of the Tenth International Conference on the
Physics of Semiconductors, Cambridge, Massachusetts,
August 17–21, 1970, pp. 407–417; CONF-700801 (August
1970)
Uniaxial stress measurements have yielded considerable information concerning location of conduction and valence
band extrema, symmetries of interband optical transitions,
electronic band deformation potentials, effective masses,
excitons, and lattice dynamical properties.

Comments on electronic transport in magnetic
semiconductors
S. von Molnar and T. Kasuya
Proceedings of the Tenth International Conference on the
Physics of Semiconductors, Cambridge, Massachusetts,
August 17–21, 1970, pp. 233–242; CONF-700801 (1970)

## Amorphous Semiconductors

Amorphous Semiconductors
David Adler
CRC Critical Revs. in Solid State Sci. 2(3):317–465 (1971)
Amorphous semiconductors are classified and their electronic structure of disordered materials is reviewed.
Experimental results on elemental, covalent, and ionic
amorphous solids are summarized, and interpretations are
presented in terms of recently proposed band models. A
brief review of the several types of switching phenomena
that characterize amorphous semiconductors is also
given. 1007 refs.

Kohn's variational method for amorphous con-
ductors
A. I. Gubanov
Phys. Stat. Sol. 47B:329–336 (1971)

Vaporization associated with surface electrical
switching in semiconducting As-Te-I and As-Te-
Ge glasses
R. T. Johnson, Jr., D. A. Northrop, and R. K. Quinn
Solid State Commun. 9:1397–1401 (1971)
Mass loss raises question of long-term device stability;
toxic vapors

A critical evaluation of a thermal method of
switching
P. W. McMillan and P. Nesvadba
J. Phys. D: Appl. Phys. 4:1401–1410 (Sept. 1971)

Electronic properties of an amorphous solid.
II. Further aspects of the theory
M. F. Thorpe and D. Weaire
Phys. Rev. 4B:3518–3527 (1971)

Basic concepts in the theory of amorphous
semiconductors
M. H. Cohen
Proceedings of the Tenth International Conference on the
Physics of Semiconductors, Cambridge, Massachusetts,
August 17–21, 1970, pp. 645–652; CONF-700801 (1970)

Amorphous and Liquid Semiconductors
Proceedings of the Third International Conference held in
Cambridge, England, September 1969
N. F. Mott, ed.
North–Holland Publishing Co., Amsterdam (1970), 638 pp.

Electronic properties of amorphous semicon-
ductors
J. Stuke
Proceedings of the Tenth International Conference on the
Physics of Semiconductors, Cambridge, Massachusetts,
August 17–21, 1970, pp. 14–22; CONF-700801 (1970)
Review on the change of the optical and electrical properties
at the transition from the crystalline to the amorphous
state

# 1. Information Centers and Other Services

Electronic Materials Information Center
   Royal Radar Establishment, St. Andrews Road, Gt. Malvern, Worcestershire, England

Electronic Properties Information Center
   Hughes Aircraft Company, Culver City, Calif. 90230, telephone 213-391-0177, extension 6596
   Collects, indexes, and abstracts the literature on the electrical and electronic properties of materials and evaluates and compiles the experimental data from that literature. The Center publishes data sheets, summary reports, thesauri, glossaries, bibliographies, and similar publications as sufficient information is evaluated and compiled.

Isotopes Information Center
   Oak Ridge National Laboratory, P. O. Box X, Oak Ridge, Tenn. 37830, telephone 615-483-8611, extension 3-1742; Federal Telecommunication System No. 615-483-1742
   Collects, evaluates, and disseminates worldwide information primarily on production and uses of radioisotopes in industry and research.

Radiation Effects Information Center
   Battelle Memorial Institute, 505 King Avenue, Columbus, Ohio 43201, telephone 614-299-3151, extension 2553
   Answers inquiries concerning technical questions, current research and development projects, and technical data or data compilations; provides engineering and reference services; publishes and disseminates state-of-the-art reviews; and conducts literature searches and prepares bibliographies and compilations on radiation effects on materials and devices, including semiconductors.

Rare-Earth Information Center
   Institute for Atomic Research, Iowa State University, Ames, Iowa 50010, telephone 515-294-2272
   Analytical, inorganic, and physical chemistry of the rare-earth elements and compounds; solid state physics and metallurgy. Publishes reviews, compilations, and bibliographies. Inquiry service by mail and telephone.

Research Materials Information Center
   Oak Ridge National Laboratory, P. O. Box X, Oak Ridge, Tenn. 37830, Telephone 615-483-8611, extension 3-1287; Federal Telecommunication System No. 615-483-1287
   Provides references on availability, purification, crystal growth, and properties of all ultrapure inorganic research specimens. Inquiry service by mail and telephone.

Cambridge Communications Corporation
   238 Main Street, Cambridge, Mass. 02142
   Publishes abstracts on preparation and properties of high-purity inorganic compounds (Solid State Abstracts).

Centre de Documentation sur les Synthèses Cristallines
   Laboratoire de Physique Moléculaire et Cristalline, Faculté des Sciences, Place Eugene-Bataillon, 34-Montpellier, France
   Identifies European crystal growers by country, installation, and material. Updated loose-leaf sheets.

Institute of Electrical and Electronics Engineers
   345 East 47th Street, New York, N. Y. 10012
   Electrical and Electronics Abstracts; also Current Papers in Electrical and Electronics Engineering.

Pierre de la Breteque, Directeur de Recherches, Société Française pour l'Industrie de l'Aluminium
   134 Chemin des Aygalades, 13 Marseille 15e, France
   Publishes annual bibliographies on gallium and compounds (including III−V, III−VI, and ternaries) covering preparation and physical properties, containing phase diagrams of new compounds.

The Selenium−Tellurium Information Service, Selenium-Tellurium Development Association, Inc.
   345 East 47th Street, New York, N. Y. 10017, telephone 212-688-2632
   Publishes periodic selenium and tellurium bibliography; inquiry service.

Joint Program on Methods of Measurement for Semiconductor Materials, Process Control and Devices Undertaken (in 1968)
   National Bureau of Standards, Electronic Technology Division, Wash., D. C.
   J. C. French is coordinator of the program; NBS-TN-472 (Dec. 1968), NBS-TN-475 (Feb. 1969), NBS-TN-488 (July 1969), NBS-TN-495 (Sept. 1969), and NBS-TN-520 (March 1970) are recent reports.

G. V. Planer Ltd.
   Windmill Road, Sunbury-on-Thames, Middlesex, England
   Monthly survey on developments in microcircuits, electrical thin/ thick films and semiconductors in Europe and the USA. Subscription service.

# 2. Journals

Archiv für Elektrotechnik
  Springer-Verlag, 175 Fifth Avenue, New York, N. Y. 10010 or
    Heidelberger Plaza 3, 1000 Berlin 33, West Germany

Archiv für technisches Messen und industrielle
Messtechnik
  R. Oldenbourg Verlag
  Rosenheimerstr. 145, Munich 8, Germany

Bell Laboratories Record
  Bell Telephone Laboratories, Inc., Murray Hill, N. J. 07974

Bell System Technical Journal
  American Telephone and Telegraph Co., 195 Broadway, New
    York, N. Y. 1007

Bulletin of the Electrotechnical Laboratory
(Tokyo)
  Ohm Sha, 1, Nishiki-cho, 3-chome, Kanda, Chiyoda-ku, Tokyo,
    Japan

Doklady Physical Chemistry
  Consultants Bureau, 227 W. 17th Street, New York, N. Y.
    10011

Ferroelectrics
  Gordon and Breach, Science Publishers, 150 Fifth Avenue,
    New York, N. Y. 10011; or 12 Bloomsbury Way, London
    W. C. 1, England
  Journal will publish original research papers on such subjects
    as theories of ferroelectricity, crystal growth, ceramic
    fabrication, and the structural, electrical, optical, and
    mechanical properties of these materials and combina-
    tions thereof

IEEE Transactions Electron Devices
  Institute of Electronic and Electrical Engineers, 345 East
    47th Street, New York, N. Y. 10017

IEEE Transactions Instrumentation and
Measurement
  Institute of Electronic and Electrical Engineers

Inorganic Materials
  Consultants Bureau, 227 West 17th Street, New York, N. Y.
    10011

International Journal of Electronics
  Taylor and Francis Ltd., 18 Red Lion Court, Fleet Street,
    London E. C. 4, England

Journal of the American Ceramic Society
  4055 North High Street, Columbus, Ohio 43214

Journal of Applied Physics
  American Institute of Physics, 335 East 45th Street, New
    York, N. Y. 10017

Journal of Crystal Growth
  North Holland Publishing Co., P. O. Box 103, Amsterdam,
    Netherlands

Journal of the Electrochemical Society
  30 East 42nd Street, New York, N. Y. 10017

Journal of Materials Science
  Chapman and Hall, 11 New Fetter Lane, London E. C. 4,
    England

Journal of the Physics and Chemistry of Solids
  Pergamon Press Ltd., Headington Hill Hall, Oxford, England

Journal of Solid State Chemistry
  Academic Press Inc., 111 Fifth Avenue, New York, N. Y.
    10003

Japanese Journal of Applied Physics
  Physical Society of Japan, Room 211, Kikai Shinko Building,
    21 Shiba Koen, Minato-ko, Tokyo, Japan

Kristall und Technik
  Akademie-Verlag GmbH, Leipziger Str. 3-4, 108 Berlin, East
    Germany

Materials Research Bulletin
  Pergamon Press, Inc., Maxwell House, Fairview Park, Elms-
    ford, New York. 10523

Physik der Kondensierten Materie
  Springer-Verlag, Postfach 1780, 69 Heidelberg 1, Germany

Radio Engineering and Electronic Physics (translation of Radiotekhnika i Elektronika, USSR)
  Institute of Electronic and Electrical Engineers, 345 East
    47th Street, New York, N. Y. 10017

Solid State Communications
  Pergamon Press, Inc., Maxwell House, Fairview Park, Elms-
    ford, N. Y. 10523

Solid State Electronics
  Pergamon Press, Inc., Maxwell House, Fairview Park,
    Elmsford, N. Y. 10523

**Journals**

Solid State Technology
  Cowan Publishing Corp., 14 Vanderventer Avenue, Port
  Washington, N Y. 11050

Sov. Physics — Crystallography
  American Institute of Physics, 335 East 45th Street, New York,
  N. Y. 10017

Sov. Physics — Semiconductors
  American Institute of Physics, 335 East 45th Street,
  New York, N. Y. 10017

Sov. Physics — Solid State
  American Institute of Physics, 335 East 45th Street,
  New York, N. Y. 10017

# 3. Methods of Crystal Growth — Books and Reviews

Crystal Growth and Epitaxy from the Vapour Phase (Proceedings of the First International Conference on Crystal Growth and Epitaxy from the Vapour Phase, Zurich, Switzerland, 23–26 September 1970)
E. Kaldis and M. Schieber, eds.,
J. Crystal Growth, 9:1–371 (May 1971)
Materials index pp. 368–371; subject index pp. 365–367

Preparation and Properties of Solid State Materials, Vol. 1: Aspects of Crystal Growth
R. A. Lefever, ed.
Marcel Dekker, Inc., New York (1971), 304 pp.

Hydrothermal Synthesis of Crystals
A. N. Lobachev
Plenum Press, New York (1971), 165 pp.

Isothermal substitutional growth of single crystals
W. Albers and J. Verberkt
Philips Res. Rept., 25:17–20 (1970)
SnS–SnSe and GaAs–GaP

Crystal growth from the melt
G. A. Chadwick
J. Sheffield Univ. Met. Soc., 9:15–22 (1970)
Review; single-phase materials

Croissance des cristaux mixtes statistique des essais et des erreurs
A. A. Chernov
Bull. Soc. Fr. Mineral. Cristallogr., 92(6):528–539 (1969–1970)

Zone Refining: A DDC Bibliography, May 1961–April 1969, Vol. 1
Defense Documentation Center, DDC-TAS-70-72-1-Vol-1; AD-712000 (Aug. 1970), 147 pp.
Purification and crystal growth

Solution growth of electronic compounds and their solid solutions
N. Hemmat, C. B. Lamport, A. A. Menna, and G. A. Wolff
Materials Engineering and Sciences Division Biennial Conference (1970), pp. 112–121

Crystal Growth in Gels
Heinz K. Henisch
The Pennsylvania State University Press, University Park and London (1970)

Investigation of nucleation sites in gels
H. K. Henisch
Pennsylvania State University, University Park, Pa., AFCRL-71-0039; AD-718 984 (Dec. 21, 1970), 25 pp.

The Growth of Single Crystals
R. A. Laudise
Solid State Physics Electronics Series (Nick Holonyak, Jr., ed.), Prentice-Hall International, Inc., London; Prentice-Hall of Australia, Pty. Ltd., Sydney (1970)

Obtention de monocristaux au sein d'un flux fondu
Y. Laurent
Rev. Chim. Miner., Fr., 6(6):1145–1186 (1969)
278 refs.

An annotated bibliography for growth of multi-element single crystals of metals and metalloids
M. J. Murtha and George Burnet
IS-2184 (Feb. 1970), 29 pp.
Includes some semiconductors

Kristallzucht aus der Gasphase
R. Nitsche
Angew. Chem., 82(1):48 (1970)

Recent developments in the study of epitaxy
D. W. Pashley
Recent Progress in Surface Science, Vol. 3 (J. F. Danielli, A. C. Riddiford, and M. D. Rosenberg, eds.) Academic Press, New York and London (1970), pp. 23–69

Controlled growth and dissolution of semiconductor crystals
E. Sirtl and T. F. Ciszek
Materials Engineering and Sciences Division Biennial Conference (1970), pp. 104–111

Handbook of Solid – Liquid Equilibria in Systems of Anhydrous Inorganic Salts, Vols. 1 and 2
N. K. Voskresenskaya, ed.
TT-55079-1; AEC-tr-6983-1 (Israel Program of Scientific Translations, Jerusalem, 1970), 869 pp.

Conference on Growth and Characterization of Electronic Materials, Chania, Greece, June 1–12, 1969
Gordon and Breach Publishing Co., 1970

Factors affecting crystal growth in molten solvents
R. W. Bartlett
High Temperature Technology (Proceedings of the Third International Symposium on High Temperature Technology held at Asilomar in Pacific Grove, California, 17–20 September, 1967) Butterworths, London (1969), pp. 173–189

Crystallization. Part II. Crystallization processes
G. D. Botsaris, E. G. Denk, G. S. Ersan, D. J. Kirwan, et al.
Ind. Eng. Chem., 61:92–101 (1969)
(Covers Spring 1967 to Spring 1969); Part I(Oct. Ind. Eng. Chem., 1969) and Part III (Dec. Ind. Eng. Chem., 1969)
Annual review

Rates of growth of crystals from solutions
Cheng T. Cheng
Thesis UCRL–19518 (Dec. 1969)

Glow discharge beam techniques
R. A. Dugdale, J. T. Maskrey, S. D. Ford, P. R. Harmer, and R. E. Lee
J. Mater. Sci., 4:323–335 (1969)
Design principles and applications in crystal growing, vapour deposition

Structure, Properties and Preparation of Perovskite-Type Compounds
F. S. Galasso, ed.
Pergamon Publishing Co., New York (1969), 218 pp.
Approximately 500 perovskite-type compounds, including the various methods of preparing powders, thin films, and single crystals

Kinetics of Reactions in Ionic Systems
Materials Science Research, Vol. 4
Proceedings of International Symposium of Special Topics in Ceramics, June 18–23, 1967
T. J. Gray and V. D. Frechette, eds.
Plenum Press, New York (1969)

On dendritic growth
Ernest Gunther Holzmann
Ph. D. thesis, Stanford University, California (1969), 339 pp.
University Microfilms, Inc., Ann Arbor, Michigan, Order No. 70–1548

Electrocristallisation des Métaux dans un Champ d'Ultrasons
A. P. Kapustin and A. N. Trofimov
Izdat. Nauka, Moscow (1969), 72 pp.

Crystallization from Solutions
E. V. Khamskii
Plenum Press, New York (1969), 110 pp.

Synthese und Kristallchemie anorganischer Stoffe bei hohen Drucken und Temperaturen
W. Kleber and K.-T. Wilke
Krist. Tech., 4(2):165–199 (1969)
Metals, semimetals, III–V, IV–V, chalcogenides; review

Hydrothermalsynthese und Struktur von Oxyden und Hydroxyden
A. Norlund Christensen
Rev. Chim. Min., 6:1187–1200 (1969)

Growing Crystals from Solution
T. G. Petrov, E. B. Treivus, and A. P. Kasatkin
Consultants Bureau, New York (1969), 106 pp.

Arc Techniques for Material Preparation and Czochralski Crystal Growth
T. B. Reed
High Temperature Technology (Proceedings of the Third International Symposium on High Temperature Technology held at Asilomar in Pacific Grove, California, 17–20 September, 1967) Butterworths, London (1969), pp. 655–664

Growth of Crystals, Vols. 1–8
N. N. Sheftal', ed.
Consultants Bureau, Plenum Press, New York (1958–1969)
A continuing series on crystal growth represents the work of Soviet scientists for the most part originally presented at the All-Union Conferences on Crystal Growth held at the Institute of Crystallography in Moscow

Semiconductor crystal manufacture
M. Sittig
Electronics Materials Review, No. 3, Noyes Dev. Corp., New Jersey (1969), 335 pp.

Possibility of using containers of refractory oxygen-free compounds for producing semiconductor single crystals
Yu. M. Smirnov, A. A. Mashnitskii, and V. A. Kuznetsov
Izv. Akad. Nauk SSSR, Ser. Fiz., 33:2005–2007 (1969)

Crystallization from the melt
R. T. Southin and G. A. Chadwick
Sci. Prog., 57:353 (1969)

Croissance de Composés Minéraux Monocristallins
J. P. Suchet, ed.
Masson et Cie., Paris (1969), 169 pp.

Contribution à l'étude des aluminates, manganites et ferrites de terres rares et d'yttrium
Gyorgy Szabo
Ph. D. thesis, Université de Lyon, France (1969)

Some factors influencing the growth of crystals in gel by the complex dilution method
A. F. Armington and J. J. O'Connor
Mat. Res. Bull., 3:923–932 (1968)

Status of crystal growth in the United States
D. S. Billington and G. C. Danielson
ORNL-RMIC-10 (Dec. 1968), 11 pp.
A survey of the demand for single crystals and the effort and training devoted to crystal growth

Techniques of Metals Research, Vol. 1, Part 2. Techniques of Materials Preparation and Handling
R. F. Bunshah, ed.
Interscience Publishers, New York (1968), 799 pp.
Crystal growth from the liquid phase, J. W. Rutter, pp. 923–990;
Crystal growth from the solid state, K. T. Aust, pp. 991–1021;
Crystal growth from the vapor phase, W. C. Ellis, pp. 1023–1068;
Growth of low-dislocation-density single crystals, F. W. Young, Jr., pp. 1133–1142;
Growth of single crystals in systems involving a solid-solid phase transformation, D. G. Westlake, pp. 1145–1153

Proceedings of the Soviet Conference on Crystal Growth (3rd)
Shou-Ch'Ing Chang
FTD-HT-23-1007-67; AD-682765 (July 19, 1968), 7 pp.
Basic theories and methods for crystal growth, testing, and determination

Fused salt mixtures: eutectic compositions and melting points
  P. V. Clark
    Supersedes and includes: SC-65-930, Vols. 1 and 2, Physical properties of fused salt mixtures: eutectic compositions and melting points, SC-R-68-1680 (Dec. 1968)
    Bibliography 1907-1968; data on eutectic compositions and melting points of fused salt mixtures are presented in two complementary tables arranged by systems and by melting points. Over 5,000 references are given through Dec. 30, 1968.

Purity and perfection of research specimens of oxides
  J. W. Cleland
    Mass Transport in Oxides (J. B. Wachtman, Jr., and A. D. Franklin, eds.), pp. 195-203, National Bureau of Standards, Wash., D. C. (1968)
    NBS-ARPA Symposium on Mass Transport in Oxides, Gaithersburg, Md. See NBS-Spec. Publ. 296; CONF-671024

Developments in melt-grown oxide crystals
  B. Cockayne
    J. Crystal Growth, 3:60-70 (1968)

Crystal growth of organic solids and mechanism of single-crystal growth from melt
  S. C. Datt and J. K. D. Verma
    J. Sci. Ind. Res., India, 27(1):11-27 (1968)

Crystal Growth, 1968, Proceedings of the Second International Conference on Crystal Growth, Birmingham, U. K., July 15-19, 1968
  F. C. Frank, J. B. Mullin, and H. S. Peiser, eds.
    North-Holland Publishing Co., Amsterdam (1968)
    J. Crystal Growth, Vol. 3, No. 4 (1968)

Crystal growth in gels
  H. K. Henisch
    Helv. Phys. Acta, 41:888-897 (1968)
    Review; 31 refs.

Rare gas crystals
  J. Hingsammer and E. Luscher
    Helv. Phys. Acta, 41:914-935 (1968)
    Review, 98 refs.

Growth of single crystals of refractory oxides
  V. K. Yanovskii
    Zh. Vses. Khim. Obshchestva D. I. Mendeleeva, SSSR, 13(2):134-142 (1968)
    142 refs.

Preparative Inorganic Reactions
  William L. Jolly, ed.
    Interscience Publishers, John Wiley and Sons, New York, Vol. 1, 1964; Vol. 2, 1965; Vol. 3, 1967; Vol. 4, 1968; and Vol. 5, 1968
    Complements the Inorganic Syntheses Series

Epitaxial growth and growth defects
  B. D. Joyce
    Electronic Compon., 9:1389-1395 (1968)
    52 refs.

Polymer crystals
  A. Keller
    Reports on Progress in Physics, Vol. XXXI, Part II, C. I. Pedersen and R. A. Cook, eds., pp. 624-704, Inst. of Physics and Physical Society, London (1968)

An investigation of the crystal growth of the heavy metal sulfides in supercritical hydrogen sulfide
  L. C. Lewis
    University Microfilms, Inc., Ann Arbor, Mich., Order No. 68-11,915
    Ph.D. thesis, Oregon State University, 1968

Growth from metal solutions
  N. P. Luzhnaya
    J. Crystal Growth, 3:97-107 (1968)

Kristallzucht aus der Gasphase
  Rudolf Nitsche
    Festkörper Probleme VIII in Referaten des Fachausschusses "Halbleiter" der Deutschen Physikalischen Gesellschaft, Berlin (1968), 42 pp.
    Zugleich Hauptvortrage des Fachausschusses "Tiefe Temperaturen" des Fachausschusses "Thermodynamik und statische Mechanik" und der Arbeitsgemeinschaft "Metallphysik" (O. Madelung, ed.), Friedr. Vieweg und Sohn, Pergamon Press, Oxford

Growth and Imperfections of Metallic Crystals
  D. C. Ovsienko, ed.
    Consultants Bureau, New York (1968), 268 pp.
    Includes alloys

Crystallization, annual review
  J. A. Palermo and G. F. Bennett
    Ind. Eng. Chem., 60(4):65 (1968)

Zur Kristallchemie und Züchtung von Laserkristallen
  Kurt Recker
    Fortschr. Miner., 45:172-213 (1968)

High-temperature solution (flux) and high-pressure solution (hydrothermal) crystal growth
  Rustum Roy and William B. White
    J. Crystal Growth, 3:33-42 (1968)

Kinetics and Mechanisms of Crystallization
  R. F. Strickland-Constable
    Academic Press, London and New York (1968), 346 pp.

Fundamental Aspects of Electrocrystallization
  J. O'M. Bockris and G. A. Razumney, eds.
    Plenum Press, New York (1967)

Crystalline synthesis
  Centre de Documentation sur les Synthèses Cristallines de Montpellier, 34, France (Commissariat à l'Energie Atomique, Centre d'Etudes Nucléaires de Saclay, 1ère Série, Mars 1967, [111]-49 feuillets)

A survey of epitaxial growth processes and equipment
  V. Y. Doo and E. O. Ernst
    IBM-TR-22.431
    Semicond. Prod., 10:31 (1967)

Growth of metallic crystals by thermal decomposition of halides in the gas phase
  Jacques Gillardeau
    CEA Bibl. No. 93, Saclay Nuclear Research Center, ORNL-tr-1888 (Sept. 1967)

Thin alloy zone crystallization
  D. T. J. Hurle, J. B. Mullin, and E. R. Pike
    J. Mater. Sci., 2:46-62 (1967)
    Crystal growing via diffusion through a thin alloy zone from a third phase

Current concepts in crystal growth from the
melt
  K. A. Jackson
  Progr. Solid-State Chem., 4:53-80 (1967)

On the nature of crystal growth from the melt
  K. A. Jackson, D. R. Uhlmann, and J. D. Hunt
  J. Cryst. Growth, 1:1-36 (1967)

The Freezing of Supercooled Liquids
  Charles A. Knight
  D. Van Nostrand Co., Inc., Princeton, N. J. (1967)

Crystal Growth, Proceedings of International
Conference on Crystal Growth, Boston, June
20-24, 1966
  H. Steffen Peiser, ed.
  Pergamon Press, New York (1967)

Energetics in Metallurgical Phenomena, Vol. 2
  William M. Mueller, ed.
  Gordon and Breach Science Publishers (1966)
  Statistical mechanics of nucleation and crystal growth

Zone melting
  W. G. Pfann
  John Wiley and Sons, Inc., New York (1966)

Growth of single crystals from non-stoichiometric
melts
  J. B. Schroeder and R. C. Linares
  Progress in Ceramic Science, Vol. 4, pp. 195-216 (J. E.
    Burke, ed.), Pergamon Press, New York (1966)

Crystallization Processes
  N. N. Sirota, F. K. Gorskii, and V. M. Varikash, eds.
  Consultants Bureau, New York (1966)

Solid State Transformations
  N. N. Sirota, F. K. Gorskii, and V. M. Varikash, eds.
  Consultants Bureau, New York (1966)

The use of non-contaminating cold crucibles for
the evaporation of refractory metals
  H. F. Sterling
  Vide, 21:121-129 (1966)

Bibliography on crystal growing (annotated
references 1960 to 1965)
  E. A. D. White
  J. Mater. Sci., 1:199-210 (1966)

Synthetic spinel flux growth survey, Appendix E
of ASE-1100
  American Science and Engineering, Inc., 11 Carleton Street,
    Cambridge, Mass. 02142, September 24, 1965

The Growth of Crystals from the Melt
  J. C. Brice
  Interscience Publishers, John Wiley and Sons, New York
    (1965)

Croissance des cristaux
  Stanislas Goldsztaub
  CEA-S4 (1965)

Crystal growth in hydrothermal systems
  J. W. Moody and R. C. Himes
  Battelle Tech. Rev., 14:3 (1965)

Condensation and Evaporation of Solids
  Emile Rutner, Paul Goldfinger, and John P. Hirth, eds.
  Gordon and Breach Science Publishers, London (1965)

Melting and Crystal Structure
  A. R. Ubbelohde
  Clarendon Press, Oxford, England (1965)

Recent advances in crystal-growing techniques
  E. A. D. White
  Brit. J. Appl. Phys., 16:1415 (1965)

The growth of oxide crystals from the fluxed
melt
  E. A. D. White
  (Hans B. Jonassen and Arnold Weissberger, eds.)
  Technique of Inorganic Chemistry, Vol. 4, p. 31, Inter-
    science Publishers, Inc., New York (1965)

Crystal Physics
  G. S. Zhdanov
  (Translated and ed. by A. F. Brown), Oliver and Boyd, Ltd.,
    Edinburgh and London (1965)

Principles of Solidification
  B. Chalmers
  John Wiley and Sons, Inc., New York (1964)

Chemical Transport Reactions
  Harald Schafer
  Academic Press, New York and London (1964)

Crystal growth techniques
  E.A.D. White
  GEC Journal, 31:43-53 (1964)

The Art and Science of Growing Crystals
  J. J. Gilman, ed.
  John Wiley and Sons, Inc., New York, London (1963), 493 pp.

Preparation and crystal synthesis of magnetic
oxides
  C. J. Kriessman and N. Goldberg
  Magnetism, Vol. III. Spin Arrangements and Crystal Struc-
    ture, Domains, and Micromagnetics (George T. Rado and
    Harry Suhl, eds.), Academic Press, New York (1963),
    553 pp.

Methods of Crystal Growth
  K. T. Wilke, ed.
  Deutscher Verlag der Wissenschaften, Germany (1963),
    612 pp.

Mechanism of growth of metal single crystals
from the melt
  D. T. J. Hurle
  Progr. Mater. Sci., 10:79-147 (1962)

Hydrothermal crystal growth
  R. A. Laudise and J. W. Nielsen
  Solid State Physics, Vol. XXI (F. Seitz and D. Turnbull, eds.),
    Academic Press, New York (1961), pp. 149-222

Reference handbook on the fusibility of systems
consisting of anhydrous inorganic salts, preface
from Vols. 1 and 2
  N. K. Voskresesskaya, N. N. Evseeva, S. I. Barul, and I. P.
    Vereshchetina
  Spravochnik po Plavkosti Sistem iz Bezvodnykh Neorganickes-
    kikh Solei, Izdatel'stvo Akademii Nauk SSSR, Moscow-
    Leningrad (1961), Vol. 1, pp. 1-7

Purification of silicon and the production of
homogeneous single (doped) crystals
  J. Goorissen and B. Okkerse
  Acta Electron., 4:379-391 (1960)
  By zone melting and pulling, review, 123 refs.

Crystals and Crystal Growing
  A. Holden and P. Singer
  Doubleday and Co., Inc., Garden City, New York (1960)

Preparation of Single Crystals
  W. D. Lawson and S. Nielsen
  Butterworths Publ. and Academic Press, London (1958)

# 4. Semiconductors — General, Reviews, and Bibliographies

Sources of information on ultrapurification and characterization
   T. F. Connolly
   Fractional Solidification (Morris Zief, ed.), Vol. III, Techniques of Ultrapurity, Marcel Dekker, New York (1972)
   Lists national and international information services, information centers, books, journals, reviews, bibliographies, and data compilations dealing with ultrapurification, activation analysis, neutron, x-ray and electron diffraction, mass and optical spectrometry, electron microprobe analysis, measurements of electrical properties, thermal analysis, resonance methods, the Mössbauer effect, and field and ion emission microscopy-all with reference to the production and characterization of ultrapure inorganic solid research specimens.

On the variation of the energy gap in substitutional semiconductor alloys
   A. Congiu
   Phys. Stat. Sol., 5A:131–35 (1971)
   II-V and III-V

Transport and free carrier electromagnetic phenomena in semiconductor boundary layers
   F. Flores, F. Garcia-Moliner, and G. Navascues
   Surface Sci., 24:61–76 (1971)
   Standard semiconductor surface transport theory rewritten in a classical Green function formulation

Electronic Properties of Materials, A Guide to the Literature, Vol. 3 (in two parts)
   D. L. Grigsby, ed.
   Plenum Publishing Corp., New York (1971)
   Part 1:1161 pp.; Part 2:756 pp.
   Semiconductors, insulator, ferroelectric dielectrics, metals, ferrites, ferromagnetics, electroluminescent materials, thermionic emitters, and superconductors

Photoconductivity and Photoconductive Materials, A Reference Guide
   John T. Milek
   (Electronic Properties Information Center, Hughes Aircraft Co., Culver City, California)
   March 1971

Chemical Vapor Phase Deposition of Electronic Materials

RCA Review, Vol. 31, No. 4 (Dec. 1970)
   Entire issue:13 papers

Physics of p-n Junctions and Semiconductor Devices
   S. M. Ryukin and Yu. V. Shmartsev, eds.
   Izdatel'stvo "Nauka," Leningrad (1969), 351 pp.
   Consultants Bureau, New York (1971)

Thermal conductivity of salt, metal, and semiconducting melts
   A. R. Regel, I. A. Smirnov, and E. V. Shadrichev
   Phys. Stat. Sol., 5A:13–57 (1971)
   242 refs.

Crystal Chemistry and Semiconduction in Transition Metal Binary Compounds
   J. P. Suchet
   Academic Press, New York (1971)

Surface Properties of Semiconductors and Dynamics of Ionic Crystals
   D. V. Skobel'tsyn, ed. (translated from Russian by Albin Tybulewicz)
   Plenum Press, New York (1971), 160 pp.

Methods for determining the distribution of imperfections in semiconductors. 2.
   H. Ahlers
   Nachrichtentechnik, 20:9–19 (Jan. 1970)
   28 refs.

Phase Diagrams Materials Science and Technology, Vol. III: The Use of Phase Diagrams in Electronic Materials and Glass Technology
   Allen M. Alper, ed.
   Academic Press, New York and London (1970)

Conduction in thin semiconductor films
   J. C. Anderson
   Advan. Phys., 19(79):311–338 (1970)

Melting enthalpies of III-V, IV-V, and V-VI compounds
   R. Blachnik and A. Schneider
   Z. Anorg. Allgem. Chem., 372(3):314–324 (1970)
   $GaSb$, $InBi$, $Bi_2Tl$, $SnAs$, $GeAs$, $GeAs_2$, $Bi_2Se_3$, $Sb_2Se_3$, $As_2Te_3$, and $As_2Se_3$

Infrared techniques for semiconductor characterization
  J. F. Black, E. Lanning, and S. Perkowitz
  Infrared Phys., 10:125-139 (1970)

Gallium: Bulletin d'information et de bibliographie
  Pierre de la Breteque Alusuisse-France, Marseilles (April 1970), 343 pp.; 7th supplement to publication of 1962.
  Physical and electrical properties of Ga, III-V, III-VI compounds, Ga carbides, halides, silicides, and germanides; phase diagrams of Ga compounds; purification; crystal growth; analysis; special supplement: 91-page survey on physical, optical, magnetic, and electrical properties of solid, liquid, and gaseous Ga

Photoconductivity of semiconductors
  R. H. Bube
  Physical Chemistry, Vol. 10, p. 765 (W. Jost, ed.), Academic Press, New York (1970)

Modulation spectroscopy of semiconductors
  Manuel Cardona
  Festkörperprobleme X: Advances in Solid State Physics (O. Madelung, ed.), Marburg, Pergamon (1970), pp. 125-173

Catalogue of semiconductor alpha-particle spectra
  R. N. Chanda and R. A. Deal
  IN-1261 (March 1970), 159 pp.

Ferroelectric Materials and Ferroelectricity; Solid State Physics Literature Guides, Vol. 1
  T. F. Connolly and Errett Turner, comp.
  IFI/Plenum, New York (1970)
  Primarily a materials bibliography which includes nonferroelectric properties of ferroelectric materials, and because of this contains many references in border-line areas such as piezoelectrics. The scope of the last two years of its coverage has been extended to include theoretical papers and materials aspects of devices, although the design and characteristics of devices per se are excluded. All types of documents-books, periodicals, reports, patents, theses-are included except that references to papers given at meetings and represented in print solely by an abstract have been deleted unless the abstract is sufficiently detailed as to constitute a précis of the paper.

Le Vide: Formation et Controle des Couches Minces
  R. David and A. Richardt
  Dunod, Paris (1970), 255 pp.

The preparation of films by chemical vapor deposition
  W. M. Feist, S. R. Steele, and D. W. Readey
  Phys. Thin Films, 5:237-322 (1969)

Techniques for melt-growth of luminescent semiconductor crystals under pressure
  A. G. Fischer
  J. Electrochem. Soc., 117(2):41C-47C (1970)
  III-V and II-VI binary crystals

Modulated piezoreflectance in semiconductors
  A. Gavini and M. Cardona
  Phys. Rev. B, Solid State, 1:672-682 (1970)
  Germanium, gallium arsenide and antimonide, indium phosphide, zinc sulfide and oxide, cadmiun telluride, selenide and oxide

Preparing semiconductor compounds
  Klaus Hein
  Neue Huette, 15(11):657-662 (1970)
  Review; 18 materials by gas phase transport reactions; in German

Superconductivity in semiconductors and semimetals
  J.K. Hulm, M. Ashkin, D. W. Deis, and C. K. Jones
  Progr. Low Temper. Phys., 6:205-242 (1970)
  90 refs.

The crystal chemistry of semiconductors with octahedral and mixed atomic coordination
  R. M. Imamov, S. A. Semiletov and Z.G. Pinsker
  Kristallografiya, 15(2):287-293 (1970)
  Sov. Phys. – Cryst., 15(2):239-244 (1970)

Main Problems in Physics of Semiconductors and Semiconductor Devices
  S. P. Kal'venas, ed.
  Institut Fiziki Poluprovodnikov Akademmi Nauk Litovskoi SSR, Vil'nyus, USSR (1969), 308 pp.

Characterization of Semiconductor Materials
  P. F. Kane and G. B. Larrabee
  McGraw-Hill, New York (1970)

Proceedings of the Tenth International Conference on the Physics of Semiconductors, Cambridge, Mass., August 17-21, 1970
  S. P. Keller, J. C. Hensel, and Frank Stern, eds.
  CONF-700801 (1970)

All-Union Conference on the Physical Properties of Semiconductors, Groups 3A-5A and 3A-6A
  I. V. Kryukova et al.
  FTD-MT-24-359-69; AD-703551 (Feb.1970); 20 pp. Transl. into English from the Russian conference held at Baku, 1965

Physical characterization of electronic materials, devices and thin films, final report, 1 Dec. 1967 - 1 Dec. 1969
  S. A. Kulin, E. V. Clougherty, E. T. Peters, and K. Kreder
  AFCRL-70-0050; AD-703275 (Jan. 1970), 43 pp.

Current Injection in Solids
  Murray A. Lampert and Peter Mark
  Academic Press, New York (1970), 354 pp.

Optical properties and electronic band structures of group V and IV – VI materials
  P. J. Lin-Chung
  J. Phys. Chem. Solids, 31:2199-2208 (1970)

Quantum size effect – present state and perspectives of experimental investigations
  V. N. Lutskii
  Phys. Stat. Sol., 1A:199-220 (1970)
  Review; thin films; Bi example

Festkörper Probleme X, Advances in Solid State Physics, Plenary Lectures of the Professional Groups "Semiconductor Physics," "Low-Temperature Physics," "Thermodynamics," "Metals Physics" of the German Physical Society, Freudenstadt, April 6-11, 1970
  O. Madelung, ed.
  Pergamon, Vieweg (1970)

Handbook for Thin Film Technology
  Leon Maissel and Reinhard Glang, eds.
  McGraw-Hill, New York (1970), 800 pp.

Survey of the Field of Magnetic Semiconductors,
  S. Methfessel
  IBM J. Res. Develop., 14:207-213 (1970)

Chalcogenide spinel magnetic semiconductors
  Kazuo Miyatani
  Kotai Butsuri, 5(4-5):1-20 (1970)
  More than 100 compounds; their magnetic, electrical, and
    optical properties reviewed

Refractive index of optical materials in the in-
frared region, data sheet
  A. J. Moses
  EPIC-DS-166 (Jan. 1970), 246 pp., 154 refs.
  Includes Si, Ge, CdTe, ZnS, ZnSe

Ionicity of the chemical bond in crystals
  J. C. Phillips
  Rev. Modern Phys., 42(3):317-356 (1970)
  Includes most tetrahedrally coordinated semiconductors

Bonds and bands in semiconductors
  J. C. Phillips
  Science, 169:1035-1042 (1970)
  Covalent bonding in crystals; energy-band spectroscopy

Spectroscopic Analysis of Cohesive Energies and
Heats of Formation of Tetrahedrally Coordinated
Semiconductors
  J. C. Phillips and J. A. Van Vechten
  Phys. Rev. 2B, 2147-2160 (1970)

A new method for measuring the ionized-impurity
concentration in high-purity materials
  A. Alberigi Quaranta, C. Canali, G. Ottaviani, and A. Taroni
  Phys. Stat. Sol., 39:315-322 (1970)
  Analysis of the current signal supplied by a reverse biased
    junction

Electrical conduction in metal oxides
  C. N. R. Rao and G. V. Subba Rao
  Phys. Stat. Sol., 1A:597-652 (1970)
  Review; 319 refs.

The physics of Schottky barriers
  E. H. Rhoderick
  J. Phys. D: Appl. Phys. 3:1153-1167 (1970)

Crystal properties as influenced by crystallo-
graphic imperfections, report for Jan. '68 -
Dec. '69
  Guenter H. Schwuttke
  Scientific 1-AFCRL-70-0110 (Mar. 2, 1970), 129 pp. (also
    AD-704 895)
  Part 1 reviews material problems related to semiconductor
    device technology

Diffusion in Semiconductors
  B. L. Sharma
  Diffusion Information Center, 22447 Lake Road 205 D, Cleve-
    land, Ohio 44116 or P. O. Box 505, CH-4500 Solothurm,
    Switzerland (1970), 200 pp.

Impurity scattering in semiconductors: Phase-
shift approach
  Om Prakash Sinha
  Ph. D. thesis, Yeshiva University, New York (1969), 85 pp.
  University Microfilms, Inc., Ann Arbor, Mich., Order No.
    69-21023

Lattice thermal conductivity of semiconductors:
a chemical bond approach
  D. P. Spitzer
  J. Phys. Chem. Solids, 31:19-40 (1970)
  Data at room temperature are compiled for more than 200
    semiconductors

Thermodynamics and calculation of the liquidus -
solidus gap in homogeneous, monotonic alloy sys-
tems
  J. Steininger
  J. Appl. Phys., 41:2713-2724 (1970)
  InAs-GaAs, InSb-InAs, InSb-GaSb, GaSb-AlSb, InSb-AlSb,
    CdTe-ZnTe, CdTe-CdSe, HgTe-CdTe, GeTe-MnTe,
    $PbBr_2$-$PbCl_2$; 53 refs.

Propriétés et applications des matériaux semi-
conducteurs magnétiques
  J. P. Suchet
  Ann. Phys., 5:67-76 (1970)
  Crystallographic and magnetic transitions, magnetic disper-
    sion, magnetoresistance, Faraday effect, Astrov effect,
    Hall effects, etc.

Measurement of Forbidden Energy Gap of Semi-
conductors by Diffuse Reflectance Techniques
  S. P. Tandon and J. P. Gupta
  Phys. Stat. Sol., 38:363-367 (1970)

Introduction aux Phénomènes de Transport
Linéaires dans les Semiconducteurs
  J. Tavernier and D. Calecki
  Masson et $C^{ie}$, Paris (1970), 237 pp.

Semiconducting thin films: an annoted bibliog-
raphy, 1969 supplements to NOLC report 712
  W. R. Turnbull
  NWCCA 613-38 (June 1, 1970)
  This document is subject to special export controls and each
    transmittal to foreign governments or foreign nationals
    may be made only with prior approval of the officer in
    charge of the Naval Weapons Center Corona Labs. (Code
    7565), Corona, Cal. 91720
  The abstracts are arranged by author under the following
    classes: (1) elemental, (2) Group III-V, (3) Group II-VI,
    (4) Group IV-VI, (5) Group V-VI, (6) amorphous films,
    (7) miscellaneous compounds, and (8) methods and tech-
    niques

SiC - mid-1965 to mid-1970
  Errett Turner
  BTL Bibliography No. 169 (Nov. 1970), 882 refs.
  All basic and general characteristics are covered thoroughly,
    along with applications based on electrical or optical prop-
    erties

Gruneisen Constant of Some Semiconductors
  J. K. D. Verma and M. D. Aggarwal
  Solid State Commun., 8:1929-1931 (1970)
  III-V and II-VI

Ferromagnetic semiconductors with exchange
effects through conduction electrons
  S. V. Vonsovskii, A. A. Samokhvalov, and A. A. Berdyshev
  Helv. Phys. Acta, 43:9-16 (1970), 27 refs.

Electronic Structure and Dielectric Properties
of Cubic Semiconductors
  J. P. Walker
  Thesis, University of California, Berkeley, UCRL-20377 (Nov.
    1970), 114 pp.

Intermetallic Semiconductor Films
  H. H. Weider
  Pergamon, Oxford, England (1970), 361 pp.
  624 refs.

Semiconductors and Semimetals. Vol. 5, Infrared Detectors
  R. K. Willardson and A. C. Beer, eds.
  Academic Press, New York (1970), 552 pp.

Semiconductors and Semimetals. Vol. 6, Injection Phenomena
  R. K. Willardson and A. C. Beer, eds.
  Academic Press, New York (1970), 361 pp.

Proceedings of the International Conference on Luminescence held at the University of Delaware, Newark, August 25-29, 1969
  F. Williams, ed.
  North-Holland Publishing Co., Amsterdam, Netherlands (1970), 975 pp.
  Alkali halides, II—VI and III—V semiconductors, molecular crystals, organic solutions, rare earth phosphors, condensed rare gases, and complex phosphors
  Reprinted from J. Luminescence, Vol. 1-2 (1970)

Temperature Coefficient of the Refractive Index of Diamond- and Zinc-Blende-Type Semiconductors
  Peter Y. Yu and Manuel Cardona
  Phys. Rev., 2B:3193-3197 (1970)

Intrinsic and extrinsic photoeffect spectra of semiconductors at liquid helium temperatures, 1964-1969
  Libraries and Information Systems Center, Bell Telephone Labs. Bibliography No. 149 (March 1970), 8 pp.

Technical Conference on Defects in Electronic Materials for Devices
  Boston, Mass., August 24-27, 1969
  Proceedings in Met. Trans., 1(3):561-748 (1970)

Report on the conference on growth and characterization of electronic materials
  Chania, Greece, 1-12 June 1969
  J. Crystal Growth, 6:116 (1969)
  Characterization of electronic materials by x-ray, electron microscopy, electron microprobe, optical, and electro-optical techniques

Theory of semimetals
  A. A. Abrikosov
  Vestn. Akad. Nauk SSSR, 39(1):45-55 (1969)

Semiconducting II—VI, IV—VI, and V—VI Compounds
  N. Kh. Abrikosov, V. F. Bankina, L. V. Poretskaya, L. E. Shelimova, and E. V. Skudnova
  Plenum Press, New York (1969), 246 pp.

Summary report of activities concerning electrical, thermal, and optical properties of semiconductors related to energy conversion. Final technical rept. 15 June 58-30 Nov. 69
  R. B. Adler and A. C. Smith
  Energy Conversion and Semiconductor Lab., Massachusetts Institute of Technology, Cambridge, Mass., Contract Nonr-184 (51), AD-693235 (August 1969), 26 pp.
  Capsule view of work performed under the contract over its ten-year period. It includes abstracts of all technical reports and theses, and bibliography of published works.

Production of semiconductor films and study of their properties
  L. N. Aleksandrov and A. V. Rzhanov
  Izv. Akad Nauk SSSR, Neorg. Mater, 5(4):652-672 (1969)
  Inorg. Mater., 5(4):555-572 (1969)

Electric conductivity of hot carriers in Si and Ge
  M. Asche and O. G. Sarbei
  Phys. Stat. Sol., 33(1):9-57 (1969)
  Review, 150 refs.

Semiconductors and Semimetals, Vol. 6
  A. M. Barnett, ed.
  Academic Press, New York (1969)

Superconductive semiconductors
  Kurt Baumann
  Acta Phys. Austr., 29:241-250 (1969)
  Review with tabulation of properties

Ternary Diamond-Like Semiconductors
  L. I. Berger and V. D. Prochukhan
  Consultants Bureau, New York (1969), 114 pp.
  Methods for the synthesis, purification, and growth of single crystals; description of their semiconducting and technical properties

Semiconductor Detectors
  G. Bertoline and A. Coche, eds.
  Interscience, New York (1969), 518 pp.
  Physics of crystals and semiconductors, behavior of lithium as a compensating material

Semiconductor Nuclear Particle Detectors and Circuits, Proceedings of conference held at Gatlinburg, Tenn. 1967
  W. L. Brown, W. A. Higinbotham, G. L. Miller, and R. L. Chase, eds.
  National Academy of Science (1969)

The evaporation of metals and elemental semiconductors (for vapour-deposition) using a work-accelerated electron-beam source
  M. St. J. Burden and P. A. Walley
  Vacuum, 19:397-402 (1969)
  Vapor deposition of 22 metallic and semiconducting materials described in detail

Tunneling Phenomena in Solids
  Elias Burstein and S. Lundquist, eds.
  Plenum Press, New York (1969), 579 pp.
  Described detailed aspects of electron tunneling in semiconductors, metals and superconductors, and atomic tunneling in solids.

Proceedings of the Meeting on Special Techniques and Materials for Semiconductor Detectors, Ispra, Italy, Oct. 1-2, 1968
  F. Cappellani and G. Restelli, eds.
  Center for information and Documentation, European Atomic Energy Community, Brussels, 1969
  Contents: pp. 13-160; Ge (Li) detectors: pp. 161-282; ion implantation: pp. 283-293; discussions.

Optical properties of some compound semi-conductors in the 36 — 150-eV region
  Manuel Cardona and Ruprecht Haensel
  DESY 69/45 (Oct. 1969)

The Physics of Selenium and Tellurium
  C. W. Cooper, ed.
  Pergamon Press, Oxford, England (1969), 380 pp.
  Proceedings of International Symposium October 12-13, 1967, Montreal, Canada

Group Theory and Electronic Energy Bands in Solids
J. F. Cornwell
North-Holland, Amsterdam; Interscience, New York (1969), 290 pp.
Selected Topics in Solid State Physics, Vol. 10.

Doping of semiconductors and semiconducting film, report bibliography Jan. 1963-Feb. 1969
(Defense Documentation Center, Alexandria, Virginia), Vol. 2, AD-853 000
Cumulative; 271 refs.

Tunneling in Solids: Solid State Physics Supplement 10
C. B. Duke, ed.
Academic Press, New York (1969), 353 pp.

Analysis of single crystal pulling methods with respect to application of crystal preparing from semiconductor compounds and alloys
Z. Dziuba
Przeglad Elektron (Poland), 10(12):581-584 (1969)
II—VI and III—V

Microwave emission from semiconductors
A. G. R. Evans (Oxford Univ., England)
Rept. 1079-69 (Contract AF-EOAR-67-33) (June 1969), 193 pp.

Proc. Conf. Intern. Advanced Study Institute: Electronic Structures in Solids, Chania, Crete, Greece, June 30-July 14, 1968. Superconducting Semiconductors
H. P. R. Frederikse
Electronic Structures in Solids
Plenum Press, Inc., New York (1969), pp. 270-282

High-vacuum vapor deposition of compound semiconductors
Helmut Freller and K. G. Guenther
Thin Solid Films, 3(6):417-438 (1969)
Review; III—V and II—VI

Heavily Doped Semiconductors
V. I. Fustul'
Moscow, 1967
Translated by Albin Tybulewicz from the Russian edition, Plenum Press, New York (1969), 420 pp.

Structure, Properties and Preparation of Perovskite-Type Compounds
F. S. Galasso
Pergamon Press, New York (1969)
Covers approximately 500 compounds; among the properties covered are electrical conductivity, ferroelectricity, ferromagnetism, optical transmittance, catalytic properties and mechanical properties

Problems concerning the spatial distribution of deep impurities in semiconductors
P. E. Gibbons
Solid State Electron., 12:989-995 (1969)

Differential thermal analysis studies of compound semiconductors
A. C. Glatz
Electron Res. Center, NASA, Cambridge, Mass., Thermal Analysis, Vol. 2, Proceedings of 2nd International Con-ference, 1968 (R. F. Schwenker, Jr., ed.), Academic Press, New York (1969), pp. 1411-1422. Equilibrium properties of semiconductors; crystal perfection

Methods of Examining the Thermoelectric Properties of Semiconductors
V. M. Glazov, A. S. Okhotin, R. P. Borovikova, and A. S. Pushkarsky
Atomizdat, Moscow (1969), 167 pp. + refs.

Physics of Thin Films
G. Haas and R E. Thun, eds.
Advances in Research and Development, Vol. 5, Academic Press, New York (1969)

Some ternary chalcogenides with a chalcopyrite structure
H. Hahn et al.
Z. Anorg. Allg. Chem. (Leipzig), 371:153-170 (1953)
NLL-RRE-Trans-219-(8036.625), July 28, 1969, 44 pp.

Ternary chalcogenides of aluminum, gallium and indium with zinc, cadmium and mercury
H. Hahn et al.
Z. Anorg. Allg. Chem. (Leipzig), 279:241-270 (1955)
NLL-RRE-Trans-220-(8036.625), July 28, 1969, 38 pp.

Electronic Structures in Solids
E. D. Haidenmenakis, ed.
Plenum Press, New York (1969)

Physics of Solids in Intense Magnetic Fields
E. D. Haidemenakis, ed.
Lectures presented at the First Chania Conf. held at Chania, Crete, July 16-29, 1967, Plenum Press, New York (1969)

Semiconductor Plasma Instabilities
Hans Hartnagel
American Elsevier, New York (1969)
Semiconductor bulk-effects

Halbleiter bei höchsten Dotierungen
W. Heywang
33. Physikertag. Karlsruhe, 1968, B. G. Teubner, Stuttgart (1969), pp. 330-349

Solid solutions in $A_3^{III}B_2^{V}-A^{II}B^{VI}$ systems
L. V. Kradinova, A. A. Vaipolin, and N. A. Goryunova
Khim. Svyaz Krist. (N. N. Sirota, ed.), Izd. "Nauka i Tekhnika," Minsk, USSR (1969), pp. 417-422 (in Russian)

Gunn effect and its applications
V. M. Kuznetsov and others
JPRS-48361, Izv. Vyssh. Ucheb. Zaved. Radiotekhtron (USSR), 2(4):319-341 (1969), July 7, 1969, 32 pp.

The preparation of pure cadmium telluride and gallium phosphide using a generally applicable procedure
W. Kwestroo, A. Huizing, and J. de Jonge
Mat. Res. Bull., 4:817-824 (1969)

Festkörper Probleme IX. Advances in Solid State Physics, Vol. 15
O. Madelung, ed.
Pergamon Press, New York (1969)
Plenary Lectures of the Professional Group "Semiconductor Physics" of the German Physical Society, Munich, March 19-22, 1969, and Invited Papers of the European Meeting of the IEEE "Semiconductor Device Research," Munich, March 24-27, 1969

Advances in Electronics and Electron Physics,
Vol. 27
  L. Marton and Claire Marton, eds.
  Academic Press, New York (1969), 358 pp.

Application of the scanning electron microscope
to semiconductors
  R. K. Matta
  Solid State Tech., 12:34–42 (1969)
  40 refs.

The 1968 supplement to NOLC Report 712, Semi-
conducting Thin Films, An Annotated Bibliography
  NWCCL TP 842 (March 1969), 156 pp., 451 refs.
  Preparation, properties, applications, and theory.

Thermal conductivity measurement techniques
for semiconductors
  P. S. Nay, J. K. D. Verma, and B. D. Nag
  Solid State Tech., 12:43–52 (1969)
    Review; 128 refs.

Infra-red absorption due to localized modes of
vibration of impurity complexes in ionic and
semiconductor crystals
  R. C. Newman
  Adv. Phys., 18(75):545–663 (1969)

Thin films of semiconductor compounds
  L. S. Palatnik and V. K. Sorokin
  Izv. Akad. Nauk, SSSR, Neorg. Mater., 5(5):822–852 (1969)
  Inorg. Mater., 5(5):699–725 (1969)

Use of thin films in determining the fundamental
optical properties of semiconductors
  William Paul
  J. Vacuum Sci. Tech., 6:483–493 (1969)
    Illustrations from recent work on the group 4, group 4–6,
    and group 2–6 semiconductors

Alloys for Semiconducting Devices
  V. V. Rozov and V. M. Soboleva
  Metallurgiya, Moscow (1969), 245 pp.

The average heat of atomization and the proper-
ties of semiconductors
  V. Sadagopan and H. C. Gatos
  Chemical Bond in Crystals (N. N. Sirota, ed.), Sciences and
    Technology Publishers, Minsk, U. S. S. R. (1969), pp. 220–
    231
    Correlating the physical properties of known semiconductors
    and predicting the properties of new ones, including the
    vitreous semiconductors and the limits of stability of var-
    ious semiconductor structures

Solid State Physics, Advances in Research and
Application, Vol. 1
  Frederick Seitz, David Turnbull, and Henry Ehrenreich, eds.
  Academic Press, New York (1969) contents:
  Insulating and metallic states in transition metal oxides, David
    Adler; The excitonic state at the semiconductor-semimetal
    transition, B. I. Halperin and T. M. Rice; Polarons, J. Ap-
    pel

Principles of doping semiconductors
  S. A. Semenkovich and Yu. P. Shishkin
  Khim. Svyaz Krist. (N. N. Sirota, ed.), Izd. "Nauka Tekhnika,"
    Minsk, USSR (1969), pp. 253–265

Introduction to Semiconductor Physics
  Yu. K. Shalabutov
  Edited by A. R. Regel; Izdat. Nauka, Leningrad (1969), 292 pp.

Semiconductor Crystal Manufacture
  Marshall Sittig
  Electronics Materials Review No. 3 (Noyes Development
    Corp., Park Ridge, N. J.), 335 pp.
  All methods; includes patent literature

Pure Chemical Elements for Semiconductors
  Marshall Sittig
  (1969), 335 pp.
  Based on U. S. patent literature since 1960; techniques in-
    clude distillation, recrystallization, electrolytic refining,
    chemical treatment, and zone refining

Semiconductors
  M. S. Sominskii
  Rept. No. FTD-HT-23-1333-68; AD-698 639, edited transla-
    tion of monograph, "Poluprovodniki," Leningrad, 1967
    (Aug. 14, 1969), 415 pp.

Thermal conductivity of semiconducting materi-
als
  E. F. Steigmeier
  Thermal Conductivity, R. P. Tye, Academic Press, New York
    (1969), Vol. 2, pp. 203–251

Cathodoluminescence of single-crystal semicon-
ductors 1963–1969
  B. A. Stevens
  Bell Telephone Laboratories Bibliography 142 (Oct. 1969), 5 pp.

Physics of Semiconductor Devices
  S. M. Sze
  Interscience Publishers, John Wiley and Sons, New York
    (1969), 812 pp.
  1000 refs. including 1968

Silicon Semiconductor Data
  Helmut F. Wolf, ed.
  Pergamon Press, New York (1969)
  International Series of Monographs on Semiconductors, Vol.
    9, 648 pp.

Applied Solid State Science, Advances in Materi-
als and Device Research
  Raymond Wolfe and C. J. Kriessman, eds.
  Academic Press, New York (1969), Vol. 1, 404 pp.
  Contents: Junction electroluminescence, P. J. Dean; Metal-
    Insulator-Semiconductor (MIS) Physics, A. Goetzberger
    and S. M. Sze; Ion implantation in semiconductors, James
    W. Mayer and Ogden J. Marsh; Electron transport through
    insulation thin films, S. R. Pollack and J. A. Seitchik

Diamagnetic excitons in semiconductors
  B. P. Zakharchenya and R. P. Seisyan
  Uspekhi Fiz. Nauk, 97:193–210 (1969)
  Sov. Phys. — Uspekhi, 12(11):70–79 (1969)

Probleme der Festkörperelektronik. Vol. 1
  Herausgegeben vom VEB Halbleiterwerk Frankfurt (Oder),
    Verlag Technik, Berlin (1969), 238 pp.

Proceedings of the third international conference
on photoconductivity
  Solid State Commun., 7(11) (June 1969)
  Conference held at Stanford, 12–25 Aug. 1969.

Analysis of High-Purity Metals in Semiconductor
Technology
  I. P. Alimarin, ed.
  Israel Program for Scientific Translations, Jerusalem, Israel
    (1968), 584 pp.
  Review of USSR work

## Electronic Properties of Semiconducting Solid Solutions
Aleksei B. Almazov
Consultants Bureau, New York (1968)

## Semiconductor Photoelectric Devices. An Introduction to Design
A. Ambroziak
Iliffe Books, Ltd., London (1968), 328 pp.

## Semiconductors
E. Antoncik
(Czechoslovak Academy Sci., Prague)
Theory of Condensed Matter, Trieste, 3 Oct.-16 Dec. 1967
(IAEA, Vienna, 1968) pp. 873-940
Aspects of the band structure of semiconductors important for optical and transport phenomena

## Physics of Electronic Conduction in Solids
Frank J. Blatt, ed.
McGraw-Hill, New York (1968), 446 pp.

## Final report (Sept. 1, 1956 - Aug. 31, 1967)
Harvey Brooks and William Paul
Harvard Univ., Cambridge Mass., Contract Nonr-1866(10), Tech. Report No. HP-23 (May 1968), 45 pp.
A general description of research on the effect of hydrostatic pressure on the properties of semiconductors carried out over a ten-year period

## Measurement of carrier lifetime in semiconductors – an annotated bibliography covering the period 1949–1967
W. Murray Bullis
NBS Tech. Note 465, November 1968, 62 pp.
Approximately 300 refs.

## Solid State Physics: Electrons in Metals, Vol. 1
J. F. Cochran, and R. R. Haering, eds.
Gordon and Breach Science Publishers, New York (1968), 572 pp.
Includes the experimental study of the electronic structure of semi-metals and degenerate semiconductors

## Isoelectronic impurities in semiconductors
R. A. Faulkner and J. J. Hopfield
Localized Excitations in Solid, Proceedings of 1st International Conference, Irvine, California, 1967, pp. 218-238, Plenum Press, New York (1968)

## Principles of the physics of semiconductor devices
Ya. A. Fedotov
1964. Translated from Russian by FTD-HT-67-194; AD-673918 (Jan. 1968), 595 pp.

## The Theory and Practice of Microelectronics
Sorab K. Ghandi, ed.
John Wiley and Sons, Inc. New York (1968), 496 pp.

## Physicochemical Principles of Semiconductor Doping
V. M. Glazov and V. S. Zemskov
Translated from Russian by Ch. Nisenbaum and B. Benny; D. Slutzkin, transl. ed.; Israel Program for Scientific Translations, Jerusalem; Davey, Hartford, Conn. (1968), 380 pp.

## Optical Properties and Band Structures of Semiconductors
David L. Greenaway and Gunther Harbeke, eds.
Pergamon Press, New York (1968), 159 pp.

## Semiconductor physics
W. Kleen and W. Heywang
Siemens Z., 42:79-95 (1968)

## Electron Spin Resonance in Semiconductors
G. Lancaster
Plenum Press, New York (1968)

## Semiconductor Physics, Devices, and Circuits
Louis H. Lenert
Merrill, Columbus, Ohio (1968), 609 pp.

## Festkörper Probleme VIII
In: Referaten des Fachausschusses "Halbleiter" der Deutschen Physikalischen Gesellschaft, Berlin (1968), zugleich Hauptvortrage des Fachausschusses "Tiefe Temperaturen" des Fachausschusses "Thermodynamik und statistische Mechanik" und der Arbeitsgemeinschaft "Metallphysik" (ed. O. Madelung); Friedr. Vieweg und Sohn; also Pergamon Press, New York (1968)

## Halbleitereigenschaften von Telluriden. X. Wärmeleitfähigkeit und elektrische Eigenschaften des Systems HgTe – CdTe
W. Markert, H. Nieke and D. Spiegler
Ann. Phys., 21:387-401 (1968)

## Experimental study of the electronic structure of semimetals and degenerate semiconductors
W. Mercouroff
NASA-TT-F-11953 (Dec. 1968), 18 pp.
Reviews the literature data on electronic structure of semimetals and degenerate semiconductors

## Semiconductor detectors of nuclear particles
Jean Messier
Commissariat à l'Energie Atomique, Saclay, France, CEA-CONF-1233; CONF-680642-6 (June 1968), 24 pp.

## Magnetic semiconductors
Siegfried Methfessel and Daniel C. Mattis
Magnetismus (H. P. J. Wijn, ed.), pp. 389-562, Vol. 18, Pt. 1 of Handbuch der Physik (S. Flugge, ed.), Springer-Verlag, Berlin, Heidelberg, New York (1968)

## Electromagnetic and space-charge wave in an active bulk semiconductor
Mark Allen Meyer
University Microfilms, Ann Arbor, Mich., Order No. 69-3656
Ph. D. thesis, University of California, Berkeley, 1968, 135 pp.

## Tables of properties of semiconductors
Brian R. Pamplin
Handbook of Chemistry and Physics, 49th edition, The Chemical Rubber Co., Cleveland, Ohio (1968), pp. E97-E102
The data is presented in 3 tables. Table I "General Properties of Semiconductors" lists the main crystallographic data and semiconducting prop. of many semiconducting materials in 3 main categories: "Tetrahedral Semiconductors" in which every atom is tetrahedrally coordinated to four nearest neighbor atoms, "Octahedral Semiconductors" in which every atom is octahedrally coordinated to six nearest neighbor atoms, and "Other Semiconductors." Table II gives more detailed information about some better-known semiconductors, while Table III gives some information about the electronic energy band structure parameters of the best known materials.

## Tabular reduction of crystal chemical information: metals and semiconductors
W. B. Pearson
Helv. Phys. Acta, 41:1070-1103 (1968)

## The Hall Effect and Semi-conductor Physics
E. H. Putley
Constable and Co., Ltd., London (1968), 236 pp.

Catalytical properties of organic semiconductors
S. Z. Roginskii and M. M. Sakharov
Zh. Fiz. Khim., 42:1331-1345 (1968)
Review, 121 refs.

Chemical Bonds in Semiconductors and Thermo-dynamics
N. N. Sirota, ed.
Consultants Bureau, New York (1968), 255 pp.

Semiconductor Handbook, 2nd Edition
Robert B. Tomer, ed.
SAMS, Indianapolis, Indiana (1968), 287 pp.

Status of diffusion data in binary compound semiconductors
D. W. Yarbrough, Solid State Tech., 11:23-54 (1968)
III−V, II−VI compounds, and SiC; review; tabulation; 118 refs.

Status of diffusion data for binary compound semiconductors
David W. Yarbrough
U. S. Govt. Res. Develop. Rept., 68:162 (1968); AD−670014,
69 pp.
III−V, II−VI compounds and SiC

IX Intern. Conf. on Physics of Semiconductors, Moscow, July 23-29, 1968. Proceedings, Vol. 1
Academy of Sciences of the USSR, Publishing House "Nauka," Leningrad Branch (1968)

Electronic transitions in semiconductors
V. G. Alekeseyeva, Yu. V. Gulyayev, and Ya. E. Pokrovskiy
Radio Eng. Electron. Phys., 12:1916-1930 (1967)
228 refs.

Comparison of solid-state photoelectronic radiation detectors
Richard H. Bube
Trans. Met. Soc. AIME, 239:291 (1967)

A review of semiconductor heterojunctions
J. T. Calow, P. J. Deasley, S. J. T. Owen, and P. W. Webb
J. Mater. Sci., 2:88-96 (1967)

High Field Transport in Semiconductors
E. M. Conwell
Academic Press, New York (1967).

Solid-state radiation detectors
G. Dearnaley
Comtemp. Phys., 8:607-626 (1967)

Diffusion furnaces for semiconductor processing
R. P. Donovan and R. M. Burger
Solid State Tech., 10:40 (1967)

Photoemission studies of the electronic band structures of gallium arsenide, gallium phosphide, and silicon
Richard C. Eden
Stanford Electronics Laboratories, Stanford Univ., Stanford, Calif.; Tech. Rept. No. 5221-1; SU-SEL-67-038; AD−675474 (May 1967), 319 pp.

Electronic Properties of Materials: A Guide to the Literature, Vol. 2
D. L. Grigsby, D. H. Johnson, M. Neuberger, and S. J. Welles, eds.
Plenum Press, New York (1967)
More than 11,000 refs.

Metallurgy in the semiconductor industry
Ralph Grubel
J. Metals, 19:13-17 (1967)

Energy structure of the bands of crystals of the $A^{IV}$, $A^{III}B^{V}$, $A^{II}B^{VI}$, and $Mg_2Si$ types
S. G. Kroitovu, V. V. Sobolev, N. N. Syrbu, and S. D. Shutov
Khimicheskaya Svyaz'v Poluprovodnikakh i Termodinamika (Chemical Bonds in Semiconductors and Thermodynamics), 1966, Nauka i Tekhnika, Minsk (1967), pp. 240-250

Energy bands in semiconductors
Donald Long, ed.
John Wiley and Sons, New York (1967), 224 pp.

Festkörperprobleme VII
O. Madelung, ed.
Pergamon Press, New York (1967), 292 pp.
Invited Papers on Semiconductor Device Research from the Joint Meeting of the Solid State Physics Committee of the German Physical Society and the European Section of the I. E. E. E., Bad Nauheim, Germany, April 17 to 22, 1967

Thermal conductivity of silicon, germanium, III − V compounds and III − V alloys
P. D. Maycock
Solid − State Electron., 10:161-168 (1967)

Electrochemistry of Semiconductors
V. A. Myamlin and Yu. V. Pleskov
Plenum Press, New York (1967), 430 pp.

Photolumineszenz und elektronische Struktur von Halbleitern
H. J. Queisser
Dtsch. phys. Gesellsch. 32. Physikertag. Berlin, 1967, Stuttgart, B. G. Teubner (1967), pp. 249-276

Electronic Processes at the Surface and in Single-Crystal Layers of Semiconductors
A. V. Rzhanov
Izdat Nauka, Novosibirsk (1967), 240 pp.

A bibliography of metal-insulator-semiconductor studies
E. S. Schlegel
Trans. IEEE Electron Devices ED − 14:728-749 (1967)

Chemical Bonds in Semiconductors and Solids
N. N. Sirota, ed.
Consultants Bureau, New York (1967)

Electronic Conduction in Solids
A. C. Smith, J. F. Janak, and R. B. Adler
McGraw-Hill, New York (1967)

Energy structure of the bands of certain compounds of the $A^{II}B^{V}$, $A^{V}B^{VI}$, and $A^{III}B^{VI}$ types
V. V. Sobolev, N. N. Syrbu, and S. D. Shutov
Khimicheskaya Svyaz'v Poluprovodnikakh i Termodinamika (Chemical Bonds in Semiconductors and Thermodynamics), 1966, Nauka i Tekhnika, Minsk (1967), pp. 221-228

Second International Conference on Solid Compounds of Transition Elements
Enschede, Netherlands, June 12-16, 1967, 140 pp.

Sixth all-union symposium on semiconductor theory
A. G. Aronov and V. S. Oskot-skii
Usp. Fiz. Nauk (SSSR), 88(1):161-177 (1966)
Soviet Physics - Acoustics, 9(1):153-167 (1966)

Photoconductivity of Solids
R. H. Bube
John Wiley and Sons, New York (1966)

Band structures and pseudopotential form factors
for fourteen semiconductors of the diamond and
zinc-blende structures
M. L. Coehn and T. K. Bergstresser
Phys. Rev., 141:789 (1966)

New horizons in semimetal alloys
L. Esaki
Trans. IEEE Spectrum, 3:74-86 (1966)

Semiconductor Physics
A. F. Gorodetskii, A. F. Kravchenko, and E. M. Samoilov
Osnovy Fiziki Poluprovodnikov i Poluprovodnikovykh Priborov,
Novosibirsk, Izdat. Nauka, Sibirskoe Otdelenie (1966), 351 pp.

Semiconducting properties of several III-V-VI
ternary materials and their metallurgical aspects
K. Jurata and T. Hirai
Solid State Electron., 9:633-640 (1966)

Solid State and Semiconductor Physics
J. P. McKelvey, ed.
New York, Evanston, London: Harper and Row; Tokyo: J.
Weatherhill (1966), 512 pp.

Optical and magneto-optical properties of some
semiconductors
Yuichiro Nishina, Kunihide Tanaka, Susumu Kurita, Masao
Yamamoto, Takehito Jimbo, Noritaka Kuroda, and Tadao
Fukuroi
Sci. Rept. Res. Inst. Tohoku Univ., 18:536-568 (1967)
Review, visible and infrared; 75 refs.

On the ternary semiconducting chalcogenides of
the $A^I B^{III} C_2^{VI}$ type

L. S. Palatnik and E. I. Rogacheva
Izv. Akad. Nauk SSSR, Neorg. Mater., 2(4):659-666 (1966)
Inorg. Mater., 2(4):568-573 (1966)

Semiconductors
D. A. Wright, ed.
London, Chapman and Hall; London, Methlei and Co. (1966),
134 pp.

Proceedings of the International Conference on
the Physics of Semiconductors, Kyoto, Japan,
1966
J. Phys. Soc. Japan, 21 (1966), 805 pp.

Electrical and optical properties of semicon-
ductors
Proceedings of Physics Institute of Academy of Sciences
Nauka Publ. House, Moscow (1966)

Symposium on Test Methods and Measurements of
Semiconductor Devices, Budapest: Scientific
Society for Telecommunication, Vols. 1 and 2
(1967)
Symposium held at Budapest, Hungary, April 25-28, 1967

Ultrasonic effects in semiconductors
N. G. Einspruch
Solid State Phys., 17:217-268 (1965)
Comprehensive review; investigation of properties of
semiconductors

Crystal chemistry of some rare earth – group
VI compounds
D. J. Haase, H. Steinfink, and E. J. Weiss
Rare Earth Research, Vol. III, Gordon and Breach, New York
(1965), pp. 535-544
Phase diagrams and crystal chemistry

Progress in Semiconductors, Vol. 9
A. F. Gibson and R. E. Burgess, eds.
Heywood Book, Temple Press Books Ltd., London (1965)
Chemical behavior of semiconductors: Etching characteris-
tics, H. C. Gatos and M. C. Lavine, pp. 1-46 (includes tab-
ulation of semiconductors, etchants, conditions)
Optical properties of semiconductors in the visible and ultra-
violet ranges, J. Tauc, pp. 87-134
Lattice bands in diamond and zinc-blende crystals, F. A.
Johnson, pp. 179-235

Chemical Physics of Semiconductors
J. P. Suchet
Van Nostrand, London (1965)

Integrated silicon device technology
Research Triangle Institute, Durham, N. C., ASD-TDR-63-316
Vol. IX (Aug. 1965): Methods and practical applications of
silicon epitaxy; Vol. X (Nov. 1965): Chemical and metal-
lurgical properties of silicon
Complete set of existing silicon binary phase diagrams

Thermomagnetic effects in semiconductors and
semimetals
R. T. Delves
Rept. on Progress in Phys. (G. B.), 28:249-289 (1965)

Ultrasonic effects in semiconductors
N. G. Finspruch
Advances in Research and Applications in Solid State Physics,
Vol. 17, Academic Press, New York (1965), pp. 217-268

Transport effects in semi-metals and narrow-gap
semiconductors
H. J. Goldsmid
Advances in Phys. (Phil. Mag. Suppl.), 14:273 (1965)

Materials Used in Semiconductor Devices
C. A. Hogarth, ed.
Interscience Publishers, John Wiley and Sons, New York (1965)
Ge, Si, Se, PbS, PbSe, PbTe, InSb, $Bi_2Te_3$, CdSb, ZnSb

Solid Semiconductors
A. K. Jonscher, ed.
Routledge and Kegan Paul, London (1965), 91 pp.

Photoelectronic Materials and Devices
Simon Larach, ed.
D. Van Nostrand Company, Inc., Princeton, New Jersey (1965)

Semiconductor Surfaces
A. Many, Y. Goldstein, and N. B. Grover, eds.
Amsterdam, North-Holland Publishing Co. (1965), 496 pp.

Photoconductivity
T. S. Moss
Reports on Progress in Physics, Vol. XXVIII (A. C. Stickland,
executive editor)
The Institute of Physics and The Physical Society, 47 Belgrave
Square, London (1965), pp. 15-59

Soviet Research in New Semiconductor Mate-
rials
D. N. Nasledov and N. A. Goryunova, eds. (translated from the
Russian by A. Tybulewicz)
Consultants Bureau, New York (1965), 121 pp. I–IV–VI, III–
VI, I–IV–V, Te, and glasses

Semiconducting compounds
 B. R. Pamplin
 Sci. Progr., 53:539-552 (1965)

Les Matériaux Semi-Conducteurs
 M. Rodot, ed.
 Dunod, Paris (1965), 288 pp.

Certain ternary compounds of the type
$A_2^I B^{IV} C_3^{VI}$ and solid solutions based on them
 G. K. Averkieva, A. A. Vaipolin, and N. A. Goryunova
 Paper from "Investigations on Semiconductors. New Semiconductor Materials," pp. 44-56 (1964)

Conference on Monocrystals at Turnove 1963
 Zpravy Vyzkumneho Ustavu Monokrystalu, Turnove (1964)

Trace Analysis of Semiconductor Materials
 J. P. Cali, ed.
 Pergamon Press, Inc., New York (1964)

Bibliography on the measurement of bulk resistivity of semiconductor materials for electron
devices
 Judson C. French
 NBS Tech. Note 232, Oct. 21, 1964

Advances in III − V and II − VI semiconductor
compounds
 Paul M. Hamilton
 Semicond. Tech., 7:15-20 (1964)

Semiconductor transducers
 C. A. Hogarth
 British J. Appl. Phys., 15:121 (1964)

Study of semiconducting properties of selected
rare earth metals and compounds
 J. F. Miller, F. J. Reid, L. K. Matson, J. W. Moody, R. D.
  Baxter, and R. C. Himes
 Battelle Memorial Institute, Columbus, Ohio, Contract AF 33
  (657)-1068, AL-TDR-64-239; AD-607082 (Feb. 1964)

Imperfections and active centres in semiconduc-
tors
 R. G. Rhodes
 International Series of Monographs of Semiconductors, Vol. 6, ,
  Pergamon Press, New York (1964), 373 pp.

Photoelectric Effect in Semiconductors
 S. M. Ryvkin, ed.
 Consultants Bureau, New York (1964), 402 pp.

Semiconductors
 Horst Teichmann
 Plenum Press, New York (1964), 139 pp.

Semiconductor Materials
 N. Kh. Abrikosov, ed. Preface by D. N. Nasledov
 Consultants Bureau, New York (1963), 139 pp.
 Proceedings of the 4th All-Union Conference on Semiconduc-
  tor Materials

Band structure of semiconductors
 L. Pincherle
 Semiconductors. Proceedings International School Phys.
  E. Fermi, Course XXII, Varenna on Lake Como, 1961,
  Academic Press, New York, London (1963), pp. 1-49

Proceedings of the AIME Technical Conference
on the Metallurgy of Semiconductor Materials
 H. Basseches, R. C. Manz, C. O. Thomas, and S. K. Tung
 Interscience Publishers, New York (1962)

Thermomagnetic Effects in Semiconductors
 I. M. Tsidil'kovskii
 Translated from Russian by A. Tybulewicz, Infosearch, London
  (1962), 333 pp.
 III−V, II−VI, IV−VI, V−VI, I−VI, Ge, Te, some oxides; ef-
  fects and methods of measurement

Conference on Monocrystals at Turnove  1961
 Zpravy Vyzkumneho Ustavu Monokrystalu, Turnove (1962)

Selected Constants Relative to Semi-Conductors,
Vol. 12 of Tables of Constants and Numerical Data
 C. Benoit à la Guillaume, R. Coehlo, O. Garreta, H. Guennoc,
  C. Sebenne, and J. Tavernier
 Pergamon Press, New York, London (1961)

# 5. I—V—VI Compounds

Electrical conductivity, thermoelectric power, and thermal conductivity of $Cu_3AsSe_4$ and $Cu_3SbSe_4$
    S. N. Aliev, G. G. Gadzhiev and Ya. B. Magomedov
    Fiz. Tekhn. Poluprovod., 3(11):1709-1711 (1969)
    Soviet Phys. —Semicond., 3(11):1437-1439 (1970)

The nonlinear optical properties of $Ag_3SbS_3$
    W. B. Gandrud, G. D. Boyd, J. H. McFee, and F. H. Wehmeier
    Appl. Phys. Letters, 16:59-61 (1970)
    Material suitable as a phase matched standard in the infrared for optical nonlinearities

Some electrical, photoelectric, and thermoelectric properties of thin films of alkali metal thio- and selenantimonites
    N. I. Gnidash, L. N. Sukhorykova, M. S. Kuznetsov, Ya. G. Finkel'shtein, S. I. Berul', N. P. Luzhnaya, and V. A. Bazakutsa
    Izv. Akad. Nauk SSSR, Neorg. Mater., 6(2):237-240 (1970)
    Inorg. Mater., 6(2):208-210 (1970)

Semiconducting compounds with the general formula $ABX_2$
    V. P. Zhuze, V. M. Sergeeva, and E. L. Shtrum
    Zh. Tekhn. Fiz., 28(10):2093-2108
    Soviet Phys.—Technical Phys., 3(10):1925-1938

Lattice thermal conductivity of semiconductors: a chemical bond approach
    D. P. Spitzer
    J. Phys. Chem. Solids, 31:19-40 (1970)

Synthetic proustite ($Ag_3AsS_3$): a summary of its properties and uses
    W. Bardsley, P. H. Davies, M. V. Hobden, K. F. Hulme, O. Jones, W. Pomeroy, and J. Warner
    Opto-Electron., 1:29-31 (1960)

Melt growth of ternary chalcogenides
    A. L. Gentile and O. M. Stafsudd
    Mat. Res. Bull., 4:869-876 (1969)

Ternary solid solutions in the system Cu—As—S
    Minael Guanaco, A. H. Clark, and A. Moraga
    Am. Mineral., 54:1269 (1969)

Thermal conductivity of the semiconductor compound $CuSbSe_2$ in the solid and liquid states
    A. U. Mal'sagov and Ja. B. Magomedov
    Teplofiz. vys. Temp., SSSR, 7:170-171 (1969)

On the crystal growth of optical quality proustite and pyrargyrite
    W. Bardsley and O. Jones,
    J. Crystal Growth, 3(4):268-271 (1968)
    $Ag_3AsS_3$, $Ag_3SbS_3$

Synthesis of optical quality proustite and pyrargyrite
    W. Bardsley and O. Jones
    Nature, 217:444-445 (1968)

Study of the microhardness and the photomechanic effect in argyrodite ($Ag_3GeS_3$) and canfieldite ($Ag_3SnS_3$)
    N. I. Bucko and I. S. Osypishin
    Ukr. Fiz. Zh., 13: 1950-1953 (1968)
    Crystal growth method given

Czochralski-grown proustite and related compounds
    A. L. Gentile and O. M. Stafsudd
    J. Crystal Growth, 3(4):272-274 (1968)
    $Ag_3AsS_3$, $Ag_3SbS_3$

The use of double decomposition reactions in melts for producing monocrystals of semiconductor compounds
    N. P. Luzhnaya
    Rost Kristallov, Vol. 6, Nauka (1965), pp. 220-225
    Growth of Crystals, Vol. 6B (N. N. Sheftal', ed.) Consultants Bureau, New York (1968), pp. 35-39

Synthetic trechmannite
    G. W. Roland
    Am. Mineral., 53:1208-1214 (1968)
    $AgAsS_2$

The system $Ag_2S - As_2S_3$ and the growth of crystals of proustite, smithite and pyrargyrite ($Ag_3SbS_3$)
    F. H. Wehmeier, R. A. Laudise, and J. W. Shiever
    Mat. Res. Bull., 3:767-778 (1968)

Investigation of some properties of silver antimony sulfide
  M. I. Butsko
  Ukr. Fiz. Zh., 9:686–688 (1967)

Synthetic proustite ($Ag_3AsS_3$): a new crystal for optical mixing
  K. F. Hulme, O. Jones, P. H. Davies, and M. V. Hobden
  Appl. Phys. Letters, 10:133 (1967)

Ricerche sul ternario $Ag_2S - Cu_2S - Sb_2S_3$
  L. Cambi, M. Elli, and I. Tangerini
  La Chimica e l'Industria, 48(6):567–568 (1966)

Phase diagrams of the pseudo-binary $Cu_2Se - Bi_2Se_3$ and $Ag_2Se - Bi_2Se_3$ systems and thermoelectric properties of $Cu_2Se - Bi_2Se_3$ solid solution
  Tadamasa Hirai, Kazuhiro Kurata, and Motohisa Hirao
  Advanced Energy Conversion, 6:195–200 (1966)

$CuGaSe_2$ and $AgInSe_2$: preparation and properties of single crystals
  Lawrence S. Lerner
  J. Phys. Chem. Solids, 27:1–8 (1966)

Processi idrotermali. Sintesi di solfosali da ossidi di metalli e metalloidi. Nota II — Cuprosol-foantimoniti
  L. Cambi and M. Elli
  La Chimica e l'Industria, 7(2):136–447 (1965)

Physical properties of single crystals of $CuSbS_2$ and $CuSbSe_2$
  G. B. Abdullaev, R. K. Nani, Y. N. Nasirov, and T. G. Osmanov
  Izv. Akad. Nauk SSSR, Ser. Fiz., 28:1096 (1964)

Preparation and electrical properties of silver antimony telluride
  R. A. Burmeister and D. A. Stevenson
  Trans. Met. Soc., AIME, 230:329 (1964)

Ternary selenides and tellurides of silver and antimony and their preparation
  Tom A. Bither, Jr.
  U. S. Patent No. 3,008,797, Nov. 14, 1961

The search for new semiconductors
  J. H. Wernick and R. Wolfe
  Bell Labs. Record 39:388–394 (1961)
  Ag, Cu(As, Sb, Bi) (S, Se)$_2$

Constitution of the $AgSbS_2$–PbS, $AgBiS_2$–PbS and $AgBiS_2$–$AgBiSe_2$ systems
  J. H. Wernick
  Am. Mineral., 45:591–598 (1960)

Ternary semiconducting compounds $A^IB^VB_2^{VI}$
  L. D. Dudkin and A. P. Ostranitsa
  Dokl. Akad. Nauk SSSR, 124:94–97 (1959)
  Abstract: Tech. Translations, 2:286 (1959)

Ternary semiconducting compounds with sodium chloride-like structure: $AgSbSe_2$, $AgSbTe_2$, $AgBiS_2$, $AgBiSe_2$
  S. Geller and J. H. Wernick
  Acta Cryst., 12:46 (1959)
  Synthesis and crystal structure

The thermal conductivity of silver antimony telluride
  E. F. Hockings
  J. Phys. Chem. Solids, 10:341–342 (1959)

Semiconducting materials and devices made therefrom
  Jack H. Wernick
  U. S. Patent No. 2,882,467, April 14, 1959
  Ag(Sb,Bi,As)Se$_2$

# 6. II–IV–V$_2$ Compounds

Semiconducting properties of glasses in the
CdGeAs$_2$ – CdSnAs$_2$ system
 V. V. Aksenov, V. M. Petrov, F. F. Kharakhorin, and B. I.
  Jurushkin
 Izv. Akad. Nauk SSSR, Neorg. Mater., 6(4):826–827 (1970)
 Bridgman method

Thermal analysis of the CdGe(As$_x$P$_{1-x}$)$_2$ system
 Z. U. Borisova, N. A. Goryunova, N. I. Kouzova, and E. O.
  Osmanov
 Vest. Leningrad. Univ., 25(4):165–167 (1970)

The preparation and growth of polycrystalline
layers of ZnSiP$_2$ in an open flow system
 B. J. Curtis and P. Wild
 Mat. Res. Bull., 5:69–72 (1970)

Growth of CdGeP$_2$ crystal by means of the meth-
od of vapor transport and properties of the
crystals
 A. I. Fedorov, A. G. Bychkov, and I. I. Tychina
 Ukr. Fiz. Zh., 15(9):1568–1569 (1970)
 In Russian

Values of the effective mass of conduction elec-
trons in CdGeAs$_2$
 L. B. Zlatkin, Yu. F. Markov, and I. K. Polushina
 Fiz. Tekhn. Poluprovod., 3(10):1590–1591 (1969)
 Soviet Phys. – Semicond., 3(10):1336–1337 (1970)

Optical properties and forbidden-band width
of ZnSnSb$_2$
 V. N. Ivakhno, L. V. Kradinova, and V. D. Prochukhan
 Fiz. Tekhn. Poluprovod., 3(7):1083–1084 (1969)
 Sov. Phys. – Semicond., 3(7):913–914 (1969)

Preparation and some optical properties of
ZnGeP$_2$
 I. M. Ivanova, E. K. Ivanov, L. B. Zlatkin, and V. D. Prochuk-
  han
 Fiz. Tekhn. Poluprovod., 3(12):1871–1873 (1969); Soviet Phys.
  Semicond., 3(12):1587–1588 (1970)

Electroreflectance spectra of some II–IV–As$_2$
compounds
 C. C. Y. Kwan and J. C. Woolley
 Can. J. Phys., 48(18):2085–2096 (1970)
 CdSnAs$_2$, CdGeAs$_2$, ZnSnAs$_2$, ZnGeAs$_2$, and ZnSiAs$_2$

Préparation et propriétés de ZnGeN$_2$
 M. Maunaye and J. Lang
 Mat. Res. Bull., 5:793–796 (1970)

Structure of the energy zones of semiconductors
with the chalcopyrite lattice, II. MgSiP$_2$, ZnGeP$_2$,
ZnSiAs$_2$, CdSiP$_2$
 A. S. Poplavnoi, Yu. I. Potygalov, and V. A. Chaldyshev
 Izv. VUZ. Fiz (USSR), 6:95–100 (1970); Soviet Phys. J. (USA),
  No. 6 (1970)

Electroreflectance study of the energy-band
structure of CdSnP$_2$
 J. L. Shay, E. Buehler, and J. H. Wernick
 Phys. Rev., 2B: 4104–4109 (1970)
 Crystals grown from Sn solution

CdSnP$_2$ emission and detection of near infrared
radiation
 J. L. Shay, R. F. Leheny, E. Buehler, and J. Wernick
 Appl. Phys. Letters, 16(9) (1970)

Lattice thermal conductivity of semiconductors:
a chemical bond approach
 D. P. Spitzer
 J. Phys. Chem. Solids, 31:19–40 (1970)

Production of ZnSiP$_2$ by crystallization from a
solution in a zinc melt
 E. E. Alekperova, Yu. A. Valov, E. A. Valov, and T. N.
  Ushakova
 Izv. Akad. Nauk SSSR, Neorg. Mater., 5(1):175–177 (1969)
 Inorg. Mater., 5(1):146–148 (1969)

Recombination radiation spectra in ZnSiP$_2$
crystals
 E. E. Alekperova, Yu. A. Valov, N. A. Goryunova, S. M.
  Ryvkin, and G. P. Shpenkov
 Phys. Stat. Sol., 32:49–54 (1969)

Electric properties of CdSiAs$_2$ crystals
 G. K. Averkieva, N. A. Goryunova, V. D. Prochukhan, et al.
 Phys. Stat. Sol., 34:5–8 (1969)

Optical and thermal carrier activation energies
in glasses of the CdGe(As$_x$P$_{1-x}$)$_2$ system
 Z. U. Borisova, N. A. Goryunova, N. I. Kouzova, E. O.
  Osmanov, and Yu. V. Rud'
 Fiz. Tekhn. Poluprovod., 2(10):1548–1549 (1968)
 Sov. Phys. – Semicond., 2(10):1292–1293 (1969)

Phase diagram of the Cd—Ge—As system
A. S. Borshchevskii and N. D. Roenkov
Zh. Neorg. Khim,, 14:2253-2258 (1969)

Behavior of CdSiP$_2$, CdGeP$_2$, CdSnP$_2$, ZnSnP$_2$,
ZnGeP$_2$ and ZnSiP$_2$ under high pressure and
temperature conditions
N. A. Goryunova, S. V. Popova, and L. G. Khvostantsev
Dokl. Akad. Nauk SSSR, 186(1-3):592-594 (1969)
Dokl. Chem., 186(1-3):401-403 (1969)

Optical properties and band structure of ZnSnP$_2$
(chalcopyrite and sphalerite modifications)
N. A. Goryunova, M. L. Belle, L. B. Zlatkin, G. V. Loshakova,
A. S. Poplavnoi, and V. A. Chaldyshev
Fiz. Tekhn. Poluprovod., 2(9):1344-1351 (1968)
Sov. Phys.—Semicond.,2(9):1126-1131 (1969)

Raman scattering in ZnSiP$_2$
I. P. Kaminow, J. H. Wernick, and E. Buehler
Bull. Am. Phys. Soc., 14:1165 (1969)

A high-pressure superconducting polymorph of
cadmium germanium diarsenide
H. Katzman, T. Donohue, W. F. Libby, and H. L. Luo
J. Phys. Chem. Solids, 30:2794-2795 (1969)

A high-pressure superconducting polymorph of
cadmium tin diarsenide
H. Katzman, T. Donohue, W. F. Libby, H. L. Luo, and J. G.
Huber
J. Phys. Chem. Solids, 30:1609-1611 (1969)

Compounds of type A$^{II}$B$^{IV}$C$_2^V$ and CuFeS$_2$ investi-
gated at high pressures and high temperatures
L. G. Khvostantsev, S. V. Popova, and N. A. Goriunova
Dokl. Akad. Nauk SSSR, 186:1365-1367 (1969)

Etude à haute t de CdGeAs$_2$
I. I. Kozhina, N. S. Boltovec, A. S. Borshchevskii, and N. A.
Goryunova
Vestn. Leningr. Univ., 24(10):93-96 (1969)

On electro-physical properties of ZnSnSb$_2$
L. V. Kradinova and T. I. Voronina
Phys. Stat. Sol., 32:K173-K174 (1969)

Preparation and phase studies of the ternary
semiconducting compounds ZnSnP$_2$, ZnGeP$_2$,
ZnSiP$_2$, CdGeP$_2$, and CdSiP$_2$
S. A. Mughal, A. J. Payne, and B. Ray
J. Mater. Sci., 4:895-901 (1969)

Studies in the adamantine family of semiconductors. IV. The Zn$_x$Cd$_{1-x}$SnAs$_2$ semiconducting
alloy system: phase diagram and preparation
B. R. Pamplin and J. S. Shah
J. Electrochem. Soc., 116:1565-1568 (1969)

Structure of the energy zones of semiconductors
with the chalcopyrite lattice. I. ZnSiP$_2$
A. S. Poplavnoi, Yu. I. Polygalov, and V. A. Chaldyshev
Izv. VUZ Fiz., 11:58-66 (1969)
English transl.: Soviet Phys. J., USA

The new semiconductor compound Zn$_2$In$_2$S$_5$ in the
system Zn—In—S
S. I. Radautsan, F. G. Donika, G. A. Kyosse, I. G. Mustya,
and V. F. Zhitar
Phys. Stat. Sol., 34:K129-K131 (1969)

Preparation and some physical properties of
ZnGeP$_2$
B. Ray, A. J. Payne, and G. J. Burrell
Phys. Stat. Sol., 35:197 (1969)

The average heat of atomization and the properties of semiconductors
Varadachari Sadagopan and Harry C. Gatos
Chemical Bond in Crystals, N. N. Sirota, ed., Science and
Technology Publishers, Minsk, USSR, (1969), pp. 220-231
Correlating the physical properties of known semiconductors
and predicting the properties of new ones including the vitreous semiconductors and the limits of stability of various
semiconductor structures

MgSiP$_2$: a new member of the II-IV-V$_2$ family of
semiconducting compounds
A. J. Spring-Thorpe and J. G. Harrison
Nature, 222:977-979 (1969)

Synthesis of A$^{II}$B$^{IV}$C$_2^V$ phosphides by chemical
transport reactions
Yu. A. Valov
Izv. Akad. Nauk SSSR, Neorg. Mater., 5(12):2115-2118 (1969)
Inorg. Mater., 5(12):1802-1805 (1969)

Some laws of ZnSiP$_2$ transfer during a chemical
transport reaction in an iodide system
Yu. A. Valov and T. N. Ushokova
Izv. Akad. Nauk SSSR, Neorg. Mater., 5(4):746-751 (1969)
Inorg. Mater., 5(4):634-637 (1969)

The Mössbauer effect in some semiconducting
compounds of the type A$^{II}$B$^{IV}$C$_2^V$
B. N. Veits, V. Ya. Grigalis, Yu. D. Lisin, E. O. Osmanov, Yu.
V. Rud, and I. M. Taksar
Latvi. PSR Zinat. Akad. Vestis, Fiz. Tekhn. Ser., No. 2, pp.
60-65 (1969)

Optical reflection of CdSnAs$_2$ single crystals
in the photon energy range 1-11 eV
L. B. Zlatkin and E. K. Ivanov
Fiz. Tekhn. Poluprovod., 3(6):926-927 (1969)
Sov. Phys.—Semicond.,3(6):781-782 (1969)

Lattice reflection and optical constants of ZnSnP$_2$
crystals with chalcopyrite and sphalerite structure
L. B. Zlatkin, J. F. Markov, A. I. Stekhanov, and M. S. Shur
Phys. Stat. Sol., 32:473-479 (1969)

Reflectivity of amorphous and polycrystalline
CdGeAs$_2$ in the 0.6 — 5.5eV region
A. Abraham, V. Vorlicek, and M. Zavetova
Czech. J. Phys., 18B:958-959 (1968)

Properties of the compound ZnSnP$_2$
A. S. Borshchevskii, A. A. Vaipolin, N. A. Goryunova, and
G. V. Loshakova
Izv. Akad. Nauk, SSSR, Neorg. Mater., 4(6):878-880 (1968)
Inorg. Mater., 4(6):772-774 (1968)

Electric properties of n — CdSnAs$_2$ in a wide range
of temperatures and impurity concentrations
V. V. Galavanov, N. A. Goryunova, N. M. Korshak, V. K.
Mityurev, and I. K. Polushina
Ukr. Fiz. Zh., 13:100-106 (1968)

Electric properties of n—CdSnAs$_2$ in a wide range
N. A. Goryunova, F. P. Kesamanly, and G. V. Loshakova
Fiz. Tekhn. Poluprovod., 1(7):1010-1012 (1967)
Sov. Phys.—Semicond., 1(7):844-846 (1968)

Investigations of some properties of vitreous and
crystalline CdGeP$_2$
N. A. Goryunova, S. M. Ryvkin, G. P. Shpenikov, I. I. Tichina,
and V. G. Fedotov
Phys. Stat. Sol., 28:489 (1968)
Recombination radiation

The structure of crystalline and amorphous CdGeP$_2$
R. Grigorovici, R. Manaila, and A. A. Vaipolin
Acta Cryst., 24B:535-541 (1968)

Properties of the band structure of semiconductors having the chalcopyrite lattice
G. F. Karavaev, A. S. Poplavnoi, and V. A. Chaldyshev
Fiz. Tekhn. Poluprovod., 2(1):113-114 (1968)
Sov. Phys. — Semicond., 2(1):93-95 (1968)

Structure of the conduction band of CdSnAs$_2$
Yu. V. Mal'tsev, T. A. Polyanskaya, G. A. Sikharulidze, V. M. Tuchkevich, Yu. I. Ukhanov, and Yu. V. Shmartsev
Fiz. Tekhn. Poluprovod., 1(10):1584-1586 (1967)
Sov. Phys.—Semicond., 1(10):1319-1321 (1968)

The structure and properties of the semiconducting compound ZnSnP$_2$
E. O. Osmanov
Phys. Stat. Sol., 29:435 (1968)

High-temperature modifications of the semiconductor compounds CdSnAs$_2$ and CdGeAs$_2$
E. O. Osmanov, Yu. V. Rud', and M. E. Stryalkovskii
Phys. Stat. Sol., 26:85-90 (1968)

Energy spectrum of electrons in ZnGeAs$_2$ and ZnSiP$_2$
A. S. Poplavnoi and G. F. Karavaev
Izv. Akad. Nauk SSSR, Neorg. Mater., 4(2):196-200 (1968)
Inorg. Mater., 4(2):161-164 (1968)

Preparation and characteristics of ZnSnP$_2$
M. Rubenstein and R. W. Ure, Jr.
J. Phys. Chem. Solids, 29:551-555 (1968)
Solution growth; optical band gap; structure; conductivity

Absorption of polarized light in CdSnAs$_2$ crystals
G. A. Sikharulidze, and Yu. I. Ukhanov
Phys. Stat. Sol., 26:K33 (1968)

Growth of some single-crystal II — IV — V$_2$ semiconducting compounds
A. J. Spring-Thorpe and B. R. Pamplin
J. Crystal Growth, 3(4):313-316 (1968)

Preparation and properties of MgSiP$_2$
R. Trykozko and N. A. Goryunova
Izv. Akad. Nauk SSSR, Neorg. Mater., 4(12):2101-2105 (1968)
Inorg. Mater., 4(12):1826-1829 (1968)

The structure and properties of the semiconducting compound ZnSnP$_2$
A. A. Vaipolin, N. A. Goryunova, L. I. Kleshchinskii, G. V. Loshakova, and E. O. Osmanov
Phys. Stat. Sol., 29:435-442 (1968)

Production of crystals of CdSiP$_2$ from the gas phase
Yu. A. Valov and R. L. Plechko
Izv. Akad. Nauk SSSR, Neorg. Mater., 4(6):993-995 (1968)
Inorg. Mater., 4(6):875-877 (1968)

Transport of ZnSiP$_2$ in chemical transport reactions in an iodide system
Yu. A. Valov and T. N. Ushakova
Izv. Akad. Nauk SSSR, Neorg. Mater., 4(7):1054-1059 (1968)
Inorg. Mater., 4(7):926-930 (1968)

Luminescence of ZnSiP$_2$ crystals
I. Kh. Akopyan, S. S. Grigor'yan, and A. S. Yakovlev
Fiz. Tverd. Tela, 8(12):3643-3646 (1966)
Sov. Phys.—Solid State, 8(12):2910-2912 (1967)

Synthesis and growth of multicomponent semiconductor crystals by chemical transport reactions
G. Averkieva, N. A. Goryunova, A. Nazarov, and V. D. Prochukhan
Krist. Tech. (German), 2:517-522 (1967)
GaAs — ZnSiAs$_2$, GaAs — ZnGeAs$_2$, and GaAs — Cu$_2$GeSe$_3$

Semiconducting A$^{II}$B$^{IV}$C$_2^V$ compounds
A. S. Borshchevskii, N. A. Goryunova, F. P. Kesamanly, and D. N. Nasledov
Phys. Stat. Sol., 21:9-55 (1967)
Review article

Die Kristallstruktur von BeSiN$_2$
P. Eckerlin
Z. Anorg. Allgem. Chem., 353:225-235 (1967)

Application to Cd$_{3-x}$Zn$_x$As$_2$ of a method of simultaneous measurement of the thermal conductivity, thermoelectric power and electric resistivity in the temperature range 80°-400°K (in French)
L. Giraudier
J. Phys., 28:667-670 (1967)

Growing and study of crystals of some solid solutions based on gallium arsenide
N. A. Goryunova, S. Mamaev, A. Nazarov, V. D. Prochukhan, and Yu. V. Rud'
Izv. Akad. Nauk Turkm. SSR, No. 5, pp. 79-83 (1967)
2GaAs — ZnGeAs$_2$ and 2GaAs — ZnSiAs$_2$

Some photoelectric properties of n−CdGeP$_2$ and p − ZnGeP$_2$ single crystals
N. A. Goryunova, I. I. Tychina, and R. Yu. Khansevarov
Fiz. Tekhn. Poluprovod., 1(1):141-142 (1967)
Sov. Phys. — Semicond., 1(1):110-111 (1967)

Growing single crystals of CdSnAs$_2$
N. A. Goryunova, A. S. Borshchevskii, Ya. Ya. Venkrbets, and N. M. Korshak
Izv. Akad. Nauk SSSR, Neorg. Mater., 3(1):180-181 (1967)
Inorg. Mater., 3(1):150-152 (1967)

Optical phenomena in p-type CdSnAs$_2$
É. L. Karaseva, G. A. Sikharulidze, V. M. Tuchkevich, Yu. I. Ukhanov, and Yu. V. Shmartsev
Fiz. Tekhn. Poluprovod., 1(2):276-280 (1967)
Sov. Phys. — Semicond., 1(2):219-222 (1967)

Diffusion of gold in crystalline and glassy samples of CdGeAs$_2$
F. F. Kharakhorin and V. V. Aksenov
Fiz. Tekhn. Poluprovod., 1(6):961-962 (1967)
Sov. Phys. — Semicond., 1(6):805-806 (1967)

Study of the effect of impurities on the properties of CdSnAs$_2$
S. Mamaev and O. Ismailov
Izv. Akad. Nauk Turkm. SSR, No. 1, pp. 17-20 (1967)

The magnetoresistance effect on n-type CdSnAs$_2$
M. Matyas
Czech. J. Phys., 17B:926-927 (1967)

Solid solutions on the 2GaAs — ZnGeAs$_2$ and 2GaAs — ZnSiAs$_2$ sections on the equilibrium diagram, and some physicochemical properties of such solid solutions
A. Nazarov, A. A. Vaipolin, V. D. Prochukhan, and N A. Goryunova
Izv. Akad. Nauk SSSR, Neorg. Mater., 3(12):1982-1984 (1967)
Inorg. Mater., 3(12):1982-1984 (1967)

Bibliography on $A^{II}B^{IV}C_2^V$ ternary compounds
B. Ray
J. Mater. Sci., 2:284–292 (1967)
Annotated references 1957 to 1966, includes a tabulation of
physical properties

Reflection spectra of CdSnAs₂
G. A. Sikharulidze, V. M. Tuchkevich, Yu. I. Ukhanov, and
Yu. V. Shmartsev
Fiz. Tekhn. Poluprovod., 1(2):309–311 (1967)
Sov. Phys. – Semicond., 1(2):254–256 (1967)

Preparation and properties of ZnSnAs₂ crystals
A. A. Vaipolin, F. P. Kesamanly and Yu. V. Rud'
Izv. Akad. Nauk SSSR, Neorg. Mater., 3(6):974–980 (1967)
Inorg. Mater., 3(6):871–875 (1967)

Study of the electrical properties of solid solutions based on In–As and of $A^{II}B^{IV}C_2^V$ compounds
A. V Voitsekhovskii, F. P. Kesamanly, and Yu. V. Rud'
Ukr. Fiz. Zh., 12:792–795 (1967)

Photoelectric properties of single crystals of ZnSiP₂
L. B. Zlatkin and B. V. Novikov
Izv. Akad. Nauk SSSR, Neorg. Mater., 3(1):78–86 (1967)
Inorg. Mater., 3(1):63–69 (1967)

Optical reflection spectra of single crystals of ZnSiP₂
I. Kh. Akopyan and L. B. Zlatkin
Dokl. Akad. Nauk SSSR, 168(3):547–549 (1966)
Sov. Phys. – Dokl., 11(5):435–437 (1966)

Optical and photoelectric properties of ZnSiP₂ single crystals
M. L. Belle, Yu. A. Valov, A N. Goryunova, L. B. Zlatkin,
A. N. Imenkov, M. M. Kozlov, and B. V. Tsarenkov
Dokl. Akad. Nauk SSSR, 163(3):606–608 (1965)
Sov. Phys. – Dokl., 10(7):641–643 (1966)

Production of the semiconducting compound CdSnAs₂ and some of its properties
A. S. Borshchevskii, N. A. Goryunova, G. A. Sikharulidze,
V. M. Tuchkevich, and Yu. V. Shmartsev
Dokl. Akad. Nauk SSSR, 171(4):830–832 (1966)
Sov. Phys. –Dokl., 11(12):1059–1058 (1967)

Some physical properties of the semiconductor compound CdSiP₂
A. G. Bychkov, R. L. Plechko, Yu. A. Valov and N. A. Goryunova
Izv. Akad. Nauk SSSR, Neorg. Mater., 2(11):2078–2079 (1966)
Inorg. Mater., 2(11):1798–1799 (1966)

Some properties of p-type CdSnAs₂
V. V. Galavanov, N. A. Goryunova, N. M. Korshak, S. Mamaev
and A. Nazarov
Fiz. Tverd. Tela, 7(12):3655–3656 (1965)
Sov. Phys. – Solid State, 7(12):2949–2950 (1966)

Solid solutions in the CdGeAs₂ system
N. A. Goryunova
Izv. Akad. Nauk Turkm. SSR, 3:29–32 (1966)

Structure of the conductivity zone of ZnSiP₂
N. A. Goryunova, S. S. Grigor'yan, and L. B. Zlatkin
Izv. Akad. Nauk SSSR, Neorg. Mater., 2(12):2125–2129 (1966)
Inorg. Mater., 2(12):1838–1841 (1966)

Preparation, lattice parameters, and phase diagram of solid solutions in the CdSnAs₂–CdGeAs₂ system
N. A. Goryunova, S. M. Mamaev, V. D. Prochukhan, and
M. Serginov
Izv. Akad. Nauk Turkm. SSR, No. 3, pp. 29–32 (1966)

Energy band-structure of some crystals belonging to the $A^{II}B^{IV}C_2^V$ groups
F. P. Kesamanly, S. G. Kroitoru, Yu. V. Rud', V. V. Sobolev,
and N. N. Syrbu
Dokl. Akad. Nauk SSSR, 163(4):868–869 (1965)
Sov. Phys. – Dokl., 10(8):743–744 (1966)

Experimental study of the band structure of the compound CdSnAs₂
P. Leroux-Hugon
J. Phys. Chem. Solids, 27:1205–1218 (1966)

Production of the semiconducting compounds ZnSnP₂ and CdSnP₂ and a study of some of their properties
G.V. Loshakova, R. L. Plechko, A. A. Vaipolin, B. V. Pavlov,
Yu. A. Pavlov and N. A. Goryunova
Izv. Akad. Nauk SSSR, Neorg. Mater., 2(11):1966–1969 (1966)
Inorg. Mater., 2(11):1702–1704 (1966)

Preparation and electronic properties of single crystals and doped crystals of ZnSnAs₂ semiconductor
Katashi Masumoto and Shigehiro Isomura
Trans. Nat. Research Inst. Metals (Japan), 8:200–209 (1966)

The preparation and properties of ZnSiAs₂, ZnGeP₂ and CdGeP₂ semiconducting compounds
K. Masumoto, S. Isomura, and W. Goto
J. Phys. Chem. Solids, 27:1939–1947 (1966)

Physical and electronic properties of Cd₃As₂–ZnSnAs₂ alloys
Katashi Masumoto, Shigehiro Isomura, and Wataru Goto
Proceedings Memorial Lecture Meeting on 10th Anniversary
of National Research Institute of Metals, Tokyo (1966),
p. 157

Galvanomagnetic effects in CdSnAs₂
T. A. Polyanskaya, G. A. Sikharulidze, V. M. Tuchkevich, and
Yu. V. Shmartsev
Fiz. Tverd. Tela, 8(6):1851–1858 (1966)
Sov. Phys.–Solid State, 8(6):1468–1473 (1966)

Optical and magneto-optical effects in CdSnAs₂
G. A. Sukharulidze, V. M. Tuchkevich, Yu. I. Ukhanov, and
Yu. V. Shmartsev
Fiz. Tverd. Tela, 8(4):1159–1164 (1966)
Sov. Phys. – Solid State, 8(4):924–928 (1966)

Diamond-like semiconductors in the glossy state
A. A. Vaipolin, E. O. Osmanov, and Yu. V. Rud'
Fiz. Tverd. Tela, 7(7):2266–2272 (1965)
Sov. Phys. – Solid State, 7(7):1833–1838 (1966)

Thermal analysis of some solid solutions
B. V. Baranov, V. D. Prochukhan, and N. A. Goryunova
Latvi. PSR Zinat. Akad. Vestis, 3:301–308 (1965)

Electrical and photoelectric properties of ZnSiP₂
A. G. Bychkov
Ukr. Fiz. Zh., 10:867–872 (1965)

The Chemistry of Diamond-like Semiconductors
N. A. Goryunova
Chapman and Hall, London (1965) pp. 142–144
Energy gaps in most $A^{II}B^{IV}C_2^V$ compounds

Some properties of $CdGeAs_2$
N. A. Goryunova
Izv. Akad. Nauk SSSR, Neorg. Mater., 1(6):885-889 (1965)
Inorg. Mater., 1(6):814-817 (1965)

Synthesis and some properties of the compound $ZnGeAs_2$
N. A. Goryunova, V. I. Sokolova, and P. H. Chiang
Zh. Prikl. Khim., 38:771-778 (1965)

Electric and photoelectric properties of $ZnSiP_2$
N. A. Goryunova, F. P. Kesamanly, D. N. Nasledov, V. V.
    Negreskul, Yu. V. Rud', and S. V. Slobodchikov
Fiz. Tverd. Tela, 7(5):1312-1314 (1965)
Sov. Phys. — Solid State, 7(5):1060-1062 (1965)

The thermo-emf and transverse Nernst-Ettings-
hausen effect in p-$ZnSnAs_2$ crystals
F. P. Kesamanly, D. N. Nasledov, and Yu. V. Rud'
Fiz. Tverd. Tela, 6(7):2187-2189 (1964)
Sov. Phys. — Solid State, 6(7):1727-1729 (1965)

Electrical properties of p-type $ZnSnAs_2$ crystals
at low temperatures
F. P. Kesamanly, D. N. Nasledov, and Yu. V. Rud'
Phys. Stat. Sol., 8:K159-K162 (1965)

On photoelectric properties of p-$ZnSiAs_2$ and
p-$CdGeAs_2$ crystals
F. P. Kesamanly, Yu. V. Rud', and S. V. Slobodchikov
Dokl. Akad. Nauk SSSR, 161(5):1065-1066 (1965)
Sov. Phys. — Dokl., 10(4):336-337 (1965)

Thermal properties of the ternary compounds
$CdSnAs_2$ and $ZnSnAs_2$
P. Leroux-Hugon and J. J Veyssie
Phys. Stat. Sol., 8:561-568 (1965)
Lattice thermal conductivity: specific heat

Effect of fast neutron irradiation on thermal
conductivity in the ternary arsenides
P. Leroux-Hugon and G. Weill
Seventh International Conference Physics Semiconductors
    (Radiation Damage), Vol. 3, Dunod, Paris (1965), pp. 73-77.

Unusual temperature dependence of the Hall
coefficient in solid solutions $CdSnAs_2$ — 2InAs
S. Mamaev and A. Allanazarov
Izv. Akad. Nauk Turkm. SSR, 3:98-99 (1965)

$CdSnAs_2$ single crystals and their electrical
properties
S. Mamaev, A. Nazarov, Ch. Doblet-Muradov, and M. Serginov
Izv. Akad. Nauk Turkm. SSR, Ser. Fiz.—Tekn. Khim.
    Geol. Nauk, 4:16-20 (1965)

The preparation and semiconducting properties
of single crystals of $ZnSnAs_2$ compound B
K. Masumoto and S. Isomura
J. Phys. Chem. Solids, 26:163-172 (1965)

The $Zn_xCd_{1-x}SnAs_2$ semiconducting alloy system
B. R. Pamplin, J. S. Shah, and R. A. L. Sullivan
J. Electrochem. Soc., 112:1249-1250 (1965)

New investigation of semiconducting mixed
crystals with special reference to phase diagrams
J. Rupprecht and R. G. Maier
Phys. Stat. Sol., 8:3-39 (1965)

New vitreous compounds
A. A. Vaipolin, N. A. Goryunova, E. O. Osmanov, and Yu. V. Rud'
Dokl. Akad. Nauk SSSR, 160(3):633-634 (1965)
Dokl. — Phys. Chem., 160(1-6):74-75 (1965)

A new method for calculating the width of the
forbidden zone
S. S. Batsanov
Russ. J. Struct. Chem., 5:862-864 (1964)
$ZnGeP_2$, $ZnGeAs_2$, $ZnSiAs_2$, and $CdGeP_2$

Preparation and some properties of $CdGeAs_2$
single crystals
N. A. Goryunova, F. P. Kesamanly and E. O. Osmanov
Fiz. Tverd. Tela, 5(7):2031-2032 (1963)
Sov. Phys. — Solid State, 5(7):1484-1485 (1964)

Electrical properties of p-type $ZnSnAs_2$ crystals
N. A. Goryunova, F. P. Kesamanly, D. N. Nasledov, and
    Yu. V. Rud'
Fiz. Tverd. Tela, 6(1):113-115 (1964)
Sov. Phys. — Solid State, 6(1):89-91 (1964)

The preparation and semiconducting properties
of single crystals of $ZnSnAs_2$ compound
K. Masumoto and S. Isomura, J. Phys. Chem. Solids, Vol. 25
    (1964)
Electrical and optical properties

Investigation of the physicochemical and electrical properties of crystals of some ternary semiconductor compounds of the $A^{II}B^{IV}C_2^V$ type
A. A. Vaipolin
Izv. Akad. Nauk SSSR, Ser. Fiz., 28:1085-1089 (1964)
$CdGeAs_2$, $ZnSnAs_2$, $CdSiAs_2$, $CdSiP_2$, and $ZnSiP_2$

Investigation of crystals of $ZnSiP_2$, $CdSiP_2$, and
$ZnSiAs_2$
A. A. Vaipolin, N. A. Goryunova, E. O. Osmanov, Yu. V. Rud',
    and D. N. Tret'yakov
Dokl. Akad. Nauk SSSR, 154(5):1116-1119 (1964)
Dokl. — Chem., 154(1-6):146-150 (1964)

The ternary semiconducting crystal $ZnSnAs_2$ and
the structure of the three-component system
Zn—Sn—As
H. Borchers and G. Maier
Metall., 17-775-780 (1963)

Pseudo-binary phase diagrams of the semiconduct-
ing crystal InAs with $ZnSnAs_2$, $ZnGeAs_2$, and
$CdGeAs_2$
Borchers and R. G. Maier
Metall., 17:1006-1010 (1963)

Band structure of $A^{II}B^{IV}C_2^V$-type semiconduc-
tor compound having the chalcopyrite structure
F. M. Gashimzade
Fiz. Tverd. Tela, 5(4):1199-1201 (1963)
Sov. Phys. — Solid State, 5(4):875-876 (1963)

Certain properties of the semiconductor
$CdSnAs_2$ — an electron analogue of indium arse-
nide
N. A. Goryunova, S. Mamaev, and V. D. Prochukhan
Dokl. Akad. Nauk SSSR, 142(3):623-626 (1962)
Dokl. — Phys. Chem., 142(1-6):91-94 (1963)

The solubility of germanium in certain ternary
semiconducting compounds
N. A. Goryunova, V. I. Sokolova, and Tsing Ping-hsi
Dokl. Akad. Nauk SSSR, 152(2):363-366 (1963)
Dokl. — Phys. Chem., 152(1-6): 808-810 (1963)

Properties of several ternary compound semicon-
ductors
P. Leroux-Hugon
Compt. Rend. Acad. Sci., 256:118-120 (1963)
Thermoelectric power, Hall effect, thermal and electrical
conductivity (CdGeAs$_2$, CdSnAs$_2$, ZnGeAs$_2$, ZnSnAs$_2$)

Thermal conductivity of the compounds CdSnAs$_2$,
CdGeAs$_2$, ZnSnAs$_2$, and ZnGeAs$_2$
P. Leroux-Hugon
Compt. Rend. Acad. Sci., 256:3991-3994 (1963)

A study of the thermo-emf and thermomagnetic
effects in alloys of the system CdSnAs$_2$ – 2InAs
D. N. Nasledov, S. Mamaev, and O. V. Emel'yanenko
Fiz. Tverd. Tela, 5(1):147-150 (1963)
Sov. Phys. – Solid State, 5(1) 104-106 (1963)

Crystal structure of ZnSnAs$_2$
H. Pfister
Acta Cryst., 16:153 (1963)

The properties of ZnSnAs$_2$ and CdSnAs$_2$
D. B. Gasson
J Phys. Chem. Solids, 23:129-302 (1962)

Study of InAs–CdSnAs$_2$ alloys
P. Leroux-Hugon
Compt. Rend. Acad. Sci., 255:662-664 (1962)
Thermoelectric power; electron effective mass

Electrical properties of semiconducting solid
solution xCdSnAs$_{2-y}$ (2InAs)
S. Mamaev, D. N. Nasledov, and V. V. Galanov
Fiz. Tverd. Tela, 3(11):3405-3413 (1961)
Sov. Phys.–Solid State, 3(11):2473-2478 (1962)

The semiconducting properties of CdSnAs$_2$
M. Matyas and P. Hoschl
Czech. J. Phys., 12:788-795 (1962)
Electrical conductivity, Hall effect, and magnetic susceptibility

New ternary semiconducting phosphides MgGeP$_2$,
CuSi$_2$P$_3$, and CuGe$_2$P$_3$
O. G. Folberth and H. Pfister
Acta Cryst., 14:325-326 (1961)

Mixed crystals of ZnSnAs$_2$–InAs and ZnGeAs$_2$–
InAs systems
G. Giesecke and H. Pfister
Acta Cryst., 14:1289 (1961)
Variation of lattice parameter with composition

Electrical and optical properties of CdSnAs$_2$
W. G. Spitzer, J. H. Wernick, and R. Wolfe
Solid State Electronics, 2:96-99 (1961)

Preparation and properties of CdSnAs$_2$
A. J. Strauss and A. J. Rosenberg
J. Phys. Chem. Solids, 17:278-283 (1961)
n-type; Hall mobility; energy gap

The crystal structure of ZnSnAs$_2$
O. G. Folberth and H. Pfister
Acta Cryst., 13:199-201 (1960)

Some problems concerning the formation of
semiconducting tetrahedral phases
N. A. Goryunova
Proceedings of the International Conference on Semiconductor Physics [in Russian], Prague, Czechoslovakia (1960),
p. 909
Mutual solubility of binary compounds with zinc-blende structure

Solid solutions in quaternary systems formed
from InAs and InSb
N. A. Goryunova and V. D. Prochukhan
Fiz. Tverd. Tela, 2(1):176-178 (1960)
Sov. Phys. – Solid State, 2(1):161-162 (1960)

Some electrical properties of quaternary alloys
based on InAs
S. Mamaev
Izv. Akad. Nauk Turkm. SSR, Ser. Fiz., 6:7-12 (1960)

Properties of CdSnAs$_2$
A. J. Rosenberg and A. J. Strauss
Bull. Am. Phys. Soc., 5:83 (1960)

The existence of tetrahedral phases
O. G. Folberth
Z. Naturforsch., 149:94-96 (1959)

The prediction of semiconducting properties of
inorganic compounds
C. H. L. Goodman
J. Phys. Chem. Solids, 6:305-314 (1958)

Crystal structure of ternary compounds of the
type A$^{II}$B$^{IV}$C$_2^V$
H. Pfister
Acta Cryst., 11:221-224 (1958)

A new group of compounds with diamond-type
(chalcopyrite) structure
C. H. L. Goodman
Nature, 179:828-829 (1957)
ZnSiAs$_2$, ZnGeP$_2$, CdGeP$_2$, ZnGeAs$_2$, CdGeAs$_2$, ZnSnAs$_2$,
CdSnAs$_2$, ZnSnP$_2$, and CdSnP$_2$

# 7. II—V Compounds

## 7.a. General, Reviews, and Bibliographies

Dielectric constants of zinc and cadmium diarsenides
N. V. Kotosonov, S. P. Artyukhov and T. A. Zyubina
Izv. Akad. Nauk SSSR, Neorg. Mater., 5(12):2207-2208 (1969)
Inorg. Mater., 5(12):1887-1888 (1969)

Energy-band structures of $Cd_3As_2$ and $Zn_3As_2$
P. J. Lin-Chung
Phys. Rev., 188(3):1272-1280 (1969)

Über ZnSb-reiche ZnSb — CdSb-Mischkristalle vom n- und p-Typ
D. Schmidt and G. Schneider
Z. Naturforsch., 24a:1586-1593 (1969)

The reflectivity spectra of zinc and cadmium diarsenides and diphosphides
V. V. Sobolev
N. N. Syrbu, Phys. Stat. Sol., 31:K51-K53 (1969)

Interband magnetoabsorption in $Cd_xZn_{3-x}As_2$
R. J. Wagner, E. D. Palik, and E. M. Swiggard
Phys. Letters, 30A:175-176 (1969)

Pair spectra in tetragonal zinc diphosphide ($ZnP_2$) and cadmium diphosphide ($CdP_2$) single crystals
W. Wardzynski, A. Wojakowski, and W. Zdanowicz
Phys. Letters, 29A:547-548 (1969)

The optical reflectivity spectra of CdSb and ZnSb single crystals
E. M. Averbakh, V. V. Sobolev, N. N. Syrbu, and Ya. A. Ugai
Phys. Stat. Sol., 30:K145-K147 (1968)

Structure et propriétés des condensats du système ternaire Cd — Sb — Zn
L. S. Palatnik, G. V. Fedorov, L. A. Kornienko, A. F. Bogdanova, and A. L. Toptygin
Izv. VUZ. Fiz., 11:48-54 (1968)

Dilatometric studies in the semiconductor system $Cd_3As_2 - Zn_3As_2$
W. Trzebiatowski, F. Krolicki, and W. Zdanowicz
Bull. Acad. Polon. Sci., Ser. Sci. Chim., 76:(7):343-346 (1968)
$Zn_3As_2$, $Cd_3As_2$, $Cd_3As_2 - Zn_3As_2$, apha—$Cd_{3-x}Zn_xAs_2$

Phase transitions of $Cd_3As_2$ and $Zn_3As_2$
S. Weglowski and K. Lukaszewicz
Bull. Acad. Pol. Sci., Ser. Sci. Chim., 16(4):177-182 (1968)

Chemistry of Semiconductor Materials and Materials for Quantum Electronics
N. P. Sazhin, N. Kh. Abrikosov, N. P. Luzhnaya, and E. G. Ippolitov
Razv. Obshch., Neorg. Anal. Khim. SSSR, 1917-1967, Akad. Nauk SSSR, Inst. Istor. Estestvozn. Tekh. (1967), pp. 214-223

Thermodynamic properties of the compounds $Zn_3As_2$ and $Cd_3As_2$
N. N. Sirota and E. M. Smolyarenko
Chemical Bonds in Semiconductors and Thermodynamics, Consultants Bureau, New York (1968), pp. 115-117
Enthalpy, entropy, and free energy of formation

Energy structure of the bands of certain compounds belonging to groups II — V, V — VI, and III — VI
V. V. Sobolev, N. N. Syrbu, and S. D. Shutov
Chemical Bonds in Semiconductors and Thermodynamics, Consultants Bureau, New York (1968), pp. 165-170
Optical reflection spectra

Halbleitendes Cadmium- und Zinkdiphospid
W. Kischio
Z. Naturforsch., 21A:1733-1734 (1966)

Antimonides of cadmium and zinc
G. R. Blackwell
Materials Used in Semiconductor Devices (C. A. Hogarth, ed.), Interscience Publishers, John Wiley and Sons, Inc., New York (1965), pp. 199-217

The preparation and properties of some II — V semiconducting compounds
G. A. Silvey, V. J. Lyons, and V. J. Silvestri
J. Electrochem. Soc., 108:653-658 (1961)
$Zn_3As_2$, $Cd_3As_2$, AlAs. $CdAs_2$, CdSb, ZnSb

Preparation of single crystals of $Zn_xCd_{1-x}Sb$ by interruption of free crystallization of the melt
K. Smirous
J. Phys. Chem. Solids, 19:170-171 (1961)
CdSb, ZnSb, and $Zn_xCd_{1-x}Sb$

29

Concerning the growth of crystals of intermetallic compounds in the Zn − Sb, Cd − Sb and In − Sb alloy systems
  V. I. Psarev
  Kristallografiya, 5(3):479–481 (1960)
  Sov. Phys. − Cryst., 5(3):459–462 (1960)

## 7.b. Zinc Compounds

### 7.b.1. Zn$_3$P$_2$

Preparation of zinc diphosphides and the low-temperature luminescence and absorption of the tetragonal polymorph
  M. Rubenstein and P. J. Dean
  J. Appl. Phys., 41:1777–1786 (1970)
  Single crystals of black monoclinic and red tetragonal ZnP$_2$
    from the elements

Transition polymorphe de Zn$_3$P$_2$ sous haute t et haute p
  J. Osugi and Y. Tanaka
  J. Chem. Soc. Japan, Pure Chem. Sect., 90(7):618–625 (1969)

Preparative, electrical, and optical characteristics of monoclinic zinc diphosphide crystals
  B. Ray and P. Burnet
  Phys. Stat. Sol., 32:K113 (1969)
  Vapor transport of the constituent elements from a charge
    containing a stoichiometric excess of phosphorus

Some electric properties of Zn$_3$P$_2$
  W. Zdanowicz and Z. Henkie
  Bull. Acad. Polon. Sci., Ser. Sci. Chim., 12 10:729 (1964)
  Purification and mxtl needles by sublimation

### 7.b.2. Zn As

The preparation and study of some of the electrical properties of single crystals of zinc arsenide
  Ya. A. Ugai and T. A. Zyubina
  Izv. Akad. Nauk SSSR, Neorg. Mater., 2(1):9–16 (1966)
  Inorg. Mater., 2(1):7–12 (1966)

The preparation and properties of some II − V semiconducting compounds
  G. A. Silvey, V. J. Lyons, and V. J. Silvestri
  J. Electrochem. Soc., 108:653–658 (1961)
  Zn$_3$As$_2$

### 7.b.3. Zn Sb

Nature of crystallization of the compounds Zn$_4$Sb$_3$ and Zn$_3$Sb$_2$
  V. I. Psarev and K. A. Dobryden'
  Izv. Akad. Nauk SSSR, Neorg. Mater., 6(2):230–236 (1970)
  Inorg. Mater., 6(2):203–207 (1970)

Thermo EMF and electroconductivity of zinc − antimony alloys
  Y. Ya. Asanovich, V. A. Kozlova, and I. T. Sryvalin
  Fiz. Metallov Metalloved., 28:373–375 (1969)

Effect of a tellurium impurity on the crystallization and electrical properties of zinc antimonide
  T. A. Kostur and V. I. Psarev
  Izv. Vyssh. Ucheb. Zaved., Fiz., 12(1):118–122 (1969)

Piezoresistance of antimonous zinc
  Z. D. Kovalyuk, E. N. Kosenkov, I. M. Rarenko, et al.
  Ukr. Fiz. Zh., 14;1215–1217 (1969)

Negative magnetische Widerstandsänderung von ZnSb
  F. Peters and G. Schneider
  Z. Naturforsch., 24a:620–628 (1969)

Anisotropy of the thermoelectric power of ZnSb single crystals
  I. M. Pilat and E. V. Osipov
  Fiz. Tekh. Poluprovod., 2(6):880–881 (1968)
  Sov. Phys.−Semicond., 2(6):730–731 (1968)

Anisotropy of electrical and thermoelectrical properties of single-crystal ZnSb
  I. M. Pilat, E. V. Osipov, and K. D. Soliichuk
  Ukr. Fiz. Zh. (USSR), 13(12):2047–2051 (Dec. 1968)
  Electrical conductivity, thermo-emf, Hall constant and trans-
    verse Nernst-Ettingshausen effect

Electrical properties of doped single crystals of the compound Zn$_4$Sb$_3$
  V. I. Psarev and N. L. Kostur
  Izv. VUZ Fizika (USSR), No. 2, 34–38 (1967)
  With Ag, Au, Ga, In, Sn, Pb, and Te; 100–490°K

Thermal and electronic transport properties of p-type ZnSb
  P. J. Shaver and John Blair
  Phys. Rev., 141:649–663 (1966)
  Lists refs. to crystal growth, electrical and optical properties,
    phase diagrams, surface etchants, bonding

Preparation of single crystals of semiconducting phases in the Zn-Sb system
  Ya. A. Ugai, E. M. Averbakh, and G. S. Kruglova
  Soviet Phys. J., 3:86–89 (May–June 1965)
  Bridgman; ZnSb, $\beta$−Zn$_4$Sb$_3$, $\epsilon$ −Zn$_3$Sb$_2$

Untersuchungen an zonengeschmolzenen ZnSb-Einkristallen
  E. Justi, W. Rasch, and G. Schneider
  Advan. Energ. Conversion, 4:27–38 (1964)

Materials for thermoelectric generators
  A. A. Machonis, J. Rivera, and I. Cadoff
  AD 425131, New York University Final Report, May 1963,
    20 pp.
  Problems producing zinc antimonide by a Bridgman technique

Single crystals
  K. Masumot and L. Roth
  Semiconductor Research Sixth Quarterly Report, January 1,
    1962, to March 31, 1962 (Purdue University, Dept. of
    Physics), p. 29, Contract DA 36-039-sc-87394, Purdue
    Research Foundation and United States Signal Corps
    PRF 2641; AD 275926
  ZnSb; Czochralski method; resistivity and Hall coefficient
    from liquid nitrogen to 297°K

Growth of ZnSb single crystals
  R. L. Eisner, R. Mazelsky, and W. A. Tiller
  J. Appl. Phys., 32:1833–1834 (1961)
  Horizontal zone leveling

On the structure of ZnSb
  K. Toman
  J. Phys. Chem. Solids, 16:160 (1960)

Constitution diagram of Zn-Sb
  Vratislav Tydlitat
  Czech. J. Phys., 9(5):638–640 (1959)

## 7.b.4. Zn Mixed Systems

Méthode de mesures simultanées et independantes
de paramètres thermiques et électriques
  L. Giraudier
  Entropie (France), No. 26, 30–51 (1969)

Über ZnSb-reiche ZnSb — CdSb-Mischkristalle vom
n- und p-Typ
  D. Schmidt and G. Schneider
  Z. Naturforsch., 24a:1586–1593 (1969)

Preparation and properties of n-type ZnSb
  G. Schneider
  Phys. Stat. Sol., 33:K133 (1969)
  And $Zn_{0.5}Cd_{0.5}Sb$

Investigation of $Zn_{0.1}Cd_{0.9}Sb$ solid-solution sin-
gle crystals
  I. M. Pilat and E. V. Osipov
  Fiz. Tekhn. Poluprovod., 2(1):64–68 (1968)
  Sov. Phys. — Semicond., 2(1):53–56 (1968)

Growing monocrystals of $Zn_xCd_{1-x}Sb$ solid solu-
tions
  I. M. Pilat, E. V. Osipov, V. Ya. Shevchenko, V. A. Skripkin,
    Ya. A. Ugai, and T. A. Marshakova
  Izv. Akad. Nauk SSSR, Neorg. Mater., 4(8):1356–1358 (1968)
  Inorg. Mater., 4(8):1190–1192 (1968)

Electric and thermoelectric effects in a solid
solution of $Zn_xCd_{1-x}Sb$
  A. Hruby, I. Kubelik, and L. Stourac
  Czech. J. Phys., 17B:426 (1967)
  Single crystals of $Zn_xCd_{1-x}Sb$ for x = 0.2 prepared by a
    modified Czochralski method

Über ZnSb und seine Mischkristalle mit CdSb
  G. Schneider
  Abh. braunschweig. wissensch. Gesellsch., 18:131–164 (1966)
  Zone-refined mixed crystals

The electrical parameters of single crystals of
$Zn_xCd_{3-x}As_2$ solid solutions
  Ya. A. Ugai, T. A. Zyubina, and E. A. Malygin
  Izv. Akad. Nauk SSSR, Neorg. Mater., 2(1):17–20 (1966)
  Inorg. Mater., 2(1):13–15 (1966)

Properties of the system ZnSb–CdSb
  Yu. P. Keloglu and A. S. Fedorko
  Kishinev. Gosudarstvennyi Universitet. Uchenye Zapiski,
    80:121–132 (1965)

Preparation of single crystals of $Zn_xCd_{1-x}Sb$
by interruption of free crystallization of the
melt
  K. Smirous
  J. Phys. Chem. Solids, 19:170–171 (1961)

## 7.c. Cadmium Compounds

### 7.c.1. $Cd_3P_2$

Optically pumped $Cd_3P_2$ laser
  S. G. Bishop, W. J. Moore, and E. M. Swiggard
  Appl. Phys. Letters, 16(11) (June 1, 1970)
  Sublimation-grown single crystals; $2.12\mu$

Photoluminescence and stimulated emission in
$Cd_3P_2$
  S. G. Bishop, W. J. Moore, and E. M. Swiggard
  Appl. Phys. Letters, 15(1):12–13 (1969)
  Crystal growth by sublimation

Photoconductivity in $Cd_3P_2$
  S. G. Bishop, W. J. Moore, and E. M. Swiggard
  Naval Res. Lab., Washington, D. C. 20390, Abstract 4.-1.,
    Proc. Third Intern. Conf. on Photoconductivity, Stanford
    Univ. (August 12-25, 1969)
  Abstracts published in Solid State Commun., 7:i–xxiii (1969).
    Sublimation-grown single crystals

Crystal structure and absolute configuration
of $\beta$-$CdP_2$
  J. Horn
  Bull. Acad. Pol. Sci., Ser. Sci. Chim., 17(2):69–74 (1969)

Field emission from $CdP_4$
  H. Neumann
  Acta Phys. Polon. (Poland), 35(3):487–488 (1969)

Anomales Verhalten des Feldemissionsstroms
aus $Cd_3P_2$
  Hans Neumann
  Solid State Commun., 7:877–878 (1969)

Production and electrical properties of cadmium
phosphide $CdP_4$
  I. S. Kovaleva, A. V. Dmitriev, and I. A. Zhizheiko
  Izv. Akad. Nauk SSSR, Neorg. Mater., 2(2):403–404 (1966)
  Inorg. Mater., 2(2):346–347 (1966)

Preparation and semiconducting properties of
cadmium phosphide ($Cd_3P_2$)
  W. Zdanowicz and A. Wojakowski
  Phys. Stat. Sol., 8:569 (1965)
  Melting point 729 ± 2°C; mxtl needles by sublimation

Some optical properties of $CdP_2$
  W. Zdanowicz and A. Wojakowski
  Phys. Stat. Sol., 10:K93 (1965)
  $CdP_2$ obtained by saturation of $Cd_3P_2$ with phosphorus vapor
    at about 700°C

Preparation and semiconducting properties of
$Cd_3P_2$
  G. Haacke and G. A. Castellion
  J. Appl. Phys., 35:2484 (1964)
  Sublimation

### 7.c.2. $Cd_3As_2$

Une deuxième bande de conduction dans $Cd_3As_2$
  M. Aubin, R. Brizard, and J. P. Messa
  Can. J. Phys., 48:2215–2220 (1970)
  Heavily doped samples

The thermal conductivity of cadmium arsenide
  D. Armitage and H. J. Goldsmid
  J. Phys. C (Solid St. Phys.), Ser. 2, 2:2138–2145 (1969)

Magneto-Seebeck and Nernst effects in cadmium
arsenide
  D. Armitage and H. J. Goldsmid
  J. Phys. C, Proc. Phys. Soc., 2:2389–2395 (1969)

Thermomagnetic effects in cadmium arsenide
  F. A. P. Blom and A. Huyser
  Solid State Commun., 7:1299–1303 (1969)

The de Haas-van Alphen effect of n-type Cd$_3$As$_2$
  H. Doi, T. Fukuroi, T. Fukase, Y. Muto, and K. Tanaka
  Sci. Rep. Res. Inst., Tohoku Univ., A 20(5-6):190-200 (1969)

On the electrical and magnetoresistance properties of Cd$_3$As$_2$ - NiAs eutectic
  C. T. Elliott and S. E. R. Hiscocks
  Brit. J. Appl. Phys. (J. Phys. D), Ser. 2, 2:1083-1087 (1969)

The electron and lattice thermal conductivity of n-Cd$_3$As$_2$ at low temperatures
  N. A. Goryunova, V. M. Muzhdaba, M. Serginov, and S. A. Shalyt
  Fiz. Tverd. Tela, 11(2):280-282 (1969)
  Sov. Phys.—Solid State, 11(2):225-227 (1969)

Physics of Solids in Intense Magnetic Fields, First Chania Conf., Crete, July 16-29, 1969
  E. D. Haidemenakis, ed.
  Plenum Press, New York (1969)
  Chapter 9: High field magnetospectroscopy and band structure of Cd$_3$As$_2$, by E. D. Haidemenakis

On the preparation, growth and properties of Cd$_3$As$_2$
  S. E. R. Hiscocks and C. T. Elliott
  J. Mater. Sci., 4:784-788 (1969)

Behavior of the Tl impurity during the growing of CdAs$_2$ single crystals
  R. A. Karieva, Sh. M. Mavlonov, and V. N. Vigdorovich
  Dokl. Akad. Nauk Tadzhik. SSR, 12(3):15-18 (1969)

Hall effect in quenched samples of Cd$_3$As$_2$
  D. R. Lovett
  Phys. Letters, 30A(2):90-91 (1969)

A refinement of the crystal structure of a''-Cd$_3$As$_2$
  A. Pietraszko and K. Lukaszewicz
  Acta Cryst., 25 B:988-990 (1969)
  Small crystals grown by sublimation

Effet Shubnikov de Haas dans Cd$_3$As$_2$: forme de la surface de Fermi et modèle non parabolique de la bande de conduction
  I. Rosenman
  J. Phys. Chem. Solids, 30:1385-1402 (1969)
  Growth of mixed crystals included

Contribution to the study of conduction bands of cadmium arsenide
  Nicole Sexer
  Ministère de l'Air, Paris, France, NT-168 (1969), 84 pp.

Thermoelectric power of cadmium arsenide
  V. Ya. Shevchenko, G. I. Goncharenko, V. F. Dvoryankin, and F. M. Gashinzade
  Fiz. Tekhn. Poluprovod., 3(6):916-917 (1969)
  Sov. Phys.—Semicond., 3(6):771-772 (1969)

Non-parabolic conduction band in Cd$_3$As$_2$
  D. Armitage and H. J. Goldsmid
  Phys. Letters, 28 A:149-150 (1968)

Thermomagnetic effects in cadmium arsenide
  D. W. G. Ballentyne and D. R. Lovett
  Brit. J. Appl. Phys. (J. Phys. D), Ser. 2, 1:585-592 (1968)

High-pressure metallic polymorph of cadmium arsenide
  H. Katzman, Terence Donohue, and W. F. Libby
  Phys. Rev. Letters, 20:442 (1968)

Preparation of a high-mobility thin film of Cd$_3$As$_2$
  H. Matsunami, M. Iwami, K. Asano, and T. Tanaka
  Japan. J. Appl. Phys., 7:444-445 (1968)

Crystal growth from the vapor phase by forced convection
  T. B. Reed and W. J. LaFleur
  Lincoln Lab., Mass. Inst. of Tech., Mass. Section III, in ESD-TR-68-17 (April 1968), pp. 20-23
  Includes Cd$_3$As$_2$

Detector investigation for 8-25 micron region, Rept. No. 26 for period July 1, 1967 to Dec. 31, 1967
  H. Shenker
  NRL Problem P01-03, RR 008-03-46,5664, ARPA Order No. 269-62, Program Code No. 8E30, NRL Memorandum Rept. 1845 (Jan. 1968), 33 pp.
  Each transmittal of this document outside the agencies of the U. S. Government must have prior approval of Director, Naval Research Lab., Washington, D. C. 20390
  Cd$_{3-x}$Zn$_x$As$_2$ and Cd$_{3-x}$Zn$_x$P$_2$ crystals grown

Thermal conductivity of cadmium arsenide from 60 to 400°K
  Donald P. Spitzer
  Natl. Bur. Stand. (U. S.), Special Publ. No. 302 (1967, Published 1968), pp. 123-130

The crystal structure of Cd$_3$As$_2$
  G. A. Steigmann and J. Goodyear
  Acta Cryst., 24B:1062 (1968)

Phase transitions of Cd$_3$As$_2$ and Zn$_3$As$_2$
  S. Weglowski and K. Lukaszewicz
  Bull. Acad. Polon. Sci., Ser. Sci. Chim., 16:177-182 (1968)

Properties of evaporated Hall elements of cadmium arsenide
  L. Zdanowicz
  Solid-State Electronics, 11:429-436 (1968)

Application to Cd$_{(3-x)}$Zn$_x$As$_2$ of a method of simultaneous measurement of the thermal conductivity, thermoelectric power and electric resistivity in the temperature range 80°-400°K
  L. Giraudier
  J. Phys. (France), 28:667-670 (Aug.-Sept. 1967)
  0 < x < 3

The Righi-Leduc effect in cadmium arsenide
  D. R. Lovett and D. W. G. Ballentyne
  Brit. J. Appl. Phys., 18:1399 (1967)
  Includes crystal growth

New semiconducting solid solution. The Cd$_3$P$_2$ - Cd$_3$As$_2$ system
  K. Masumoto and S. Isomura
  Trans. Japan. Inst. Metals, 8:139-149 (1967)

Thermodynamic properties of the compounds Zn$_3$As$_2$ and Cd$_3$As$_2$
  N. N. Sirota and E. M. Smolyarenko
  Chemical Bonds in Semiconductors and Thermodynamics, Consultants Bureau, New York (1968), pp. 115-117

Preparation and electric properties of thin Cd$_3$As$_2$-films deposited by thermal evaporation in vacuum
  L. Zdanowicz
  Acta Phys. Polon., 31:1021-1040 (1967)

Some optical properties of thin evaporated
$Cd_3As_2$ films
  L. Zdanowicz
  Phys. Stat. Sol., 20:473 (1967)

High-field magneto-optical studies of cadmium
arsenide
  E. D. Haidemenakis, M. Balkanski, E. D. Palik, and J.
    Tavernier
  Proc. Internat. Conf. on Physics of Semiconductors, Kyoto,
    Japan (1966), pp. 189-192

Observation of interband transitions in $Cd_3As_2$
  E. D. Haidemenakis, J. G. Mavroides, M. S. Dresselhaus,
    and D. F. Kolesar
  Solid State Commun., 4:65-68 (1966)

Formation de monocristaux d'arséniure de
cadmium ($Cd_3As_2$)
  B. Koltirine and M. Chaumereuil
  Phys. Stat. Sol., 13:K1 (1966)

Shubnikov-de Haas effect and electron band struc-
ture of cadmium arsenide
  I. Rosenman
  Proc. Internat. Conf. on Physics of Semiconductors, Kyoto,
    Japan (1966), pp. 370-373

Conduction bands in cadmium arsenide
  N. Sexer
  Phys. Stat. Sol., 14:K43 (1966)

Anomalous thermal conductivity of $Cd_3As_2$ and
$Cd_3As_2 - Zn_3As_2$ alloys
  D. P. Spitzer, G. A. Castellion, and G. Haacke
  Bull. Am. Phys. Soc., 11:401 (A) (1966)

Effet Shubnikov-de Haas dans l'arséniure de
cadmium
  Izio Rosenman
  Compt. Rend., 259:2621-2624 (1964)

Some semiconducting properties of thin evapora-
ted $Cd_3As_2$ films
  L. Zdanowicz
  Phys. Stat. Sol., 6:K153 (1964)

Semiconducting properties of $Cd_{3-x}Zn_xAs_2$-type
solid solutions
  L. Zdanowicz and W. Zdanowicz
  Phys. Stat. Sol., 6:227 (1964)
  n-type for $0 \leq x \leq 1.35$, and p-type for $1.5 \leq x \leq 3$

La magnétorésistance de l'arséniure de cadmium
entre 1.6 et 300°K
  V. Zhdanovich
  Acta Phys. Polon., 25:663-673 (1964)

The preparation and properties of some II-V
semiconducting compounds
  G. A. Silvey, V. J. Lyons, and V. J. Silvestri
  J. Electrochem. Soc., 108:653-658 (1961)
  Crystal growth - $Cd_3As_2$ and $CdAs_2$

Physical properties of several II-V semicon-
ductors
  W. J. Turner, A. S. Fischler, and W. E. Reese
  Phys. Rev., 121:759 (1961)
  Optical band gap

Physical chemistry of the II - V semiconductors
CdSb and $CdAs_2$
  V. J. Lyons, V. J. Silvestri, and G. A. Silvey
  IBM Corp., 1959 Fall Mtg. Electrochem. Soc.
  Purification and crystal growth

A noncubic semiconductor with unusually high
electron mobility
  A. J. Rosenberg and T. C. Harman
  J. Appl. Phys., 30:1621-1622 (1959)
  Includes crystal growth

Standard x-ray diagrams of several selenides,
tellurides, arsenides, and sulfides of copper,
silver, zinc, cadmium, gallium, and indium
  V. A. Kotovich and V. A. Frank-Kamenetskii
  Uchenye Zapiski Leningradskogo Gosudarstvennogo Ordena
    Lenina Universiteta imeni A. A. Zhdanova, Ser.
    Geologicheskikh Nauk, No. 8,135-156 (1957)

## 7.c.3. CdSb, $Cd_3Sb_2$

Hole effective masses in highly doped CdSb
determined from infrared-plasma-reflectivity
measurements
  B. Rheinlander
  Phys. Stat. Sol., 38:193-202 (1970)
  Czochralski grown; Ag-doped

Photoconductivity of p-CdSb in the millimeter
band
  V. M. Afinogenov, G. I. Goncharenko, V. I. Trifonov, and
    V. Ya. Shevchenko
  Zh. Eksp. Teor. Fiz. Pis'ma, 10(8):370-372 (1969)
  JETP Letters, 10(8):234-236 (1969)

Thermal irreversibility of the electrical prop-
erties of cadmium antimonide
  L. I. Anatychuk, V. D. Iskra, O. Ya. Luste, and I. M. Rarenko
  Izv. Akad. Nauk SSSR, Neorg. Mater., 5(9):1501-1507 (1969)
  Inorg. Mater., 5(9):1275-1280 (1969)

Elastic moduli of CdSb
  L. I. Anatychuk, V. D. Iskra, I. M. Rarenko, and B. M. Sharlai
  Fiz. Tverd. Tela, 10(11):3419-3420 (1968)
  Sov. Phys. - Solid State, 10(11):2702-2703 (1969)

Electrical properties of doped CdSb crystals at
low temperatures
  I. K. Andronik, E. K. Arushanov, O. V. Emel'yanenko, and
    D. N. Nasledov
  Fiz. Tekhn. Poluprovod., 2(9):1248-1252 (1968)
  Sov. Phys. - Semicond., 2(9):1049-1052 (1969)
  Electrical conductivity, Hall effect, and magnetoresistance;
    Cu, Ag, In, Sn, and Ge

The Peltier coefficient at the solid-liquid phase
boundary in antimony and some of its semicon-
ducting compounds
  N. V. Kolomiets and N. I. Strekopytova
  Fiz. Tverd. Tela, 11(4):866-868 (1969)
  Sov. Phys. - Solid State, 11(4):711-713 (1969)
  $Sb_2Te_3$, CdSb, $AgSbTe_2$

Gravitational effect in the segregation of thallium
in cadmium antimonide
  A. A. Kuliev, V. N. Vigdorovich, Sh. Mavlonov, and F. Gulmova
  Izv. Akad. Nauk SSSR, Neorg. Mater., 5(2):248-251 (1969)
  Inorg. Mater., 5(2):206-209 (1969)

New infrared detector based on n-type cadmium
antimonide
    V. K. Malyutenko, V. A. Romanov, and I. M. Rarenko
    Ukr. Fiz. Zh., 14:1570-1572 (1969)

Electrical properties of undoped p-CdSb at low
temperatures
    Hiroyuki Matsunami, Yoshikazu Nishihara, and Tetsuro Tanaka
    J. Phys. Soc. Japan, 27:1507-1516 (1969)

Segregation of Se, S, Tl and Fe in the growth of
cadmium antimonide and thallium telluride single
crystals from melt
    Sh. Mavlonov
    Rost Kristallov, Vol. 7, Izv. Akad. Nauk (1967), pp. 165-170
    Growth of Crystals, Vol. 7 (A. V. Shubnikov and N. N. Sheftal',
       eds.) Consultants Bureau, New York (1969), pp. 139-144

Magnetic susceptibility of CdSb
    K. D. Tovstyuk, E. I. Slyn'ko, and I. M. Rarenko
    Ukr. Fiz. Zh., 14:21-29 (1969)

The effective mass of holes in CdSb from reflectivity measurements
    M. Zavetova and V. Vorlicek
    Czech. J. Phys., 19B:677-680 (1969)

Anisotropy of electroconductivity in CdSb
    L. I. Anatychuk, V. D. Iskra, O. J. Luste, I. M. Rarenko, and
       L. I. Zarubin
    Phys. Stat. Sol., 27:101 (1968)

Anisotropy of the thermoelectric power of CdSb
in the intrinsic conduction region
    L. I. Anatychuk and O. Ya. Luste
    Fiz. Tekhn. Poluprovod., 3(2):432-433 (1968)
    Sov. Phys. Semicond., 3(2):354-355 (1968)

Electrical properties of indium- and copper-
doped cadmium antimonide crystals
    I. K. Andronik and E. K. Arushanov
    Fiz. Tekhn. Poluprovod., 2(6):869-872 (1968)
    Sov. Phys. − Semicond., 2(6):719-722 (1968)

Growth of cadmium antimonide dendrites from
melts doped with surface-active and surface-in-
active impurities
    M. Ya. Dashevskii and A. N. Poterukhin
    Izv. Akad. Nauk SSSR, Neorg. Mater., 4(9):1478-1482 (1968)
    Inorg. Mater., 4(9):1292-1295 (1968)

Conditions of the crystallization of the metastable compound $Cd_3Sb_2$
    K. A. Dobryden' and V. I. Psarev
    Izv. Akad. Nauk SSSR, Neorg. Mater., 4(7):1036-1039 (1968)
    Inorg. Mater., 4(7):911-913 (1968)

Doping of intermetallic crystals using melts
    N. L. Kostur and V. I. Psarev
    Rost Kristallov, Vol. 6B, Izv. Akad. Nauk (1965), pp. 288-295
    Growth of Crystals, Vol. 6B (A. V. Shubnikov and N. N. Sheftal',
       eds.) Consultants Bureau, New York (1968), pp. 96-102

Relationships between crystallization processes
and surface phenomena in alloyed cadmium antimonide melts
    V. B. Lazarev, M. D. Korsakova, and A. V. Pershikov
    Dokl. Akad. Nauk SSSR, 179(1-3):133-136 (1968)
    Dokl.−Phys. Chem., 179(1-3):156-159 (1968)

Magnetische Suszeptibilität des legierten und
flüssigen CdSb
    M. Matyas
    Helv. Phys. Acta, 41:1032-1035 (1968)
    Mn, Fe, Ni, and La doped

Magnetic susceptibility of CdSb doped with Mn,
Fe, Ni and La
    M. Matyas and M. Kligl
    Czech. J. Phys., 18B:376 (1968)

Segregation of gold and copper during normal
crystallization of cadmium antimonide
    Sh. Mavlonov, A. A. Kuliev, and A. Sadiev
    Izv. Akad. Nauk SSSR, Neorg. Mater., 4(10):1804-1805 (1968)
    Inorg. Mater., 4(10):1576-1577 (1968)

Field emission of CdSb:Ag
    H. Neumann
    Z. Naturforsch., 23:1240 (Aug. 1968)

Crystallization and structural stabilization of
CdSb and $Cd_3Sb_2$ in Cd−Sb melts
    V. I. Psarev and K. A. Dobryden'
    Rost Kristallov, Vol. 6B, Izv. Akad. Nauk (1965), pp. 247-255
    Growth of Crystals, Vol. 6B (A. V. Shubnikov and N. N. Sheftal',
       eds.) Consultants Bureau, New York (1968), pp. 60-65

Production and properties of alloyed single
crystals of CdSb
    G. V. Rakin
    Izv. Akad. Nauk SSSR, Neorg. Mater., 4(7):1032-1035 (1968)
    Inorg. Mater., 4(7):908-910 (1968)

Phase transitions in the $Zn_4Sb_3−Cd_4Sb_3$ system
    V. Ya. Shevchenko, V. A. Skripkin, Ya. T. Ugai, and T. A.
       Marshakova
    Izv. Akad. Nauk SSSR, Neorg. Mater., 4(8):1359-1360 (1968)
    Inorg. Mater., 4(8):1193-1194 (1968)

The properties of cadmium antimonide in strong
electric fields
    L. I. Zarubin, V. K. Malyutenko, and I. Yu. Nemish
    Ukr. Fiz. Zh. (USSR), 13(11):1911-1913 (Nov. 1968)
    Hall coefficient, mobility and specific resistance

The magnetic susceptibility of p-type CdSb
    M. Matyas
    Czech. J. Phys., 17B:227 (1967)

Phase diagram of the cadmium − antimony semiconductor system
    Ya. A. Ugai, T. A. Marshakova, K. B. Aleinkova, and N. P.
       Demina
    Izv. Akad. Nauk SSSR, Neorg. Mater., 3(8):1360-1369 (1967)
    Inorg. Mater., 3(8):1188-1194 (1968)

Stable and metastable equilibrium diagrams for
the cadmium − antimony system
    K. A. Dobryden' and V. I. Psarev
    Zh. Fiz. Khim., 40(11):2894-2896 (1966)

Galvanomagnetic effects in cadmium antimonide
    Tatsuo Kawasaki and Tetsuro Tanaka
    J. Phys. Soc. Japan, 21:2475 (1966)

Absorption edge of CdSb
    M. Zavetova
    Czech. J. Phys., 14(8):615-621 (1964)

The preparation and properties of some II − V
semiconducting compounds
    G. A. Silvey, V. J. Lyons, and V. J. Silvestri
    J. Electrochem. Soc., 108:653 (1961)

Anisotropy of electrical properties of single crystals of cadmium antimonide
  I. K. Andronik and M. V. Kot
  Fiz. Tverd. Tela, 2(6):1128-1133 (1960)
  Sov. Phys. — Solid State, 2(6):1022-1026 (1960)

Concerning the growth of crystals of intermetallic compounds in the Zn—Sb, Cd—Sb and In—Sb alloy systems
  V. I. Psarev
  Kristallografiya, 5(3):479-481 (1960)
  Sov. Phys. — Cryst., 5(3):456-458 (1960)

# 8. II—VI Compounds

## 8.a. General, Reviews, and Bibliographies

A survey of optical and electrical properties of
thin films of II — VI semiconducting compounds
R. Ludeke
J. Vacuum Sci. Tech., 8:199-209 (1971)

Electronic core levels of the IIB-VIA com-
pounds
C. J. Vesely and D. W. Langer
Phys. Rev., 4B:451-462 (1971)

Injection electroluminescence in II — VI semi-
conducting compounds, final report, 1 Oct.
1968 - 30 Sept. 1969
Manuel Aven, D. T. F. Marple, and H. H. Woodbury
General Electric Co., Schenectady, N. Y., AFML-TR-70-17;
AD-702897 (March 1970), 104 pp.

The rare-earth doping of binary and ternary
chalcogenides
M. R. Brown and W. A. Shand
J. Mater. Sci., 5:790-795 (1970)

Growth and structure of monocrystalline films
of $A^{II}B^{VI}$ compounds
I. P. Kalinkin, K. K. Muravyeva, L. A. Sergeyewa, V. B.
Aleskowsky, and N. S. Bogomolov
Kristal. Tech., 5(1):51-59 (1970)

Etude des vibrations de ressau des composés
semiconducteurs II — VI par spectrometrie in-
frarouge
R. Le Toullec
Ph. D. thesis, University of Paris, 1968, CNRS No. 4403 (1970)

Growth and electrophysical properties of monocrystalline films of cadmium and zinc chalcogenides
K. K. Muravjeva, I. P. Kalinkin, and V. B. Aleskovskii
Thin Solid Films, 5:7-14 (1970)

Study of the growth and structure of single-
crystal films of cadmium zinc chalcogenides
K. K. Murav'eva, I. P. Kalinkin, L. A. Sergeeva, V. B.
Aleskovskii, and N. S. Bogomolov
Izv. Akad. Nauk SSSR, Neorg. Mater., 6(3):434-440 (1970)
Inorg. Mater., 6(3):381-386 (1970)

Melting points of II—VI compounds under argon
pressure
K. Narita, H. Watanabe, and M. Wada
Japan. J. Appl. Phys., 9:1278 (1970)

Cadmium telluride, a bibliographic up-date
M. Neuberger
Electronic Properties Information Center, Hughes Aircraft
Co., Culver City, Calif. (May 28, 1970)
Up-date to cadmium telluride and the cadmium telluride—
mercury telluride system, DS-157 (1967)

Cadmium telluride—mercury telluride system, a
bibliographic up-date
M. Neuberger
Electronic Properties Information Center, Hughes Aircraft
Co., Culver City, Calif. (July 1, 1970)
Up-date to cadmium telluride and the cadmium telluride—mer-
cury telluride system, DS-157 (1967)

Research on improved II — VI crystals, final tech-
nical report - 1 June 1968 to 31 May 1970
L. R. Shiozawa and J. M. Jost
Clevite Corp., Electronic Research Div., 540 East 105th St.,
Cleveland, Ohio 44108, Contract F33615-68-C-1601 (August
1970), 108 pp.
Vapor-grown; thermodynamic properties; phase diagrams;
equilibrium constants for II—VI sublimation

Physics of stimulated emission in II — VI semiconducting compounds
H. F. Taylor
Naval Electronics Lab. Center (San Diego, Calif.), NELC-TR-
1713; AD-710988 (1970), 41 pp.

Semiconducting II — VI, IV — VI, and V — VI
Compounds
N. Kh. Abrikosov, V. F. Bankina, L. V. Poretskaya, L. E.
Shelimova, and E. V. Skudnova
Plenum Press, New York (1969), 246 pp.

Etude de l'effet faraday dans les composés semi-
conducteurs du groupe II — VI
Emile Amzallag
Ph. D. thesis, Université de Paris, France (1969)
To identify effective mass of free carriers and interband tran-
sitions

37

Impurity vibration modes in group II—VI crystals
R. Beserman
Ph.D. thesis, Paris, 1968, CNRS No. 3237 (April 1969)

Optical properties of some compound semiconductors in the 36-150 eV region
Manuel Cardona and Ruprecht Haensel
Deutsches Elektronen-Synchrotron, Hamburg, Germany, DESY 69/45 (Oct. 1969)
ZnS, CdS, ZnSe, CdSe, ZnTe, CdTe, PbS, and PbTe

Diffusion and luminescence of rare earth in cadmium sulfide, Interim Tech. Rept.
Dexter G. Girton
Ph.D. thesis, Ohio State Univ. Research Foundation, Columbus, RF-2419-TR-69-1; AROD-6835-2-E; AD-693887 (June 1969), 154 pp.
RE doping of CdS and ZnSe

Determination and systematization of the properties of binary compounds of group II metals with group VI elements
S. D. Gromakov and A. I. Partala
Zh. Fiz. Khim., 43:267-270 (1969)

Growth of the II—VI semiconductor compounds ZnS, ZnSe and ZnTe
H. Hartmann
Rost Kristallov, Vol. 7, Izv. Akad. Nauk (1967), pp. 252-257
Growth of Crystals, Vol. 7 (A. V. Shubnikov and N. N. Sheftal', eds.) Consultants Bureau, New York (1969), pp. 220-223

II—VI semiconducting compounds - data tables
Meta Neuberger
Electronic Properties Information Center, Hughes Aircraft Company, Culver City, Calif. 90230, Contract F 33615-68-C-1225, EPIC S-11 (October 1969), 156 pp., 386 refs.
Every property; mechanical, crystallographic, physical, thermal, magnetic, electronic, and optical of each of the II—VI binary semiconducting compounds

II—VI Compounds
Brian Ray
Vol. 2 of International Series of Monographs in the Science of the Solid State (B. R. Pamplin, ed.)
Pergamon Press, Oxford, New York, London (1969)
262 pp. + index

The preparation of graded-band-gap single crystals of II—VI compounds
P. Reimers
Phys. Stat. Sol., 35:707-716 (1969)

Imperfections ponctuelles dans les composés II—VI
M. Rodot, H. Rodot, G. Rouy, Y. Marfaing, and R. Triboulet
Rapp. final D. G. R. S. T., Action concertée: composants circuits micromin., Contrat No. 67 00 807 (May 1969), 57 pp.

Energy bands of hexagonal II—VI semiconductors
U. Rossler
Phys. Rev., 184:733-738 (1969)

Solid solutions in the pseudobinary (III—V) — (II—VI) systems and their optical energy gaps
W. Michael Yim
J. Appl. Phys., 40:2617-2624 (1969)

Technology, structural defects, and properties of semiconductor compounds of $A^{II}B^{VI}$-type single crystals
Jozef Zmija
Biul. Wojsk. Akad. Tech., 18(8):Suppl. 5-262 (1969)
134 refs.

Electroluminescence in II—VI compounds
Albert G. Fischer
RCA Lab., Princeton, N. J., Proceedings of International Conference on Luminescence 1966, Vol. 2 (G. Szigeti, ed.), Budapest, Hungary (1968), pp. 1765-1781

Optical transitions in II-VI compounds containing transition elements (survey)
H. E. Gumlich
Optical Properties of Dielectric Films (Norman H. Axelrod, ed.)
The Electrochem. Soc., New York (1968), pp. 45-77

Optical properties of II—VI wurtzite-type semiconductors in the band-edge region
E. Gutsche
Humboldt-Univ., Berlin, Germany
Proc. 9th Intern. Conf. Phys. Semiconductors
Nauka, Leningrad, USSR (1968), pp. 1157-1171

Structure of $A^{II}B^{VI}$ epitaxial films
I. P. Kalinkin, L. A. Sergeeva, K. K. Murav'eva, K. K. Khristoforov, Ya. G. Pevzner, and V. B. Aleskovskii
Protsessy Rosta Strukt. Monokrist. Sloev Poluprov., Tr. Simp. 1966 (Publ. 1968), Vol. 1, pp. 355-363 (L. N. Aleksandrov, ed.)
Izv. "Nauka" Sib. Otd., Novosibirsk, USSR

An investigation of the crystal growth of the heavy metal sulfides in supercritical hydrogen sulfide
Leroy C. Lewis
Ph.D. thesis, Oregon State University, 1968; University Microfilms, Inc., Ann Arbor, Mich., Order No. 68-11,915

Preparation and analysis of $A^{II}B^{VI}$ semiconducting compounds
I. B. Mizetskaya
Izv. Akad. Nauk SSSR, Neorg. Mater., 4(8):1220-1224 (1968)
Inorg. Mater., 4(8):1072-1075 (1968)

Photometric methods of determining microimpurities in compounds of A(II) B(VI) type
R. P. Pantaler et al.
Aztec School of Languages, Inc., Maynard, Mass., Research Translation Division, Rept. 179-69-405-RULL (May 1969), 10 pp.
Tr. Komis. Po Analit. Khim. Akad. Nauk SSSR (Moscow), 16:24-29 (1968)

Etudes des propriétés électriques des états de surface dans le silicium et les composés II—VI
C. Sebenne
Rapp. final D. G. R. S. T., Action concertée: composants circuits micromin., Contrat No. 6700761 (1968), 75 pp.

Physical—chemical properties and preparation of $A^{II}B^{VI}$ semiconductor compounds
A. V. Vanyukov
Mosk. Inst. Stali Splavov, No. 52, 244-261 (1968)
Critical review

Cadmium Telluride
B. M. Vul, ed.
Nauka, Moscow (1968), 142 pp.

Status of diffusion data for binary compound semiconductors
David W. Yarbrough
U. S. Govt. Res. Develop. Rept., 68:162 (1968); AD-670014, 69 pp.
III-V, II—VI compounds, and SiC

39

Physics and Chemistry of II—VI Compounds
M. Aven and J. S. Prener, eds.
North-Holland Publishing Co., Amsterdam (1967), 850 pp.

Band structure of II—VI compounds
M. Balkanski
J. Phys., 28, Suppl. No. 5-6: C3-36-42 (May–June 1967)

Chemistry of semiconductor materials and materials for quantum electronics
N. P. Sazhin, N. Kh. Abrikosov, N. P. Luzhnaya, and E. G. Ippolitov
Razv. Obshch., Neorg. Anal. Khim. SSSR, 1917–1967, Akad. Nauk SSSR, Inst. Istor. Estestvozn. Tekh. (1967), pp. 214–223

Energy structure of bands of $A^{II}B^{VI}$-type solid solutions
V. V. Sobolev
Opt. Mekh. Svoistva Poluprov. Dielek., Akad. Nauk Mold. SSR, Inst. Prikl. Fiz. (1967), pp. 27–34

II—VI Semiconducting Compounds
D. G. Thomas, ed.
W. A. Benjamin, New York (1967), 1489 pp.
Intern. Conf., Providence, Rhode Island (Sept. 6–8, 1967)

Methods of activating and recrystallizing thin films of II—VI compounds
A. Vecht
Phys. Thin Films, 3:165–210 (1966)
Doping; simultaneous deposition of host and dopant; annealing; 87 refs.

Review of the physical properties of some new semiconductors of groups II, IV, and VI
Jozef Wojas
Postepy Fiz., 17(3):317–329 (1966)
Preparation, purification, and semiconducting properties of HgSe and HgTe

Research on II—VI compound semiconductors
L. R. Shiozawa, J. L. Barrett, G. P. Chotkevys, et al.
Clevite Corporation, Cleveland, Ohio, final report, Contract AF 33(616)-6865, Project 7021, Task 70846, ARL 62-365 (June 1962)
Two years of research on the preparation, purification, crystal growth, and measurement of the fundamental bulk properties of CdS, CdSe, ZnTe, and CdS–CdSe mixed crystals are summarized. Purification of component elements by distillation and zone–refining are described

Properties of cadmium sulfide, zinc sulfide, and mercuric sulfide, Parts I-III, an annotated bibliography
Helen M. Abbott, comp.
Lockheed Aircraft Corp., Missiles and Space Division, Sunnyvale, Cal., SRB-61-2, Vol. 1 (March 1961)

Properties of cadmium sulfide, zinc sulfide, and mercuric sulfide, Part IV, an annotated bibliography
Helen M. Abbott, comp.
Lockheed Aircraft Corp., Missiles and Space Div., Sunnyvale, Calif., SRB-61-2, Vol. II (March 1961)

Crystal growth of CdS and other II—IV compounds
P. Flogel
Z. Anorg. Allgem. Chem., 370 (1/2):16–30 (1960)

## 8.b. Zinc Compounds

### 8.b.1. ZnO

#### Preparation and Properties

A mechanism for the formation of dislocations in ZnO single crystals
K. A. Jones
J. Cryst. Growth, 8:63–68 (1971)

Vapor-condensed single crystals of ZnO
A. Billmann
Compt. Rend. B, Sci. Phys., 270:170–173 (1970)

Growth mechanism of ZnO single crystal by oxidation of $ZnI_2$
Masami Hirose and Yoshio Furuya
Japan. J. Appl. Phys., 9:423–424 (1970)

Growth of ZnO single crystal from $ZnBr_2$
M. Hirose, Y. Furuya, and I. Kubo
Japan. J. Appl. Phys., 9:726–727 (1970)

Zinc oxide films by oxidation of zinc selenide films
Masaru Ohnishi, Sumiaki Ibuki, and Michio Yoshizawa
Mitsubishi Electric Corp., Ger. Offen. 1,948,376 (April 9, 1970), 21 pp.

Structure and properties of single-crystal zinc oxide films
R. A. Rabadanov, S. A. Semiletov, and Z. A. Magomedov
Fiz. Tverd. Tela, 12(5):1431–1436 (1970)
Sov. Phys. — Solid State, 12(5):1124–1127 (1970)
Grown from gas phase on cleaved mica

High-resistivity transparent ZnO thin films
D. L. Raimondi and E. Kay
J. Vacuum Sci. Tech., 7:96–99 (1970)

Lattice parameters of ZnO from 4.2° to 296°K
R. R. Reeber
J. Appl. Phys., 41:5063–5066 (1970)

Crystal structure of zinc oxide made by oxidizing thin films of ZnS
K. V. Shalimova, V. A. Dmitriev, and N. M. Satybaev
Kristallografiya, 15:200–201 (1970)
Sov. Phys. — Cryst., 15:167–168 (1970)

Densification and grain growth in hot-pressed zinc oxide
S. K. Dutta and R. M. Spriggs
Mat. Res. Bull., 4:797–806 (1969)

Crystallographic orientation of zinc-oxide films deposited by triode sputtering
N. F. Foster
J. Vacuum Sci. Tech., 6:111–114 (1969)

Growth of ZnO single crystal by oxidation of $ZnI_2$
M. Hirose and I. Kubo
Japan J. Appl. Phys., 8:402 (1969)

Preparation and properties of thin films of ZnO
N. L. Kenigsberg and A. N. Chernets
Fiz. Tverd. Tela, 10:2834–2836 (1968)
Sov. Phys.–Solid State, 10:2235 (1969)

Pyrolytic growth of semiconductor layers of ZnO
in an inert medium
V. F. Korzo, P. S. Kireev, and G. A. Lyashchenko
Izv. Akad Nauk SSSR, Neorg. Mater., 5(2):367-368 (1969)
Inorg. Mater., 5(2):306-307 (1969)

Crystallization kinetics of zincite under hydro-
thermal conditions
I. P. Kuz'mina
Kristallografiya, 13(5):920-921 (1968)
Sov. Phys. — Cryst., 13(5):803-804 (1969)

Intake of iron in hydrothermal zincite crystals
M. M. Lukina and L. A. Chernyaev
Kristallografiya, 13:1113-1114 (1968)
Soviet Phys. — Cryst., 13:979 (1969)

Epitaxial thin films of ZnO on CdS and sapphire
G. A. Rozgonyi and W. J. Polito
25th National Vacuum Symp., Pittsburgh (Oct. 30 - Nov. 1, 1968)
J. Vacuum Sci. Technol., 6(1):115-119 (1969)

Optical properties of impurities-doped hydrother-
mally grown zinc oxide
N. Sakagami and S. Hasegawa
J. Ceram. Soc. Japan, 77(9):309-312 (1969)

Vapour growth of Cu- and Li-doped single-crystal zinc oxide in the resistivity range 50 to 10³
ohm-cm
J. A. Savage and E. M. Dodson
J. Mater. Sci., 4:809-813 (1969)

Selective doping of piezoelectric crystals by
ion implantation, Semiannual report; 1 January
1969 through 30 June 1969
G. A. Shifrin, K. R. Zanio, D. M. Jamba, W. R. Jones, O. J.
Marsh, and R. G. Wilson
Hughes Research Labs., Malibu, California, Contract N00014-
69-C-0171 (August 1969), 48 pp.

Preparation of ZnO thin-film transducers by
vapor transport
R. F. Belt and G. C. Florio
J. Appl. Phys., 39:5215 (1968)

Method of forming zinc oxide infrared trans-
mitting optical element
Edward Carnall, Jr., and Donald W. Roy
Eastman Kodak Co., U. S. Patent 3,415,907 (Dec. 17, 1968)

Vapour growth of single-crystal zinc oxide
E. M. Dodson and J. A. Savage
J. Materials Sci., 3:19-25 (1968)

Growth of ZnO single crystals by the vapor-
phase reaction method
K. F. Nielsen
J. Crystal Growth, 3:141-145 (1968)

Hydrothermalsynthese von Zinkit
D. Rykl and J. Bauer
Kristall u. Tech., Dtsch., 3(3):375-384 (1968)

Optical and electrical properties of zinc oxide
single crystals
Prafulla Chandra, V. B. Tare, and A. P. B. Sinha
Indian J. Pure Appl. Phys., 5:6 (1967)

Localized cooling in flux crystal growth
A. B. Chase and J. A. Osmer
J. Am. Ceram. Soc., 50:325 (1967)

Investigation of evaporation and condensation
mechanism
F. H. Cocks, B. N. Das, C. B. Lamport, A. A. Menna, E. A.
Trickett, et al.
Final Report (15 May 1966 - 31 August 1967), Tyco Labs., Inc.,
Waltham, Mass., Contract AF 33(615)-5060, AFML-TR-
67-385; AD-666 073 (Dec. 1967), 149 pp.

On the morphology and polarity of the ZnO crys-
tal
H. Iwanaga and N. Shibata
Japan J. Appl. Phys., 6:415 (1967)

The hydrothermal growth of low-carrier-concen-
tration ZnO at high water and hydrogen pressures
E. D. Kolb, A. S. Coriell, R. A. Laudise, and A. R. Hutson
Mat. Res. Bull., 2:1099-1106 (1967)

Zinc-oxide film microwave acoustic transducers
R. M. Malbon, D. J. Walsh, and D. K. Winslow
Appl. Phys. Letters, 10:9 (1967)

Growth of ZnO single crystals
Y. S. Park and D. C. Reynolds
J. Appl. Phys., 38:756 (1967)

Hydrothermal growth of impurities-doped ZnO
single crystals
N. Sakagami, S. Hasegawa, and G. Ohara
Rep. Res. Inst. Elec. Commun. Tohoku Univ., 19:199-213 (1967)

Hydrothermal growth of lithium and copper-doped
ZnO single crystals
N. Sakagami, S. Hasegawa, and G. Ohara
J. Ceram. Assoc. Japan, 75:255-257 (1967)

Vapor phase growth of ZnO single crystals
E. A. Weaver
J. Crystal Growth, 1:320-333 (1967)

Hydrothermally grown ZnO crystals of low and
intermediate resistivity
E. D. Kolb and R. A. Laudise
J. Am. Ceram. Soc., 49:302 (1966)

Crystallization of zincite under hydrothermal
conditions
I. P. Kuz'mina and V. F. Antonova
Rost Kristallov, Vol. 4, Izv. Akad. Nauk (1964), pp. 151-156
Growth of Crystals, Vol. 4 (A. V. Shubnikov and N. N. Sheftal',
eds.), Consultants Bureau, New York (1966), pp. 125-128

Growth and morphology of hydrothermally crys-
tallized zinc oxide
R. A. Laudise
Acta Cryst., 21:Pt. 7, Suppl. A262 (Dec. 1966)
Seventh International Congress and Symp. Internat'l Union
of Crystallography, Moscow (1966)

Etude par spectrométrie γ de l'imprégnation de
l'oxyde de zinc par le cobalt et l'argent
G. Mesnard and M. Ladous
Phys. Stat. Sol., 14:K59 (1966)

Preparation and properties of noncrystalline
zinc-oxide films
R. A. Mickelsen and W. D. Kingery
J. Appl. Phys., 37:3541 (1966)

Preparation of ZnO thin films by sputtering of
the compound in oxygen and argon
G. A. Rozgonyi and W. J. Polito
Appl. Phys. Letters, 8:220 (1966)

Vapor reaction growth of ZnO single crystal
  T. Takahashi, A. Ebina, and A. Kamiyama
  Japan. J. Appl. Phys., 5:560 (1966)

Schmelzen von Zinkoxyd durch Hochfrequenzer-
hitzung
  J. Burmeister
  Phys. Stat. Sol., 10:K1 (1965)

Flux growth of ZnO using a temperature gradient
  A. B. Chase and J. A. Osmer
  Aerospace Corp., Contract AF 04(695)-669, TDR-669(9230-02)-
    1; SSD-TR-65-155 (Nov. 1965)

Kinetische Untersuchung der Festkörperreaktion
ZnS + CdO = ZnO + CdS
  H. Kleykamp
  G.-M. Schwab and R. Sizmann
  Z. Physik. Chem. (Frankfurt), 44:15 (1965)

Growth of zinc-oxide single crystals by hydroly-
sis of zinc fluoride
  I. Kubo
  Japan. J. Appl. Phys., 4:225 (1965)

Crystal growth in hydrothermal systems
  Jerry W. Moody and Richard D. Himes
  Battelle Tech. Rev., 14:3 (1965)

The activation energy for oxygen desorption
from zinc-oxide surfaces
  H. Watanabe, M. Wada, and T. Takahashi
  Japan. J. Appl. Phys., 4:945 (1965)

Growth of ZnO single crystals by a traveling
solvent zone technique
  G. A. Wolff and H. E. LaBelle, Jr.
  J. Am. Ceram. Soc., 48:441 (1965)

The growth of crystals by solvent zone tech-
niques
  G. A. Wolff and A. I. Mlavsky
  Colloq. Intern. Centre Natl. Rech. Sci. (Paris) No. 152, Ad-
    sorption et Croissance Cristalline (1965), p. 711

Über die Homogenität dotierter Zinkoxide
  H. L. Gruber and R. Ulrich
  Monatsch. Chem., 95:1026 (1964)

Hydrothermal growth of large sound crystals of
zinc oxide
  R. A. Laudise, E. D. Kolb, and A. J. Caporaso
  J. Am. Ceram. Soc., 47:9 (1964)

Raw materials for refractory oxide ceramics
  Hayne Palmour, III, Jerry M. Waller, R. Douglas McBrayer,
    Dong M. Choi, and Lawrence D. Barnes
  N. Carolina State College, ML-TDR-64-110; AD-605804 (July
    1964)
  Contracts AF 33(616)-7288 and AF 33(657)-8741

Optical and electrical properties of ZnO crystals
  H. Watanabe, M. Wada, and T. Takahashi
  Japan. J. Appl. Phys., 3:617 (1964)

The solubility of zincite in basic hydrothermal
solvents
  R. A. Laudise and E. D. Kolb
  Am. Mineralogist, 48:642 (1963)

Crystal growth of zinc-oxide by chemical reac-
tion of zinc-fluoride with air
  Ikumaro Kubo
  J. Phys. Soc. Japan, 16:2358 (1961)

Hydrothermal synthesis of zinc oxide and zinc
sulfide
  R. A. Laudise and A. A. Ballman
  J. Phys. Chem., 64:688 (1960)

Electronic processes in zinc oxide
  G. Heiland and E. Mollwo
  Solid State Phys., 8:193 (1959)

The preparation of zinc-oxide crystals having
definite additives
  G. Bogner and E. Mollwo
  J. Phys. Chem. Solids, 6:136 (1958) (in German)
  AEC-tr-4661

Electroluminescence
  F. Pribyl
  Slaboproudy Obzor., 19:239-246 (1958)

Concentration of hydrogen and semiconductivity
in ZnO under hydrogen-ion bombardment
  J. J. Lander
  J. Phys. Chem. Solids, 3:87 (1957)

## Electrical Properties

Etude de la résistance électrique de l'oxyde de
zinc en poudre en fonction de la pression et de
la température
  Robert Guillien, Wolfgang Palz, and Edouard Yvroud
  Compt. Rend., Ser. B, 270:101-104 (1970)

Polar surfaces of zinc-oxide crystals
  G. Heiland and P. Kunstmann
  Surface Sci., 1:72-83 (1969)

Untersuchung von Raumladungsschichten an
Zinkoxid-Oberflächen mit Hilfe der Elektrore-
flexion
  Bernd Hoffmann
  Z. Physik, 219:354-363 (1969)

Surface photo-voltage of zinc oxide powder
layer
  A. Kondo, Y. Takahashi, and S. Murakawa
  Kogyo Kagaku Zasshi, 72:140-145 (1969)

Differential capacity of a ZnO:Li single crystal
between asymmetric contacts
  R. Meaudre and G. Mesnard
  Compt. Rend., Ser. B, Sci. Phys., 268:293-296 (1969)

Photoconductivity and polarization by laser-
radiation in the volume of ZnO-crystals
  E. Mollwo and G. Pensl
  Z. Physik, 228:193-221 (1969)

Surface states associated with alcohols in zinc
oxide
  S. Roy Morrison
  Bull. Am. Phys. Soc., 14:340 (1969)

Surface states associated with chemisorbed
species on zinc oxide
  S. R. Morrison
  Surface Sci., 13:85-98 (1969)

Contact of ZnO thin films with Rhodamine B dye
  Ikuo Niikura, Hideo Watanabe, and Masanobu Wada
  Japan. J. Appl. Phys., 8(6): 755-758 (1969)

Excited terminal states of a bound exciton—donor
complex in ZnO
  D. C. Reynolds and T. C. Collins
  Bull. Am. Phys. Soc., 14:429 (1969)

Energy bands of hexagonal II—VI semiconductors
U. Rossler
Phys. Rev., 184:733–738 (1969)

Transmission of 35 GHz microwaves through acousto-electrically excited ZnO
M. Bruun and N. I. Meyer
Solid State Commun., 6:359–362 (1968)

Effect of chemisorbed oxygen, carbon, carbon monoxide and carbon dioxide on some electrical properties of zinc oxide single crystal
Herschel Clopper
Ph.D. thesis, Rice University, Houston, Texas (1968), 191 pp.
Univ. Microfilms, Ann Arbor, Michigan, Order No. 68–15610

Dielectric, piezoelectric, and electromechanical coupling constants of zinc-oxide crystals
D. F. Crisler, J. J. Cupal, and A. R. Moore
Proc. IEEE, 56:225–227 (1968)

Deposition, structure, and performance of thin-film piezoelectric transducers
N. F. Foster
Bell Telephone Lab., Allentown, Pa., Sendai Symp. Acousto-electron., 6th (1968), pp. 11–20

A study of some of the electrical properties of zinc-oxide
William Andrew Keenan
Ph.D thesis, Princeton Univ., New Jersey (1968), 93 pp.
University Microfilms, Ann Arbor, Mich., Order No. 68–9683

Conductivity of oriented layers of carbon-doped zinc oxide
V. F. Korzo, L. A. Ryabova, Ya. S. Savickaya, and G. A. Lyashchenko
Radiotekh. i Elektron. SSSR, 13:2041–2043 (1968)

Transverse conductivity of oriented layers of carbon-doped zinc oxide (15–320°C)
V. F. Korzo, L. A. Ryabova, Ya. S. Savitskaya, and others
Radio Eng. Electron. Phys., 13:1789–1791 (Nov. 1968)

Gigahertz acousto-electric oscillations in zinc oxide
J. D. Maines, F. G. Marshall, E. G. S. Paige, and R. A. Stuart
Phys. Letters A, 26:388–389 (1968)

The effect of water vapour on the photoconductivity of zinc oxide
J. McK. Nobbs
J. Phys. Chem. Solids, 29:439–450 (1968)

The effect of doping on the non-stoichiometry of zinc oxide
V. J. Norman
Australian J. Chem., 21:299–305 (1968)
Influence on electrical conductivity

Estimation of the spontaneous polarization of hexagonal ZnS, CdS and ZnO crystals
Tomoya Ogawa
J. Phys. Soc. Japan, 25:1126–1128 (1968)

Piezoelectric crystals in audioelectronics
Wincenty Pajewski
Przegl. Elektron., 9:417–423 (1968)

Current saturation and electron drift mobility in ZnO
Andreas Rannestad
Norwegian Defence Research Establishment, Kjeller, AD–689184 (Dec. 1968), 18 pp.

Studies in zinc-oxide photoconductivity
Howard Saltsburg
Gulf General Atomic, San Diego, California, Final Tech. Report, Contract DAAK02–67–C–0115, GA–8431; AD–673836 (August 1968), 46 pp.

Electronic drift mobilities and space-charge-limited currents in lithium-doped ZnO
M. A. Seitz
Northwestern University, 1966, Dissertation Abstr. 27, 3545B (1966–67)
J. Phys. Chem. Solids, 29:1033–1049 (1968)

Piezoelectric materials
Lebo R. Shiozawa
Clevite Corp., U. S. Patent 3,409,464 (Nov. 5, 1968) applied April 29, 1964, 4 pp.

Zinc oxide is made electroconductive by absorbing volatile halogen compound $XMY_3$ onto nonconducting ZnO and heating at 400–900°C, where M is Si, C, Ti, or Ge, X is halogen, hydrogen, or alklyl group of one to three carbon atoms and Y is halogen
St. Joseph Lead, British Patent, 1,114,509; appl. (US) 13 Jan 66 and 23 Nov 66; publ. 14 June 68

The electrical and optical properties of refractory oxides, final rept., Aug. 1966–July 1968
Walter C. Tripp and Edward T. Rodine
Systems Research Labs, Inc., Dayton, Ohio, Contract AF 33(615)–2765, SRL–11580 ARL–68–0221 (December 1968), 147 pp.

Electron multiplication and surface charge on zinc-oxide single crystals
R. Williams and A. Willis
J. Appl. Phys., 39:3731–3736 (1968)

Faraday rotation in ZnO: Determination of the electron effective mass
W. S. Baer
Phys. Rev., 154:785 (1967)

Conductivité électrique des oxydes de zinc purs et dopés, catalyseurs d'hydrogénation de l'éthylène. II. Adsorption chimique d'hydrogène et d'éthylène par l'oxyde de zinc
F. Bozon-Verduraz, B. Arghiropoulos, and S. J. Teichner
Bull. Soc. Chim. Fr., 8:2854–2861 (1967)

Effect of oxygen chemisorption on the electrical conductivity of ZnO single crystals
P. Chandra, V. B. Tare, and A. P. B. Sinha
Indian J. Pure Appl. Phys., 5:313–317 (1967)

Polycrystalline zinc-oxide dielectrics
R. A. Delaney and H. D. Kaiser
J. Electrochem. Soc., 114:833 (1967)

High-electric-field effects in ZnO single crystals
N. I. Meyer, E. Mosekilde, and M. H. Jorgensen
II—VI Semiconducting Compounds, 1967 Intern. Conf., Providence, R. I. (Sept. 6–8, 1967), pp. 950–962
W. A. Benjamin, Inc., New York (1967)

Adsorption of hydrogen atoms on ZnO and the influence on its electro-conductivity
I. A. Myasnikov and I. N. Pospelova
Zh. Fiz. Khim., 41:567–575 (1967)

The effect of variations in the concentration of atmospheric constituents on the electrical conductivity of zinc oxide
J. McK. Nobbs
J. Phys. Chem. Solids, 28:205 (1967)

Acoustoelectric effects in solids
Andreas Rannestad
Norwegian Defence Research Establishment, Kjeller, Progress report (1 Apr. 1966 – 31 Mar. 1967) Contract AF 61(052)-958, PR-4; AFCRL-67-0636; AD-666 452 (June 1967), 7 pp.

Optical and electrical properties of ZnO monocrystals with excessive Zn
E. Scharowsky
Royal Radar Establishment, Malvern, England, RRE-TRANS-165; TIL/OT/8701 (1967), 14 pp.

Photoconductive effect on ZnO-Cu crystals
M. Sumita
Japan. J. Appl. Phys., 6:418 (1967)

Electrical properties of ZnO:Ni and ZnO:Co crystals
Minoru Sumita
Japan. J. Appl. Phys., 6:1469-1470 (1967)

Slow ambient insensitive traps in lithium-doped zinc oxide
H. Van Hove, D. Bohrmann, and A. Luyckx
Surface Sci., 7:474-477 (1967)

L'influence de l'anhydride maléique adsorbé sur la photoconduction de ZnO microcristallin
T. Yamaguchi, H. Kokado, and E. Inoue
J. Chem. Soc. Japan, Pure Chem., Sect. 88:144-148 (1967)

Contribution à l'étude de la variation de résistance de quelques semiconducteurs en poudre, et particulièrement de l'oxyde de zinc, en fonction de la pression et de la température
E. Yvroud
Ph.D. thesis, Univ. Nancy, Oudart et Clement, Nancy (1967), 72 pp.

Variation en fonction de la température de la résistivité de l'oxyde de zinc fritté contenant Cu-ZnO en impureté
N. A. Dereberya
Izv. Vysshikh Uchebn. Zaveden., Fiz., 2:177-179 (1966)

Pyroelectricity of zinc oxide
G. Heiland and H. Ibach
Solid State Commun., 4:353-356 (1966)

Electronic structure of zinc oxide and related crystals
Frank Herman
Materials Sciences Lab., Lockheed Missiles and Space Co., Palo Alto, Calif., Quarterly Status Report, 1 May – 31 July 1966, QSR-1; AD-638482 (July 1966), Contract AF33(615)-5072, 6 pp.

Non-ohmic behaviour and oscillation phenomena in ZnO single crystals
N. I. Meyer, M. H. Jorgensen, and E. Mosekilde
Proc. of International Conf. on Physics of Semiconductors, Phys. Soc. Japan, Tokyo (1966), pp. 406-410

Non-ohmic behaviour in ZnO single crystals
N. I. Meyer and M. H. Jørgensen
Phys. Letters, 20:450 (1966)

Photoconductivity decay of ZnO crystals in oxygen
H. Van Hove and A. Luyckx
Solid State Commun., 4:603 (1966)

Effect of oxygen on the electrical conductivity of zinc-oxide powder
H. Watanabe, M. Wada, and T. Takahaski
ATS-23T98J (April 1967), 14 pp.
Denshi Shashin, Japan, 7:2-9 (1966)

Influence of ambient oxygen on the electronic conductivity of zinc oxide powder
H. Watanabe, M. Wada, and T. Takahashi
Rept. Res. Inst. Elec. Commun. Tohoku Univ., 18:159-202 (1966)

Propriétés photoconductrice de ZnO modifié par $H_2S$
T. Yamanouchi, E. Ochiai, and F. Iwai
J. Chem. Soc. Japan, Ind. Chem. Sect., 69:179 (1966)

Influence of illumination on surface equilibria in ZnO
J. Haber and A. Kowalska
Bull. Acad. Polon. Sci., 13:463 (1965)

Hall-Effekt und Thermokraft von gesintertem Zinkoxyd
H. Heinrich
Z. Naturforschung, 20a:99 (1965)

Hall-Effekt und Photoleitung an Nickel – dotiertem zinkoxyd
H. Heinrich and H. Preier
Z. Naturforschung, 20a:249 (1965)

Hypersensitization of photoconduction in microcrystalline zinc oxide
E. Inoue, H. Kokado, and T. Yamaguchi
J. Phys. Chem., 69:767 (1965)

Surface barriers on ZnSe and ZnO
C. A. Mead
Phys. Letters, 18:218 (1965)

Electroluminescence and current oscillations in ZnO single crystals excited by short pulses
N. I. Meyer, M. H. Jørgensen, and I. Balslev
Solid State Commun., 3:393 (1965)

Piezoelectric measurements on zinc-oxide crystals
C. Solbrig
Z. Physik – West Germany, 184:293-298 (1965)
TT-65-14162

The study of Seebeck effect on the semiconductor system $ZnO-Al_2O_3$
I. Ursu, F. Puskas, and V. Cristea
Rev. Roumaine Phys., 10:223-228 (1965)

Sensitization of photoconduction in a zinc oxide film by eosin
S. J. Dudkowski and L. I. Grossweiner
J. Opt. Soc. Am., 54:486 (1964)

Etude de la conductivité électrique de l'oxyde de zinc
I. Manikowska
Rocz. Chem. Polska, 38:1641 (1964)

Piezoelektrische Potentialfelder um Stufenversetzungen beliebiger Richtung in piezoelektrischen Kristallen mit elastischer Isotropie
L. Merten
Z. Naturforschung, 19a:1161 (1964)

Propriétés électroniques de monocristaux d'oxyde de zinc contenant des éléments "accepteurs"
Guy Mesnard and Claude Eymann
Compt. Rend., 258:3672 (1964)

Der Einfluss einer Glühbehandlung auf das Leitfähigkeitsverhalten dünner ZnO-Schichten
H. Preier
Z. Naturforschung, 19a:1431 (1964)

Electronic structure of copper impurities in ZnO
R. E. Dietz, H. Kamimura, M. D. Sturge, and A. Yariv
Phys. Rev., 132:1559 (1963)

Polar properties of zinc oxide
G. Heiland, P. Kunstmann, and H. Pfister
Z. Physik - West Germany, 176:485–497 (1963)
TT-65-11534

Electrophotographic properties of zinc oxide dispersed in dielectric resin
Eiichi Inoue
(Tokyo Institute of Technology, Meguroku, Tokyo, Japan)
Proc. Tech. Assoc. of the Graphic Arts - Fourteenth Annual Meeting (June 11-13, 1962), p. 175

Measurements of the electrical conductivity and the Hall effect of ZnO-crystals and their interpretation in terms of impurity bands
G. Bogner
Physics and Chemistry of Solids, 19:235-250 (1961)

Photoconductivity and chemisorption kinetics in sintered zinc-oxide semiconductor
D. B. Medved
J. Phys. Chem. Solids, 20:255 (1961)

The effect of illumination on the contact potentials of some semiconductors
I. A. Akimov
Dokl. Akad. Nauk SSSR, 128(4):691-694 (1959)
Sov. Phys. — Dokl., 4(5):1046-1049 (1960)

Piezoelectricity and conductivity in ZnO and CdS
A. R. Hutson
Phys. Rev. Letters, 4:505 (1960)

Pressure dependence of the resistivity of zinc oxide
A. R. Hutson, W. Paul, et al.
Z. Physik, 158:151 (1960)

Reactions of lithium as a donor and an acceptor in ZnO
J. J. Lander
J. Phys. Chem. Solids, 15:324 (1960)

Untersuchungen der elektrischen Leitfähigkeit von Zinkoxyd
J. Deren, J. Haber, and T. Wilkowa
Z. Physik, 155:453 (1959)

Electronic properties of ZnO
A. R. Hutson
J. Phys. Chem. Solids, 8:467 (1959)

Surface potential, field-effect mobility, and surface conductivity of ZnO crystals
H. J. Krusemeyer
Phys. Rev., 114:655 (1959)

Diffusion of excess zinc in zinc-oxide crystals
R. Pohl
Z. Physik, 155:120 (1959)

The effect of Cu, Zn, and Li impurities on the dielectric properties of zinc oxide
M. Blanchard and M. Martin
J. Phys. Radium, 19:677 (1958)

Photoconduction and surface effects with zinc-oxide crystals
R. J. Collins and D. G. Thomas
Phys. Rev., 112:388 (1958)

Feldeffekt und Photoleitung an ZnO-Einkristallen
Gerhard Heiland
J. Phys. Chem. Solids, 6:155 (1958)

The dielectric constant of zinc oxide over a range of frequencies
N. H. Langton and D. Matthews
British J. Appl. Phys., 9:453 (1958)

Concentration and mobility of electrons in zinc-oxide single crystals with selected additives
H. Rupprecht
Phys. Chem. Solids, 6:144-154 (1958), TT-65-17076

Photoproperties of zinc oxide with ohmic and blocking contacts
H. J. Gerritsen, W. Ruppel, and A. Rose
Helv. Phys. Acta, 30:504 (1957)

Seebeck effect in zinc oxide
A. R. Hutson
Bull. Am. Phys. Soc., 2:56 (1957)

Hall-effect studies of doped zinc-oxide single crystals
A. R. Hutson
Phys. Rev., 108:222 (1957)

Investigation of the connection between the electro-conductivity and the adsorption properties of zinc oxides
I. A. Myasnikov
Zhur. Fiz. Khim., 31:1721 (1957)
AEC-tr-3301, 19 pp.

Influence of impurities of the photoconductance of zinc oxide
H. A. Papazian, P. A. Flinn, and D. Trivich
J. Electrochem. Soc., 104:84 (1957)

Surface conductivity produced on zinc oxide by zinc and hydrogen
D. G. Thomas and J. J. Lander
J. Phys. Chem. Solids, 2:318 (1957)

Catalytic and electrical properties of semiconductors. II. Semiconductors
J. Decrue
Helv. Phys. Acta, 39:812 (1956)

Semiconducting properties of ZnO
A. R. Hutson
Bull. Am. Phys. Soc., 1:381 (1956)

Electrical and optical properties of ZnO
E. Mollwo
Photoconductivity Conference, held at Atlantic City (Nov. 4-6, 1954), pp. 509-528, John Wiley and Son, Inc., New York (1956)

Hydrogen as a donor in zinc oxide
D. G. Thomas and J. J Lander
J. Chem. Phys., 25:1136 (1956)

Conductivity and Hall effect of ZnO at low temperatures
S. E. Harrison
Phys. Rev., 93:52 (1954)

Work function and energy levels in insulators
D. A. Wright
Proc. Phys. Soc. (London), 60:13 (1948)

## Optical Properties

Absorption and emission in copper-doped zinc oxide
R. Dingle
Bull. Am. Phys. Soc., 15:280 (1970)

Relative signs of nonlinear optical coefficients of polar crystals
Robert C. Miller and William Z. Nordland
Appl. Phys. Letters, 16:174-176 (1970)

Ultraviolet reflectivity spectra of ZnO
Y. S. Park, T. S. Wagner, M. Skibowski, and R. Klucker
Bull. Am. Phys. Soc., 15:43 (1970); also DESY 70/9 (Feb. 1970)

X-ray spectroscopic studies of the chemical bond in oxides
H.-U. Chun
Angew. Chem., 81(10):400-401 (1969)

Comparison of ultraviolet reflectivity spectrum and electron energy loss measurements of ZnO
R. L. Hengehold and F. L. Pedrotti
Bull. Am. Phys. Soc., 14:304 (1969)

Infrared electroreflectance of zinc oxide
H. Luth
Phys. Stat. Sol., 33:267-275 (1969)

Infrared electroabsorption of zinc oxide in the multi-phonon region
H. Luth
Solid State Commun., 7:585-588 (1969)

The absorption and reflection character of infrared bands of powdered inorganic materials
Conrad M. Phillippi
Air Force Materials Lab., Wright-Patterson Air Force Base, Ohio, Project 7367, Task No. 736702, AFML-TR-68-317 (Jan. 1969), 59 pp.

Mechanisms governing the emission processes in nickel-doped zinc oxide
M. L. Reynolds, W. E. Hagston, and G. F. J. Garlick
Phys. Stat. Sol., 33:579 (1969)

Spatial dispersion of excitons in ZnO
T. Skettrup
Solid State Commun., 7:869 (1969)
Erratum, Solid State Commun., 7:xiii (1969)

Refractive indices of ZnO, ZnS, and several thin-film insulators
J. C. Burgiel, Y. S. Chen, F. Vratny, and G. Smolinsky
J. Electrochem. Soc., 115:729 (1968)

Effect of a stoichiometric excess of zinc on the properties of zinc oxide. I. Luminescence
M. N. Danchevskaya, G. P. Panasyuk, and N. I. Kobozev
Russ. J. Phys. Chem., 42:1512-1514 (Nov. 1968)

Superstoichiometric zinc effect on zinc oxide properties. I. Luminescence
M. N. Danchevskaya, G. P. Panasyuk, and N. I. Kobozev
Zh. Fiz. Khim., 42:2843 (1968)

Thermoluminescence of ZnO powder
D. De Muer and W. M. Van Der Vorst
Physica, 39:123-132 (1968)

Ultraviolet emission spectrum of ZnO
I. Filinski and T. Skettrup
Solid State Commun., 6:233-237 (1968)

Optical transitions in II-VI compounds containing transition elements
H.-E. Gumlich
In: Optical Properties of Dielectric Films (Norman N. Axelrod, ed.), The Electrochem. Soc., New York (1968), pp. 45-77

Optical absorption and conduction due to $Co^{2+}$ in ZnO crystals
Y. Kanai
J. Phys. Soc. Japan, 24:956 (1968)

Transmission spectra of ZnO single crystals
W. Y. Liang and A. D. Yoffe
Phys. Rev. Letters, 20:59 (1968)

Thermoreflectance in semiconductors
E. Matatagui, A. G. Thompson, and Manuel Cardona
Phys. Rev., 176:950-960 (1968)

Zinc oxide crystals for electron-beam pumped lasers
F. H. Nicoll
J. Appl. Phys., 39:4469-4470 (1968)

Index of refraction of ZnO
Y. S. Park and J. R. Schneider
J. Appl. Phys., 39:3049 (1968)

Valence band symmetry and deformation potentials of ZnO
J. E. Rowe, Manuel Cardona, and F. H. Pollak
Solid State Commun., 6:239-242 (1968)

Decay times of the ultraviolet and green emission lines in ZnO
T. Skettrup and L. R. Lidholt
Solid State Commun., 6:589-592 (1968)

Electroluminescence of current-carrying ZnO single crystals
T. Skettrup and N. I. Meyer
Phys. Kondens. Materie, 7:97-106 (1968)

The electrical and optical properties of refractory oxides, final rept., August 1966 — July 1968
Walter C. Tripp and Edward T. Rodine
(Systems Research Labs, Inc., Dayton, Ohio), Contract AF 33(615)-2765, SRL-11580 ARL-68-0221 (December 1968) 147 pp.

Contribution of excitons to the edge luminescence in zinc oxide
R. L. Weiher and W. C. Tait
Phys. Rev., 166:791-796 (1968)

Effect of preparation conditions on the formation of electron traps in luminescent ZnO
A. A. Bundel and G. V. Zhukov
Optik. Spektrosk., 22:103-106 (1967)

Electroreflectance at a semiconductor-electrolyte interface
M. Cardona, K. L. Shaklee, and F. H. Pollak
Phys. Rev., 154:696-720 (1967)

Präparation und Lumineszenzeigenschaften von Zinkoxyd
D. Hahn, R. Nink, and Karl Tobisch
Phys. kondens. Materie, 6:229–235 (1967)

Multiphonon infrared absorption in zinc oxide
G. Heiland and H. Luth
Solid State Commun., 5:199 (1967)

Anisotropy in lattice vibrations in zinc oxide
E. C. Heltemes and H. L. Swinney
J. Appl. Phys., 38:2387 (1967)

Reflectance and surface conductivity of zinc-oxide crystals
B. Hoffmann
Solid State Commun., 5:61 (1967)

"Elektroreflexion" durch Gasadsorption am Zinkoxyd
Bernd Hoffmann
Z. Physik, 206:293–308 (1967)

Spectre d'absorption de l'ion Cu$^{2+}$ place en substitution dans un cristal de ZnO
Maurice Kobler and Françoise Calendini
J. Phys., 28:878 (1967)

Absorption "edges" in CdS and ZnO
B. Segall
International Conference on II—VI Semiconducting Compounds, Providence, Rhode Island (September 6–8, 1967), pp. 327–336, W. A. Benjamin, Inc., New York (1967)

Infrared study of oxygen adsorption on impure zinc oxide
D. M. Smith and R. P. Eischens
J. Phys. Chem. Solids, 28:2135–2142 (1967)

Der Einfluss von Temperatur und uniaxialer Verspannung auf das Linienspektrum von Zinkoxyd-Kristallen
Ch. Solbrig and E. Mollwo
Solid State Commun., 5:625 (1967)

Features of the radical-recombination luminescence spectra of ZnO and ZnS
V. V. Styrov and Sokolov
Izv. Vysshikh Uchebn. Zaveden. Fiz., 6:135–139 (1967)

Influence of preparation conditions and predecomposition phenomena on thermoluminescence curves and luminescence spectra of ZnS, ZnO and ZnS—Cu$_2$S solid solutions
A. A. Bundel et al.
Izv. Akad. Nauk SSSR, Ser. Fiz., 30:637–643 (1966)

Raman effect in zinc oxide
T. C. Damen, S. P. S. Porto, and B. Tell
Phys. Rev., 142:570–574 (1966)

Luminescence spectra of zinc oxide
F. F. Gavrilov and F. I. Vergunas
Zh. Eksperim. Teor. Fiz., Moscow, 18:224 (1948)
NASA-TT-F-10454 (Nov. 1966), 7 pp.

Der Einfluss der Löcherhaftstellen auf den Leuchtmechanismus des Zinkoxyds
D. Hahn and R. Nink
Symp. International Luminescence. Phys. Chim. Scintillateurs. Munich, 1965. Munich, Verlag Karl Thiemig (1966), pp. 345–351

Zur grünen Lumineszens des Zinkoxyds, II. Spontanlumineszenz
Dietrich Hahn and Reinhard Nink
Phys. Kondens. Materie, 4:336–348 (1966)

Electrooptic light modulators
I. P. Kaminow and E. H. Turner
Proc. IEEE, 54:1374 (1966)

Photoluminescence radicalaire de l'oxyde de Zn autoactive
V. G. Kornin and A. N. Gordan'
Opt. i Spektrosk. SSSR, 21:390–392 (1966)

Edge emission of n-type conducting ZnO and CdS
W. Lehmann
Solid-State Electron., 9:1107 (1966)

Exciton spectrum of ZnO
Y. S. Park, C. W. Litton, T. C. Collins, and D. C. Reynolds
Phys. Rev., 143:512 (1966)

Current and light storage effects in lithium- and sodium-doped crystals of ZnO
Y. S. Park and C. W. Litton
Aerospace Research Labs., Wright-Patterson AFB, Ohio, AEDC Proc. of the 13th Ann. AF Sci. and Eng. Symp. Vol. II (1966), 25 pp.

Sharp-line emission due to preferential pairing in ZnO crystals
D. C. Reynolds, C. W. Litton, Y. S. Park, and T. C. Collins
J. Phys. Soc. Japan, 21 Suppl., 143–147 (1966)
Proceedings Eighth International Conf. on the Physics of Semiconductors, Kyoto, 1966

Photoconductivity decay of ZnO crystals in oxygen
H. Van Hove and A. Luyckx
Solid State Commun., 4:603–606 (1966)

Optical properties of free electrons in ZnO
R. L. Weiher
Phys. Rev., 152:736 (1966)

Luminescence, photoconductivity and photochemical reactions of zinc oxide
M. Yamamoto
Dissertation Abstr., B27:1872 (1966)

Influence of small amounts of vanadium pentoxide on the luminescence of zinc oxide
S. L. Berdnikov and Ya. M. Zelikin
Opt. i Spektroskopiya, 19:611–615 (1965) (in Russian)

Measurement of the refractive indices of several crystals
W. L. Bond
J. Appl. Phys., 36:1674 (1965)

Optical properties of CdSe and ZnO single crystals
John Shermann DeWitt
(Master's thesis, Air Force Inst. of Tech., Wright-Patterson AFB, Ohio), Report SP/PH/65-8; AD-617923 (May 1965), 104 pp.

On the green luminescence of zinc oxide. Part I. Thermoluminescence
D. Hahn and R. Nink
Phys. Kondens. Materie, 3:311–322 (1965)

Optical absorption edges of ZnO and CdS
W. Lehmann
J. Electrochem. Soc., 112:1150 (1965)

The green luminescence of ZnO
W. Maenhout-Van der Vorst and F. Van Craeynest
Phys. Stat. Sol., 9:749 (1965)

Some optical properties of group II—VI semiconductors (I)
D. C. Reynolds, C. W. Litton, and T. C. Collins
Phys. Stat. Sol., 9:645 (1965)

Infrared absorption of copper impurities in II—VI semiconducting compounds
R. E. Dietz and H. Kamimura
Congr. Internation. Phys. Semiconducteurs, Paris (1964), pp. 62–63

Ruby-laser-induced photocurrents and luminescence in ZnO
Donald R. Hotchkiss
J. Appl. Phys., 35:2455 (1964)

Optical second-harmonic generation in piezoelectric crystals
Robert C. Miller
Appl. Phys. Letters, 5:17 (1964)

Optical and electrical properties of ZnO crystals
H. Watanabe, M. Wada, and T. Takahashi
Japan. J. Appl. Phys., 3:617 (1964)

Effect of copper activation of the fluorescent excitation spectra of zinc oxide in the visible region
M. L. Blanchard and G. Monod-Hersen
Compt. Rend., 256:4189 (1963)

Properties of ZnO phosphors doped with Li, Ni, and Cu
Arnold Pfahnl
J. Electrochem. Soc., 110:381 (1963)

The diffuse reflectance spectra of zinc oxide and zinc peroxide
Audrey L. Companion
J. Phys. Chem. Solids, 23:1685 (1962)

The infrared spectrum of hydrogen chemisorbed on zinc oxide
R. P. Eischens, W. A. Pliskin, and M. J. D. Low
J. Catalysis, 1:180–191 (1962)

Optical absorption study of Co-doped oxide systems
R. Pappalardo, D. L. Wood, and R. C. Linares, Jr.
J. Chem. Phys., 35:2041 (1961)

Optical absorption spectra of Ni-doped oxide systems. Part I
R. Pappalardo, D. L. Wood, and R. C. Linares, Jr.
J. Chem. Phys., 35:1460 (1961)

The exciton spectrum of zinc oxide
D. G. Thomas
J. Phys. Chem. Solids, 15:86 (1960)

Absorption of ZnO single crystals in the infrared
R. Arneth
Z. Physik, 155:595 (1959)

Infrared reflectivity of zinc oxide
R. J. Collins and D. A. Kleinman
J. Phys. Chem. Solids, 11:190 (1959)

The low-temperature luminescence of zinc oxide in the red region of the spectrum
V. V. Osiko
Opt. Spectroscopy, 7:454 (1959)

Infrared absorption in zinc-oxide crystals
D. G. Thomas
J. Phys. Chem. Solids, 10:47 (1959)

The existence for zinc oxide, of several distinc mechanisms for dipolar Debye absorption
M. Blanchard
Compt. Rend., 244:767 (1957)

On the electrothermoluminescence effect
D. Hahn
J. Phys. Radium, 17:748 (1956)

On the emission mechanism of zinc-oxide phosphors
H. Gobrecht, D. Hahn, and K. Scheffler
Z. Physik, 139:365 (1954)

Dispersion, absorption and thermal emission of zinc-oxide crystals
E. Mollwo
Z. Angew. Phys., 6:257 (1954)

Luminescence of the electronic semiconductor–zinc oxide
F. I. Vergunas and G. A. Konovaloz
Zh. Eksp. i Teoret. Fiz., 23:712–719 (1952)
AEC-tr-1636, 17 pp.

### Physical Properties and Structure

Lattice dynamics of ZnO and BeO
A. W. Hewat
Solid State Commun., 8:187–189 (1970)

Dispersive properties of the fundamental Rayleigh wave mode $(M_{11})$ in a heteroepitaxial zinc-oxide film on a sapphire substrate
T. C. Lim, E. A. Kraut, and B. R. Tittman
Appl. Phys. Letters, 17:34–36 (1970)

Iron cyanide as a surface state to prevent ZnO photolysis in vacuum
S. Roy Morrison
J. Vacuum Sci. Tech., 7:84–89 (1970)

Remeasurement of the structure of hexagonal ZnO
S. C. Abrahams and J. L. Bernstein
Acta Cryst., 25 B:1233 (1969)

Increase of flow stress in ZnO under illumination
Lennart Carlsson and Christer Svensson
Solid State Commun., 7:177–179 (1969)

Ultrasonic attenuation in ZnO
L. I. Claiborne, R. B. Hemphill, and N. G. Einspruch
J. Acoust. Soc. Am., 45:1352–1355 (1969)

Uni-directional channeling and blocking: a new technique for defect studies
L. C. Feldman and B. R. Appleton
Appl. Phys. Letters, 15(9) (Nov. 1, 1969)

Feldemission aus ZnO-Kristallen
K. P. Frohmader
Solid State Commun., 7:1543–1549 (1969)

Determination and systematization of the properties of binary compounds of Group II metals with Group VI elements
  S. D. Gromakov and A. I. Partala
  Zh. Fiz. Khim., 43 (1):267–270 (1969)

Thermal expansion of silicon and zinc oxide (I)
  H. Ibach
  Phys. Stat. Sol., 31:625 (1969)

Thermal expansion of silicon and zinc oxide (II)
  H. Ibach
  Phys. Stat. Sol., 33:257–265 (1969)

Thermal expansion anisotropy of oxides and oxide solid solutions
  Henry P. Kirchner
  J. Am. Ceram. Soc., 52:379–386 (1969)

Polarity of ZnO crystal (I) ion-bombarding behavior
  Ikumaro Kubo and Yoshinobu Tokita
  Japan. J. Appl. Phys., 8:626–627 (1969)

Polarity of ZnO crystal (I) ion-bombarding behavior
  Ikumaro Kubo, Mitsuhiro Fujii, and Masami Hirose
  Japan. J. Appl. Phys., 8:627–628 (1969)

Nonequilibrium vaporization rates of single-crystal zinc-oxide basal faces
  R. B. Leonard and A. W. Searcy
  J. Chem. Phys., 50:5419–5420 (1969)

Surface states associated with chemisorbed species on zinc oxide
  S. Roy Morrison
  Surface Sci., 13:85–98 (1969)

Electron capture by ions at the ZnO/solution interface
  S. Roy Morrison
  Surface Sci., 15:363–379 (1969)

Diffusion of aluminum and gallium in zinc oxide
  V. J. Norman
  Australian J. Chem., 22:325–329 (1969)

Determination of the lattice constant of cubic ZnO
  O. E. Radczewski and R. F. Schicht
  Naturwissenschaften, 56:514 (Oct. 1969)

Wurtzite z parameter for beryllium oxide and zinc oxide
  T. M. Sabine and S. Hogg
  Acta Cryst., 25B:2254–2256 (1969)

Crystal structure of thin films of zinc oxide deposited by cathode sputtering
  K. V. Shalimova, A. F. Botnev, V. A. Dmitriev, and N. M. Satybaev
  Kristallografiya, 13(4):679–682 (1968)
  Sov. Phys. — Cryst., 13(4):576–579 (1969)

Temperature dependence of the $C_{33}$ elastic parameter for insulating ZnO in the temperature range 25–100°K
  V. Tarnow
  British J. Appl. Phys., 2:1383–1387 (1969)

The free energy of formation of ZnO(s) for the temperature range 420° to 908°C
  Thomas C. Wilder
  Trans. AIME, 245:1370 (1969)

Low-energy electron diffraction studies of [0001] surfaces of CdS and ZnO
  B. D. Campbell
  (Brown University), Dissertation Abstr., 28B:3843 (1967–68)

Leed studies of the polar (001) surfaces of the II—VI compounds CdS, CdSe, ZnO, and ZnS
  B. D. Campbell (Los Alamos Scientific Lab., N. M.), C. A. Haque (Bell Telephone Labs., Murray Hill, N. J.), and H. E. Farnsworth (Brown Univ., Providence, R. I.)
  Contract W-7405-eng-36, LA-DC-9377; CONF-680611-3 (1968), 23 pp.

The Voigt-Reuss-Hill (VRH) approximation and the elastic moduli of polycrystalline ZnO, $TiO_2$ (rutile), and $\alpha - Al_2O_3$
  D. H. Chung and W. R. Buessem
  J. Appl. Phys., 39:2777 (1968)

Study of surface properties of atomically-clean metals and semiconductors
  H. E. Farnsworth, M. F. Chung, and C. A. Haque
  Contract DA-28-043 AMC-02511 (E), Semi-annual report, 1 March 1968 - 31 Aug. 1968, ECOM-02511-4 (Dec. 1968), 15 pp. (140)
  This document is subject to special export controls and each transmittal to foreign governments or foreign nationals may be made only with prior approval of the Commanding General, U. S. Army Electronics Command, Fort Monmouth, N. J., Attn: AMSEL-XL-E

Chemical reactions involving anodic processes on a single-crystal zinc-oxide catalyst
  W. P. Gomes, T. Freund, and S. R. Morrison
  Surface Sci., 13:201–208 (1968)

Self-diffusion in simple oxides (a bibliography)
  P. J. Harrop
  J. Materials Sci., 3:206–222 (1968)

X-ray determination of thermal expansion of ZnO
  A. A. Khan
  Acta Cryst., 24A:403 (1968)

Linear compressibilities of II—VI compound single crystals
  R. A. Montalva and D. W. Langer
  Bull. Am. Phys. Soc., 13:367 (1968)

Characterization of point defects in oxides
  John E. Wertz
  Mass Transport in Oxides (J. B. Wachtman, Jr., and A. D. Franklin, eds.), National Bureau of Standards Special Publ. 296 (August 1968), p. 11. Proceedings of Symposium held at Gaithersburg, Md. (Oct. 1967)

Low energy electron diffraction studies of (0001) surfaces of CdS and ZnO
  B. D. Campbell
  Thesis from Brown University (1967), 120 pp.
  Univ. Microfilms, Ann Arbor, Michigan, Order No. 68-1443

Study of surface properties of atomically-clean metals and semiconductors
  H. E. Farnsworth and B. D. Campbell
  Brown Univ., Providence, Rhode Island, Semi-annual report 1 Sept. 1966 - 28 Feb. 1967, Contract DA-28-043-AMC-02511 (E), ECOM-02511-1 (April 1967), 21 pp.

Zur Gitterenergie vom Zinkoxyde
  Rudolf Hoppe
  Naturwissenschaften, 54:587–588 (1967)

On the morphology and polarity of the ZnO crystal
H. Iwanaga and N. Shibata
Japan. J. Appl. Phys., 6:415 (1967)

Microwave acoustic loss mechanisms in crystals
R. W. Kedzie, M. Kestigian, D. H. McMahon, A. B. Smith, and
E. W. Prohofsky
(Sperry Rand Research Center, Sudbury, Mass.), Final Rept.
(1 May 1966 - 31 July 1967), Contract AF 33(615)-3937,
SRRC-CR-67-47; AFML-TR-67-361 (Nov. 1967), 153 pp.

Polar character of zinc-oxide crystals. 2: Etch patterns on prism faces
Ansgar Klein
(Royal Radar Establishment, Malvern, England), RRE-TRANS-
174 (May 1967), 11 pp.; Z. Physik, 188:352-360 (1965)

The distribution of interstitial zinc in zinc oxide
V. J. Norman
Australian J. Chem., 20:85 (1967)

The solubility of transition metal oxides in zinc oxide and the reflectance spectra of $Mn^{3+}$ and $Fe^{3+}$ in tetrahedral fields
C. H. Bates, W. B. White, and R. Roy
J. Inorg. Nucl. Chem., 28:397 (1966)

Magnetic susceptibility of group II, group VI semiconductor ZnO
R. B. Lal
Solid State Commun., 4:529 (1966)

Oxygen chemisorption on zinc oxide
J. Haber and A. Kowalska
Bull. Acad. Polon. Sci., 13:419 (1965)

Vapor pressure and evaporation coefficient studies of stannic oxide, zinc oxide, and beryllium nitride
Clarence L. Hoenig
Dissertation Abstr., 65-3007. 134 pp.

Vaporization and thermodynamic properties of zinc oxide
Donald F. Anthrop
Dissertation Abstr., 64-2018. 45pp.

Differential isotopic method for investigating the adsorbent surface
G. Bliznakov and Z. Karamanova
C. R. Acad. Bulg. Sci., 15:527-530 (1964)
$CO_2$, NiO, and ZnO

A lattice parameter study of defective zinc oxide, I. Zinc excess and distortions in pure ZnO
A. Cimino, G. Mazzone, and P. Porta
Z. Physik. Chem. (Frankfurt), 41:154 (1964)

Microhardness of single crystals of BeO and other wurtzite compounds
C. F. Cline and J. S. Kahn
J. Electrochem. Soc., 110:773 (1963)

Structural change of $TiO_2$ and ZnO by means of mechanical grinding
T. Kubo, M. Kato, Y. Mitarai, J. Takahaski, and K. Ohkura
Kogyo Kagaku Zasshi, 66:318 (1963)

Crystallographic polarity of ZnO crystals
A. N. Mariano and R. E. Hanneman
J. Appl. Phys., 34:384 (1963)

Elastic moduli of single-crystal zinc oxide
T. B. Bateman
J. Appl. Phys., 33:3309 (1962)

Alteration of the work function of oxide semiconductors by the introduction of additives
É. Kh. Enikeev, L. Ya. Margolis, and S. Z. Roginskii
Dokl. Akad. Nauk SSSR, 130 (4):807-809 (1960)
Dokl. — Phys. Chem., 130 (1-6):99-102 (1960)

The effect of pressure on zinc blende and wurtzite structures
A. L. Edwards, T. E. Slykhouse, and H. G. Drickamer
J. Phys. Chem. Solids, 11:140 (1959)

Diffusion of zinc and oxygen in zinc oxide
W. J. Moore and E. L. Williams
Discussions Faraday Soc., 28:86 (1959)

An experimental study of anomalous x-ray scattering by zinc sulfide and zinc oxide
J. R. Townsend, G. A. Jeffrey, and G. N. Panagis
Z. Krist., 112:150 (1959)

Creep of silver atoms along the surface of zinc-oxide crystals
A. G. Shekhter, A. I. Echeistova, and I. I. Tret'yakov
Zhur. Fiz. Khim., 24:202-206 (1950)
AERE-Trans-11/3/5/157; AEC-tr-1096; WO-25490

## 8.b.2. ZnS

The mechanism of polytype formation in vapour-phase grown ZnS crystals
E. Alexander, Z. H. Kalman, S. Mardix, and I. T. Steinberger
Phil. Mag., 21:1237-1246 (1970)

Some physical properties of ZnS crystals grown from the vapour phase
G. Alzetta, G. Chella, I. Chudacek, and R. Scarmozzino
Czech. J. Phys., B20:808-815 (1970)
An improved flow method

Preparation of single crystals from zinc sulfide
R. Dimitrov
Compt. Rend. Acad. Bulgare Sci., 22(6):619-622 (1969)
Needles by sublimation of single crystals

Heteroepitaxial overgrowth of ZnS on GaS single crystals
M. Harsy and E. Lendvay
J. Mater. Sci., 5:988-991 (1970)

The epitaxial growth of zinc sulphides on silicon by forced vapour transport in hydrogen flow
P. Lilley, P. L. Jones, and C. N. W. Litting
J. Mater. Sci., 5:891-897 (1970)

The epitaxy of zinc sulphide on silicon
T. G. R. Rawlins
J. Mater. Sci., 5:881-890 (1970)

Electrooptical properties of cubic ZnS crystals grown by the hydrothermal method
V. A. Shamburov, V. A. Kuznetsov, A. N. Lobachev, I. V. Kharitonova, and V. G. Soshnikov
Kristallografiya, 15(2):302-307 (1970)
Sov. Phys. — Cryst., 15(2):252-255 (1970)

Growth of zinc sulfide by iodine transport
P. N. Dangel
Ph. D. thesis, Dept. of Metallurgy and Materials Science,
Massachusetts Institute of Technology (Sept. 1969)

Growth and electrical properties of zinc–cadmium sulfide graded-band-gap crystals
Indra Dev, L. J. Van Ruyven, and Ferd Williams
J. Appl. Phys., 39:3344 (1968)

Growth of single-crystal layers of zinc sulphide by a gas transport method
M. F. Kuznetsov and P. E. Ramazanov
Izv. Vysshikh Uchebn. Zaveden., Fiz., 4:132–134 (1969)

Kristallzucht aus der Gasphase
Rudolf Nitsche
Festkörper Probleme 8 (O. Madelung, ed.)
Pergamon Press, New York (1969)

Molten flux growth of cubic zinc sulfide crystals
S. G. Parker and J. E. Pinnell
J. Crystal Growth, 3:490–495 (1968)

Solution growth of some II—VI compounds using tin as a solvent
M. Rubenstein
J. Crystal Growth, 3:309–312 (1968)

Process for growing single crystals of sulfides, selenides, and tellurides of metals of groups II and III of periodic system
Leonid A. Sysoev, Leonid V. Konvisar, and Emmanuil K. Raiskin,
U. S. Patent 3,414,387 (Dec. 3, 1968)

Hydrothermal crystal growth of $Y_2O_3$ and ZnS
S. Yamaguchi and T. Katsurai
Z. Anorg. Allgem. Chem., 356:327–328 (1968)

Structure of zinc-sulfide single crystals grown from the melt under pressure
G. V. Anan'eva, K. K. Dubenskii, A. I. Ryskin, and G. I. Khil'ko
Fiz. Tverd. Tela, 10(6):1800–1806 (1968)
Sov. Phys. — Solid State, 10(6):1417–1421 (1968)

Growth and structural properties of CdS and ZnS crystal platelets
R. J. Caveney
J. Phys. Chem. Solids, 29:851–853 (1968)

Research in purification and single crystal growth of II-VI compounds, final report, April 15, 1965–April 14, 1968
Richard H. Fahrig, George N. Webb, and Lloyd W. Brown
Contract F33615-67-C-1575, ARL-68-0096; AD-679 636, 146 pp.

The epitaxial growth of zinc sulphide on silicon by vacuum evaporation
P. L. Jones, C. N. W. Litting, D. E. Mason, and V. A. Williams
Brit. J. Appl. Phys., Ser. 2, 1:283 (1968)

Infrared absorption due to donor states in ZnS crystals
H. Kukimoto, S. Shionoya, T. Koda, and R. Hioki
J. Phys. Chem. Solids, 29:935–944 (1968)

Growth of cubic zinc sulfide from molten lead chloride
R. C. Linares
Trans. AIME, 242:441–443 (1968)

Contribution à l'étude des propriétés électriques du sulfure de zinc en couches minces évaporées sous ultra-vide
G. Marchal
Doct. Sci. Phys. Thèse, University of Nancy, Nancy, France (1969), 102 pp.
Importance of preparation methods

Preparation of single crystals and purifying or doping solids
Michael Avinor
N. V. Philips Gloeilampenfabrieken, German patent 1,240,816 (May 24, 1967)
CdS, CdSe, ZnS, ZnSe, SiC, and As

A study of the transformation of sphalerite to wurtzite in atmospheres of zinc and sulfur
E. R. J. Bank
Dissertation Abstracts, B28(3):823 (1967)

Growth of ZnS and CdS crystals from Ga, In, Tl and Sn melts
M. Harsy
Krist. Tech., 2:447–449 (1967)

The growth of homogeneous mixed crystals of zinc-cadmium sulfide
R. Hill and R. B. Lauer
Materials Res. Bull., 2:861–864 (1967)

Growth of ZnS single crystals from the melt at 1850°C under argon pressure of 50 atm
Marian Kozielski
J. Crystal Growth, 1:293–296 (1967)

Luminescent chalcogen
VEB Leuchtstoffwerk Bad Liebenstein
French Patent 1,480,576 (May 12, 1967)

Formation of zinc sulphide polytypes by spiral growth around dislocation clusters
S. I. Ben-Abraham and J. Regev
J. Materials Sci., 1:212 (1966)

Parameters of growth by the flow method of ZnS crystals
I. Dev
Brit. J. Appl. Phys., 17:761 (1966)

Parameters of growth from the vapor phase of ZnS crystals
Indra Dev
Mat. Res. Bull., 1:173 (1966)

Growth parameters of doped ZnS crystals
Indra Dev and R. B. Lauer
Mat. Res. Bull., 1:185 (1966)

Preparation of rare-earth-activated zinc sulfide single crystals
I. I. Kisil', V. L. Levskin, L. A. Sysoev, S. A. Fridman, and V. V. Shchaenko
Izv. Akad. Nauk SSSR, Ser. Fiz., 30(9): 1500–1503 (1966)
Bull. Acad. Sci. USSR, Phys. Ser., 30(9):1564–1566 (1966)

Über die epitaktische Abscheidung von Zinksulfid und Bleisulfid mit Hilfe von chemischen Transportreaktionen und durch Sublimation
W. Kleber and I. Meusel
Z. Physik. Chem., 231:191 (1966)

Luminescence of rare-earth-activated zinc sulfide
W. W. Anderson, S. Razi, and D. J. Walsh
J. Chem. Phys., 43:1153 (1965)

Electroluminescence of evaporated ZnS and ZnSe films
G. A. Antcliffe
Brit. J. Appl. Phys., 16:1467 (1965)

Dendritic growth of ZnS crystals
I. Bertoti, E. Lendvay, M. Farkas-Jahnke, M. Harsy, and P. Kovacs
Phys. Stat. Sol., 12:K1 (1965)

Study of the influence of preparation conditions on the formation of electron traps in self-activated zinc sulfide
A. A. Bundel and G. V. Zhukov
Optika i Spektroskopiya, 18:479 (1965)
Optics and Spectroscopy, 18:270 (1965)

Production of crystals of zinc, cadmium, and lead sulfides, selenides, and tellurides
Bernard Kopelman
U. S. Patent 3,174,823 (Cl. 23-50) (March 23, 1965)

On the diffusion method of crystal growth
Odon Lenvay
Air Force Cambridge Research Labs., L. G. Hanscom Field, Bedford, Mass. 01730, AFCRL-69-0275, Translations No. 51
Magyar Fiziikai Folyoirat, 8(3):231-349 (1965)

Growth spirals on ZnS crystals
E. Lendvay and P. Kovacs
Phys. Stat. Sol., 8:K125 (1965)

Crystal growth of ZnS and CdS by the high-pressure melting method
Shoji Makishima and Shigeo Shionoya
Gijutsu Kenkyu, 17:286 (1965)

Crystal growth in hydrothermal systems
Jerry W. Moody and Richard C. Himes
Battelle Tech. Rev., 14:3 (1965)

Growth of single crystals at high gas pressures
A. Ya. Preobrazhenskii and V. A. Stepanov
Pribory i Tekhn. Eksperim., 10(2):196-198 (1965)

The relation of the morphology of ZnS crystals to the conditions of growth
Yu. M. Rumyantsev, F. A. Kuznetsov, and S. A. Stroitelev
Kristallografiya, 10(2):263-264 (1965)
Sov. Phys.—Cryst., 10(2):212-213 (1965)

Polarization of the green-copper luminescence in hexagonal ZnS single crystal
Shigeo Shionoya, Yoshitomo Kobayashi, and Takao Koda
J. Phys. Soc. Japan, 20:2046 (1965)

Single crystals of sulfides, selenides, and tellurides of zinc, cadmium, gallium, and indium
L. A. Sysoev, L. V. Konvisar, and E. K. Raiskin
German Patent 1,289,034 (Feb. 13, 1969, Appl. Nov. 23, 1965), 4 pp.
ZnS

Solid transport rate in the vapor-solvent growth system ZnS:I
F. Jona and G. Mandel
J. Phys. Chem. Solids, 25:187 (1964)

Piezoelectric materials
Lebo R. Shiozawa
U. S. Patent 3,409,464 (Nov. 5, 1968, Appl. April 29, 1964), 4 pp.
ZnS

Crystal morphology in evaporation, equilibrium and growth in the vapour phase
G. A. Wolff and Heitanen
Condensation and Evaporation of Solids, Gordon and Breach, New York and London (1964), pp. 451-469

On the luminescent properties of ZnS phosphors prepared at room temperature
E. Lendvay, J. Schanda, K. Richter, P. Kovacs, and M. Somogyi
Czech. J. Phys., B13:142 (1963)

Hollow crystals of hexagonal ZnS
E. J. Soxman
J. Appl. Phys., 34:948 (1963)

Very pure hexagonal semiconductor materials of CdSe, Cd$_2$SeS, and ZnS
Roland Weisbeck
German Patent 1,280,232 (Oct. 17, 1968, Appl. Nov. 22, 1963), 4 pp.
ZnS

Crystallization of the sulfides of lead and zinc from aqueous solutions of the chlorides
L. V. Bryatov and I. P. Kuz'mina
Rost Kristallov, Izd. Akad. Nauk, SSSR (1961), pp. 416-420
Growth of Crystals, Vol. 3 (A. V. Shubnikov and N. N. Sheftal', eds.) Consultants Bureau, New York (1962), pp. 294-297

Kinetics of vapor-solvent growth in the system ZnS:HCl
F. Jona
J. Phys. Chem. Solids, 23:1719 (1962)

Methods for the preparation of ZnS single crystals
P. Kovacs and J. Szabo
Acta Phys. Acad. Sci. Hung., 14:131-144 (1962)

Growth of zinc sulphide single crystals from flux
Yo Mita
J. Phys. Soc. Japan, 17:784 (1962)

Growth of cubic ZnS single crystals by a chemical transport process
Harold Samelson
J. Appl. Phys., 33:1779 (1962)

Some effects of electric fields on the cathodoluminescence of zinc-sulphide single crystals
D. W. Satchell
Brit. J. Appl. Phys., 13:589 (1962)

Properties of cadmium sulfide, zinc sulfide, and mercuric sulfide, Parts I-III, an annotated bibliography
Helen M. Abbott, comp.
SRB-61-62 (March 1961)

Crystal growth by chemical transport reactions. I. Binary, ternary, and mixed-crystal chalcogenides
R. Nitsche, H. U. Bolsterli, and M. Lichtensteiger
J. Phys. Chem. Solids, 21:199 (1961)

Vapor-phase growth of single crystals of II-VI
compounds
  W. W. Piper and S. J. Polich
  J. Appl. Phys., 32:1278 (1961)

Vapor phase growth and properties of zinc-sul-
fide single crystals
  H. Samelson
  J. Appl. Phys., 32:309 (1961)

Synthesis and crystallography of structurally
pure cubic and hexagonal single crystals of ZnS
  Harold Samelson and Vincent A. Brophy
  J. Electrochem. Soc., 108:150-154 (1961)

Some properties of zinc-sulfide crystals grown
from the melt
  A. Addamiano and M. Aven
  J. Appl. Phys., 31:36 (1960)

Study of improved single crystals of zinc-sul-
fide
  Charles B. Jordan
  EOS Rept. 300-Final (Electro-Optical Systems, Inc., Pasa-
    dena, California), May 20, 1960, Contract DA-04-495-ORD-
    1520

The growing of electroluminescent and photo-
conductive single crystals of zinc sulphide
  F. Karel
  Cs. Cas. Fys., 10:316 (1960)

Growth and heat treatment of zinc-sulfide single
crystals
  A. Kremheller
  J. Electrochem. Soc., 107:422 (1960)

Hydrothermal synthesis of zinc oxide and zinc
sulfide
  R. A. Laudise and A. A. Ballman
  J. Phys. Chem., 64:688 (1960)

Electroluminescence in zinc sulfide single crys-
tals
  Shin-ichiro Narita
  J. Phys. Soc. Japan, 15:126 (1960)

Preparation and properties of ZnS-type crys-
tals from the melt
  Albrecht G. Fischer
  J. Electrochem. Soc., 106:838 (1959)

Monocrystals of ZnS and the spectrum of their
absorption region at low temperatures
  E. F. Gross and L. G. Suslina
  Opt. and Spectros. USSR, 6:70 (1959)

Synthesis of large CdS and ZnS crystals
  J. Nishimura and Y. Tanabe
  J. Phys. Soc. Japan, 14:850-851 (1959)

The synthesis of single crystals of the sulphide
of zinc, cadmium and mercury and of mercuric
selenide by vapour-phase methods
  D. R. Hamilton
  Brit. J. Appl. Phys., 9:103 (1958)

Preparation of zinc-sulfide single crystals
  T. Matsumura, H. Fujisaki, and Y. Tanabe
  Sci. Rep. Res. Inst. Tohoku Univ., A10:459 (1958)

Zinc-sulfide single crystals for phosphor re-
search
  Afred Kremheller
  Sylvania Technologist, 8(1) (January 1955)

Growth of zinc-sulfide single crystals
  William W. Piper
  J. Chem. Phys., 20:1343 (1952)

### 8.b.3. ZnSe

Croissance épitaxiale du séléniure de zinc sur
l'arséniure de gallium
  G. Bougnot, D. Etienne, J. Chevrier, and C. Bohe
  Mat. Res. Bull., 6:145-152 (1971)

Zinc selenide: preparation and enthalpy of for-
mation
  Claude Charlot, Nicolas Tikhomiroff, and Marc Laffitte
  Bull. Soc. Chim. Fr., 2:459-463 (1970)

Chemical precipitation of thin films of zinc
selenide
  G. A. Kitaev and T. P. Sokolova
  Zh. Neorg. Khim., 15(2):319-323 (1970)

Hydrothermal crystallization of zinc selenide
  E. D. Kolb and R. A. Laudise
  J. Crystal Growth, 7:199-202 (1970)

Morphology of ZnSe crystals grown by means of
a chemical transport reaction
  A. A. Simanovskii
  Kristallografiya, 14:1098-1100 (1969)
  Sov. Phys.— Cryst., 14:960-962 (1970)

Growing zinc selenide crystals from a melt
  K. K. Dubenskii, V. A. Sokolov, and G. A. Anan'in
  Opt. Mekh. Prom., 36:30-33 (1969)

Study of semiconductor heterojunctions of zinc
selenide, gallium arsenide, and germanium
  D. L. Feucht and A. G. Milnes
  Contract NGR-39-087-002, NASA-CR-101139 (April 1969),
    36 pp.

Synthesis and melt growth of doped ZnSe crys-
tals
  W. C. Holton, R. K. Watts, and R. D. Stinedurf
  J. Crystal Growth, 6:97-100 (1969)

The epitaxy of ZnSe on Ge, GaAs, and ZnSe by
an HCl close-spaced transport process
  H. J. Hovel and A. G. Milnes
  J. Electrochem. Soc., 116:843-847 (1969)

Growth and transport properties of ZnSe crys-
tals by chemical transport
  Sidney G. Parker and Jack E. Pinnell
  Trans. AIME, 245:451 (1969)

Production of ZnSe single crystals via a trans-
port reaction
  A. A. Simanovskii
  Rost.Kristallov, Izd. Akad. Nauk SSSR (1967), pp. 258-263
  Growth of Crystals, Vol. 7 (A. V. Shubnikov and N. N. Sheftal',
    eds.) Consultants Bureau, New York (1969), pp. 224-229

Kinetics of the vapor growth of II — VI compound
crystals. II. Zinc selenide
  Masaharu Toyama and Tetsuo Sekiwa
  Japan. J. Appl. Phys., 8:855 (1969)

A technique for vapor phase growth of zinc sele-
nide
  Paul Vohl
  Mat. Res. Bull., 4:689-698 (1969)

The growth and electrical characteristics of epitaxial layers of zinc selenide on p-type germanium
  J. T. Calow, S. J. T. Owen, and P. W. Webb
  Phys. Stat. Sol., 28:295 (1968)

Preparation of single-crystal zinc selenide by direct fusion under high inert gas pressure
  Osamaru Eguchi and Masadazu Fukai
  Nat. Tech. Rept. (Matsushita Elec. Ind. Co., Osaka), 14(1): 71-76 (1968)

Determination of the parameters of recombination centers in ZnSe single crystals
  I. Ya. Gorodetskii, K. K. Dubenskii, V. E. Lashkarev, A. V. Lyubchenko, and M. K. Sheinkman
  Fiz. Tekhn. Poluprovod., 1(11):1666-1673 (1967)
  Sov. Phys.— Semicond., 1(11):1383-1388 (1968)

Growth and properties of ZnSe crystals by chemical transport
  S. G. Parker and J. E. Pinnell
  AIME Conference on Preparation and Properties of Electronic Materials: Optical and Nuclear Radiation, Chicago (Aug. 12-14, 1968)

Solution growth of some II — VI compounds using tin as a solvent
  M. Rubenstein
  J. Crystal Growth, 3:309-312 (1968)

Process for growing single crystals of sulfides, selenides, and tellurides of metals of groups II and III of periodic system
  Leonid A. Sysoev, Leonid V. Konvisar, and Emmanuil K. Raiskin
  U. S. Patent 3,414,387 (Dec. 3, 1968)

Preparation of single crystals and purifying or doping solids
  Michael Avinor
  N. V. Philips Gloeilampenfabrieken, German Patent 1,240,816 (May 24, 1967)

Improved method for growing II — VI crystals from the vapor phase
  W. M. DeMeis and A. G. Fischer
  Mat. Res. Bull., 2:465 (1967)

Crystal growth and defect structure of zinc sulfide and zinc selenide platelets
  A. G. Fitzgerald, M. Mannami, E. H. Pogson, and A. D. Yoffe
  J. Appl. Phys., 38:3303 (1967)

Synthesis and crystal growth of cadmium selenide, zinc telluride, and zinc selenide
  A. Libicky
  II—VI Semiconductor Compounds, Intern. Conf. Brown Univ. (1967), pp. 389-401 (D. Thomas, ed.), G. W. A. Benjamin, Inc., New York

Preparation of thin films of ZnSe, ZnTe, and $ZnSe_xTe_{1-x}$ by flash evaporation
  Suguru Nakamura and Masakazu Fukai
  Japan. J. Appl. Phys., 6:1473-1474 (1967)

Crystal syntheses and growth in strong acid solutions under hydrothermal conditions
  H. Rau and A. Rabenau
  Solid State Commun., 5:331 (1967)

Luminescent chalcogen
  VEB Leuchtstoffwerk Bad Liebenstein
  French Patent 1,480,576 (May 12, 1967)

Research in purification and single-crystal growth of II — VI compounds
  R. H. Fahrig, L. W. Brown, and G. N. Webb
  Qtrly. Rept. No. 4 (Jan. 15-Apr. 14, 1966), Contract AF 33(615)- 2947, AD-634 591 (1966)

Semiconductor heterojunction structure studies
  D. L. Feucht and A. G. Milnes
  Interim Rept. (Dec. 15, 1965 - June 14, 1966), Contract AF 19(628)-5811, AD-636 806, AFCRL-66-447

Growth of zinc-selenide single crystals from the vapor phase
  H. Hartmann
  Krist. Tech., 1(4):569-580 (1966)

Excitons and the absorption edge in ZnSe
  G. E. Hite et al.
  General Electric Res. and Dev. Center, Schenectady, N. Y., Rept. 66-C-374 (Oct. 1966)

Preparation and some properties of ZnSe diodes
  K. Nojima and S. Ibuki
  Japan. J. Appl. Phys., 5:253 (1966)

Growth of single crystals of ZnSe by chemical transport reaction
  J. Shirogane, Y. Onodera, and M. Fukai
  J. Japan. Inst. Metals, 30:574-579 (1966)
  With $I_2$; temperature of evaporation 1025°C; temperature of crystallization 825°C; crystals have about 180 ppm $I_2$

Preparation of a large single crystal of ZnSe from the melt
  Y. Tsujimoto, Y. Onodera, and M. Fukai
  Japan. J. Appl. Phys., 5:636 (1966)

Electro-optic effect in ZnSe crystals
  I. I. Adrianova, G. V. Dreiden, K. K. Dubenskii, Y. V. Popov, and V. A. Sokolov
  Opt. i Spektrosk. (USSR), 19:142 (1965)

Electroluminescence of evaporated ZnS and ZnSe films
  G. A. Antcliffe
  Brit. J. Appl. Phys., 16:1467 (1965)

Epitaxial growth of ZnSe on GaAs
  Alexander Baczewski
  J. Electrochem. Soc., 112:577 (1965)

Preparation and properties of p-type ZnSe:Cu
  J. H. Haanstra and J. Dieleman
  J. Electrochem. Soc., 112:60C (1965)

Nucleation and growth of single crystals by chemical transport — II. Zinc selenide
  E. Kaldis
  J. Phys. Chem. Solids, 26:1701 (1965)

Recrystallization of evaporated films of ZnSe
  A. Vecht
  Nature, 201:486 (1964)

Crystal morphology in evaporation, equilibrium and growth in the vapour phase
  G. A. Wolff and Heitanen
  Condensation and Evaporation of Solids, Gordon and Breach, New York and London (1964), pp. 451-469

Triangular growth patterns on the surfaces of ZnSe crystals
  T. Yamanaka, T. Shiraishi, and H. Mitsuda
  J. Phys. Soc. Japan, 18:463 (1963)

Crystal growth by chemical transport reactions
—I. Binary, ternary, and mixed-crystal chalcogenides
R. Nitsche, H. U. Bolsterli, and M. Lichtensteiger
J. Phys. Chem. Solids, 21:199 (1961)

Certain peculiarities of the structure of crystals of zinc selenide
Y. Cruceanu and Y. D. Chistyakov
Kristallografiya, 5(3):364–368 (1960)
Sov. Phys. — Cryst., 5(3):344–348 (1960)

Pressure of saturated vapor of solid zinc and cadmium selenide
I. Korneyeva, V. V. Sokolov, and A. V. Novoselova
Zh. Neorg. Khim., 5:241 (1960)

Preparation of pure zinc selenide
J. Horak and J. Klikorka
Sb. Vedeckych Praci, Vysoka Skola Chem.-Tech., Pardubice
    (1959), p. 67

Electroluminescent phosphors based on sulphides and selenides
O. N. Kazankin, F. M. Pekerman, and L. N. Petoshina
Opt. i Spektroskopiya, 6:672 (1959)

Investigation of carrier injection electroluminescence
A. G. Fischer and A. S. Mason
AD-270 128, AFCRL-979 (Nov. 15, 1951)

### 8.b.4. ZnTe

An investigation of electronic defects in ZnTe
Ted LeRoy Larsen
Ph. D. thesis, Stanford University, 1970, 163 pp.
**Available from University Microfilms, Ann Arbor, Mich.,
    Order No. 71-2789**
*A technique was developed for growing large crystals from Te-,
Te(Al)-, and In-rich solutions at temperatures well below the
congruent melting temperature*

Liquid phase epitaxy on zinc telluride
R. Widmer, D. P. Bortfeld, and H. P. Kleinknecht
J. Crystal Growth, 6:237–240 (1970)
5-mm-diameter layers of ZnTe on ZnTe substrates

Vapor growth of high-resistivity ZnTe
A. S. Jordan and L. Derick
J. Electrochem. Soc., 116:1424–1430 (1969)

Growing zinc telluride on zinc selenide by reaction diffusion
N. N. Koren and N. N. Sirota
Vestsi Akad. Navuk BSSR, Ser. Fiz. Mat. Navuk, No. 4, 134–136
    (1969)

Transparent furnace for vapor crystal growth
T. B. Reed
ESD-TR-69-12, Lincoln Laboratory, Massachusetts Institute
    of Technology (April 1969), 24 pp.

Light-emitting semiconductor device
Billy L. Crowder, Frederick F. Morehead, and Peter R. Wagner
U. S. Patent 3,366,819 (January 30, 1968)

Growing zinc telluride single crystals from the gas phase and a study of some of the properties of the compound
I. Godau, B. F. Ormont, and V. Kiss
Izv. Akad. Nauk SSSR, Neorg. Mater., 4(10):1798–1800 (1968)
Inorg. Mater., 4(10):1568–1570 (1968)

Preparation of ZnTe epitaxial films on CdS by vapor transport method
Toshiyuki Ido, Shunzo Oshima, and Manabu Saji
Japan. J. Appl. Phys., 7:1141–1142 (1968)

The growth of ZnTe single crystals in a Zn atmosphere
R. T. Lynch
J. Crystal Growth, 2:106–108 (1968)

Crystal growth from the vapor phase by forced convection
T. B. Reed and W. J. LaFleur
ESD-TR-68-17, April 1968, p. 20 (Lincoln Lab., Mass. Institute of Technology)

Solution growth of some II-VI compounds using tin as a solvent
M. Rubenstein
J. Crystal Growth, 3:309–312 (1968)

Research on improved II — VI crystals, Final Tech. Rept., March 8, 1965, to May 7, 1968
L. R. Shiozawa, J. M. Jost, and G. A. Sullivan
Electronic Div., Clevite Corp., Contract AF 33(615) -2708
    S/A 2(67–3242) (August 9, 1968)

Thermodynamic properties and defect structure of intermetallic compounds, Progr. Rept. May 1, 1967, to May 31, 1968, SU-283-6, June 1968
D. A. Stevenson
Stanford Univ., Contract AT (04-3) 283

Process for growing single crystals of sulfides, selenides, and tellurides of metals of groups II and III of periodic system
Leonid A. Sysoev, Leonid V. Konvisar, and Emmanuil A. Raiskin
U. S. Patent 3,414,387 (Dec. 3, 1968)

Solution growth of (Zn, Hg) Te and Ga(P, As) crystals
G. A. Wolff, H. E. LaBelle, Jr., and B. N. Das
Trans. AIME, 242:436 (1968)

Improved method for growing II — VI crystals from the vapor phase
W. M. DeMeis and A. G. Fischer
Mat. Res. Bull., 2:465 (1967)

Synthesis and crystal growth of cadmium selenide, zinc telluride, and zinc selenide
A. Libicky
II-VI Semicond. Compounds, Intern. Conf., Brown Univ., 1967,
    pp. 389–401 (D. Thomas, ed.), G. W. A. Benjamin, Inc., New York

Preparation of thin films of ZnSe, ZnTe, and $ZnSe_xTe_{1-x}$ by flash evaporation
Suguru Nakamura and Masakazu Fukai
Japan. J. Appl. Phys., 6:1473–1474 (1967)

Optical properties of zinc telluride in the infrared
Shin-ichiro Narita, Hiroyuki Harada, and Keigo Nagasaka
J. Phys. Soc. Japan, 22:1176 (1967)

The crystalline structure of ZnTe thin films
I. Spinulescu-Carnaru
Phys. Stat. Sol., 18:769 (1966)

The orientation of crystallites in ZnTe thin films
I. Spinulescu-Carnaru
Phys. Stat. Sol., 23:157 (1967)

The preparation of single-phase single crystals of zinc telluride
W. Albers and A. C. Aten
Philips Res. Repts., 20:556 (1965)

Reverse biased electroluminescence in alloyed ZnTe diodes
N. Watanabe
Japan. J. Appl. Phys., 4:343 (1965)

A new method for growing ZnTe crystals
T. Yamanaka and T. Shiraishi
Japan. J. Appl. Phys., 4:826 (1965)

Preparation and properties of n-type ZnTe
A. G. Fischer, J. N. Carides, and J. Dreaner
Solid State Commun., 2:157-159 (1964)

Crystal morphology in evaporation, equilibrium and growth in the vapour phase
G. A. Wolff and Heitanen
Condensation and Evaporation of Solids, Gordon and Breach, New York and London (1964), pp. 451-469

Growth and decoration of ZnTe crystals
R. T. Lynch, D. G. Thomas, and R. E. Dietz
J. Appl. Phys., 34:706 (1963)

Growth of zinc telluride from vapor phase
P. Klose
Semiconductor Research 6th quarterly report (Jan. 1 to Mar. 31, 1962), p. 29. Contract DA 36-039-sc-87394. AD-275926

Research on new high-temperature semiconducting materials
S. S. Devlin, J. M. Jost, and L. R. Shiozawa
Contract AF 33(616)-3923. WADD-TR 60-11; AD-240875 (June 1960)
Armed Services Technical Information Agency Bulletin No. U60-4-1

Photoconductivity of zinc telluride
Y. Simon and J. Bok
Compt. Rend., 248:2176 (1959)

Semiconducting materials
T. J. Gray
Annual Rept., Nov. 1957, pp. 3-6
AD-154800

Zinc telluride, a semiconducting compound
J. Horak, M. Machovec, and F. Kosek
Czech. J. Phys., 7:468-475 (1957)

Preparation and characteristics of zinc telluride crystals
E. L. Lind and R. H. Bube
Bull. Am. Phys. Soc., 1:110 (1956)

## 8.b.5. Zn Mixed Systems

Growth and structure of $Zn_xHg_{1-x}Te$ and $Se_xTe_{1-x}Hg$ crystals produced from solutions
E. Cruceanu and N. Nistor
Rost Kristallov, Izd. Akad. Nauk SSSR (1967), pp. 264-267
Growth of Crystals, Vol. 7 (A. V. Shubnikov and N. N. Sheftal', eds.) Consultants Bureau, New York (1969), pp. 230-232

New solid-state device concepts
M. Aven, R. N. Hall, W. Garwacki, and J. R. Richardson
AFCRL-68-0048, January 1968, Contract AF 19(628)-4976
$ZnSe_xTe_{1-x}$

Growth of ZnSe-ZnTe solid solutions in $Te_2$ atmosphere
M. Aven, R. B. Hall, and J. S. Prener
J. Electrochem. Soc., 115:846-850 (1968)

Crystal growth from the vapor phase by forced convection
T. B. Reed and W. J. LaFleur
ESD-TR-68-17 (April 1968), p. 20
$ZnTe_{1-x}Se_x$

Growth of single crystals of ZnTe and $ZnTe_{1-x}Se_x$ by temperature gradient solution zoning
Jacques Steininger and Robert E. England
Trans. AIME, 242:444 (1968)

Photodiodes
Manuel Aven and Walter Garwacki
German Patent 1,248,164, August 24, 1967
ZnSe:ZnTe

Research in purification and single-crystal growth of II-VI compounds
R. H. Fahrig, L. W. Brown, and G. N. Webb
Qtr. Prog. Rept. No. 7, Oct. 15, 1966-Jan. 14, 1967, Contract AF 33(615)2947, AD-651 140, 31 pp.

Preparation of thin films of ZnSe, ZnTe, and $ZnSe_xTe_{1-x}$ by flash evaporation
Suguru Nakamura and Masakazu Fukai
Japan. J. Appl. Phys., 6:1473 (1967)

Preparation of $ZnSe_xTe_{1-x}$ crystals by vapor transport
Yoshinobu Tsujimoto, Tatsunori Nakajima, Yoshinori Onodera, and Masakazu Fukai
Japan. J. Appl. Phys., 6:1014 (1967)

New solid-state device concepts
M. Aven, R. N. Hall, L. M. Rosenberg, and H. H. Woodbury
AFCRL-65-896, December 1965, Contract AF 19(628)-4976
$ZnSe_xTe_{1-x}$

Synthesis and transport properties of ZnSe — ZnTe mixed crystals in n- and p-type form
M. Aven and W. Garwacki
Appl. Phys. Letters, 5:160 (1964)

## 8.c. Cadmium Compounds

### 8.c.1. CdS

The CdS crystal synthesis from vapours of the component elements
B. M. Bulakh and G. S. Pekar
J. Cryst. Growth, 8:99-103 (1971)

A simple versatile method to grow cadmium-chalcogenide single crystals
A. Corsini-Mena, M. Elli, C. Paorici, and L. Pelosini
J. Cryst. Growth, 8:297-298 (1971)

Growth of large cadmium-sulfide single crystals by coalescence
R. M. Mikulyak
J. Cryst. Growth, 8:149-152 (1971)

Growth of hollow crystals of cadmium sulphide
S. D. Sharma and L. K. Malhotra
J. Cryst. Growth, 8:285-287 (1971)

Heteroepitaxial growth of cadmium-sulfide films
S. A. Aitkhozhin, G. B. Bokii, V. F. Dvoryankin, I. M.
    Kotelyanskii, and V. V. Panteleev
Soviet Phys. Dokl. (USA), 14(10):1032-1033 (1970)

Production of CdS single crystals from the gaseous phase
B. M. Bulakh
Izv. Akad. Nauk SSSR, Neorg. Mater., 5(8):1357-1361 (1969)
Inorg. Mater., 5(8):1159-1162 (1969)

Growth of boule CdS single crystals from the vapour phase
B. M. Bulakh
J. Crystal Growth, 7:196-198 (1970)

Single-crystal formation in boule CdS crystal growth from the vapour phase
B. M. Bulakh and G. S. Pekar
J. Crystal Growth, 7:375-376 (1970)

The epitaxial growth of cadmium sulphide on gallium-arsenide substrates
B. J. Curtis and H. Brunner
J. Crystal Growth, 6:269-277 (1970)

Rf-sputtered cadmium-sulfide "thin crystals"
I. Lagnado and M. Lichtensteiger
J. Vacuum Sci. Tech., 7(2):318-321 (1970)

Etude de la croissance des monocristaux de sulfure de cadmium
M. Moulin
Rev. Tech. Thomson - C.S.F., 1(3):365-376 (1969)
Vapor transport

Production of epitaxial CdS films by gas-transport reactions
S. A. Aitkhozhin, I. M. Kotelyanskii, G. B. Bokii, et al.
Kristallografiya, 14:373-374 (1969)
Soviet Phys.—Cryst., 14:309-311 (1969)

Growth of single-crystal platelets of cadmium sulfide
Lodewijk van den Berg
Master's thesis, Dept. of Physics, University of Delaware,
    Newark, Del., AD-688 903, June 1969, 90 pp.

Synthesis of cadmium-sulfide single crystals from the gas phase
B. M. Bulakh
Izv. Akad. Nauk SSSR, Neorg. Mater., 5(1):20-24 (1969)
Inorg. Mater., 5(1):16-19 (1969)

The real conditions of CdS single-crystal growth from the vapour phase
B. M. Bulakh
J. Crystal Growth, 5:243-250 (1969)

The structure and growth of hollow conical single crystals of cadmium sulphide
M. N. Chandrasekharaiah and P. Krishna
J. Crystal Growth, 5:213-215 (1969)

Crystal growth of cadmium sulfide and other group II — IV compounds. III. Equilibrium between selenium and hydrogen at 1000°
Peter Floegel
Z. Anorg. Allgem. Chem., 370(12):16-30 (1969)

Growth hillocks and crystal structures in heteroepitaxial CdS deposited on GaAs
Osamu Igarashi
Japan. J. Appl. Phys., 8:642-651 (1969)

Crystal growth and growth rates of CdS by sublimation and chemical transport
E. Kaldis
J. Crystal Growth, 5:376-390 (1969)

p-Type cadmium-sulfide crystalline films
M. Lichtensteiger, I. Lagnado, and H. C. Gatos
Appl. Phys. Letters, 15:418-420 (1969)
rf sputtering in an argon atmosphere

Iodine-doped hollow CdS crystals
C. Paorici
J. Crystal Growth, 5:315-316 (1969)

Liquidus solubilities of CdS in a metals solvent
Martin Rubenstein
Trans. AIME, 245:457 (1969)

Formation of epitaxial films by vacuum deposition
S. A. Semiletov, V. A. Vlasov, and Z. A. Magomedov
Rost Kristallov, Izd. Akad. Nauk SSSR (1968), pp. 184-187
Growth of Crystals, Vol. 8 (A. V. Shubnikov and N. N.
    Sheftal', eds.) Consultants Bureau, New York (1969),
    pp. 150-152

Research on improved II — VI crystals, summary report, June 1, 1968, to May 31, 1969
L. R. Shiozawa and J. M. Jost
Electronic Res. Div., Clevite Corp., Cleveland, Ohio, Contract F33615-68-C-1601, June 25, 1969

Epitaxy of CdS on SrF$_2$
W. H. Strehlow and E. L. Cook
Phys. Rev., 188:1256-1266 (1969)

Growing semiconducting single crystals of sulfides and selenides of groups IIA and IIIA
L. A. Sysoev, L. V. Konvisar, and E. K. Raiskin
U.S.S.R. Patent 177,844 (June 20, 1969); from Otkrytiya,
    Izobret., Prom. Obraztsy, Tovarnye Znaki, 46:169 (1969)
From melt under pressure

The production of homogeneous monocrystals of the CdS type with given structures and orientations
R. V. Bakradze, L. A. Sysoev, E. K. Raiskin, and L. V.
    Konvisar
Rost Kristallov, Izd. Akad. Nauk SSSR (1965), pp. 261-265
Growth of Crystals, Vol. 6B (A. V. Shubnikov and N. N.
    Sheftal', eds.) Consultants Bureau, New York (1968),
    pp. 72-77

Melt growth and properties of CdS crystals
  L. M. Belaev, A. B. Gil'varg, V. P. Panova, I. M. Sil'ves-
    trova, and S. P. Smirnov
  Rost Kristallov, Izd. Akad. Nauk SSSR (1965), pp. 255-260
  Growth of Crystals, Vol. 6B (A. V. Shubnikov and N N.
    Sheftal', eds.), Consultants Bureau, New York (1968),
    pp. 66-71

Epitaxial growth of II − VI compounds
  R. J. Caveney
  J. Crystal Growth, 2:85-90 (1968)

Growth and structural properties of CdS and
ZnS crystal platelets
  R. J. Caveney
  J. Phys. Chem. Solids, 29:851-853 (1968)

Growth of single crystals of cadmium sulphide
  L. Clark and J. Woods
  J. Crystal Growth, 3:127-130 (1968)

Optical properties of tellurium as an isoelec-
tronic trap in cadmium sulfide
  J. D. Cuthbert and D. G. Thomas
  J. Appl. Phys., 39:1573 (1968)

Electron-beam crystallization of silicon, ger-
manium and cadmium sulfide
  John C. Evans, Jr.
  NASA-TN-D-4522, April 1968

Research in purification and single-crystal
growth of II − VI compounds, final report,
April 15, 1965-April 14, 1968
  Richard H. Fahrig, George N. Webb, and Lloyd W. Brown
  Eagle-Picher Industries, Inc., Miami, Okla., Contract
    F33615-67-C-1575, ARL-68-0096; AD-679 636, 146 pp.

Growth of cadmium-sulphide single crystals of
controlled composition from the vapour phase
  P. D. Fochs, W. George, and P. D. Augustus
  J. Crystal Growth, 3:122-125 (1968)

Structure and electronic properties of chem-
ically grown CdS films
  G. Hecht, J. Herberger, and C. Weissmantel
  Thin Solid Films, 2:293-304 (1968)

Préparation de couches minces orientées de
CdS obtenues par voie chimique en phase va-
peur
  J. C. Heyraud and L. Capella
  J. Cryst. Growth, 2:405-410 (1968)

Nonstoichiometry of CdS crystals grown by dif-
ferent methods
  L. Hildrisch
  J. Crystal Growth, 3:131-134 (1968)

Electrical and optical properties of CdS-MnS
single crystals
  Mitsusuke Ikeda, Kohji Itoh, and Hisanao Sato
  J. Phys. Soc. Japan, 25(2):455-460 (1968)

Growth of cadmium-sulphide crystals from re-
acting solutions
  P. A. Jackson
  J. Crystal Growth, 3:395-399 (1968)

Growth of strip and whisker crystals of cad-
mium sulfide
  L. S. Palatnik, M. N. Naboka, and N. T. Gladkikh
  Kristallografiya, 12(4):684-687 (1967)
  Sov. Phys.−Cryst., 12(4):594-596 (1968)

Sodium-doped hollow CdS crystals
  Carlo Paorici
  J. Crystal Growth, 2:324-325 (1968)

Effect of self-purification and polarity on sur-
face photoconductivity of CdS crystals
  Kazuo Sasaki
  Japan. J. Appl. Phys., 7:584 (1968)

Epitaxial growth of CdS single-crystal layers
on ZnS platelets
  H. Seiler, P. Reimers, and Indra Dev
  Mat. Res. Bull., 4:119-124 (1969)

Préparation et étude de couches de CdS évapo-
rées par bombardement électronique
  J. P. Sorbier, J. P. Legre, and S. Martinuzzi
  Vide, 133:32-39 (1968)

Process for growing single crystals of sulfides,
selenides, and tellurides of metals of groups II
and III of periodic system
  Leonid A. Sysoev, Leonid V. Konvisar, and Emmanuil A.
    Raiskin
  U. S. Patent 3,414,387, Dec. 3, 1968

Epitaxy and crystal growth of CdS thin films
  R. Ueda and T. Inuzuka
  Rost Kristallov, Izd. Akad. Nauk SSSR (1968), pp. 171-176
  Growth of Crystals, Vol. 8 (A. V. Shubnikov and N. N.
    Sheftal', eds.) Consultants Bureau, New York (1968),
    pp. 211-218

Growth mechanism of cadmium-sulfide single
crystals
  J. Chikawa and T. Nakayama
  Tech. J. Japan Broadcasting Corp., 19(2):7-16 (1967)
  Vapor phase in argon

An improved furnace for the growth and treatment of cadmium-sulfide single-crystal platelets
  L. C. Greene and C. R. Geesner
  J. Appl. Phys., 38:3662 (1967)

Growth of ZnS and CdS crystals from Ga, In, Tl
and Sn melts
  M. Harsy
  Krist. Tech., 2:447-449 (1967)

Closed-system vapor growth of bulk CdS crys-
tals from the elemental constituents
  Naim Hemmat and Martin Weinstein
  J. Electrochem. Soc., 114:851 (1967)

Growth of CdS from liquid Cd solution
  Naim Hemmat and Martin Weinstein
  J. Electrochem. Soc., 114:403 (1967)

Studies of the growth of single-crystal layers
of semiconductors on dielectric substrates
  D. C. Jillson, V. A. Russell, and C. A. Neugebauer
  AFML-TR-67-313, December 1967, Contract AF 33(615)-
    1539
  CdS

Dislocations and their origin in cadmium-sul-
fide crystals grown from the vapour phase
  W. Mohling
  PTI 67-2, Physikalisch-Technisches Institut der Deutschen
    Akademie der Wissenschaften zu Berlin, Abt. Kristall-
    wachstum und Realstruktur, 108 Berlin (1967)
  Also in: J. Crystal Growth, 1:115-124 (1967)

Croissance des plaques de CdS par transport en phase vapeur
  Michel Y. Moulin
  Compt. Rend., 265:933–935 (1967)

Physical properties of vapor- and melt-grown cadmium-sulfide and cadmium-selenide crystals
  M. Onuki and M. Amemiya
  J. Phys. Chem. Solids, 28 (Suppl. 1):225–227 (1967)

Crystal syntheses and growth in strong acid solutions under hydrothermal conditions
  H. Rau and A. Rabenau
  Solid State Commun., 5:331 (1967)

Growing CdS crystals for ultrasonic amplification purposes
  L. A. Sysoev, B. L. Timan, A. S. Gershun, E. K. Raiskin, L. V. Konvisar, and V. K. Komar'
  Kristallografiya, 11(6):933–935 (1966)
  Sov. Phys.—Cryst., 11(6):790–792 (1967)

Luminescent chalcogen
  VEB Leuchtstoffwerk Bad Liebenstein, French Patent 1,480,576 (May 12, 1967)

Evaporated and recrystallized CdS layers
  K. W. Boer, A. S. Esbitt, and W. M. Kaufman
  J. Appl. Phys., 37:2664 (1966)

Effects of substrate on films of chemical-spray-deposited CdS
  R. R. Chamberlin
  Am. Ceram. Soc. Bull., 45:698 (1966)

Research on high-temperature dielectric materials
  F. Chernow, W. B. Westphal, and R. E. Newnham
  Lab. for Insulation Research, Mass. Inst. of Tech., Cambridge, Mass., Summary Tech. Rept. No. 2, Nov. 1965 to Nov. 1966, Contract AF 33(615)-2199, AFML-TR-67-27 (Nov. 1966), 38 pp.

The preparation of large single crystals of cadmium sulphide
  L. Clark and J. Woods
  Brit. J. Appl. Phys., 17:319 (1966)

Acoustic wave propagation in cadmium sulfide
  B. Das, N. Hemmat, and M. Weinstein
  Tyco Laboratories, Inc., Waltham, Mass., Final report, Contract DA 49-186-AMC-239 (D) (May 1966). 97 pp.

Effects of growth conductions on microstructures in CdS:Cu single crystals
  Arthur Dreeben
  J. Electrochem. Soc., 113:1275 (1966)

Synthesis and characterization of electronically active materials, RCA Labs., Princeton, N. J.
  E. F. Hockings and H. W. Leverenz, eds.
  Tech. Report No. 5 covering period Nov. 16, 1965, to May 15, 1966 (June 15, 1966)
  Contract SD-182

Production, structure, and photoelectric properties of cadmium-sulfide and cadmium-selenide single-crystal films
  I. P. Kalinkin, L. A. Sergeeva, and V. B. Aleskovskii
  Izv. Akad. Nauk SSSR, Neorg. Mater., 2(12):2110–2115 (1966)
  Inorg. Mater., 2(12):1824–1829 (1966)

Preparation of CdS single crystals and their structural and electrophysical properties
  I. D. Konozenko, E. O. Muzalev'skii, A. I. Rovna, O. P. Galushka, G. G. Shmatke, et al.
  Ukr. Fiz. Zh., Kiev, 11:171–176, 1966, NASA-TT-F-10277 (Sept. 1966), 16 pp.

Contact properties and related conduction phenomena in insulating cadmium sulphide
  Eric Leon and Anne Courtens
  Massachusetts Inst. of Tech., Cambridge, Mass., Tech. Report, 1 Oct. 1964 to 1 March 1966, Contract AF 33(615)-2199, AFML-TR-66-251 (Dec. 1966), 195 pp.

Mobility of electrons in pure CdS crystal grown from the melt
  M. Onuki, M. Amemiya, and N. Hase
  J. Phys. Soc. Japan, 21:811 (1966)

Structural defects arising during the growth and heat treatment of cadmium-sulfide single crystals
  M. Ya. Skorokhod and L. I. Datsenko
  Kristallografiya, 11(2):300–304 (1966)
  Sov. Phys.—Cryst., 11(2):272–275 (1966)

Crystallization of cadmium sulfide from solutions in cadmium halides
  L. A. Sysoev, Ya. A. Obukhovskii, and D. S. Bidnaya
  Rost Kristallov, Izd. Akad. Nauk SSSR (1964), pp. 157–159
  Growth of Crystals, Vol. 4 (A. V. Shubnikov and N. N. Sheftal', eds.) Consultants Bureau, New York (1966), pp. 130–131

Whisker-crystal growth and structural changes in CdS films
  B. Tooper, L. Cartz, and A. P. van den Heuvel
  Acta Cryst., 21, Pt. 7, Suppl., A283 (1966)
  Seventh Intern. Congr. and Symp. Intern. Union of Crystallography, Moscow (1966)

Kinetics of the vapor growth of II — VI compounds crystals
  Masaharu Toyama
  Japan. J. Appl. Phys., 5:1204 (1966)

Photoconduction du sulfo-séléniure de cadmium en couche évaporée
  S. Asano and N. Yamashita
  Japan. J. Appl. Phys., 4:839 (1965)

Melt growth and properties of CdS crystals
  L. M. Belaev, A. B. Gil'varg, V. P. Panova, I. M. Sil'vestrova, and S. P. Smirov
  Rost Kristallov, Izd. Akad. Nauk SSSR (1965), pp. 255–260
  Growth of Crystals, Vol. 6B (A. V. Shubnikov and N. N. Sheftal', eds.) Consultants Bureau, New York (1968), pp. 66–72

Investigation of growth of single-crystal films on dielectric substrates
  J. M. Blank, E. C. Henry, K. K. Reinhartz, and V. A. Russell
  General Electric Co., Syracuse, New York., Tech. Documentary Report (August 25, 1965), 70 pp.

Incorporation of Cd-interstitial double donors into CdS single crystals
  R. Boyn, O. Goede, and S. Kuschnerus
  Phys. Stat. Sol., 12: 57 (1965)

Epitaxies de cristaux de sulfure de plomb et de sulfure de cadmium sur différents supports monocristallins
  Lucien Capella and Jean-Claude Heyraud
  Compt. Rend., 261: 4053 (1965)

New cadmium-sulfide evaporation source
  R. T. Galla
  Rev. Sci. Instr., 36: 403 (1965)

Some properties of CdS single crystals grown by the zone sublimation method
  A. P. Galushka, I. B. Yermolovich, N. E. Korsunskaya, I. D. Konozenko, and M. K. Sheinkman
  Ukr. Fiz. Zh., 10(7): 808-809 (1965)

A fused salt solvent for growing cadmium-sulphide crystals
  D. T. Haworth and D. P. Lake
  Chem. Commun., 21: 553 (1965)

Crystal growth of ZnS and CdS by the high-pressure melting method
  S. Makishima and S. Shionoya
  Gijutsu Kenkyu, 17: 286-289 (1965)

Growth of single crystals at high gas pressures
  A. Ya. Preobrazhenskii and V. A. Stepanov
  Pribory i Tekhn. Eksperim., 10: 196-198 (1965)

Crystallization of cadmium sulfide from cadmium-halide solutions
  L. A. Sysoev, Ya. A. Obukhovskii, and D. S. Bidnaya
  Rost Kristallov, Izd. Akad. Nauk SSSR (1964), pp. 157-159
  Growth of Crystals, Vol. 4 (A. V. Shubnikov and N. N. Sheftal', eds.) Consultants Bureau, New York (1965), pp. 130-131

The growth of wurtzite CdTe and sphalerite-type CdS single-crystal films
  Martin Weinstein, G. A. Wolff, and B. N. Das
  Appl. Phys. Letters, 6:73 (1965)

Dislocation structure and growth mechanism of cadmium-sulfide crystals
  Jun-ichi Chikawa and T. Nakayama
  J. Appl. Phys., 35: 2493 (1964)

Synthesis and characterization of electronically active materials, RCA Laboratories, Princeton, New Jersey
  E. F. Hockings and H. W. Leverenz, eds.
  Tech. Report No. 2 covering period February 16, 1964, to November 15, 1964 (December 1964), Contract SD-182

Acoustic amplification and electron mobility in lithium- and sodium-doped cadmium sulfide
  John A. Hubbard
  M. S. thesis, Ohio School of Engineering, GNE/PHYS/64-11; AD-603611 (June 1964), 42 pp.

Effect of light on the evaporation and oxidation of CdS single crystals
  G. A. Somorjai
  Solid Surfaces. Proc. International Conf. Phys. Chem. Solid Surfaces, Providence, R. I. (1964), pp. 298-306
  North-Holland Publishing Co., Amsterdam

Purification of CdS by partial sublimation in a gas flow
  A. Vecht, B. W. Ely, and A. Apling
  J. Electrochem. Soc., 111: 666 (1964)

Einfaches Verfahren zur Herstellung von hoch-reinem hexagonalem Cadmiumsulfid
  Roland Weisbeck
  Chem.-Ing.-Tech., 36: 442 (1964)

Structures of evaporated films of zinc sulphide and cadmium sulphide
  P. S. Aggarwal and A. Goswami
  Indian J. Pure Appl. Phys., 1: 366 (1963)

Photoconductivity performance in large single crystals of cadmium sulfide
  A. B. Dreeben and R. H. Bube
  J. Electrochem. Soc., 110: 456 (1963)

Einige physikalische Eigenschaften grosser CdS-Einkristalle
  J. Eckstein, A. Kuhn, J. Jindra, and M. Holas
  Czech. J. Phys., B13:182 (1963)

Hollow cadmium-sulfide crystals
  D. H. Mash and F. Firth
  J. Appl. Phys., 34: 3636 (1963)

Purification of II—VI compounds by solvent extraction
  M. Aven and H. H. Woodbury
  Appl. Phys. Letters, 1:53 (1962)

Preparing crystals of CdS at high pressures
  L. M. Belyaev, G. P. Shaknovskoi, S. P. Smirnov, and I. P. Kuz'mina
  Kristallografiya, 6(4): 641-643 (1961)
  Sov. Phys.—Cryst., 6(4): 516-518 (1962)

Aftertreatment of CdS single crystals grown by vapour transport with iodine
  J. A. Beun, R. Nitsche, and H. U. Bolsterli
  Physica, 28:184 (1962)

Introduction of microimpurities into single crystals of CdS during their growth and some characteristics of the alloyed samples
  T-U-3; TT 65-63977 (Sept. 1964), 9 pp.
  Trans. by B. M. Bulakh and I. B. Mizetska
  Ukr. Fiz. Zh. (USSR), 7:1125-1127, 1962

Properties of cadmium sulfide, zinc sulfide and mercuric sulfide. Parts I-III, an annotated bibliography
  Helen M. Abbott, comp.
  Lockheed Aircraft Corp., Missiles and Space Div., Sunnyvale, Calif., SRB-61-2 (Vol. 1) (March 1961), 169 pp.

Crystal growth by chemical transport reactions-1. Binary, ternary, and mixed-crystal chalcogenides
  R. Nitsche, H. U. Bolsterli, and M. Lichtensteiger
  J. Phys. Chem. Solids, 21:199 (1961)

Vapor-phase growth of single crystals of II—VI compounds
  W. W. Piper and S. J. Polich
  J. Appl. Phys., 32:1278 (1961)

Research on new high-temperature semiconducting materials
  S. S. Devlin, J. M. Jost, and L. R. Shiozawa
  Electronic Research Div., Clevite Corp., WADD-TR 60-11; Ad-240875 (June 1960); Contract AF 33(616)-3923

An improved method of growing CdS crystals
from the vapor phase
  P. D. Fochs
  J. Appl. Phys., 31:1733 (1960)

Crystal growth of CdS in the vertical furnace
  H. Fujisaki and Y. Tanabe
  J. Phys. Soc. Japan, 15:204 (1960)

Production of large cadmium-sulfide monocrys-
tals for gamma field dosimetry
  I. D. Konozenko and V. I. Ustyarov
  Ukr. Fiz. Zh., 5:606-614 (Sept.-Oct. 1960)

Vaporization — crystallization method for grow-
ing CdS single crystals
  D. R. Boyd and Y. T. Sihvonen
  J. Appl. Phys., 30:176 (1959)

The achievement of maximum photoconductivity
performance in cadmium sulfide crystals
  R. H. Bube and L. A. Barton
  RCA Rev., 20:564 (1959)

On the crystal growth of cadmium sulphide
  S. Ibuki
  J. Phys. Soc. Japan, 14:1181 (1959)

Growth of cadmium sulphide crystals by coevaporation of cadmium and sulfur in a
vacuum
  R. J. Miller
  Syracuse Univ., Dissertation Abstr., 19:2368 (March 1959)

Synthesis of large CdS and ZnS crystals
  J. Nishimura and Y. Tanabe
  J. Phys. Soc. Japan, 14:850-851 (1959)

The synthesis of single crystals of the sulphides of zinc, cadmium and mercury and of mercuric selenide by vapour phase methods
  D. R. Hamilton
  Brit. J. Appl. Phys., 9:103 (1958)

High-pressure, high-temperature growth of cad-
mium-sulfide crystals
  W. E. Medcalf and R. H. Fahrig
  J. Electrochem. Soc., 105:719 (1958)

Production of cadmium-sulfide crystals by co-
evaporation in a vacuum
  R. J. Miller and C. H. Bachman
  J. Appl. Phys., 29:1277 (1958)

Crystal-growth mechanism in cadmium sulfide
crystals
  D. C. Reynolds and L. C. Greene
  J. Appl. Phys., 29:559 (1958)

### 8.c.2. CdSe

Cadmium-selenide and cadmium-sulphide crys-
tals grown in the presence of arsenic
  F. Brunetti and C. Paorici
  Phys. Stat. Sol., A1:K13-K15 (1970)

Growth of CdSe single crystals from the melt
under an atmospheric pressure of argon
  K. Narita, H. Watanabe, and M. Wada
  Japan. J. Appl. Phys., 9:1275 (1970)

Epitaxy of CdSe during sublimation in vacuum
  L. V. Al'tman
  Izv. Akad. Nauk SSSR, Neorg. Mater., 5(5):868-871 (1969)
  Inorg. Mater., 5(5):740-742 (1969)

Growing cadmium-selenide single crystals with
p-type conductivity
  R. Baubinas, J. Vaitkus, and J. Viscakas
  Izv. Akad. Nauk SSSR, Neorg. Mater., 5(6):1005-1007 (1969)
  Inorg. Mater., 5(6):855-856 (1969)

Analysis of the growth conditions of CdSe single
crystals in the static sublimation method
  R. Baubinas, Yu. Vishchakas, and A. Smilga
  Litov. Fiz. Sbornik (USSR), 9(4):795-805 (1969)

Growth of vapour phase deposits of CdSe and
CdTe on single-crystal substrates
  N. G. Dhere and A. Goswami
  Indian J. Pure Appl. Phys., 7:398-402 (1969)

Effect of thermal treatment in Cd or Se atmo-
sphere on electrical properties of CdSe
  Ming-Pan Hung, N. Ohashi, and K. Isaki
  Japan. J. Appl. Phys., 8:652-659 (1969)

Epitaxy of cadmium selenide on $CaF_2$
  L. V. Al'tman, E. N. Vorontsova, Yu. V. Ruban, and G. P.
    Tikhomirov
  Kristallografiya, 12(4):694-697 (1967)
  Sov. Phys. — Cryst., 12(4):601-604 (1968)

Electrical properties of n-type CdSe single
crystals prepared under a nitrogen pressure
  P. Hoschl and S. Kubalkova
  Czech. J. Phys., B18, 897-899 (1968)

Growth of CdSe single crystals by temperature
gradient solution zoning in excess Se
  Jacques Steininger
  Mat. Res. Bull., 3:595-598 (1968)

Process for growing single crystals of sulfides,
selenides, and tellurides of metals of groups II
and III of periodic system
  Leonid A. Sysoev, Leonid V. Konvisar, and Emmanuil A.
    Raiskin
  U. S. Patent 3,414,387 (Dec. 3, 1968)

Epitaxial growth of CdSe films evaporated on
sodium chloride
  Yukio Yasuda
  Japan. J. Appl. Phys., 7:1171-1180 (1968)

p-Type CdSe single crystals
  R. Baubinas, A. Sakalas, A. Smilga, and J. Viscakas
  Phys. Stat. Sol., 24:K91-93 (1967)

Electrical properties of n-type CdSe
  R. A. Burmeister, Jr., and D. A. Stevenson
  Phys. Stat. Sol., 24:683 (1967)

Synthesis and crystal growth of cadmium sele-
nide, zinc telluride and zinc selenide
  A. Libicky
  II-VI Semicond. Compounds, Intern. Conf., Brown Univ., 1967,
    pp. 389-401 (D. Thomas, ed.), G. W. A. Benjamin, Inc., New
    York

Growth and optical properties of wurtzite and
sphalerite CdSe epitaxial thin films
  R. Ludeke and W. Paul
  Phys. Stat. Sol., 23:413 (1967)

Physical properties of vapor- and melt-grown
cadmium-sulfide and cadmium-selenide crystals
    M. Onuki and M. Amemiya
    J. Phys. Chem. Solids, 28:Suppl. 1, 225-227 (1967)

Production, structure, and photoelectric properties of cadmium-sulfide and cadmium-selenide single-crystal films
    I. P. Kalinkin, L. A. Sergeeva, and V. B. Aleskovskii
    Izv. Akad. Nauk SSSR, Neorg. Mater., 2(12):2110-2115 (1966)
    Inorg. Mater., 2(12):1824-1829 (1966)

The optical absorption edges of cadmium-selenide single crystals
    P. Manca, G. Mula, and F. Raga
    Rend. Seminario Fac. Sci. Univ. Cagliara 34:221 (1964)

Effect of the partial pressure of selenium on
the electrical conductivity of single crystals
of cadmium selenide
    S. Stonkus and J. Viscakas
    Lietuvos Fiz. Rinkinys, 4:263-266 (1964)

A study of the solid state – gaseous phase equilibrium in CdS: and its application to the growing of single crystals from the gaseous phase
by Frerich's method
    P. Hoschl and C. Konak
    Czech. J. Phys., 13:364 (1963)

Preparation of photosensitive crystals
    Otto Weinreich
    Tung-Sol Electric, Inc., U. S. Patent 2,868,736, Jan 13,
    1959

Photoconductivity in cadmium selenide
    H. J. Dirksen
    Univ. of Delft, Dissertation (1958) 98 pp.

Growth and some properties of a large single
crystal of cadmium selenide
    D. M. Heinz and E. Banks
    J. Chem. Phys., 24:391 (1956)

## 8.c.3. CdTe

Improvements in the purification of cadmium
telluride by zone refining
    A. Cornet, P. Siffert, and A. Coche
    J. Crystal Growth, 7:329-332 (1970)

Zone melting of cadmium telluride
    S. V. Prokof'ev
    Izv. Akad. Nauk SSSR, Neorg. Mater., 6(6), 1077-1080 (1970)

The conditions of crystallization of a chemical
compound from a melt enriched with a volatile
component
    S. V. Prokof'ev and Yu. V. Rud'
    J. Crystal Growth, 6:187-189 (1970)

Research and development on single-crystal
high-resistivity cadmium telluride for use as
a gamma-ray spectrometer, interim progress
report Jan. 1, 1968, through Dec. 31, 1968
    Kenneth R. Zanio
    Hughes Res. Labs., Malibu, California, SAN-549-5
    (June 1970)

Growth of vapour-phase deposits of CdSe and
CdTe on single-crystal substrates
    N. G. Dhere and A. Goswami
    Indian J. Pure Appl. Phys., 7:398-402 (1969)

Growing cadmium-telluride single crystals
    O. A. Matveev, S. V. Prokof'ev, and Yu. V. Rud'
    Izv. Akad. Nauk SSSR, Neorg. Mater., 5(7):1175-1180 (1969)
    Inorg. Mater., 5(7):1000-1004 (1969)

Vertical zone-recrystallization of cadmium telluride
    O. A. Matveev, Yu. V. Rud', and K. V. Sanin
    Izv. Akad. Nauk SSSR, Neorg. Mater., 5(9):1650-1652 (1969)
    Inorg. Mater., 5(9):1398-1400 (1969)

On the formation of the radial inhomogeneity in
cadmium-telluride ingots
    Y. V. Rud' and K. V. Sanin
    Mater. Sci. Eng., 4(2-3):186-187 (1969)
    Control of furnace thermal gradients

Preparation of CdTe crystals and Te whiskers
by vapour-phase reaction
    C. H. Chung, Y. Chea, and K. B. Kim
    J. Korean Phys. Soc., 1:112-114 (1968)

Matériaux semiconducteurs en couches minces,
Final report D.G.R.S.T., Action concertée:
Electron. (Composants circuits microminiatur.)
    C. Guillaud, M. Rodot, C. Paparoditis, and Y. Marfaing
    C. N. R. S., Lab. Magnet. Phys. Solide, 92-Meudon-Bellevue,
    Contract No. 63 FR 0 12 (April 1968)

Change in the amount of the deviation from
stoichiometry of cadmium telluride crystals
during heat treatment
    S. N. Maksimovskii, S. A. Medvedev, and V. A. Rukavishnikov
    Tellurid Kadmiya (B. M. Vul, ed.), pp. 25-31
    Izd. "Nauka," Moscow, USSR (1968)

Synthesis and directed crystallization of cadmium telluride in the presence of a control-
lable deviation from the stoichiometry
    S. A. Medvedev, S. N. Maksimovskii, Yu. V. Klevkov, and
    P. V. Shapkin
    Izv. Akad. Nauk SSSR, Neorg. Mater., 4(11):2025-2027 (1968)
    Inorg. Mater., 4(11):1759-1761 (1968)

Process for growing single crystals of sulfides, selenides, and tellurides of metals of
groups II and III of periodic system
    Leonid A. Sysoev, Leonid V. Konvisar, and Emmanuil A.
    Raiskin
    U. S. Patent 3,414,387 (Dec. 3, 1968)

Recrystallization of thin CdTe layers by an
electron beam
    V. Tolutis and V. Jasutis
    Tonkie Plenki Ikh Primen. (1968), pp. 29-30

The crystalline structure of the CdTe thin
films
    I. Dima, G. Vasiliu, and D. Borsan
    Ann. Univ. Bucuresti, Sti, Nat., Fiz., 16:47-53 (1967)
    Conditions for obtaining cubic structure

Some properties of a compound semiconductor:
CdTe
    Reiji Ishikawa and Takeshi Mitsuma
    Denki-Kagaku, 35:115-120 (1967)

Crystal growth and orientation of vacuum-deposited films of CdTe
Jin-ichi Matsuno and Morio Inoue
Japan. J. Appl. Phys., 6:297 (1967)

Zone melting
Hermann Schildknecht
Transl. by Express Translation Service, London; Verlag Chemie, GmbH, Weinheim/Bergstr.; Academic Press, New York and London (1966)

Preparation of hexagonal cadmium telluride films
K. V. Shalimova, O. S. Bulatov, E. N. Voronkov, and V. A. Dmitriev
Kristallografiya, 11(3): 480-482 (1966)
Sov. Phys. — Cryst., 11(3):431-432 (1966)

Study of growth of gallium phosphide and cadmium telluride on gallium arsenide in gas transport reactions
Zh. I. Alferov, V. I. Korol'kov, I. P. Mikhajlova-Mikheeva, V. N. Romanenko, and V. M. Tuchkevich
Fiz. Tverd. Tela, 6(8):2353-2357 (1964)
Sov. Phys.—Solid State, 6(8):1865-1869 (1965)

Research on CdTe
R. E. Halsted, D. T. F. Marple, and B. Segall
General Electric Co., Research Lab., Schenectady, N. Y. Final Report May 1961 - May 1965, Contract AF 33(616)8264, ARL-65-205; AD-626941 (Oct. 1965), 48 pp.

Les propriétés semiconductrices des tellures. IV. La conductivité thermique, le pouvoir thermoélectrique et la conductivité électriques des monocristaux de tellure de cadmium
R. Horch and H. Nieke
Ann. Physik, 16:289-299 (1965)

Formation of single-crystal layers of cadmium and mercury tellurides
Yu. E. Maronchuk, E. A. Krivorotov, and A. P. Sherstyakov
Vychisl. Sistemy, Novosibirsk, Sb., No. 15, 67-75 (1965)

Partial pressures and Gibbs free energy of formation for congruently subliming CdTe (c)
R. F. Brebrick and A. J. Strauss
J. Phys. Chem. Solids, 25:1441 (1964)

Cadmium-telluride devices
C. G. Kirkpatrick
Autonetics, 3370 Miraloma Ave., Anaheim, California, Final report Contract DA 36-039 AMC-02188 (E), EM-1164-167 (1964)

Growing of CdTe single crystals by sublimation in Cd vapours
P. Hoschl and C. Konak
Czech. J. Phys., 13:785 (1963)

Growing of CdTe single crystals by static sublimation on a cool wall under the pressure of one of the components
P. Hoschl and C. Konak
Czech. J. Phys., 13:850 (1963)

High-purity CdTe by sealed-ingot zone refining
M. R. Lorenz and R. E. Halsted
J. Electrochem. Soc., 110:343-344 (1963)

Investigation relating to the development of cadmium-telluride energy converters
B. J. Robinson, M. Scott, O. G. Brandt, A. R. Moser, and A. P. van den Heuvel
Technology Center, IIT Research Institute, Chicago, III., Final report, July 10, 1962 - Aug. 31, 1963, Contract NASw-455, NASA-CR-56836; IITRI-1220 (Sept. 1963), 18 pp.

Vapour growth patterns of CdTe crystals
Iwao Teramoto
Phil. Mag., 8:357 (1963)

Vapour growth of cadmium-telluride crystals in the ⟨111⟩ polar directions
Iwao Teramoto and Morio Inoue
Phil. Mag., 8:1593 (1963)

Preparation of CdTe crystals from near-stoichiometric and Cd-rich melt compositions under constant Cd pressure
M. R. Lorenz
J. Appl. Phys., 33:3304-3306 (1962)

Phase equilibria in the system Cd — Te
M. R. Lorenz
J. Phys. Chem. Solids, 23:939 (1962)

The solid-vapor equilibrium of CdTe
M. R. Lorenz
J. Phys. Chem. Solids, 23:1449-1451 (1962)

Vapor growth of cadmium-telluride single crystals
R. T. Lynch
J. Appl. Phys., 33:1009-1011 (1962)

On the electrical and optical properties of n-type cadmium-telluride crystals
S. Yamada
J. Phys. Soc. Japan, 17:645 (1962)

Study of CdTe single crystals (Part 1) Preparation of high-purity CdTe single crystals
Masami Yokozawa and Iwao Teramoto
National Tech. Rept., 8:428 (1962)

Twinning of CdTe crystals
Tomislav Simecek
Czech. Phys. J., 10:180-181 (1960)

Investigation of the photoconductivity of CdTe
Fotoelektricheskie i opticheskie iavleniia v poluprovodnikakh (A. D. Schneider, ed.), Academy of Science of the Ukrainian SSR, Kiev (1959), pp. 107-110

Photoconductivity in cadmium telluride
Clyde E. McNeilly
New York State College of Ceramics, Alfred Univ., Contract Nonr-1503(01), AD-202 407; PB 144 263 (July 1958), 98 pp.

### 8.c.4. Cd Ternaries

The preparation and crystallography of cadmium zinc sulfide solid solutions
Paul Cherin, E. L. Lind, and E. A. Davis
J. Electrochem. Soc., 117:233-236 (1970)

The phase diagram of $Zn_xCd_{1-x}Te$ solid solutions
S. I. Radautsan, A. E. Tsurkan, and O. G. Maksimova
Phys. Stat. Sol., 37:K9-K11 (1970)

Phase diagram of the CdTe—CdSe pseudobinary system
A. J. Strauss and Jacques Steininger
J. Electrochem. Soc., 117:1420-1426 (1970)

Growth and properties of $Cd_xHg_{1-x}Te$ crystals
B. E. Bartlett, J. Deans, and P. C. Ellen
J. Mat. Sci., 4:266-270 (1969)

Photoelectric properties of the $CdS_{0.9}Te_{0.1}$ solid solution
Eugen Cruceanu and A. Dimitrov
Fiz. Tverd. Tela, 11(6):1715-1716 (1969)
Sov. Phys. — Solid State, 11(6):1389-1390 (1969)

Electroluminescent material and device
John D. Cuthbert and David G. Thomas
Bell Telephone Labs., Inc., U. S. Patent 3,462,630 (Aug. 19, 1969), 3 pp.
CdS:Te crystals

CdTe — HgTe heterostructures
G. S. Almasi and A. C. Smith
J. Appl. Phys., 39:233 (1968)

Matériaux semiconducteurs en couches minces, Final report D.G.R.S.T., Action concertée: Electron. (Composants circuits microminiatur.)
C. Guillaud, M. Rodot, C. Paparoditis, and Y. Marfaing
C. N. R. S., Lab. Magnet. Phys. Solide, 92-Meudon-Bellevue, Contract No. 63 FR 0 12 (April 1968)

Hydrothermal crystallization of some II — VI compounds
E. D. Kolb, A. J. Caporaso, and R. A. Laudise
J. Crystal Growth, 3:422-425 (1968)
CdSe and CdTe

Effect of annealing on the electrical properties of $Cd_xHg_{1-x}Te$ single crystals
R. V. Lutsiv, M. V. Pashkovskii, L. G. Svekolkina, and B. A. Sus
Izv. Akad. Nauk SSSR, Neorg. Mater., 4(5):778-779 (1968)
Inorg. Mater., 4(5):681-682 (1968)

Crystal structure of a cadmium-sulfide — cadmium-selenide heterojunction grown from the gas phase
E. A. Muzalevskii and A. P. Ostranitsa
Protsessy Rosta Strikt. Monokrist. Sloev Poluprov., Tr. Simp., 1966 (publ. 1968), Vol. 1, pp. 351-354 (L. N. Aleksandrov, ed.), Izd. "Nauka" Sib. Otd., Novosibirsk, USSR

The preparation of CdS — CdSe graded single crystals
P. Reimers and W. Ruppel
Phys. Stat. Sol., 29:K31 (1968)

Solution growth of some II — VI compounds using tin as a solvent
M. Rubenstein
J. Crystal Growth, 3:309-312 (1968)
CdS, CdSe, CdTe, ZnS, ZnSe, and ZnTe

Research on improved II — VI crystals, Final Tech. Rept., March 8, 1965, to May 7, 1968
L. R. Shiozawa, J. M. Jost, and G. A. Sullivan
Electronic Div., Clevite Corp., Contract AF 33(615)-2708 S/A 2(67-3242) (August 9, 1968)
CdS, CdSe

Semiconducting thin films: An annotated bibliography, 1967 Supplement
W. R. Turnbull
Naval Weapons Center Corona Labs., Corona, Calif., NOLC Report 745 (1 March 1968), 155 pp.
CdS, CdSe, CdTe

Very pure hexagonal semiconductor materials of CdSe, $Cd_2SeS$, and ZnS
Roland Weisbeck
Farbenfabriken Bayer A.-G., German Patent 1,280,232, Oct. 17, 1968

Preparation of single- and multilayer monocrys-doping solids
Michael Avinor
N. V. Philips' Gloeilampenfabrieken holds W. German Patent 1,240,816, May 24, 1967
CdS, CdSe, ZnS, ZnSe, SiC, and As

The structure of $CdS_x \cdot CdSe_{1-x}$ films
N. I. Dovgoshei and I. A. Gryadil'
Kristallografiya, 12(1):148-149 (1967)
Sov. Phys. — Cryst., 12(1):123-124 (1967)

Preparation, structure, and some photoelectric properties of single- and multilayer monocrystalline chalcogenide films of the types $A^{II}B^{VI}$ and $A^{IV}B^{VI}$
L. A. Sergeeva, I. P. Kalinkin, and V. B. Aleskovskii
Kristallografiya, 12(1):113-118 (1967)
Sov. Phys. — Cryst., 12(1): 90-94 (1967)

Sublimation of cadmium telluride and cadmium selenide under a vapour pressure of one of their components and the equilibrium form of crystal growth
P. Hoschl and C. Konak
Phys. Stat. Sol., 9:167 (1965)

Production of crystals of zinc, cadmium, and lead sulfides, selenides, and tellurides
Bernard Kopelman
U. S. Patent 3,174,823 (March 23, 1965)

The growth of wurtzite CdTe and sphalerite-type CdS single-crystal films
Martin Weinstein, G. A. Wolff, and B. N. Das
Appl. Phys. Letters, 6:73 (1965)

Crystal morphology in evaporation, equilibrium and growth in the vapour phase
G. A. Wolff and Heitanen
Condensation and Evaporation of Solids, Gordon and Breach, New York and London (1964), pp. 451-469
CdSe, CdS, and CdTe

Preparation, doping and electrical properties of $Cd_{0.1}Hg_{0.9}Te$
R. R. Galazka
Acta Phys. Polon., 24:791-800 (1963)

Photoconductivity of cadmium sulfide with selenium
H. Okimura and Y. Sakai
Denki Gakkai Zasshi, 83:10-18 (1963)

Some electrical measurements of single-crystal Hg — CdTe
  Yu. K. Tovpentsev and P. V. Sharavskii
  Dokl., Nauch. Konf., Leningrad. Inzh. -Stroit. Inst., 21st, Leningrad (1963), pp. 25-28

Characteristics of CdS·CdSe and CdSe·CdTe crystals growth from the vapor state
  N. I. Vitrikhovskii and I. B. Mizetskaya
  Rost Kristallov, Izd. Akad. Nauk SSSR (1961), pp. 345-350
  Growth of Crystals, Vol. 3 (A. V. Shubnikov and N. N. Sheftal', eds.), Consultants Bureau, New York (1962), pp. 247-250

Crystal growth by chemical transport reactions — I. Binary, ternary, and mixed-crystal chalcogenides
  R. Nitsche, H. U. Bolsterli, and M. Lichtensteiger
  J. Phys. Chem. Solids, 21:199 (1961)
  CdS and CdSe

Effect of the conditions of growth on certain physical properties of mixed CdS·CdSe single crystals
  N. I. Vitrikhovskii and I. B. Mizetskaya
  Fiz. Tverd. Tela, 3(5):1581-1587 (1961)
  Sov. Phys. — Solid State, 3(5):1148-1151 (1961)

Preparation of mixed CdS·CdSe monocrystals from the vapor phase and some of their properties
  N. I. Vitrikhovskii and I. B. Mizetskaya
  Fiz. Tverd. Tela, 1(3):397-402 (1959)
  Sov. Phys. — Solid State, 1(3):358-363 (1959)

Preparation of mixed CdS·CdTe monocrystals and some of their properties
  N. I. Vitrikhovskii and I. B. Mizetskaya
  Fiz. Tverd. Tela, 1:996 (1959)
  Soviet Phys. — Solid State, 1:912-914 (1959)

Photoconductors consisting of microcrystalline cadmium sulfide and cadmium selenide
  T. Damokos
  Periodica Polytech., 2:235-240 (1958)

Optical absorption of CdS-CdSe mixed crystals prepared by solid state diffusion
  E. T. Handelman and W. Kaiser
  Bell Telephone Labs., Inc., Murray Hill, New Jersey

## 8.d. Mercury Compounds

### 8.d.1. HgS

Recrystallization by shifting the equilibrium of chemical complexes — the growth of cinnabar
  A. F. Armington and J. J. O'Connor
  J. Crystal Growth, 6:278-280 (1970)

Production of thin films of the $\alpha$-modification of mercury sulfide
  B. F. Bilen'kii, M. V. Miliyanchuck, M. V. Pashkovskii, and I. V. Savitskii
  Kristallografiya, 14:111-112 (1969)
  Soviet Phys. — Cryst., 14:975-976 (1970)

Growth of cinnebar (HgS) from sodium sulfide-sulfur fluxes
  R. W. Garner and W. B. White
  Pennsylvania State Univ., University Park, Pa., preprint rec'd. RMIC June 12, 1970, submitted to: J. Crystal Growth

Crystal growth of HgS from Hg-rich solutions
  E. Cruceanu and N. Nistor
  J. Crystal Growth, 5:206 (1969)

Photoelectronic properties of synthetic mercury-sulphide crystals
  G. G. Roberts, E. L. Lind, and E. A. Davis
  J. Phys. Chem. Solids, 30:833-844 (1969)

Hydrothermal growth of single crystals of cinnabar (red HgS)
  S. D. Scott and H. L. Barnes
  Mater. Res. Bull., 4:897-904 (1969)

Croissance par voie hydrothermale du sulfure mercurique HgS sous forme de cinabre
  Yves Toudic and Roger Aumont
  Compt. Rend., 269:74-77 (1969)

Epitaxial growth of II — VI compounds
  R. J. Caveney
  J. Crystal Growth, 2:85-90 (1968)

Electrical properties of black HgS
  Kunio Takei and Rikuo Tominaga
  ORNL-tr-2000 translated from Denki Kagaku, 36:734-739 (1968)

Ore solution chemistry II. Solubility of HgS in sulfide solutions
  H. L. Barnes, S. B. Romberger, and M. Stemprok
  Economic Geology, 62:957-982 (1967)

The growth of HgS and $Hg_3S_2Cl_2$ single crystals by a vapor-phase method
  Ernest H. Carlson
  J. Crystal Growth, 1:271-277 (1967)

Crystal syntheses and growth in strong acid solutions under hydrothermal conditions
  H. Rau and A. Rabenau
  Solid State Commun., 5:331-332 (1967)

Growth of $\alpha$-HgS crystals from a melt and study of their properties
  P. M. Koval's'kii, V. M. Mel'nik, and A. D. Shneider
  ORNL-tr-1680 translated from Ukr. Fiz. Zh., 11(8):921-922 (1966)

[Title Not Given]
  V. V. Ribalka
  Ukr. Fiz. Zh., 11(6):684-686 (1966); 11(6):682-684 (1966); 11(4):382 (1966)

Raising-crucible method for single-crystal growing
  Hiromu Sasaki and Shunkichi Kisaka
  Japan. J. Appl. Phys., 3:170-171 (1964)

Obtention de cristaux de HgS cubique, leur stabilisation et leur transformation en cinabre
  Philippe Terrée and Jean-Claude Monier
  Compt. Rend., 250:3990-3992 (1964)

Certain physical properties of mercuric − sulfide crystals
M. V. Pashkovskii, I. V. Savitskii, and V. V. Rybalka
Izv. Leningr. Elektrotekhn. Inst. 1963 (51)

Conduction processes in single crystals of mercury sulfide
M. V. Pashkovskii et al.
Fiz. Tverd. Tela, 4(7):1970−1971 (1963)
Sov. Phys. − Solid State, 4(7):1443−1444 (1963)

Effect of a phase transformation on the vapor phase growth of single-crystal HgS
O. L. Curtis, Jr.
J. Appl. Phys., 33:2461−2463 (1962)

Research in purification of cadmium-sulfide crystals and other II − VI compounds
K. E. Bean, W. E. Medcalf, R. H. Fahrig, J. E. Powderly, and J. S. Roderique
ARL-62-319 (March 1962)

Effect of conditions during the synthesis of mercury sulfide on its electric properties
M. V. Pashkovskii, I. V. Savitskii, and V. K. Zelenyuk
Visn. L'vivs'k. Univ., Ser. Fiz. (1962), pp. 90−96

Properties of cadmium sulfide, zinc sulfide and mercuric sulfide, Parts I − III, an annotated bibliography
Helen M. Abbott, comp.
SRB-61-2, March 1961

Electrical properties of $\alpha$-HgS crystals
M. V. Pashkovskii, V. V. Ribalka, and I. V. Savitskii
Dopovidi L'vivs'k. Derzh. Univ., No. 9, Pt. 2, 40−42 (1961)

Equilibria of red HgS (cinnabar) and black HgS (metacinnabar) and their saturated solutions in the systems $HgS - Na_2S - H_2O$ and $HgS - Na_2S - Na_2O - H_2O$ from 25°C to 75°C at 1 atmosphere pressure
F. W. Dickson and George Tunell
Am. J. Sci., 256:654−679 (Nov. 1958)

The synthesis of single crystals of the sulphides of zinc, cadmium and mercury and of mercuric selenide by vapour-phase methods
D. R. Hamilton
Brit. J. Appl. Phys., 9:103 (1958)

The saturation curves of cinnabar and metacinnabar in the system $HgS - Na_2S - H_2O$ at 25°C
Frank W. Dickson and George Tunell
Science, 119:467 (April 9, 1954)

## 8.d.2. HgSe

Single-crystal preparation of mercury selenide
Charles R. Whitsett and Donald A. Nelson
J. Crystal Growth, 6:26−28 (1969)

Growth of monocrystals of $A_2B_6$ compounds with narrow forbidden gaps
I. M. Rarenko and N. P. Gavaleshko
Rost Kristallov, Izd. Akad. Nauk SSSR (1965), pp. 267−270
Growth of Crystals, Vol. 6B (A. V. Shubnikov and N. N. Sheftal', eds.) Consultants Bureau, New York (1968), pp. 78−81

A new process of crystal growth: evaporation-diffusion under isothermal conditions
Y. Marfaing, G. Cohen-Dolal, and F. Bailly
J. Phys. Chem. Solids, 28, Suppl. No. 1, 549−552 (1967)

Growth of HgSe crystals from solutions and some of their properties
E. Cruceanu, N. Nistor, and D. Niculescu
Kristallografiya, 11:305−310 (1966)
Sov. Phys. − Cryst., 11:276−279 (1966)

Preparation of HgTe single crystals
V. P. Shchastlivii and A. V. Vanyukov
Izv. Akad. Nauk SSSR, Neorg. Mater., 2(8):1378−1382 (1966)
Inorg. Mater., 2(8):1177−1180 (1966)

Constitutional supercooling and two-liquid growth of HgTe alloys
R. T. Delves
Brit. J. Appl. Phys., 16:343 (1965)

Properties of mercury chalcogenides
T. C. Harman
Physics and chemistry of II−VI Compounds (M. Aven and J. S. Prener, eds.), North-Holland Publishing Company, Amsterdam (1967), pp. 769−816.
Also as ESD-TR-67-248, March 1965, Contract AF 19(628)-5167

Some properties of HgTe crystals grown from solutions
E. Cruceanu, D. Niculescu, N. Nistor, N. Stamatescu, and S. Ionescu-Buzhor
Fiz. Tverd. Tela, 7(6):1808−1812 (1965)
Sov. Phys. − Solid State, 7(6):1456−1459 (1965)

Formation of single-crystal layers of cadmium and mercury tellurides
Yu. E. Maronchuk, E. A. Krivorotov, and A. P. Sherstyakov
Vychisl. Sistemy, Novosibirsk, Sb. No. 15, 67−75 (1965)

Growth of mercury-selenide and telluride single crystals from the gas phase and their study
E. Cruceanu, D. Niculescu, and A. Vancu
Kristallografiya, 9:537 (1964)
Sov. Phys. − Cryst., 9:445 (1964)
Also available as TT-65-10741

Electrical and optical properties of mercury selenide
H. Gobrecht, U. Gerhardt, B. Peinemann, and A. Tausend
J. Appl. Phys., 32:2246−2250 (1961)

Method of growing single crystals of mercury telluride and selenide
M. V. Kury'k and M. P. Gavaleshko
Nauk. Zap. Chernivets'k. Univ., 53:32−36 (1961)

Preparation and properties of HgTe and mixed crystals of HgTe − CdTe
W. D. Lawson, S. Nielsen, and A. S. Young
Paper from: Solid State Physics in Electronics and Telecommunications. Proc. Intern. Conf., Brussels (June 2-7, 1958), Academic Press, New York (1960), Vol. 2, Pt. 2, pp. 830−835

Band structure and dispersion mechanisms in single-crystal telluride and selenide of mercury
Michel Rodot and Huguette Rodot
Compt. Rend., 250:1447−1449 (1960)

Preparation and electrical properties of mercury selenide
T. C. Harman
1959 Fall Meeting of Electrochemical Society (paper)

The synthesis of single crystals of the sulphides of zinc, cadmium and mercury and of mercuric selenide by vapour-phase methods
D. R. Hamilton
Brit. J. Appl. Phys., 9:103 (1958)

The crystalline perfection of some semiconductor single crystals
R. L. Bell
J. Electronics and Control, 3:487–493 (1957)

Semiconducting materials
T. J. Gray
Alfred University, Annual Report, pp. 3–6 (Nov. 1957);
AD 154 800

### 8.d.3. HgTe

Preparation and transport properties of epitaxial HgTe films
G. A. Antcliffe and H. Kraus
J. Phys. Chem. Solids, 30: 243–248 (1969)

Croissance épitaxique de HgTe sur CdTe en présence d'un gradient de température
Francis Bailly, Gerard Cohen-Solal, Ludvik Svob, and Yves Marfaing
Compt. Rend. 269C:465 (1969)

Zone melting of mercury telluride
V. I. Ivanov-Omskii, B. T. Kolomiets, V. K. Ogorodnikov, and K. P. Smekalova
Izv. Akad. Nauk SSSR, 5(3): 487–491 (1969)
Inorg. Mater., 5(3): 406–409 (1969)

Growth direction and cleavage anisotropy of HgTe single crystals
E. E. Kolosiv
Izv. Akad. Nauk SSSR, Neorg. Mater., 5: 1285–1286 (1969)
Inorg. Mater., 5: 1093–1094 (1969)

Growth of monocrystals of $A_2B_6$ compounds with narrow forbidden gaps
I. M. Rarenko and N. P. Gavaleshko
Rost Kristallov, Izd. Akad. Nauk, SSSR (1965), pp. 267–270
Growth of Crystals, Vol. 6B (A. V. Shubnikov and N. N. Sheftal', eds.) Consultants Bureau, New York (1968), pp. 78–81

Growth of mercury telluride crystals by the Bridgman method
A. V. Vanyukov, V. P. Schastlivy, E. L. Polisar, and G. V. Indenbaum
Izv. Akad. Nauk, Neorg. Mater., 3(7): 1271–1272 (1967)
Inorg. Mater., 3(7): 1122–1123 (1967)

Epitaxial growth of composite semiconductors by evaporation-diffusion in an isothermal system
G. Cohen-Solal, Y. Marfaing, and F. Bailly
Rev. Phys. Appl., 1: 11–17 (1966)

Preparation of high-purity HgTe
Z. Dziuba
Acta Phys. Polon., 26: 897–903 (1964)

Preparation of n- and p-type HgTe single crystals and study of their electric properties
L. A. Osnach and P. V. Sharavskii
Dokl. Nauch. Konf., Leningrad. Insh.-Stroit. Inst., 21st, Leningrad (1963), pp. 16–19

Preparation and physical properties of crystals in the HgTe – CdTe solid-solution series
J. Blair and R. Newnham
Metallurgy of Elemental and Compound Semiconductors, Interscience Publishers, New York (1961), Vol. 12, pp. 393–402

Method of growing single crystals of mercury telluride and selenide
M. V. Kury'k and M. P. Gavaleshko
Nauk. Zap. Chernivets'k. Univ., 53;32–36 (1961)
HgTe–HgSe

Preparation and properties of HgTe and mixed crystals of HgTe – CdTe
W. D. Lawson, S. Nielsen, and A. S. Young
Solid State Physics in Electronics and Telecommunications, Proc. Intern. Conf., Brussels (June 2–7, 1958), Academic Press, Inc., New York (1960), Vol. 2, Part 2, Semiconductors, pp. 830–835

### 8.d.4. Hg Ternaries

The liquidus curve and crystal growth in the Hg:Te system
E. Z. Dziuba
J. Cryst. Growth, 8:221–222 (1971)

Hydrothermal crystallization of HgSe and HgTe
A. Pajaczkowska
J. Cryst. Growth, 8:137–138 (1971)

Profile of the fundamental absorption edge of mercury selenide
V. V. Volkov, V. M. Bezborodova, V. P. Dmitriev, A. V. Vanyukov, and P. S. Kireev
Sov. Phys. – Semicond., 4(8):1234–1239 (1971)

Controlled growth of crystalline layers of ternary compounds from group II and VI elements
Donald R. Carpenter, Gerald W. Manley, Philip S. McDermott, and Ralph J. Riley
IBM, Ger. Offen. 1,947,382 (April 16, 1970), 18 pp.

Effect of ordered vacancies on ultrasonic wave propagation in some mercury – indium tellurides
G. A. Saunders and T. Seddon
J. Phys. Chem. Solids, 31:2495–2504 (1970)

Phase transformation studies in the mercury – selenide – mercury-sulfide system
H. I. Andrews, II
Ph. D. thesis, Dept. of Metallurgy and Materials Science, Massachusetts Institute of Technology (May, 1969)

Growth and properties of $Cd_xHg_{1-x}Te$ crystals
B. E. Bartlett, J. Deans, and P. C. Ellen
J. Mater. Sci., 4: 266–270 (1969)

Growth and structure of $Zn_xHg_{1-x}Te$ and $HgSe_xT_{1-x}$ crystals produced from solutions
E. Cruceanu and N. Nistor
Rost Kristallov, Izd. Akad. Nauk, SSSR (1967), pp. 264–267
Growth of Crystals, Vol. 7 (A. V. Shubnikov and N. N. Sheftal', eds.) Consultants Bureau, New York (1969), pp. 230–232

Growth and structure of crystals obtained from solutions in systems based on mercury selenide and mercury telluride
N. Nistor and E. Cruceanu
Krist. Tech., 4(3): 337-344 (1969)

Growth and properties of $Hg_{1-x}Cd_xTe$ epitaxial layers
O. N. Tufte and E. L. Stelzer
J. Appl. Phys., 40: 4559-4568 (1969)

Kinetic properties of $HgS_xSe_{1-x}$
Ya. S. Budzhak, R. V. Lutsiv, O. P. Mikhailyuk, M. V. Pashkovskii, and O. R. Shelyug
Ukr. Fiz. Zh. (Ukr. Ed.), 13(8): 1337-1344 (1968)

Study of solid solutions in the HgSe — HgS system
F. F. Kharakhorin, M. V. Pashkovskii, R. V. Lutsiv, V. M. Petrov, L. G. Svekolkina, and G. P. Mikhailyuk
Izv. Akad. Nauk SSSR, Neorg. Mater., 4(1): 39-43 (1968)
Inorg. Mater., 4(1): 30-33 (1968)

Phase diagram of $Hg_{1-x}Cd_xTe$
J. L. Schmit and C. J. Speerschneider
Infrared Phys., 8: 247-253 (1968)

Solution growth of (Zn, Hg) Te and Ga(P, As) crystals
G. A. Wolff, H. E. LaBelle, Jr., and B. N. Das
Trans. AIME, 242: 436 (1968)

Optical properties of mercuric-sulfide — mercuric-selenide solid solutions
B. F. Bilen'kii, R. V. Gerasimchuk, V. M. Korolishin, R. V. Lutsiv, and M. V. Pashkovskii
(L'vivs'k. Derzhuniv. im. Franka, Lvov), Ukr. Fiz. Zh., 12: 1198-1200 (1967)

Optical and photoelectric properties of the $HgS_xSe_{1-x}$ system in the region of the semi-metal-semiconductor transition
F. F. Kharakhorin and V. M. Petrov
Fiz. Tekh. Poluprovod, 9: 143-145 (Jan. 1967)
Soviet Phys.—Semicon., 9(1) (July 1967)

Preparation of homogeneous HgTe — ZnTe alloys
E. Cruceanu and N. Nistor
J. Electrochem. Soc., 113: 955 (1966)

The electrical properties of some $HgTe - In_2Te_3$ alloys
J. E. Lewis and D. A. Wright
Brit. J. Appl. Phys., 17: 783 (1966)

Constitutional supercooling and two-liquid growth of HgTe alloys
R. T. Delves
Brit. J. Appl. Phys., 16:343 (1965)
(Hg, Mn) Te

# 9. Boron

Die Kristallzüchtung von β-rhomboedrischem Bor
W. Borchert, W. Dietz, and H. Kolker
Z. Angew. Phys., 29(5):277-281 (1970)

Crystalline modifications of boron deposited
on boron substrates
Z. Olempska, A. Badzian, K. Pietrzak, and T. Niemyski
Rost Kristallov, Izd. Akad. Nauk, SSSR (1968), pp. 193-197
Growth of Crystals, Vol. 8 (A. V. Shubnikov and N. N. Sheftal',
    eds.) Consultants Bureau, New York (1969), pp. 157-160

Vapor transport of boron, boron phosphide and
boron arsenide
A. F. Armington
J. Cryst. Growth, 1:47 (1967)

Boron (data sheets)
J. T. Milek and S. J. Welles
DS-151 (Feb. 1967), 251 pp.
Electronic Properties Information Center, Hughes Aircraft
    Company, Culver City, Calif. 90232

Electrolytic growth and preparations of tran-
sition-metal compound single crystals
Aaron Wold
(Brown Univ., Providence, R. I.)
AD-660615 (1967), 41 pp.

Vapour growth of boron crystals
S. Mierzejewska and T. Niemyski
J. Less-Common Metals, 10:33 (1966)

Purification and crystallization of amorphous
boron powder by zone refining. Pre-purifica-
tion of boron before zoning
J. R. Musgrave and R. J. Starks
Eagle-Picher Co., TID-22221 (Feb. 1966). Contract AT-
    (40-1)-3327

Crystalline modifications of boron deposited
on boron substrates
Z. Olempska, A. Badzian, K. Pietrzak, and T. Niemyski
J. Less-Common Metals, 11:351 (1966)

Preparation and morphology of boron filamentary crystals grown by the vapor-liquid-solid
mechanism
J. P. Sitarik and W. C. Ellis
J. Appl. Phys., 37:2399 (1966)

Synthesis of red, α-rhombohedral boron
E. Amberger and W. Dietze
Boron Preparation, Properties, and Applications, Vol. 2
    (Gerhart K. Gaulé, ed.), p. 1
Papers Presented at the 1964 International Symp. on Boron,
    Paris, Plenum Press, New York (1965)

Tetragonal boron and borides with similar struc-
tures
Hermann J. Becker
Boron, Preparation, Properties, and Applications, Vol. 2
    (Gerhart K. Gaulé, ed.), Plenum Press, New York (1965) p. 89

Method for the extreme high temperature and
pressure production of single crystals (carbon,
boron, boron carbide or nitride, silicon carbide)
R. Daheberg
German Patent 1,198,322 (1965), 3 pp.

On the birefringence of simple rhombohedral
boron
F. H. Horn, E. A. Taft, and D. W. Oliver
Boron, Preparation, Properties, and Applications, Vol. 2
    (Gerhart K. Gaulé, ed.), Plenum Press (1965), pp. 231

Preparation and characterization of single-crys-
tal boron
I. R. King, F. E. Wawner, Jr., G. R. Taylor, Jr., and C. P.
    Talley
Boron, Preparations, Properties, and Applications (Gerhart
    K. Gaulé, ed.), Plenum Press, New York (1965), p. 45

Studies on crystallization of pure boron
T. Niemyski, I. Pracka, R. Szczerbinski, and Z. Frukacz
Boron, Preparations, Properties, and Applications (Gerhart
    K. Gaulé, ed.), Plenum Press, New York (1965), p. 35

The preparation and structure of thin films of
boron on silicon
E. T. Peters and W. D. Potter
Trans. AIME, 233:473 (1965)

The preparation of high-purity boron via the
iodide
A. F. Armington, G. F. Dillon, and R. F. Mitchell
Trans. AIME, 230:350 (1964)

Single-crystal boron films on silicon
  A. F. Armington, W. D. Potter, and L. E. Tanner
  J. Appl. Phys., 35:730 (1964)

[Title not given]
  R. C. Keezer
  Sci. Tech. Aerospace Rept., 2:749 (1964)

Sur la préparation des monocristaux du bore
  T. Niemyski, I. Pracka, R. Szczorbinski, and Z. Frukacz
  Congr. International Phys. Semiconducteurs, Paris (1964),
    pp. 105-106

The structure of boron crystals grown from
the melt
  F. N. Tavadze, I. A. Bairamashvili, G. V. Tsagareishvili,
    K. P. Tsomaya, and N. A. Zoidze
  Kristallografiya, 9 (6):918-920 (1964)
  Sov. Phys. – Cryst., 9 (6):768-770 (1965)

Preparation and characterization of high-purity
single-crystal boron
  I. R. King, G. R. Taylor, Jr., F. E. Wawner, Jr., and C. P
    Talley
  ASD-TDR-62-427; AD-297801 (Jan. 1963)

Elements
  R. H. Wentorf, Jr.
  The Art and Science of Growing Crystals, John Wiley and
    Sons (1963), pp. 176-193

Floating zone refining of boron using electron
beam melting
  C. B. Hood and M. O. Thurston
  J. Electrochem. Soc., 109:66 (1962)

Die Herstellung von einkristallinem Tellur
  R. Link
  Z. Physik. Chem. (Leipzig), 221:274 (1962)

Preparation of pure crystalline boron
  T. Niemyski, Z. Olempska, and I. Pracka
  Report of the International Conf. on The Physics of Semicon-
    ductors held at Exeter (July 1962), p. 722

Research investigations in the physical chemistry and metallurgy of semiconducting materials
  R. J. Starks
  Eagle-Picher AD-277480 (May 1962), pp. 22-27

Züchtung von Tellur-Einkristallen mit einer
zur c-Achse senkrechten Ziehrichtung
  R. von Kujawa
  Phys. Status Solidi, 1:K34 (1961)

The preparation of high-purity boron with crys-
tal growth by a crucible-free method
  W. E. Medcalf, K. E. Bean, and R. J. Starks
  Metallurgy of Elemental and Compound Semiconductors,
    AIME-Interscience (1961), pp. 381-392

Preparation of single-crystal boron
  C. P. Talley
  J. Appl. Phys., 32:1787 (1961)

Preparation on new crystalline modification of
boron and notes on the synthesis of boron tri-
iodide
  L. V. McCarty and D. R. Carpenter
  J. Electrochem. Soc., 107:38 (1960)

Bigger boron crystal produced by floating zone
melting and new "pressed-bar" technique
  E. S. Greiner
  Bell Labs. Record, 37:476 (1959)

# 10. III—V Compounds

## 10.a. General, Reviews, and Bibliographies

Contribution aux études théoriques du transport en phase vapeur des composés III — V: GaAs, GaP, GaSb
  G. Bougnot, J. Chevrier, D. Etienne, and C. Bohe
  Mat. Res. Bull., 6:137-144 (1971)

The band structure of several zinc-blende semiconductors from a self-consistent pseudopotential approach
  David Brust
  Solid State Commun., 9:481-85 (1971)
  Tabulated for seven III—V compounds

A review of the electrical and optical properties of III — V compound semiconductor films
  H. H. Wieder
  J. Vacuum Sci. Tech., 8:210-223 (1971)

Charge-related properties of III — V compounds
  A. E. Attard
  J. Phys. C: Solid State Phys., 3:184-89 (1970)

Core transitions and density of conduction states in the III — V semiconductors
  M. Cardona, W. Gudat, E. E. Koch, M. Skibowski, B. Sonntag, and P. Y. Yu
  Phys. Rev. Letters, 25:659-661 (1970)

Single-crystal electroluminescent materials
  H. C. Casey, Jr., and F. A. Trumbore
  Mater. Sci. Eng., 6:69-109 (1970)
  Review paper; 155 refs.; crystal growth; properties; theory

Evaluation of electronic-energy band structures of GaAs and GaP
  A. Marcus Gray
  Phys. Stat. Sol., 37:11-28 (1970)
  Review; 57 refs.

Ionicities, effective charges, dielectric constants, and atomic polarizabilities of III — V semiconductors
  K. Hubner
  Phys. Letters, 31A:365-366 (1970)
  Tabulation

On the preparation of epitaxial films of III — V compounds
  V. K. Jain and S. K. Sharma
  Solid-State Electron., 13:1145-1162 (1970)
  Relative merits of the important techniques

A review of bulk and process-induced defects in GaAs semiconductors
  E. D. Jungbluth
  Met. Trans., 1:575-585 (1970)

Heteroepitaxy of III — V compound semiconductors on insulating substrates, final report
  H. M. Manasevit and A. C. Thorsen
  Autonetics, Anaheim, Calif., NASA-CR-86408 (Jan. 1970), 102 pp.

Gallium arsenide — a bibliography supplement
  M. Neuberger
  Electronics Properties Information Center, Hughes Aircraft Co., Culver City, Calif., DS-144 (Suppl.6), Oct. 1970

Negative differential conductivity effects in semiconductors
  B. R. Pamplin
  Contemp. Phys., 11(1):1-19 (1970)
  Gunn effect basic physics

Vapor-phase growth technique and system for several III — V compound semiconductors
  J. J. Tietjen, R. Clough, R. Enstrom, M. Ettenberg, H. P. Maruska, et al.
  Radio Corp. of America, Princeton, N. J., ISR-3; NASA-CR-110194 (April 1970), 34 pp.

Lattice mobility of holes in III — V compounds
  J. D. Wiley and M. DiDomenico
  Phys. Rev., B, Solid State, 2:427-433 (1970)

Group IIIB — VB compounds
  Robert K. Willardson, Worth P. Allred, and James E. Cook
  Bell and Howell Co., U. S. Patent 3,496,118 (Feb. 17, 1970) 8 pp.
  Czochralski method; low lattice defects

Lattice dynamics of III — V compounds
  R. Banerjee and Y. P. Varshni
  Canad. J. Phys., 47(4):451-462 (1969)
  36 refs.

Microzone recrystallization of semiconductor compound films
  A. R. Billings
  J. Vacuum Sci. Tech., 6(4):757-765 (1969)

Experimental investigations of superconductivity in degenerate semiconductors
  N. M. Builova and V. B. Sandomirskii
  Usp. Fiz. Nauk, 97:119-192 (1969)
  Sov. Phys.—Usp., 12(1):64-69 (1969)

Energy-loss measurements on III—V compounds
  C. V. Festenberg
  Z. Phys., 227:453-481 (1969)

Effet électrooptique et charges effective dans les composés III—V
  C. Flytzanis
  Compt. Rend., B269(8):325-328 (1969)

Second-order optical susceptibilities of III—V semiconductors
  Chr. Flytzanis and J. Ducuing
  Phys. Rev., 178:1218-1228 (1969)

Gunn effect bibliography supplement
  T. K. Gaylord, P. L. Shah, and T. A. Tabson
  Trans. IEEE Electron Devices, ED-16:490-494 (1969)

Gallium Arsenide Lasers
  C. H. Gooch
  Wiley-Interscience, New York (1969), 348 pp.

Development of techniques for the growth of bulk single crystals of several 3-5 compound semiconductors
  J. J. Haggerty and J. F. Wenckus
  Arthur D. Little, Inc., Cambridge, Mass.
  Contract NAS12-2020, NASA-CR-86308 (June 1969), 67 pp.

An investigation of the preparation and properties of some IIIa—Vb compounds
  S. E. R. Hiscocks and J. B. Mullin
  J. Mater. Sci., 4:962-973 (1969)
  Preparation reviewed; preparative problems systematically analyzed; thin films of PrAs, NdAs, SmAs, GdAs, DyAs, TmAs, YbAs, PrSb, SmSb, YbSb, and SmP

The exciton-binding energy of III—V semiconductor compounds
  J. K. Kubler
  Phys. Stat. Sol., 35:189-195 (1969)

Effects of anisotropy in properties on impurity segregation for growing semiconductor crystals
  U. M. Kulish and A. P. Vyatkin
  Rost Kristallov, Izd. Akad. Nauk, SSSR (1968), pp. 72-77
  Growth of Crystals, Vol. 8 (A. V. Shubnikov and N. N. Sheftal', eds.) Consultants Bureau, New York (1969), pp. 59-62

The use of metal-organics in the preparation of semiconductor materials. I. Epitaxial gallium-V compounds
  H. M. Manasevit and W. I. Simpson
  J. Electrochem. Soc., 116:1725 (1969)

Thermochemical constants of $A^{III}B^V$ compounds
  L. I. Marina and A. Ya. Nashelskii
  Russ. J. Phys. Chem., 43:963-966 (1969)

III-V semiconducting thin films, bibliography
  M. Neuberger
  Hughes Aircraft Company, Culver City, California, Contract F33615-68-C-1225, Report No. S-13 (Dec. 1969), 92 pp. 394 refs.
  With annotations or abstracts

Preparation of single crystals of GaS—InSe and their physical properties
  P. G. Rustamov, Z. D. Melikova, Ya. N. Nasirov, and M. A. Alidzhanov
  Izv. Akad. Nauk SSSR, Neorg. Mater., 5(5):881-884 (1969)
  Inorg. Mater., 5(5):750-752 (1969)

Gallium-phosphide electroluminescence
  D. G. Thomas
  Brit. J. Appl. Phys., 2:637-654 (1969)

Vapor-phase growth technique and system for several III—V compound semiconductors (interim scientific report)
  J. J. Tietjan, R. Clough, A. B. Dreeben, R. Enstrom, and D. Richman
  RCA, Princeton, New Jersey, Contract NAS12-538, NASA-CR-86192; ISR-2 (March 1969), 29 pp.
  GaN and $In_{1-x}Ga_xP$ and $GaAs_{1-x}Sb_x$ and $GaAl_{1-x}As$

High-purity metals for the electronics industry
  J. E. Wardill and D. J. Dowling
  Chem. Britain, 5(5):226-228 (1969)
  GaAs, GaP, and InP

Solid solutions in the pseudobinary (III—V)—(II—VI) systems and their optical energy gaps
  W. Michael Yim
  J. Appl. Phys., 40:2617-2624 (1969)

Chemical bonding and structure of the semiconductor compounds
  R. J. Caveney
  Phil. Mag., 17:943-950 (1968)

Radiative and nonradiative recombination mechanisms in the III—V compound semiconductors (survey; 94 refs.)
  P. J. Dean
  Trans. AIME, 242:384-400 (1968)

Gallium arsenide — a review
  J. Frey
  Atomic Energy Research Establishment, Harwell, England, AERE-R 5757 (March 1968)

Gunn effect bibliography
  T. K. Gaylord, P. L. Shah, and T. A. Tabson
  Trans. IEEE Electron Devices, ED-15:777-788 (1968)

Zone refining
  A. Lawley
  Techniques of Metals Research, Vol. 1, Part 2, Techniques of Materials and Handling (R. F. Bunshah, ed.), Interscience Publishers, New York (1968), pp. 8451-8920
  High-melting-point metals, low-melting-point metals, semiconductors, the group III—V compounds, and alkali halides; 162 refs.

Liquid encapsulation crystal pulling at high pressures
  J. B. Mullin, R. J. Heritage, C. H. Holliday, and B. W. Straughan
  J. Crystal Growth, 3(4):281-285 (1968)
  The practical application of the technique to the growth of good-quality InP and GaP

# General, Reviews, and Bibliographies

**Indium antimonide data sheets**
M. Neuberger
Electronic Properties Information Center, Hughes Aircraft
Co., Culver City, Calif., Contract AF 33(616)-8438, Project 7381, Task 738103, DS-121 (Feb. 1963)
2nd edition, DS-121 (Dec. 1965); also 2nd edition, Suppl. 1
(June 1968)

**Semiconductors and Semimetals, Vol. 4: Physics of III − V Compounds**
R. K. Willardson and A. C. Beer, eds.
Academic Press, New York (1968), 511 pp.
Fourth volume of a series, of which the first appeared in 1966

**Status of diffusion data for binary-compound semiconductors**
David W. Yarbrough
U. S. Govt. Res. Develop. Rept. 68:162 (1968); Ad-670014, 69 pp.
III−V, II−VI compds., and SiC

**Preparation and properties of III − V compounds for radiative processes**
Louis G. Bailey
Trans. AIME, 239:310 (1967)

**Photoelectronic analysis**
R. H. Bube
Semiconductors and Semimetals. III. Optical Properties of III−V Compounds (R. K. Willardson and A. C. Beer, eds.),
Academic Press, New York (1967), pp. 461–495

**Croissance des cristaux non-centrosymétriques suivant une direction polaire, rôle de la polarisation superficielle**
R. Cadoret and J. C. Monier
J. Crystal Growth, 1:59–66 (1967)

**Chemistry of semiconductor materials and materials for quantum electronics**
N. P. Sazhin, N. Kh. Abrikosov, N. P. Luzhnaya, and E. G. Ippolitov
Razv. Obshch., Neorg. Anal. Khim. SSSR, 1917-1967, Akad.
Nauk SSSR, Inst. Istor. Estestvozn. Tekh., 214-223 (1967)

**Bibliography vapor growth of the arsenides and phosphides of group III metals by chemical transport reactions**
R. C. Taylor
IBM Corp., Yorktown Heights, N. Y., NC-684 (Jan. 1967)

**Systematic classification of the properties of A(III) − B(V) semiconducting compounds on the basis of Mendeleev's periodic table**
S. D. Gromakov, Z. M. Latypov, and P. S. Kirilyuk
Russian J. Phys. Chem., 40(6):677 (1966)

**Epitaxial growth of III − V materials**
E. W. Mehal and G. R. Cronin
Electrochem. Tech., 4:540 (1966)
Basic considerations of chemical transport are presented, together with typical reactor and operating procedure

**Evaluated bibliography on aluminum antimonide, nitride, phosphide, arsenide and bismuthide**
M. Neuberger
Electronic Properties Information Center, Hughes Aircraft
Co., Culver City, Calif., EPIC Interim Report IR-23
(March 1966)

**Investigation of the Gunn effect in gallium arsenide**
Alois A. Slepicka
Master's thesis, Naval Postgraduate School, Monterey, Calif.,
AD-800479 (May 1966), 74 pp.

**Gallium antimonide; properties and perspectives of applications in semiconducting devices**
Ireneusz Wojcik
Przegl. Elektron., 7(12):584-595 (1966)

**Gallium arsenide**
C. Hilsum
Progress in Semiconductors, Vol. 9 (A. F. Gibson and R. E.
Burgess, eds.), Heywood Book, Temple Press Books Ltd.,
London (1965), pp. 135-178

**Photoconductivity, section 3.2, group III − V compounds**
T. S. Moss
Reports on Progress in Physics, Vol. XXVIII (A. C. Stickland,
ed.), The Institute of Physics and the Physical Society,
47 Belgrave Square, London (1965), pp. 41-51

**A fundamental study of epitaxy by flash evaporation, final report, Oct. 1, 1964 − March 30, 1965**
John L. Richards
Philco Corp., Blue Bell, Penn., Contract AF 19(628)-2937,
AFCRL-65-412; A049-Final Sc Rpt (April 30, 1965), 80 pp.
Conditions under which films of III−V compounds can be evaporated and deposited stoichiometrically onto a number of substrates

**Emittance studies of III − V compound semiconductors**
D. L. Stierwalt and R. F. Potter
Naval Ordnance Lab., Corona, Calif., Task R360-FR-104/211-1/R001-01-01, NASA-W-11-400-B, Rept. No.
NOLC-630; AD-626411 (Nov. 1965), 32 pp.

**The growth of semiconductor crystals from solution using the twin-plane reentrant-edge mechanism**
J. W. Faust, Jr., and H. F. John
J. Phys. Chem. Solids, 25:1407-1415 (1964)

**The growth of crystals from compounds with volatile components**
G. R. Cronin, Morton E. Jones, and O. Wilson
J. Electrochem. Soc., 110:582-584 (1963)
Czochralski

**Determination of thermodynamic properties of semiconductor materials**
M. J. Pool, R. W. Sullivan, and C. E. Lundin
Denver Research Institute, Univ. of Denver, Colorado, Contract AF 19(604)-7222, Project 4608, AFCRL-63-156 (May 1963)

**Properties of GaAs alloy diodes**
R. H. Rediker and T. M. Quist
Solid-State Electron., 6:657-665 (1963)

**Gallium**
Pierre de la Breteque
Société Française pour l'Industrie de l'Aluminium, Marseille
and Aluminium Suisse S A, Zurich et Neuhausen, Switzerland (1968)
6th Suppl. to bibl. published in 1962

Indium antimonide — a review of its preparation, properties and device application
K. F. Hulme and J. B. Mullin
Solid-State Electron., 5:211-247 (1962)

Compound Semiconductors — Preparation of III—V Compounds
R. K. Willardson and H. L. Goering
(Reinhold, 1962), 553 pp.
Contains bibliography by R. C. Sangster, pp. 509-548, 1172 refs., catalogued by compound

III—V compound growth and solubility considerations
Metallurgy of Elemental and Compound Semiconductors, Proceedings of a Technical Conference, Boston, Massachusetts, August 29-31, 1960, Vol. 12
(Ralph O. Grubel, ed.), Interscience Publishers, New York (1961), pp. 303-380

Semiconducting III—V Compounds, Vol. 1
C. Hilsum and A. C. Rose-Innes
Pergamon Press (1961), 239 pp.

Materials research on GaAs and InP
L. R. Weisberg, F. D. Rosi, and P. G. Herkart
Properties of Elemental and Compound Semiconductors, Vol. 5, Interscience Publishers, New York (1960), pp. 25-67
Large single crystals by four different growth techniques; purification methods

Binding and semiconductor properties of $A^{III}B^V$ compounds
O. G. Folberth and H. Welker
J. Phys. Chem. Solids, 8:14-20 (1959)

Intermetallic semiconductors
H. T. Minden
Semicond. Prod., 2:30-42 (Feb. 1959)
InSb, InAs, InP, GaAs, HgTe, CdTe, and $Bi_2Te_3$; 50 refs

Bibliography on $A^{III}$-$B^V$ intermetallic compound semiconductors
M. E. Tanner, H. C. Gorton, and R. K. Willardson
Battelle Memorial Institute, 505 King Avenue, Columbus, Ohio, Technical Memorandum No. 1
Consists of refs. to pertinent literature through 1957

## 10.b. Boron Compounds

### 10.b.1. BN

Preparation and properties of Ba phosphides
K. E. Maass
Z. Anorg. Allgem. Chem., 374(1):1-10(1970)

Sectorial growth of cubic boron—nitride crystals
N. E. Filonenko, T. P. Nikitina, and N. M. Kamentseva
Dokl. Akad. Nauk, 187(3):569-570 (1969)

Crystallization of cubic nitride and of synthetic diamond
G. N. Bezrukov, V. P. Butuzov, T. P. Nikitina, et al.
Dokl. Akad. Nauk SSSR, 179:1326-1328 (1968)

Study of the evaporation of boron, aluminum and gallium nitride
A. S. Bolgar, S. P. Gordienko, E. A. Ryklis, and V. V. Fesenko
Chemistry and Physics of the Nitrides, Naukova Dumka, Kiev (1968), pp. 151-156

Etching of crystal faces of cubic boron nitride
N. E. Filonenko, G. M. Zaretskaya, N. M. Kamentseva, et al.
Dokl. Akad. Nauk SSSR, 179(1):88-89 (1968)

Refractory materials from silicon and boron nitride
V. K. Kazakov
Chemistry and Physics of the Nitrides, Naukova Dumka, Kiev (1968), pp. 121-129

Preparation and properties of thin film boron
M. J. Rand and J. F. Roberts
J. Electrochem. Soc., 115:423 (1968)

Production of dense modifications of boron nitride by high pressures and shear stresses
L. F. Vereshchagin, E. V. Zubova, L. N. Burenkova, and N. I. Revin
Dokl. Akad. Nauk SSSR, 178(1):72-73 (1968)
Sov. Phys. — Dokl., 13(1):25-26 (1968)

The chemical vapor deposition of pyrolytic boron nitride
C. E. Frahme
(Rutgers State Univ., New Brunswick, N. J.), Dissertation Abstr., 28(3):885 (1967)

Heats of formation of BN and $BF_3$
P. Gross, C. Hayman, and M. C. Stuart
Proc. Brit. Ceram. Soc., 8:39-50 (1967)

Chemical vapor deposition of ceramic compounds
R. L. Heestand, D. W. Short, and W. C. Robinson
(Oak Ridge National Lab., Tenn.), Proceedings of the Conference on Chemical Vapor Deposition of Refractory Metals, Alloys, and Compounds (A. C. Schaffhauser, ed.), Gatlinburg, Tenn., Sept. 12-14, 1967, American Nuclear Soc., Inc., Hinsdale, Ill. (1967), pp. 175-191
CONF-670905

The deposition of boron nitride in the gas-phase reaction
H. Tagawa and K. Ishii
Denki Kagaku, 35(1):50-53 (1967)

Some factors affecting crystal growth of cubic boron nitride
K. Kudaka, H. Konno, and T. Matoba
J. Chem. Soc. Japan, 69:365 (1966)

Production of ultrafine boron nitride by arc vaporization
W. E. Kuhn
Electrochem. Tech., 4:166 (1966)

Preparation of chemical vapor-deposited materials for use in field-enhanced electron-emission studies
David G. McMaster
U. S. Army Electronics Command, Fort Monmouth, New Jersey, Tech. Rept. ECOM-2681 (March 1966)

Chemically vapor-deposited boron nitride
P. C. Li, A. J. Capriulo, and M. P. Lepie
Vol. 1A, Proc. of OSU-RTD Symposium on Electromagnetic Windows, Dept. of Electrical Engineering, Ohio State University, Columbus, Ohio (June 1964)

Process for the manufacture of very pure crystalline carbides, nitrides, or borides
Wacker-Chemie GmbH
British Patent 968,590 (Sept. 2, 1964)

Direct transformation of hexagonal boron nitride to denser forms
F. P. Bundy and R. H. Wentorf, Jr.
J. Chem. Phys., 38:1144 (1963)

Preparation of semiconducting boron nitride
R. H. Wentorf, Jr.
J. Chem. Phys., 36:1990 (1962)

Synthesis of the cubic form of boron nitride
R. H. Wentorf, Jr.
J. Chem. Phys., 34:809 (1961)

Transformation of cubic BN to a graphitic form of hexagonal BN
H. J. Milledge, E. Nave, and F. H. Weller
Nature, 184:715 (1959)

BN
V. D. Romanov, G. V. Samsonov, and D. I. Nikitin
USSR Patent 120,509 (June 19, 1959)

Synthesis of boron phosphide and nitride
R. C. Vickery
Nature, 184:268 (1959)

Improvements in or relating to processes for fusing powdered semi-conductor materials
Siemens and Halske
British Patent 806,697 (Dec. 31, 1958)

BN
A. R. Globus
United International Research, Inc., U. S. Patent 2,812,240 (Nov. 5, 1957)

Hot-pressed BN
K. M. Taylor
Ind. Eng. Chem., 47:2506 (1955)

BN — an unusual refractory
G. R. Finlay and G. H. Fetterley
Am. Ceram. Soc. Bull., 31:141 (1952)

Reaction between HCl and nitrides
I. C. Montemartini and L. Losana
Giorn. Chim. Ind. Ed. Appl., 6:323 (1924)

Producing nitrides in an electric furnace
A. R. Lindblad
U. S. Patent 1,311,568 (July 29, 1919)

## 10.b.2. BP

Crystals and epitaxial layers of boron phosphide
T. L. Chu, J. M. Jackson, A. E. Hyslop, and S. C. Chu
J. Appl. Phys., 42:420–424 (1971)

Chemical transport reactions of boron monophosphide and sulfur
Z. S. Medvedeva, J. H. Greenberg, and E. G. Zhukov
Krist. Tech., 4(4):487–493 (1969)

Vapor transport of boron, boron phosphide, and boron arsenide
Alton F. Armington
J. Cryst. Growth, 1:47–48 (1967)

Production of coarse-crystalline boron phosphide from solution in a melt with $Cu_3P$
B. V. Baranov, V. D. Prochukhan, and N. A. Goryunova
Izv. Akad. Nauk SSSR, Neorg. Mater., 3(9):1691–1692 (1967)
Inorg. Mater., 3(9):1477–1478 (1967)

Use of the chemical transport reaction for growing boron-phosphide single crystals
Z. S. Medvedeva, Ya. Kh. Grinberg, E. G. Zhukov, and N. P. Luzhnaya
Krist. Tech., 2(4):523–534 (1967)

Boron-phosphide single crystals from the BP–$Cu_3P$ system by using temperature programming for speeding the melt as well as growth
I. A. Goryunova, B. V. Baranov, and V. D. Prochukhan
A. F. Ioffe Phys. Tech. Inst., USSR Patent 185,087 (July 30, 1966)

Physical evaluation of thin films of solid state materials, final scientific rept., Dec. 1, 1962–Nov. 30, 1965
Edward T. Peters, Irita Grierson, and S. A. Kulin
AFCRL-66-132 (Feb. 1966), Contract AF19(628)-2438

Production of boron-phosphide single crystals from vapor
Ya. Kh. Grinberg, Z. S. Medvedeva, A. A. Eliseev, and E. G. Zhukov
Dokl. Akad. Nauk SSSR, 160(2):337 (1965)
Sov. Phys. — Dokl., 10(1):6 (1965)

Obtaining single crystals of boron phosphide
Ya. Kh. Grinberg, Z. S. Medvedeva, and L. A. Klinkova
Izv. Akad. Nauk SSSR, Neorg. Mater., 1(4):478–479 (1965)
Inorg. Mater., 1(4):440–441 (1965)

Preparation and properties of the boron phosphides
James L. Peret
J. Am. Ceram. Soc., 47:44 (1964)

Preparation, optical properties and band structure of boron monophosphide
C. C. Wang, M. Cardona, and A. G. Fischer
RCA Review, 25(2):159–167 (1964)

Production of boron monophosphite
G. V. Samsonov and Y. B. Tithov
Russian J. Appl. Chem., 36:669 (1963)

Production of single-crystal boron phosphide
Bobbie D. Stone and Robert A. Ruehrwein
Monsanto Chemical, U. S. Patent 3,073,670 (Jan. 15, 1963)

[Title Not Given]
B. D. Stone
U. S. Patent 3,009,780 (1961)

The floating zone melting of boron and the properties of boron and its alloys
E. S. Greiner
Boron, Vol. 2, Synthesis, Structure, and Properties
Proceedings of the Conference on Boron sponsored by the Institute for Exploratory Research, U. S. Army Signal Research and Development Laboratory, Fort Monmouth, N. J. (J. A. Kohn, W. F. Nye, and G. K. Gaulé, eds.), Plenum Press, New York (1960)

The preparation and properties of boron phosphides and arsenides
F. V. Williams and R. A. Ruehrwein
J. Am. Chem. Soc., 82:1330 (1960)

Preparation and semiconducting properties of single crystal boron phosphide
B. Stone and D. Hill
Bull. Am. Phys. Soc., Ser. II, 4:408(A) (Nov. 27, 1959)

Synthesis of boron phosphide and nitride
K. Vickery
Nature, 184:268 (July 25, 1959)

Boron phosphide, a III — V compound of zinc-blende structure
P. Popper and T. A. Ingles
Nature, 179:1075 (May 25, 1957)

### 10.b.3. BAs

Vapor transport of boron, boron phosphide and boron arsenide
Alton F. Armington
J. Crystal Growth, 1:47-48 (1967)

Preparation and properties of boron arsenides and boron-arsenide — gallium-arsenide mixed crystals
S. M. Ku
J. Electrochem. Soc., 113:813 (1966)
Vapor-phase technique

Process for preparing crystalline boron arsenide
F. V. Williams
Monsanto Company, U. S. Patent 3,413,092 (July 21, 1958)

## 10.c. Aluminum Compounds

### 10.c.1. AlN

Properties of aluminum nitride derived from AlCl₃·NH₃
D. W. Lewis
J. Electrochem. Soc., 117:978-982 (1970)

Energy band structure of AlN
B. Hejda and K. Hauptmanova
Phys. Sta. Sol., 36:K95-99 (1969)

Preparation of aluminum nitride from the gaseous phase
M. D. Lyutaya, I. G. Chernysh, and Z. A. Yaremenko
Izv. Akad. Nauk SSSR, Neorg. Mater., 5(11):1929-1932 (1969)
Inorg. Mater., 5(11):1642-1644 (1969)

Aluminum nitride and the nitriding process
1957-1968, Bibliography No. 130, Bell Telephone Laboratories, Murray Hill, N. J. (April 1969)

Deposition, structure, and performance of thin-film piezoelectric transducers
N. F. Foster
6th Sendai Symp. Acoustoelectron. (1968), pp. 11-20

Formation of aluminum nitride by use of a transferred plasma torch
Osamu Matsumoto
J. Electrochem. Soc. Japan, 36:207-212 (1968)

Epitaxial growth of aluminum nitride
T. L. Chu, D. W. Ing, and A. J. Noreika
Solid-State Electron., 10:1023-1027 (1967)

Lattice vibration spectra of aluminum nitride
A. T. Collins, E. C. Lightowlers, and P. J. Dean
Phys. Rev., 158:833 (1967)

On the preparation, optical properties and electrical behaviour of aluminium nitride
G. A. Cox, D. O. Cummins, K. Kawabe, and R. H. Tredgold
J. Phys. Chem. Solids, 28:543 (1967)

Electrical and optical properties of AlN — a thermostable semiconductor
K. Kawabe, R. H. Tredgold, and Y. Inuishi
Electron. Eng. (Japan), 87:62-70 (1967)
Large single crystals grown

Laser induced photoconduction in AlN single crystals
K. Kawabe, K. Yoshino, and Y. Inuishi
J. Phys. Soc. Japan, 21:1604 (1966)

Development on high-temperature insulation materials, Part 1. Pyrolytic deposition of aluminum and silicon nitrides
D. W. Lewis, D. E. Sestrich, J. N. Esposito, T. W. Dakin, and D. Berg
Westinghouse Research Labs., Pittsburgh, Pa., Annual summary report - June 1965 to June 1966, Contract AF 33(615)-2782, AFML-TR-66-320, Pt. 1 (July 1966)

Strength of aluminium-nitride whiskers
T. J. Davies and P. E. Evans
Nature, 207:254-255 (1965)

Space-charge conduction and electrical behaviour of aluminum-nitride single crystals
J. Edwards, K. Kawabe, G. Stevens, and R. H. Tredgold
Solid State Commun., 3:99-100 (1965)

Epitaktisches Aufwachsen von AlN-Schichten auf SiC- und Si-Einkristallen in der Gasentladung
J. Pastrnak and L. Roskovcova
Phys. Stat. Sol., 9:K73 (1965)

Materials with special physical properties in the aluminum — boron — nitrogen system
L. I. Prikhod'ko
Vestn. Kiev Politekhn. Inst., Mekhan. -tekhnol., 2:59-63 (1965)

Pure elements and physicochemical analysis of microimpurities
N. M. Sisakyan
Vestn. Akad. Nauk SSSR, 3:52-53 (1965)

Some physical properties of aluminum nitride
T. V. Andreeva, I. G. Barantseva, E. M. Dudnik, and V. L. Yupko
High Temp. (English Transl.), 2:742-744 (1964)

Morphology and growth mechanism of AlN single crystals
J. Pastrnak and L. Roskovcova
Phys. Stat. Sol., 7:331 (1964)

New materials suitable for the fabrication of space-charge amplifiers
R. H. Tredgold
NASA Accession No. N65-16148, AD-451467 (Sept. 1964)

Preparation and properties of aluminum nitride
A. M. Lejus, J. Thery, J. C. Gilles, and R. Collongues
Compt. Rend., 257:157 (1963)

Preparation of aluminum and gallium nitride
A. Rabenau
Compound Semiconductors — Preparation of III—V Compounds
(R. K. Willardson and H. L. Goering, eds.), Reinhold Pub-
lishing Co., New York (1962), pp. 174–180

On the preparation of the nitrides of aluminum
and gallium
Arrigo Addamiano
J. Electrochem. Soc., 108:1072 (1961)

Growth and morphology of AlN-single crystals
W. Kleber and H. D. Witzke
UCRL-Trans-882(L); trans. from: Z. Krist., 116:126–133 (1961)

Synthesis of AlN crystals
T. Matsumura and Y. Tanabe
J. Phys. Soc. Japan, 15:203 (1960)

Some properties of aluminum nitride
K. M. Taylor and Camille Lenie
J. Electrochem. Soc., 107:308 (1960)

Synthesis of AlN monocrystals
J. A. Kohn, P. G. Cotter, and R. A. Potter
Am. Mineralogist, 41:355–359 (1956)

Growth and perfection of aluminum nitride
crystals
Charles M. Drum
Dissertation Abstr., 64-703, 195 pp., University Microfilms,
Ann Arbor, Mich.

## 10.c.2. AlP

Phase and thermodynamic properties of the
Ga—Al—P system: solution epitaxy of $Ga_xAl_{1-x}P$
and AlP
M. B. Panish, R. T. Lynch, and S. Sumski
Trans. Met. Soc. AIME, 245:559 (1969)

Preparation of aluminum phosphide by solution-
growth method
Hajimu Sonomura and Takeshi Miyauchi
Japan. J. Appl. Phys., 8:1263 (1969)

High-temperature-materials study
R. D. Baxter, E. P. Stambaugh, and F. J. Reid
Battelle Mem. Inst., Columbus, Ohio, Electronic Res. Center,
NASA-CR-86043 (Jan. 1968), Contract NAS 12-107

Vapor-phase growth and properties of aluminum
phosphide
D. Richman
J. Electrochem. Soc., 115:945–947 (1968)

Development of injection electroluminescent
materials, final rept.
M. Rubenstein and R. Mazelsky
AD-667 249, March 28, 1968, Contract NObs-94326 (Westinghouse Res. Labs., Pittsburgh, Pa.)

Synthesis and solution growth of aluminum phos-
phide, I.
Sylvan Z. Beer
Trans. Met. Soc. AIME, 242:424 (1968); Part II, Trans. Met
Soc. AIME, 242:428 (1968)

Vapor growth of AlP single crystals
F. J. Reid, S. E. Miller, and H. L. Goering
J. Electrochem. Soc., 113:467 (1966)

## 10.c.3. AlAs and Mixed Systems

Isothermal solution mixing growth of thin
$Ga_{1-x}Al_xAs$ layers
J. M. Woodall
J. Electrochem. Soc., 118:150–152 (1971)

Investigation of heterojunctions and p—n junc-
tions in the AlAs—GaAs system using a scan-
ning electron microscope with a probe micro-
analyzer
Zh. I. Alferov, V. M. Andreev, V. I. Murygin, and V. I.
Stremin
Fiz. Tekhn. Poluprovod., 3(10):1470–1477 (1969)
Sov. Phys. — Semicond., 3(10):1234–1239 (1970)

$Al_xGa_{1-x}As_{1-x'}P_{y'}GaAs_{1-y}P_y$ heterostructure laser
and lamp junctions
R. D. Burnham, N. Holonyak, Jr., and D. R. Scifres
Appl. Phys. Letters, 17:455–457 (1970)

Phase and thermodynamic properties of the Ga—
Al—P system: solution epitaxy of $Ga_xAl_{1-x}P$ and
AlP
M. B. Panish, R. T. Lynch, and S. Sumski
Trans. Met. Soc. AIME, 245:559 (1969)
An approximate value of 2.49 ± 0.05 eV for the band gap of
AlP at 300°K

Growing single crystals of aluminum antimonide
A. A. Pavlyuk
Izv. Akad. Nauk SSSR, Neorg. Mater., 5(7):1103 (1969)
Inorg. Mater., 5(7):1295 (1969)

Vapor-phase growth technique and system for
several III—V compound semiconductors
J. J. Tietjen, R. Clough, R. E. Enstrom, and M. Ettenberg
Radio Corp. of America, Princeton, N. J., Contract NAS12-538
NASA-CR-86306 (Sept. 1969), 14 pp.

Research and development study on wide band-
gap semiconductor films
A. J. Noreika et al.
Westinghouse Electric Corp., Pittsburgh, Pa., Interim Rept.
June 1967-Nov. 1968, Contract NAS12-568, N69 22975 (Dec.
1968), 127 pp.

Growing aluminum antimonide crystals at the
free surface of the melt
A. A. Pavlyuk and N. N. Vertoprakhov
Izv. Sibirsk. Otd. Akad. Nauk SSSR, Khim. Nauk, No. 5, 130–
132 (1968)

Croissance des cristaux non-centrosymétriques
suivant une direction polaire; rôle de la polarisation superficielle
R. Cadoret and J. C. Monier
J. Crystal Growth, 1:59–66 (1967)

Preparation of alloyed AlAs—GaAs heterojunc-
tions and investigation of their electrical prop-
erties
T. D. Dzhafarov, M. I. Stepanova, and V. K. Subashiev
Fiz. Tekhn. Poluprovod., 1(9):1306–1310 (1967)
Sov. Phys. — Semicond., 1(9):1087–1090 (1968)

Preparation and properties of AlAs — GaAs mixed crystals (halide-vapor disproportionation and melt-growth)
　J. F. Black and S. M. Ku
　J. Electrochem. Soc., 113:249 (1966)

Study of the preparation and electrical properties of semiconducting thin films of aluminum antimonide
　J. P. David, L. Capella, L. Laude, and S. Martinuzzi
　Rev. Phys. Appl., 1:172 (1966)

Certain characteristics of the growth of single crystals of compounds in the $A^{III} B^V$ group
　M. S. Mirgalovskaya, E. V. Skudnova, and I. A. Strel'nikova
　Solid State Transformations (N. N. Sirota, F. K. Gorskii, and V. M. Varikash, eds.), Consultants Bureau, New York (1966), pp. 77–81

Electrical properties of n-type aluminum arsenide
　J. Whitaker
　Solid-State Electron., 8:649–652 (1965)

Wide band gap semiconductors
　D. E. Hill and R. I. Stearns
　Monsanto Research Corp., Everett, Mass., AD 610 629 (June 1964)
　Films; vapor deposited by reduction of the chloride on (100) oriented gallium-arsenide substrates

Über Aluminiumarsenid
　W. Kischio
　Z. Anorg. Allgem. Chem., 328:187–193 (1964)
　Preparation, properties and some reactions of AlAs crystals

Preparation of aluminum arsenide by a vapour-phase transport reaction
　D. E. Bolger and B. E. Barry
　Nature, 199:1287 (1963)

Preparation of alloyed monocrystals and p — n junctions in aluminum antimonide
　M. S. Mirgalovskaya and I. A. Strel'nikova
　Sov. Phys. — Solid State 3:332 (1961)
　Synthesis; Czochralski growth

The preparation and properties of aluminum antimonide
　W. P. Allred, W. L. Mefford, and R. K. Willardson
　J. Electrochem. Soc., 107:117–122 (1960)
　Oxide eliminated by a vacuum heat-treating technique which also removes volatile impurities

Preparation and properties of AlSb
　W. P. Allred, W. L. Mefford, and R. K. Willardson
　Battelle Memorial Institute; 1959 Spring Meeting Electrochem. Soc.

Zone-melting and crystal-pulling experiments with AlSb
　W. P. Allred, Bernard Paris, and M. Genser
　J. Electrochem. Soc., 105:93–96 (1958)

Zur Isomorphie der Verbindungen des Typs $A^{III}B^V$
　Werner Koster and Werner Ulrich
　Stuttgart, Z. Metallk., 49:365–367 (1958)
　Preparation, mixed crystals, GaSb— InSb, AlSb—GaSb and AlSb— InSb

Einkristalle und pn-Schichtkristalle aus Aluminiumantimonid (Czochralski)
　Hans Achim Schell
　Z. Metallk., 49:140–144 (1958)

## 10.d. Gallium Compounds

### 10.d.1. GaN

Gallium nitride formed by vapour deposition and by conversion from gallium arsenide
　K. R. Faulkner, D. K. Wickenden, B. J. Isherwood, B. P. Richards, and I. H. Scobey
　J. Mater. Sci., 5:308–313 (1970)
　Thin films were prepared by reacting $GaCl_3$ and $NH_3$ and depositing onto single-crystal silicon carbide substrates; bulk gallium nitride was prepared by the conversion of single crystals of gallium arsenide using an intermediate oxide phase

Preparation of single-phase gallium nitride from single-crystal gallium arsenide
　B. J. Isherwood and D. K. Wickenden
　J. Mater. Sci., 5:869–872 (1970)

Optical studies of the phonons and electrons in gallium nitride
　D. D. Manchon, Jr., A. S. Barker, Jr., P. J. Dean, and R. B. Zetterstrom
　Solid State Commun., 8:1227–1231 (1970)
　Growth of needle crystals

Synthesis and growth of single crystals of gallium nitride
　R. B. Zetterstrom
　J. Mater. Sci., 5:1102–1104 (1970)

The preparation and properties of vapor-deposited single-crystalline GaN
　H. P. Maruska and J. J. Tietjen
　Appl. Phys. Letters, 15:327 (1969)
　Vapor-phase growth; the first reported specimens suitable for electrical and optical evaluation; also doped with Ge, Zn, Mg, Hg, and Si

Vapor-phase growth technique and system for several III — V compound semiconductors (interim scientific report)
　J. J. Tietjen, R. Clough, A. B. Dreeben, R. Enstrom, and D. Richman
　RCA, Princeton, N. J., Contract NAS12-538, NASA-CR-86192; ISR-2 (March 1969), 29 pp.
　Single crystalline, colorless layers of GaN were grown with sufficient size to permit good electrical and optical characterization

Preparation of aluminum and gallium nitride
　A. Rabenau
　Compound Semiconductors — Preparation of III— V Compounds, Reinhold, New York (1962)
　Crystal growth reviewed; 23 refs.

### 10.d.2. GaP

The growth of epitaxial gallium phosphide from the vapor phase by halogen transport
　A. Mottram, A. R. Peaker, and P. D. Sudlow
　J. Electrochem. Soc., 118:318–324 (1971)

Properties of GaP single crystals grown by
liquid encapsulated pulling
S. F. Nygren, C. M. Ringel, and H. W. Verleur
J. Electrochem. Soc., 118:306-312 (1971)

Epitaxial growth of gallium phosphide on
cleaved and polished (111) calcium fluoride
A. Y. Cho
J. Appl. Phys., 41(2):782-786 (1970)

Epitaxial growth and optical evaluation of gallium-phosphide and gallium-arsenide thin films
on calcium-fluoride substrate
A. Y. Cho and Y. S. Chen
Solid State Commun., 8:377-379 (1970)

Effects of the dendritic-growth mechanism on
impurity nonuniformity in GaP crystals
V. M. Grachev and L. I. Marina
Kristallografiya, 15(2):392-393 (1970)
Sov. Phys. – Cryst., 15(2):337-338 (1970)

High-efficiency red-emitting GaP diodes grown
by single epitaxy on solution-grown ($\eta \simeq 6\%$)
and Czochralski ($\eta \simeq 2\%$) substrates
W. H. Hackett, Jr., R. H. Saul, H. W. Verleur, and S. J. Bass
Appl. Phys. Letters, 16(12) (June 15, 1970)

Alloying of GaP crystals in the process of ver-
tical crucibleless zonal melting
Iu. L. Il'in, V. S. Sorokin, and D. A. Iaskov
Izv. Akad. Nauk, SSSR Neorg. Mater., 6(3):447-451 (1970)
Inorg. Mater., 6(3):392-395 (1970)

Bulk growth of GaP by halogen vapor transport
L. C. Luther
Met. Trans., 1:593-601 (1970)

Distribution of impurities in Zn,O-doped GaP
liquid-phase epitaxy layers
R. H. Saul and W. H. Hackett
J. Electrochem. Soc., 117:921-924 (1970)

The synthesis and epitaxial growth
of GaP by fused salt electrolysis
J. J. Cuomo and R. J. Gambino
J. Electrochem. Soc., 116:874 (1969)

Oxygen doping of solution-grown GaP
L. M. Foster and J. Scardefield
J. Electrochem. Soc., 116:494-498 (1969)

Effects of impurities and growth conditions on
the growth of platy gallium-phosphide crystals
N. A. Goryunova, S. L. Pyshkin, A. S. Borshchevskii, S. I.
Radautsian, et al.
Rost Kristallov, Izd. Akad. Nauk, SSSR (1968), pp. 84-89
Growth of Crystals, Vol. 8 (A. V. Shubnikov and N. N. Sheftal',
eds.) Consultants Bureau, New York (1969), pp. 68-72

Photoluminescence of epitaxial GaP
R. S. Ignatkina, I. A. Kucherenko, S. S. Meskin, V. N. Ravich,
B. V. Tsarenkov, and E. G. Shevchenko
Fiz. Tekhn. Poluprovod., 2(9):1259-1265 (1968)
Sov. Phys. – Semicond., 2(9):1057-1061 (1969)

Properties of GaP red-emitting diodes grown by
liquid-phase epitaxy. I. Effect of oxygen and
tellurium concentrations in the epitaxial n layer
Akinobu Kasami, Makoto Naito, Masaharu Toyama, and Keiji
Maeda
Japan. J. Appl. Phys., 8:1469-1480 (1969)

The use of metal-organics in the preparation of
semiconductor materials. I. Epitaxial gallium-
V compounds
H. M. Manasevit and W. I. Simpson
J. Electrochem. Soc., 116:1725-1732 (1969)

The synthesis of bulk GaP from Ga solutions
T. S. Plaskett
IBM Research Div., Yorktown Heights, N. Y., RC 2529 (No.
12165), July 2, 1969
Starting material for large single crystals
Also: J. Electrochem. Soc., 116:1722-1725 (1969)

Effect of a $GaAs_xP_{1-x}$ transition zone on the per-
fection of GaP crystals grown by deposition onto
GaAs substrates
R. H. Saul
J. Appl. Phys., 40:3273 (1969)

Growth of single-crystal GaP from organometal-
lic sources
R. W. Thomas
J. Electrochem. Soc., 116:1449 (1969)

Electron mobility and impurity concentration
in n-GaP crystals grown by slow cooling of Ga
solution
Masaharu Toyama, Makoto Naito, and Akinobu Kasami
Japan. J. Appl. Phys., 8:358 (1969)

Preparation and properties of epitaxial gallium
phosphide grown by HCl-gas transport
J. A. W. van der Does de Bye and R. C. Peters
Philips Res. Repts., 24:210-230 (1969)

Crystallization of GaP from a stoichiometric
melt (horizontal gradient freeze method) and
in gallium solution (thermal gradient)
J. P. Besselere and J. M. Leduc
Mater. Res. Bull., 3:797-806 (1968)

Research on the preparation of pure metals
Walter Brenner and Yoshiyuki Okamoto
New York University, New York, N. Y., AD-674766 (1968),
232 pp.

Optical properties of the group IV elements car-
bon and silicon in gallium phosphide
R. J. Dean, C. J. Frosch, and C. H. Henry
J. Appl. Phys., 39:5631 (1968)

Growth of epitaxial GaP
Ferranti, Ltd., England, AD-845 517 (Dec. 1968), 54 pp.

Anomalous electrical properties of solution-
grown p-type GaP
L. M. Foster, J. F. Woods, and J. E. Lewis
International Business Machines, Research Div., Yorktown
Heights, N. Y., preprint RC 2227 (No. 11071), Oct. 2, 1968

Behavior of tellurium and zinc during the nor-
mal crystallization of nonstoichiometric galli-
um – phosphorus melts
L. I. Marina, A. Ya. Nashel'skii, and B. A. Sakharov
Izv. Akad. Nauk SSSR, Neorg. Mater., 4(9):1467-1470 (1968)
Inorg. Mater., 4(9):1283-1285 (1968)

Close-spaced epitaxial growth of $GaAs_xP_{1-x}$
from powder GaAs and GaP
R. K. Purohit
J. Materials Sci., 3:330-332 (1968)

The electrical and optical properties of vapor-grown GaP
R. C. Taylor, J. F. Woods, and M. R. Lorenz
IBM Corp., Research Div., Yorktown Heights, New York, RC-2093 (#10628), May 15, 1968
J. Appl. Phys., 39:5404 (1968)

Solution growth of zinc-doped GaP with tin or germanium added to promote incorporation of zinc into the lattice
W. Westerveld and W. P. De Graaf
North American Philips Co., U. S. Patent 3,394,085 (appl. Neth., Aug. 29, 1964; publ. July 23, 1968)

Intrinsic and extrinsic edge luminescence in epitaxial GaP
D. R. Wight
J. Phys. C (Proc. Phys. Soc.), Ser. 2, 1:1759-1767 (1968)

Preparation and properties of III—V compounds for radiative processes
Louis G. Bailey
Trans. AIME, 239:310 (1967)
Reviews some of the key developments in the synthesis of the III—V compound semiconductors and the associated progress in obtaining high-quality material for device development; 120 refs.

Preparation of GaP by zone melting under pressure
Yu. L. Il'in and D. A. Yaskov
Izv. Akad. Nauk SSSR, Neorg. Mater., 3(2):296-300 (1967)
Inorg. Mater., 3(2):264-266 (1967)

Mécanismes de croissance et qualité cristalline des composés du type diamant et des cristaux des groupes III—V et II—VI
R. Kern and B. Simon
Acta Met., 15:911-920 (1967)

A crystal field analysis of cobalt impurities in GaP
Douglas Henry Loescher
Stanford Univ., California, Ph. D. thesis (1967), 72 pp.
University Microfilms, Ann Arbor, Mich., No. 67-7945
Epitaxial single crystals of GaP doped with either sulfur or zinc

Fundamental studies of the metallurgical, electrical, and optical properties of GaP
G. L. Pearson
Stanford University, Solid State Electronic Lab., Quarterly Prog. Rept., N67-40312 (Sept. 30, 1967), 20 pp.

Kristallwachstum mit alternierendem Temperaturgradienten
H. Scholz
Philips Tech. Rev., 28:220 (1967)

Vapor-transport thermodynamics of the gallium phosphide—chlorine—hydrogen system in an open tube
H. Seki and H. Araki
Japan. J. Appl. Phys., 6:1414-1418 (1967)

A method for rapid removal of metallic Ga from GaP crystals
Peter R. Wagner
Electrochem. Tech., 5:64 (1967)

Growth of GaP films of the p-type (on GaAs by sealed-system vapor transport) and their electrical properties
Zh. I. Alferov, V. I. Korolkov, M. K. Trukan, and S. P. Chashchin
Krist. Tech., 1(2):205-211 (1966)

Production of epitaxial films
J. W. Burd and W. O. Groues
Monsanto Co., French Patent 1,509,254 (Jan. 12, 1968); U. S. appl. Jan. 3, 1966 - Dec. 16, 1966, 23 pp.

Semiconductor crystal for use in apparatus for thermoelectric cooling (doped with Ge, Cu, and Zn)
O. G. Folberth
Siemens-Schuckertwerke AG, German Patent 1,219,241 appl. July 1960; publ. June 16, 1966, 2 pp.

Growth of GaP crystal platelets from solutions of phosphorus in gallium
N. A. Goryunova, A. S. Borshchevskii, G. A. Kalyuzhnaya, A. D. Smirnova, and N. K. Takhtareva
Krist. Tech., 1(1):85-91 (1966)

Epitaxial growth of III—V materials
E. W. Mehal and G. R. Cronin
Electrochem. Tech., 4:540 (1966)

The effect of impurities and crystallization conditions on the growth of platelet crystals of gallium phosphide
S. L. Pyskin, A. S. Borscevskij, S. I. Radaucan, G. A. Kaljuznaja, and Ju. I. Maksimov
Acta Cryst. (Internat.) 21, Pt. 7, Suppl., A242 (Dec. 1966)
Seventh International Congress and Symposium International Union of Crystallography, Moscow (1966)

Epitaxial materials
B. Schwartz
Vapor Deposition, John Wiley and Sons, Inc., New York (1966), p. 612

Optical properties of gallium phosphide
H. Sonomura, N. Yamamoto, and T. Miyauchi
Bull. Univ. Osaka Prefecture, Ser. A, 15: 91 (1966)

Luminescence due to the isoelectronic substitution of bismuth for phosphorus in gallium phosphide
F. A. Trumbore, M. Gershenzon, and D. G. Thomas
Appl. Phys. Letters, 9:4 (1966)

Transport reaction in closed-tube process
T. Arizumi and T. Nishinaga
J. Appl. Phys., 4:165 (1965)

Synthesis and characterization of electronically active materials (epitaxial growth of GaP and GaP—GaAs from vapor)
E. F. Hockings and H. W. Leverenz
RCA Labs., David Sarnoff Research Center, Princeton, N. J., Tech. Rept. No. 4; AD-475 116 (1965), 95 pp.

Chemical transport reactions
A. Rabenau
Philips Tech. Rev., 26:117 (1965)

Synthesis of GaP in a bismuth melt
Ya. A. Ugai and O. Ya. Gukov
Izv. Akad. Nauk SSSR, 1(6):857-860 (1965)
Inorg. Mater., 1(6):787-789 (1965)

Optical and electrical properties of single-crystal GaP vapor grown on GaAs substrate
H. Flicker, B. Goldstein, and P. A. Hoss
J. Appl. Phys., 35:2959 (1964)

Structural defects in GaP crystals and their electrical and optical effects
M. Gershenzon and R. M. Mikulyak
J. Appl. Phys., 35:2132 (1964)
Floating-zone

Development of improved single-crystal gallium phosphide solar cells
W. O. Groves and A. S. Epstein
Monsanto Research Corp., Dayton, Ohio, Qtr. Rept. No. 3, Dec. 1963-Apr. 1964, NASA Contract NAS3-2776, NASA CR-53986 (1964), 52 pp.
Vapor transport and epitaxial deposition

Preparation and properties of GaP – GaAs p – n junction lasers (halogen-vapor transport)
N. Holonyak
Trans. AIME, 230:276-281 (1964)

On the morphology and structure of GaAs and GaP platelets grown from the gas phase
W. Krieglstein, H. Pfister, and G. Ziegler
Z. Angew. Phys., 17:295 (1964)

Preparation of GaP single crystals
A. Ya. Nashelskii, L. I. Marina, and S. V. Yakobson
U. S. S. R. Patent 163,362 (appl. Feb. 1962; publ. July 1964)

Preparation and properties of solution-regrown gallium phosphide
A. Pfahnl
J. Electrochem. Soc., 111:58C (1964)

Pair spectra and "edge" emission in gallium phosphide
D. G. Thomas, M. Gershenzon, and F. A. Trumbore
Phys. Rev., 133:A269 (1964)

The reaction of GaP(s) with $H_2O$(g) and the range of stability of GaP(s) under pressures of $Ga_2O$ and $P_2$
C. D. Thurmond and C. J. Frosch
J. Electrochem. Soc., 111:184 (1964)

Growth of GaP crystal and p – n junctions by a traveling solvent method
M. Weinstein and A. I. Mlavsky
J. Appl. Phys., 35:1892-1894 (1964)

Growth of gallium phosphide from the melt
L. R. Weisberg and E. A. Miller
Synthesis and characterization of electronically active materials, RCA Labs. Tech. Rept. No. 1 covering period May 15, 1963, to Feb. 15, 1964, Contract SD-182, ARPA Order 446. L. R. Weisberg and H. W. Leverenz, eds. (March 15, 1964), pp. 21-25

Epitaxial vapor growth of III – V compounds
J. F. Gibbons and P. C. Prehn
Solid-State Electronics Lab., Stanford Univ., Calif., Interim Tech. Rept. No. 4711-1, Oct. 1961-Aug. 1963, Contract AF 33(616)-7726, SEL-63-105; RTD-TDR-63-4238; AD-434756 (Oct. 1963), 52 pp.
Open-tube method

Dissociation pressures of GaAs, GaP and InP and the nature of III – V melts
D. Richman
J. Phys. Chem. Solids, 24:1131-1139 (1963)

A flow synthesis of gallium phosphide and some properties of gallium phosphide powder layers
L. J. Bodi
J. Electrochem. Soc., 109:497 (1962)

Preparation and properties of GaP single crystals (direct combination, crystallization from gallium solution)
S. Iijima and M. Kikuchi
Kogyo Kagaku Zasshi, 65:1722-1725 (1962)
Chem. Abstr., 58:9854B/C

Preparation of gallium phosphide
J. F. Miller
Compound Semiconductors - Preparation of III–V Compounds, Reinhold (1962), pp. 194-206

Preparation of solid solutions of GaP and GaAs by a gas-phase reaction
F. A. Pizzarello
J. Electrochem. Soc., 109:226 (1962)

High-temperature semiconductor research (epitaxial GaP on GaAs)
J. Blanc, R. H. Bube, R. D. Gold, B. Goldstein, et al.
RCA Labs., Princeton, N. J., AD-260 767 (1961), 55 pp.

Some properties of p – n junctions in GaP
H. G. Grimmeiss, A. Rabenau, and H. Koelmans
J. Appl. Phys., 32:2123-2127 (1961)
Vapor method using an iodine carrier gas

GaP single crystals (zone melting under 10-15 atmosphere phosphorus pressure)
M. Lorant
Chem. Rundschau, 14:469 (1961)

On the preparation of the phosphides of aluminum, gallium and indium
Arrigo Addamiano
J. Am. Chem. Soc., 82:1537 (1960)
Synthesis; metal and ZnS

Gallium phosphide for power rectifiers
R. E. Davis
Properties of Elemental and Compound Semiconductors, Interscience, 1960, pp. 295-302
Preparation and electrical properties

Preparation of InAs, InP, GaAs, and GaP by chemical methods
D. Effer and G. R. Antell
J. Electrochem. Soc., 107:252 (1960)

Preparation and properties of gallium phosphide
G. J. Frosch, M. Gershenzon, and D. F. Gibbs
1960 Spring Mtg. Electrochem. Soc.

High-temperature semiconductor research
L. R. Weisberg et al.
RCA, Sci. Rep. No. 2, pp. 1-22, 28-38 (June 1960), AD 243 363
Bridgman

### 10.d.3. GaAs

Pits and hillocks on epitaxial GaAs grown from the vapor phase
  H. T. Minden
  J. Cryst. Growth, 8:37-44 (1971)

The effect of growth orientation on the crystal perfection of horizontal Bridgman-grown GaAs
  T. S. Plaskett, J. M. Woodall, and A. Segmuller
  J. Electrochem. Soc., 118:115-117 (1971)

Liquid-encapsulated Czochralski growth of GaAs
  M. E. Weiner, D. T. Lassota, and B. Schwartz
  J. Electrochem. Soc., 118:301-306 (1971)

Vapor-phase transport in the epitaxial growth of GaAs
  H. Araki, G. Iwane, and T. Aoki
  Rev. Electrical Commun. Lab., 18(9-10):608-617 (1970)

Thermodynamic and experimental aspects of gallium-arsenide vapor growth
  A. Boucher and L. Hollan
  J. Electrochem. Soc., 117:932-936 (1970)

Morphology of epitaxial growth of GaAs by a molecular beam method: the observation of surface structures
  A. Y. Cho
  J. Appl. Phys., 41:2780 (1970)

Some problems in studying the growth of ribbon-like dendrites in semiconducting materials
  M. Ya. Dashevsky and B. A. Sakharov
  Izv. Akad. Nauk SSSR, Neorg. Mater., 6(9):1557-1560 (1970)
  In Russian

Inhomogeneity in GaAs crystals grown from non-stoichiometric melts
  F. A. Gimel'farb, B. D. Lainer, M. G. Mil'vidskii, and V. I. Fistul'
  Kristallografiya, 15:201-202 (1970)
  Sov. Phys. — Cryst., 15:169-170 (1970)

Epitaxial growth of GaAs on insulating substrates using $HCl - H_2$ vapor transport
  W. A. Gutierrez
  Solid-State Electron., 13:1199-1200 (1970)

A review of bulk and process-induced defects in GaAs semiconductors
  E. D. Jungbluth
  Met. Trans., 1:575-585 (1970)
  Effects of nonstoichiometry, dislocations, segregation, and precipitation

Crystallization of nonstoichiometric melts of decomposable compounds
  B. D. Lainer, A. V. Berkova, M. G. Mil'vidskii, and V. V. Rakov
  Sov. Phys. — Cryst., 15:743-746 (1970)

Epitaxial gallium-arsenide p - n junctions grown in a closed iodide system with varying iodine concentrations
  L. G. Lavrent'eva, M. D. Vilisova, S. P. Gaidareva, and O. M. Ivleva
  Izv. Vyssh. Ucheb. Zaved., Fiz., 13(2):31-35 (1970)

Gallium diethyl chloride: a new substance in the preparation of epitaxial gallium arsenide
  K. Lindeke, W. Sack, and J. J. Nickl
  J. Electrochem. Soc., 117(10):1316-1318 (1970)

Appareil pour le tirage de monocristaux de GaAs par la methode de Czochralski
  M. Moldovanova, P. Gladkov, J. Georgiev, and B. Arnaudov
  God. Sofii. Univ., Fiz. Fak., Bulg., 62:161-166 (1967-1968) (pub. 1970)

Thermal instability of the growth interface in the horizontal solidification of GaAs
  M. Ohyama
  J. Phys. Soc. Japan, 29:706-710 (1970)

Ga — As — Si: phase studies and electrical properties of solution-grown Si-doped GaAs
  M. B. Panish and S. Sumski
  J. Appl. Phys., 41:3195-3196 (1970)

Preparation and properties of GaAs single crystals
  E. Papp, S. Zsindely, T. Legat, and B. Podor
  Acta Tech. Hung., 68(1-2):245-253 (1970)
  Modified Bridgman method

Germanium-doped gallium arsenide
  F. E. Resztoczy, F. Ermanis, I. Hayashi, and B. Schwartz
  J. Appl. Phys., 41:264-270 (1970)
  On GaAs seeds from Ga solution

GaAs, The Institute of Physics Conference Series No. 9. (A Comprehensive Report of the Proceedings of the Third Gallium Arsenide Symposium held in Aachen, October 1970)
  Dawsons of Pall Mall, Cannon House, Park Farm Road, Folkestone, Kent, England (April 1971)

Initial stage of epitaxial growth in thin films
  L. N. Alexandrov
  Krist. Tech., 4:K17-K24 (1969)

Epitaxie liquide de GaAs
  E. André and J. M. LeDuc
  R. T. C. Radiotechnique-Compelec, 14-Caen, Rapp. final
  D. G. R. S. T., Action concertée: Electron., Contrat No. 6600141 (1969), 8 pp.

Perfection of materials technology for producing improved Gunn-effect devices, interim scientific report
  R. D. Baxter, J. F. Miller, H. L. Leonard, and A. C. Beer
  Battelle Memorial Inst., Columbus, Ohio, NASA-Cr-86277; ISR-2 (Dec. 1969), 26 pp.
  Chemical-vapor deposition of GaAs and the preparation of $Ga_xIn_{1-x}Sb$ ingots

Vapor growth of GaAs for Gunn oscillators
  A. E. Blakeslee and B. K. Bischoff
  IBM, Yorktown Heights, N. Y., RC 2448 (11873) (April 24, 1969)

Etude radiocristallographie de couchés épitaxiales de germanium et d'arséniure de gallium obtenues à partir d'une phase liquide
  L. Castet, L. Mayet, and G. Mesnard
  Rev. Phys. Appl., 4:431-435 (1969)

Epitaxy by periodic annealing
  A. Y. Cho
  Surface Sci., 17(2):494-503 (1969)

Supercooling of melts of $A^{II}B^V$ compounds
M. Ya. Dashevskii and A. N. Poterukhin
Izv. Akad. Nauk SSSR, Neorg. Mater., 5(11):2012-2014 (1969)
Inorg. Mater., 5(11):1713-1715 (1969)

Relation of the surface properties of a gallium-
antimonide melt to crystal growth
M. Ya. Dashevskii, G. V. Kukuladze, V. B. Lazarev, and
M. S. Mirgalovskaya
Rost Kristallov, Izd. Akad. Nauk, SSSR (1968), pp. 104-107
Growth of Crystals, Vol. 8 (A. V. Shubnikov and N. N.
Sheftal', eds.) Consultants Bureau, New York (1969), pp.
85-90

Liquid-phase epitaxial growth of gallium arse-
nide
A. R. Goodwin, C. D. Dobson, and J. Franks
Gallium Arsenide, Proc. 2nd International Symposium 1968
(C. I. Pedersen, ed.), Institute of Physics and the Physics
Society, London, England (1969), pp. 36-40

Homogeneous solution grown epitaxial GaAs by
tin doping
J. S. Harris and W. L. Snyder
Solid-State Electron., 12:337-340 (1969)

High-purity GaAs by liquid-phase epitaxy
H. G. B. Hicks and D. F. Manley
Solid State Commun., 7:1463-1465 (1969)

Preparation of bulk gallium arsenide material
Kenneth L. Klohn and Lothar Wandiner
Army Electronics Command, Fort Monmouth, N. J., AD-
700134; ECOM-3198 (Nov. 1969), 29 pp.
Review; Czochralski, floating zone, Bridgman, and gradient
freeze

Techniques for characterization and evaluation
of zero-gravity grown gallium arsenide
A. P. Kulshreshtha, T. Mookherji, and G. M. Arnett
National Aeronautics and Space Administration, Marshall
Flight Center, Huntsville, Ala., in its Space Processing
and Manufacturing Meeting (Oct. 21, 1969), pp. 306-322

Effect of melt stoichiometry on the crystalliza-
tion of gallium arsenide
B. D. Lainer, V. V. Rakov, and M. G. Mil'vidskii
Izv. Akad. Nauk SSSR, Neorg. Mater., 5(1):25-29 (1969)
Inorg. Mater., 5(1):20-23 (1969)

Crystallization of gallium arsenide
B. D. Lainer, V. V. Rakov, and M. G. Mil'vidskii
Izv. Akad. Nauk SSSR, Neorg. Mater., 5(8):1370-1372 (1969)
Inorg. Mater., 5(8):1169-1171 (1969)

Distribution of a volatile impurity ahead of the
crystallization front in the melt
B. D. Lainer, V. V. Voronkov, M. G. Mil'vidskii, and V. V.
Rakov
Kristallografiya, 14(3):537-538 (1969)
Sov. Phys. — Cryst., 14(3):451-453 (1969)

The use of metal-organics in the preparation
of semiconductor materials. I. Epitaxial gallium-V compounds
H. M. Manasevit and W. I. Simpson
J. Electrochem. Soc., 116:1725-1732 (1969)

Recent advances in gallium-arsenide materials
technology
H. T. Minden
Solid State Tech., 12(4):25-35 (1969)

Production et propriétés des monocristaux de
GaAs
E. Papp, B. Podor, S. Zsindely, and T. Legat
Hiradastechnica Magyar, 20(12):368-373 (1969)

Behavior of iron in the crystallization of gal-
lium arsenide
O. V. Pelevin, B. G. Girich, and M. G. Mil'vidskii
Izv. Akad. Nauk SSSR, Neorg. Mater., 5(7):1200-1202 (1969)
Inorg. Mater., 5(7):1021-1023 (1969)

Epitaxial gallium arsenide from trimethyl gal-
lium and arsine
P. Rai-Choudhury
J. Electrochem. Soc., 116:1745-1746 (1969)

Arsenic losses during production of GaAs sin-
gle crystals
V. V. Rakov, B. D. Lainer, and M. G. Mil'vidskii
Inorg. Mater., 5:1242-1243 (1969)

Losses of arsenic during the growing of mono-
crystals GaAs
V. V. Rakov et al.
Izv. Akad. Nauk SSSR 5: 1458-1459 (1969)
Translation: National Lending Library for Science and Tech.,
Boston Spa, England, NLL-REE-Trans-280-(8036.625)

Temperature dependence of the rate of epitaxial
growth in chemical crystallization. The gallium arsenide — iodine — hydrogen system
Yu. M. Rumyantsev and F. A. Kuznetsov
Izv. Sibersk. Otd. Akad. Nauk SSSR, 5(12):66-72 (1969)

Preparation of oxygen-doped GaAs in a three-
zone Bridgman furnace
Takashi Shimoda and Shin-ichi Akai
Japan. J. Appl. Phys., 8:1352 (1969)

Epitaxial growth of GaAs(I)
H. Teshima, Y. Tarui, and O. Takeda
Bull. Electrotech. Lab. (Japan), 33(6):692-699 (1969)

Faults in epitaxial layers formed on strongly
doped GaAs substrates
V. N. Vasilevskaya, L. I. Dacenko, and L. N. Mikhajlov
Ukr. Fiz. Zh., 14(5):867-870 (1969)

Growth bands in undoped gallium arsenide sin-
gle crystals
E. N. Zhelikhovskaya and L. A. Borisova
Izv. Akad. Nauk SSSR, Neorg. Mater., 5(12):2205-2206 (1969)
Inorg. Mater., 5(12):1885-1886 (1969)

Epitaxial deposition of gallium arsenide from
the vapor phase 1961-1968
Bell Telephone Laboratories (Public Relations and Publications Div., Technical Information Libraries), Bibliography
No. 122 (Nov., 1968)

Production of epitaxial films
John W. Burd and Warren O. Groues
Monsanto Co., French Patent 1,509,254 (Jan. 12, 1968) 23 pp.

Doping of gallium-arsenide crystals grown from
gallium solution
C. Constantinescu, P. Mihailovici, I. Petrescu-Prahova, and
G. Popovici
Rev. Roum. Phys., 13(5):489-490 (1968)

Growth of gallium-arsenide dendrites and study
of their properties
M. Ya. Dashevskii and A. V. Zakharova
Izv. Akad. Nauk SSSR, Neorg. Mater., 4(9):1471-1473 (1968)
Inorg. Mater., 4(9):1286-1288 (1968)

Gallium arsenide, a review
J. Frey
Electronics and Appl. Phys. Div., U. K. A. E. A. Research Group, Atomic Energy Research Establishment, Harwell, England, AERE-R 5757 (March 1968)

Impurity transfer in GaAs vapor growth and carrier-concentration profiles of the grown films
Fumio Hasegawa and Takeshi Saito
Japan. J. Appl. Phys., 7:1342 (1968)

High-peak power microwave devices, quarterly report No. 1, July 1, 1968, to Sept. 30, 1968
Bert Jeppsson and L. F. Eastman
Cayuga Associates, Inc., Cornell Research Park, Ithaca, New York 14850, ECOM-0397-1 (U. S. Army Electronics Command, Fort Monmouth, N. J. 07703) (Dec. 1968), 18 pp.

Single-crystal gallium arsenide on insulating substrates
Harold M. Manasevit
Appl. Phys. Letters, 12:156 (1968)

Evaluation of bulk and epitaxial GaAs by means of x-ray topography
Eugene S. Meieran
Trans. AIME, 242:413 (1968)
Effects of methods of crystal growing are reviewed

Behavior of zinc during growth of doped single crystals of gallium arsenide
O. V. Pelevin and M. G. Mil'vidskii
Izv. Akad. Nauk SSSR, Neorg. Mater., 4(11):1864-1868 (1968)
Inorg. Mater., 4(11):1625-1628 (1968)

Mechanism for the generation of dislocations in heavily doped GaAs during growth from the melt
T. S. Plaskett, J. M. Woodall, and A. Segmuller with A. H. Parsons, and W. C. Wuestenhoefer
IBM, Yorktown Heights, N. Y., RC-2052 (No. 10495) (April 10, 1968)

Preferred growth directions of gallium-arsenide crystals
V. A. Presnov, V. A. Selivanova, and S. S. Khludkov
Rost Kristallov, Izd. Akad. Nauk SSSR (1965), pp. 275-280
Growth of Crystals, Vol. 6B (A. V. Shubnikov and N. N. Sheftal', eds.) Consultants Bureau, New York (1968), pp. 86-90

Germanium-doped gallium arsenide
F. E. Resztoczy, F. Ermanis, I. Hayashi, and B. Schwartz
Bull. Am. Phys. Soc., 13:375 (1968)

Single-crystal GaAs without dislocations
A. Steinemann, H. R. Winteler, and U. Zimmerli
Helv. Phys. Acta, 41:1210-1220 (1968)

Theoretical analysis of requirements for crystal growth from solution
W. A. Tiller
J. Cryst. Growth, 2:62-79 (1968)

Dislocation-free gallium arsenide (Versetzungsfreies galliumarsenid)
Ulrich Zimmerli
Ph. D. thesis, Eidgnössische Technische Hochschule, Zurich, Switzerland (1968), DISS-4072, 41 pp.
By vertical growth from solution and in closed systems with magnetic seed suspension

Preparation and properties of III—V compounds for radiative processes
Louis G. Bailey
Trans. AIME, 239:310 (1967)

Properties of gallium-arsenide crystals produced by liquid encapsulation pulling
S. J. Bass and P. E. Oliver
Proc. Internat. Symposium on Gallium Arsenide, Reading, September 1966 (1967), pp. 41-45

Growth of gallium-arsenide bulk single crystals (1957-1966)
Bell Telephone Laboratories (Technical Information Libraries, Public Relations and Publication Div., Murray Hill, N. J.), Bibliography No. 109 (March 1967)

Preparation and characteristics of gallium arsenide
D. E. Bolger, J. Franks, J. Gordon, and J. Whitaker
Proc. Internat. Symp. on Gallium Arsenide, Reading, Sept. 1966 (1967), pp. 16-22

Mass spectrometric studies of impurities in gallium-arsenide crystals
J. C. Brice, J. A. Roberts, and G. Smith
J. Mater. Sci., 2:131-138 (1967)

Microwave oscillations in bulk semiconductors
N. Braslau, J. M. Woodall, C. Lanza, and A. E. Blakeslee
(IBM, Yorktown Heights, N. Y. 10598), 8th Quart. Progr. Rept. — April 1, 1967, to June 30, 1967; 7th Quart. Rept. — Jan. 1, 1967, to March 31, 1967; 5th Quart. Rept. — July 1 to Sept. 30, 1966 (1967)
Each transmittal of these documents outside the Dept. of Defense must have prior approval of Commanding General, U. S. Army Electronics Command, Fort Monmouth, N. J., AMSEL-KL-SM

Croissance des cristaux non-centrosymétriques suivant une direction polaire, rôle de la polarisation superficielle
R. Cadoret and J. C. Monier
J. Cryst. Growth, 1:59-66 (1967)

The preparation and properties of epitaxial gallium arsenide
D. V. Eddolls, J. R. Knight, and B. L. H. Wilson
Proc. Internat. Symp. on Gallium Arsenide, Reading, Sept. 1966 (1967), pp. 3-9

Recent progress in the preparation of very pure gallium arsenide
J. L. Fertin, J. Lebailly, and E. Deyris
Proc. Internat. Symp. on Gallium Arsenide, Reading, Sept. 1966 (1967), pp. 46-51

Preparation of single crystals of gallium arsenide by the Czochralski method under flux cover
A. Hruban
Bul. Wojsk. Akad. Tech., 16:89-95 (1967)

Pyramid formation in epitaxial gallium-arsenide layers
B. D. Joyce and J. B. Mullin
Proc. Internat. Symp. on Gallium Arsenide, Reading, Sept. 1966 (1967), pp. 23-26

Preparation and properties of high-purity epitaxial GaAs grown from Ga solution
C. S. Kang and P. E. Greene
Appl. Phys. Letters, 11:171-173 (1967)

Preparation of GaAs — Ge and InAs — GaAs heterojunctions in a closed-tube system using iodine process
Hiroyuki Kasano and Shinya Iida
Japan. J. Appl. Phys., 6:1038 (1967)

The growth of semiconductors on single-crystal metallic substances
J. E. Knappet and S. J. T. Owen
J. Metals and Mat. (Nov. 1967), p. 369

Observation of the growth process of gallium arsenide on tungsten
J. E. Knappet and S. J. T. Owen
Phys. Stat. Sol., 21:K99 (1967)

Production of epitaxial gallium-arsenide films by the sandwich method
A. V. Kovda and S. A. Semiletow
Kristallografiya, 12(3):535-536 (1967)
Sov. Phys. — Cryst., 12(3):468-469 (1967)

Preparation and properties of GaAs — Si heterojunctions by solution growth method
Takao Nakano
Japan. J. Appl. Phys., 6:854 (1967)

Ternary condensed phase diagram of the gallium — arsenic — tellurium system
M. B. Panish
J. Electrochem. Soc., 114:91-95 (1967)

Gallium-arsenide epitaxial technology
D. W. Shaw, R. W. Conrad, E. W. Mehal, and O. W. Wilson
Proc. Internat. Symp. on Gallium Arsenide, Reading, Sept. 1966 (1967), pp. 10-15

Growth rate of crystalline layers of gallium arsenide in the iodide process
N. N. Sheftal' and Kh. A. Magomedov
Kristallografiya, 12(1):152-153 (1967)
Sov. Phys. — Cryst., 12(1):129-131 (1967)

Gallium Arsenide
Proc. Internat. Symp., Reading, England, September 1966 (A. C. Strickland, ed.), Institute of Physics and Physical Society, London (1967)

Bibliography vapor growth of the arsenides and phosphides of Group III metals by chemical transport reactions
R. C. Taylor
IBM, Yorktown Heights, N. Y., NC-684 (Jan. 19, 1967)

Vapor-phase growth technique and system for several III — V compound semiconductors, quarterly technical report
J. J. Tietjen, R. Clough, and D. Richman
RCA, Princeton, N. J., NASA-CR-90110; QTR-1 (June 1967), 16 pp.

The observation of defects in GaAs using photoluminescence at 20°K
E. W. Williams and D. M. Blacknall
Trans. AIME, 239:387 (1967)

Structural effects in epitaxial gallium arsenide
F. V. Williams
Proc. Internat. Symp. on Gallium Arsenide, Reading, Sept. 1966 (1967), pp. 27-31

Donor and carrier distributions in oxygen-grown GaAs
J. M. Woodall
Trans. AIME, 239:378 (1967)

Dislocations and precipitates in GaAs injection lasers
M. S. Abrahams and C. J. Buiocchi
J. Appl. Phys., 37:1974 (1966)

Recent problems of impurities in gallium-arsenide single crystal
I. Akasaki
Cyo Butsuri, 35(8):597 (1966)

Preparation and properties of aluminum-arsenide — gallium arsenide mixed crystals (melt growth and halide vapor disproportionation)
J. F. Black and S. M. Ku
J. Electrochem. Soc., 113:249-254 (1966)

Epitaxial growth of bulk-quality gallium arsenide on gallium arsenide and germanium substrates
L. C. Bobb, H. Holloway, K. H. Maxwell, and E. Zimmerman
J. Appl. Phys., 37:3909 (1966)

Oriented growth of semiconductors — II. Homoepitaxy of gallium arsenide
L. C. Bobb, H. Holloway, K. H. Maxwell, and E. Zimmerman
J. Phys. Chem. Solids, 27:1679-1685 (1966)

The influence of growth conditions on the structure of GaAs crystals grown from super-cooling melts
E. N. Bojko, L. A. Borisova, and K. E. Mironov
Acta Cryst., 21, Pt. 7, Suppl., A258 (Dec. 30, 1966)
Seventh Intern. Congr. Symp. Intern. Union of Crystallography, Moscow, 1966

Préparation. Etude des propriétés fondamentales et applications des couches minces d'arséniure de gallium evaporées sous vide
P. Bourgeois
Le Vide, No. Special A. B. I. SEM (Oct., 1966), p. 34

Effect of arsenic pressure on dislocation densities in melt-grown gallium arsenide
J. C. Brice and G. D. King
Nature, 209:1346 (1966)

Inhomogeneities in the electrical properties of gallium-arsenide crystals
J. C. Brice, R. E. Hunt, G. D. King, and H. C. Wright
Solid-State Electronics, 9:853-857 (1966)

Twinning and morphology of epitaxial layers of GaAs on Ge
Yu. M. Chashchinov and V. A. Mokievskii
Fiz. Tverd. Tela, 8(5):1511-1516 (1966)
Sov. Phys. — Solid State, 8(5):1201-1205 (1966)

Incorporation of zinc into epitaxial GaAs using diethyl zinc
R. W. Conrad and R. W. Haisty
J. Electrochem. Soc., 113:199 (1966)

Single-crystal compounds (gallium arsenide coeled under vapor equilibrium in double boat, crystal forming in inner one)
W. R. Derby and A. M. Herzog
Monsanto Co., U. S. Patent 3,240,568 (March 15, 1966), 7 pp.

Transport-reaction produced high-purity GaAs single crystal
H. J. Dersin and E. Sirtl
Z. Naturforsch., 21a: 332 (1966)

Silicon contamination of GaAs crystals grown in glassy carbon boats
Tohru Hara and Isamu Akasaki
Japan. J. Appl. Phys., 5:1255-1256 (1966)

The preparation of epitaxial semi-insulating gallium arsenide by iron doping
P. L. Hoyt and R. W. Haisty
J. Electrochem. Soc., 113:296-297 (1966)

GaAs-whisker crystals containing germanium core
S. Iida and Y. Sugita
Appl. Phys. Letters, 8:77 (1966)

Vapor-solvent growth of semiconducting crystals, final report Dec. 1, 1962-May 31, 1966
Franco P. Jona
IBM, Yorktown Heights, N. Y., AFOSR-68-2887; AD 680 049 (1966), 18 pp.

The chemical vapor deposition of single-crystals films
B. A. Joyce
The Use of Thin Films in Physical Investigations, Academic Press, New York, 1966, p. 87

Growth "pyramids" in epitaxial GaAs
B. D. Joyce and J. B. Mullin
Solid State Commun., 4:463 (1966)

Method for preparation of thin, oriented GaAs crystals
Peter Knoll and Rainer Zuleeg
J. Appl. Phys., 37:5006-5007 (1966)

Preparation and properties of boron-arsenide—gallium-arsenide mixed crystals
S. M. Ku
J. Electrochem. Soc., 113:813-816 (1966)

Vapor growth parameters and impurity profiles on n-type films grown on $N^+$ GaAs by the hydrogen—water vapor process
K. L. Lawley
J. Electrochem. Soc., 113:240 (1966)

The preparation of high-purity gallium arsenide by the Czochralski method
W. K. Liebmann and G. Kampschulte
Solid-State Electronics, 9:828-830 (1966)

Epitaxial growth of III—V materials
E. W. Mehal and G. R. Cronin
Electrochem. Tech., 4:540 (1966)

GaAs epitaxial technology for integrated circuits
E. W. Mehal, R. W. Haisty, and D. W. Shaw
Trans. AIME, 236:263 (1966)

The substrate orientation effect on impurity profiles of epitaxial GaAs films
R. R. Moest
J. Electrochem. Soc., 113:141 (1966)

Production of monocrystals of gallium arsenide
K. K. Ozolis and E. Yu. Kokorish
Rost Kristallov, Izd. Akad. Nauk SSSR (1964), pp. 181-202
Growth of Crystals, Vol. 4 (A. V. Shubnikov and N. N. Sheftal', eds.) Consultants Bureau, New York (1966), pp. 150-166

Epitaxial synthesis of GaAs using a flow system
M. Rubenstein and E. Myers
J. Electrochem. Soc., 113:365 (1966)

Some characteristics of GaAs — Ge epitaxy
R. G. Schulze
J. Appl. Phys., 37:4295 (1966)

Epitaxial materials
B. Schwartz
Vapor Deposition, John Wiley and Sons, Inc., New York (1966), p. 612

Selective epitaxial deposition of gallium arsenide in holes
Don W. Shaw
J. Electrochem. Soc., 113:904 (1966)

Silicon contamination of gallium arsenide grown in nonsilica boats
Robert I. Stearns and James B. McNeely
J. Appl. Phys., 37:933-934 (1966)

Vapor-phase growth of gallium arsenide microwave diodes
J. J. Tietjen, G. Kupsky, and H. Gossenberger
Solid-State Electronics, 9:1049-1053 (1966)

Dislocations in gallium arsenide grown from gallium by the traveling solvent method
M. Weinstein, H. E. Labelle, and A. I. Mlavsky
J. Appl. Phys., 37:2913-2914 (1966)

Preparation of high-purity epitaxial GaAs
C. M. Wolfe, T. M. Quist, and A. J. Strauss
ESD-TR-66-207, Lincoln Laboratory, Mass. Inst. of Technology, Cambridge, Mass., pp. 7-8

Preparation of $0.5-10^3$ $\Omega$-cm GaAs by acceptor precipitation during heat treatment of oxygen-grown crystals
J. M. Woodall and J. F. Woods
Solid State Commun., 4:33-36 (1966)

Cristobalite formation in vitreous silica boats and the relation to gallium-arsenide crystal growth
M. Yamaguchi, Y. Mizushima, S. Hirota, and H. Noake
J. Electrochem. Soc., 113:294-296 (1966)

A syringe crystal puller for materials having a volatile component
E. M. N. Baldwin, J. C. Brice, and E. J. Millett
J. Sci. Instr., 42:883-884 (1965)

Critical cooling rate in a vapor deposition process to form blade-like semiconductor compound crystals
H. R. Barkemeyer, W. J. McAleer, and P. I. Pollak
Merck and Co., U. S. Patent 3,206,406 (Sept. 14, 1965)

Whisker crystals of gallium arsenide and gallium phosphide grown by the vapor — liquid — solid mechanism
R. L. Barns and W. C. Ellis
J. Appl. Phys., 36:2296-2301 (1965)

Studies on high-power GaAs lasers
S. E. Blum, M. Pilkuhn, T. S. Plaskett, H. Rupprecht, R. S.
Title, J. M. Woodall, and J. F. Woods
IBM, Yorktown Heights, N. Y., RC-1351 (Feb. 9, 1965)

Distribution of impurities in gallium-arsenide
single crystals grown by the Czochralski
method
Yu. M. Burdukov, I. T. Voronina, O. V. Emel'yanenko, and
T. S. Lagunova
Izv. Akad. Nauk SSSR, Neorg. Mater., 1(9):1459-1461 (1965)
Inorg. Mater., 1(9):1332-1334 (1965)

Preparation of monocrystalline gallium arse-
nide
B. Ciszewski, K. Wieczffinski, A. Kalinowski, S. Himmel,
J. Luty, and H. Ziencik
Biul. Wojskowej Akad. Tech., 14(157):79-107 (1965)
Chem. Abstr., 65(6):8094-8095

Epitaxial growth of doped and pure GaAs in an
open flow system
D. Effer
J. Electrochem. Soc., 112:1020 (1965)

Crystal growth
W. C. Ellis, W. G. Pfann, and R. S. Wagner
Western Electric Company, Belg. Patent 658,975 (May 17,
1965), 18 pp.

Preparation of single crystals of multicompo-
nent alloys (of gallium arsenide and copper (2)
germanium (1) selenium (3))
N. A. Goryunova, G. K. Averkieva, and A. A. Vaipolin
Fiz. Dokl. K 23-El Nauchn.Konf. Leningr. Inzh.-Stroit. Inst.
Leningrad. Sb. (1965), pp. 52-53
Chem. Abstr., 64:11957-11958

Transpiration in an open tube GaAs/HI/H₂
system
G. Hellbardt
J. Electrochem. Soc., 112:443 (1965)

Gallium arsenide
C. Hilsum
Progress in Semiconductors, Vol. 9 (Alan F. Gibson and
R. E. Burgess, eds.), A Heward Book, Temple Press
Books, Ltd., London (1965), pp. 135-178

Production of single crystals
Hitachi, Ltd., Netherlands Appl. 291,059 (Jan. 25, 1965), 7 pp.

Die Verteilung von Cd in Czochralski-gezoge-
nen GaAs-Einkristallen
W. Hoffmeister, G. Kampschulte, and W. K. Liebmann
Phys. Vehr. D. P. G., Dtsch., 5(4):130 (1965)

Oriented growth of semiconductors. I. Orientations in gallium arsenide grown epitaxially on
germanium
H. Holloway, K. Wollmann, and A. S. Joseph
Phil. Mag., 11:263 (1965)

The preparation of high-purity gallium arsenide
by vapor-phase epitaxial growth
J. R. Knight, D. Effer, and P. R. Evans
Solid-State Electronics, 8:178 (1965)

Production of single crystalline semiconductors containing p-n junctions (vapor growth)
W. Krieglstein and B. Reiss
Siemens-Schuckerwerke AG, U. S. Patent 3,168,423 (Feb. 2,
1965), 6 pp.

Synthesis of GaAs by vapor transport reaction
H. R. Leonhardt
J. Electrochem. Soc., 112:237-240 (1965)

Concentration gradient apparatus for growing
crystals
V. J. Lyons
IBM Corp., U. S. Patent 3,198,606 (Aug. 3, 1965)

Growth mechanism and faults of epitaxial GaAs
films
Kh. A. Magomedov and N. N. Sheftal'
Izv. Akad. Nauk SSSR, Neorg. Mater., 1(12):2113-2119 (1965)
Inorg. Mater., 1(12):1911-1917 (1965)

Some trends in the growth of epitaxial layers
of GaAs
Kh. A. Magomedov and N. N. Sheftal'
Kristallografiya, 9(6):902-909 (1964)
Sov. Phys.—Cryst., 9(6):756-760 (1965)

Growth faults in epitaxial films of GaAs
Kh. A. Magomedov and Yu. N. Yarmukhamedov
Izv. Akad. Nauk SSSR, Neorg. Mater., 1(12):2120-2127 (1965)
Inorg. Mater., 1(12):1918-1924 (1965)

Epitaxial growth of GaAs through cracks in
SiO₂ masks
M. Michelitsch
J. Electrochem. Soc., 112:747 (1965)

Etching and crystal growth in the GaAs—Sn
system
M. F. Millea and W. R. Wilcox
J. Electrochem. Soc., 112:872-874 (1965)

Determination of the distribution coefficients
of volatile impurities in the growth of gallium-
arsenide crystals by oriented crystallization
M. G. Mil'vidskii and O. V. Pelevin
Izv. Akad. Nauk SSSR, Neorg. Mater., 1(9):1454-1458 (1965)
Inorg. Mater., 1(9):1328-1331 (1965)

Liquid encapsulation techniques—use of an
inert liquid in suppressing dissociation during
the melt growth of indium-arsenide and gallium-arsenide crystals
J. B. Mullen, B. W. Straughan, and W. S. Brickell
J. Phys. Chem. Solids, 26:782-784 (1965)

A fundamental study of epitaxy by flash evaporation, final report, Oct. 1, 1964—March 30,
1965
John L. Richards
Philco Corp., Blue Bell, Penn., AFCRL-65-412 (April
30, 1965)

GaAs laser diodes, third quarterly progress
report, Jan. 1 through March 31, 1965
R. A. Sehr
Contract DA 28-043 AMC-00235 (E) (U. S. Army Elec-
tronics Command, Fort Monmouth, N. J., AMSEL-RD-
PFM) (July 1965), 11 pp.

Production of ribbon-shaped dendrites of semiconductor material
Siemens and Halske AG, British Patent 993,701 (June 2,
1965)

A novel crystal-growth phenomenon: single-
crystal GaAs overgrowth onto silicon dioxide
F. W. Tausch, Jr., and A. G. Lapierre, III
J. Electrochem. Soc., 112:706 (1965)

The growth of crystals by solvent zone techniques, a review
G. A. Wolff and A. I. Mlavsky
Colloques Internationaux du Centre National de la Rècherche Scientifique, No. 152, Adsorption et Croissance Cristalline, Nancy, 6–12 juin 1965, Editions du Centre National de la Recherche Scientifique, 15, quai Anatole-France, Paris 7e (1965), pp. 711–722

X-ray thickness determination for epitaxial films of GaAs
B. G. Zakharov
Kristallografiya, 10(3):442–443 (1965)
Sov. Phys. — Cryst., 10(3):366–367 (1965)

Purification of gallium arsenide by zone melting
G. Ziegler
Siemens-Schuckertwerke AG, German Patent 1,201,319 (Sept. 23, 1965), 3 pp.

Vapor-phase equilibria in the gallium − chlorine and gallium − arsenide − chloride systems
M. A. Zuegel
J. Electrochem. Soc., 112:1153 (1965)

Gallium arsenide single-crystal program
M. F. Amsterdam, R. H. Moss, and S. O'Hara
Westinghouse Electric Corp., AS-606891 (May 1964)

The preparation of semi-insulating gallium arsenide by chromium doping
G. R. Cronin and R. W. Haisty
J. Electrochem. Soc., 111: 874–877 (1964)

Growth and properties of GaAs and GaP dendrites from the gas phase
H. J. Dersin and E. Sirtl
Z. Metallk., 55:536–543 (1964)

Influence of vapor composition on the growth rate and morphology of gallium-arsenide epitaxial films
R. E. Ewing and P. E. Greene
J. Electrochem. Soc., 111:1266 (1964)

Growth of semiconductor crystals from solution using the thin-plane reentrant-edge mechanism
J. W. Faust and H. F. John
J. Phys. Chem. Solids, 25:1407–1415 (1964)

The transport of gallium arsenide in the vapor phase by chemical reaction
R. R. Ferguson and T. Gabor
J. Electrochem. Soc., 111:585 (1964)

Epitaxial growth of gallium arsenide on germanium substrates. I. The relationship between fault formation in GaAs films and the surface of their germanium substrate
T. Gabor
J. Electrochem. Soc., 111: 817 (1964)

Epitaxial growth of gallium arsenide on germanium substrates. II. Deterioration of the (111) surface of Ge at 570° −850°C
T. Gabor
J. Electrochem. Soc., 111:821 (1964)

Epitaxial growth of gallium arsenide on germanium substrates. III. Deposition on high index planes and curved surfaces
T. Gabor
J. Electrochem. Soc., 111:825 (1964)

Crystal growing process
P. G. Herkart and L. R. Weisberg
RCA, Belg. Patent 641,009 (April 1, 1964)

Epitaxial growth of GaAs in a closed-tube system
A. Ito, M. Ishii, and M. Ito
Mitsubishi Denki Lab. Rept., 5: 501 (1964)

Epitaxial growth of gallium arsenide with ammonium halide as transporting agents
He Bong Kim and Richard Longini
Carnegie Inst. of Tech., Pittsburgh, Penn., NASA-CR-76429 (1964)

On the morphology and structure of GaAs and GaP platelets grown from the gas phase
W. Kriegstein, H. Pfister, and G. Ziegler
Z. Angew. Phys., 17:295 (1964)

Physics of III − IV Compounds
O. Madelung
John Wiley and Sons, New York (1964)
Contains section on introduction and removal of impurities and on semi-insulating gallium arsenide

Reactions of gallium arsenide with water vapor and hydrogen-chloride gas
M. Michelitsch, W. Kappallo, and G. Hellbardt
J. Electrochem. Soc., 111:1248 (1964)

Crucible-free preparation of gallium-arsenide rods
W. Miederer, G. Ziegler, and R. Doetzer
Siemens-Schuckertwerke AG, German Patent 1,176,102 (Aug. 20, 1964)

Etching and crystal growth in GaAs − Sn system
M. F. Millea and W. R. Wilcox
Aerospace Corp., El Segundo, Calif., SSD-TDR-64-263; TDR-469 (9230-03) -1; AD-455135 (Dec. 30, 1964)

Apparatus for floating zone melting
Philips Electrical Industries, Ltd., British Patent 958,870 (May 27, 1964)

Electron theory of crystallization of semiconducting compounds of the type $A^{III}B^V$
V. A. Presnov and V. A. Selivanova
Pverkhn. i Kontaktn Yavleniya v Poluprov, Sibirsk Fiz. Tekhn. Nauchn. Issled. Inst. Pro. Tomskom. Gos. Univ. (1964) (in Russian), pp. 495–503
Experimental, zone method

Rapid growth of gallium-arsenide crystals by vapor transport
D. Richman
Synthesis and characterization of electronically active materials, tech. rept. No. 2 (RCA, Princeton, N. J.) (Dec. 15, 1964), pp. 17–20

Chemical Transport Reactions
H. Schafer
Academic Press, New York (1964), p. 59

Incorporation of zinc in vapor-grown gallium arsenide
V. J. Silvestri and F. Fang
J. Electrochem. Soc., 111:1164-1167 (1964)

Studies on high-power GaAs lasers
R. S. Title
IBM, Yorktown Heights, N. Y., Third semiannual tech. summary rept., May 31-Dec. 31, 1964 (1964)

GaAs laser materials study, second semiannual tech. summary report, Jan. 1-May 31, 1964
W. J. Turner and H. Rupprecht
IBM, Yorktown Heights, N, Y., AD-442598 (1964), 33 pp.

The effect of orientation on the electrical properties of epitaxial gallium arsenide
F. V. Williams
J. Electrochem. Soc., 111:886 (1964)

Heavy doped n-type crystal epitaxial layers of gallium arsenide
Forrest V. Williams
Solid-State Electronics, 7:833-834 (1964)

Silicon-free gallium arsenide (laser crystals by using an atmosphere of Ga₂O to suppress reaction of Ga with the SiO₂ container)
J. M. Woodall
IBM Corp., French Patent 1,440,448, May 27, 1966; British Patent 1,049,356, Nov. 23, 1966
Chem. Abstr., 66(8):33006; 66(10):41726

High-resistivity gallium arsenide
J. M. Woodall
Electrochem. Technol., 2:167-169 (1964)

Semiconductor device concepts, scientific report No. 9 and final report—August 1964 to Oct. 1964
Hugh H. Woodbury, Manuel Aven, Robert N. Hall, Richard Baertsch, and Frederick K. Heumann
General Electric Research Laboratory, AFCRL-64-1007 (Nov. 1964)

The growth of single-crystal GaAs layers on Ge substrates
Single-Crystal Films, Macmillan Company, New York (1964), p. 283

The growth of single-crystal gallium-arsenide layers on germanium and metallic substances
James A. Amick
RCA Rev., 24:555-573 (1963)

Gallium arsenide dendrite single-crystal program, final report, Mar.-Dec. 1963
M. F. Amsterdam, R. H. Moss, S. O'Hara, R. K. Riel, H. F. John, et al.
Air Force Materials Lab., Wright-Patterson AFB, Ohio, MLTDR-64-129; AD-606891 (May 1964), 138 pp.

The growth of crystals from compounds with volatile components
G. R. Cronin, Morton E. Jones, and O. Wilson
J. Electrochem. Soc., 110:582-584 (1963)

High-purity indium phosphide and gallium arsenide
E. Enk, H. Jacob, and J. Nickl
Wacker-Chemie GmbH, U. S. Patent 3,077,384 (Feb. 12, 1963)

Epitaxial vapor growth of III − V compounds, interim tech. report No. 4711-1, Oct. 1961-Aug. 1963
James F. Gibbons and Preben C. Prehn
Solid State Electronics Lab., University of Stanford, Calif., SEL-63-105; RTD-TDR-63-4238; AD-434756 (Oct. 1963), 52 pp.

Growth rates of epitaxial gallium arsenide
N. Goldsmith
J. Electrochem. Soc., 110:588 (1963)

Vapor-phase synthesis and epitaxial growth of gallium arsenide
N. Goldsmith and W. Oshinsky
RCA Rev., 24:546-554 (1963)

Epitaxial growth of GaAs using water vapor
G. Eugene Gottlieb and John F. Corboy
RCA Rev., 24:585-595 (1963)

A quartz ampoule for growing single crystals
A. P. Izergin, V. N. Chernigovskaya, and A. P. Kalashnikov
Foreign Tech. Div., Air Force Systems Command, Wright-Patterson AFB, Ohio, FTD-HT-66-675; AD-647725 [trans. Russian Patent No. 169064 (Appl. No. 833269/23-4, 26 Apr. 1963]

Single-crystal growing of gallium arsenide by the zone-melting method
A. P. Izergin, V. A. Selivanova, and V. N. Chernigovskaya
Izv. Vyssh. Ucheb. Zaved. Fiz., 3;23-26 (1963)

Growth of gallium-arsenide single crystals by free-surface method
Kazuhiro Kurata, Junji Shirafuji, and Takao Endo
Japan. J. Appl. Phys., 2:64-65 (1963)

Constitution supercooling and facet formation of GaAs
Charlotte Z. LeMay
J. Appl. Phys., 34:439 (1963)

Growth of GaAs from gas by traveling solvent method
A. I. Mlavsky and M. Weinstein
J. Appl. Phys., 34:2885-2892 (1963)

Epitaxial growth from the liquid state and its application to the fabrication of tunnel and laser diodes
H. Nelson
RCA Rev., 24:603-615 (1963)

Vapor growth of GaAs in the polar direction
Takasi Okada
Japan. J. Appl. Phys., 2:206 (1963)

Kinetics of vapor growth in the system GaAs − I₂
T. Okada, T. Kano, and S. Kikuchi
Japan. J. Appl. Phys., 2:780 (1963)

Chemical transport and epitaxial deposition of gallium arsenide
F. A. Pizzarello
J. Electrochem. Soc., 110:1059 (1963)

Gas-phase equilibria in the system GaAs-I₂
D. Richman
RCA Rev., 24, 596-602 (1963)

Transport of gallium arsenide by a close-spaced technique
  P. H. Robinson
  RCA Rev., 24: 574–584 (1963)

The "sandwich method," a new method for the preparation of epitaxially grown semiconductor films
  E. Sirtl
  J. Phys. Chem. Solids, 24:1285 (1963)

Growth peculiarities of gallium-arsenide single crystals
  A. Steinemann and U. Zimmerli
  Solid-State Electronics, 6:597–604 (1963)

Traveling solvent method (TSM) of crystal growth
  M. Weinstein, M. A. Wright, L. B. Griffiths, and A. I. Mlavsky
  NASA Report N63-20412 (1963), 72 pp.

Gallium arsenide dendrite single-crystal program
  Westinghouse Elec. Corp., quarterly report (May 1963)
    (U. S. Gov. Res. Rep., 38 (20):125)

Role of oxygen in reducing silicon contamination of gallium arsenide during crystal growth
  J. T. Woods and N. G. Ainslie
  J. Appl. Phys., 34:1469–1475 (1963)

Process and apparatus for pulling crystals from a bath
  N. V. Philips Gloeilampenfabrieken, Belg. Patent 625,139 (May 21, 1963) 11 pp.

Crystal growth of III—V compounds (Czochralski)
  W. P. Allred
  Conf. on the Ultrapurification of Semiconductor Materials, Boston, 1961, Proc. (1962), pp. 550–567

Vapor-phase reactions
  G. R. Antell
  Compound Semiconductors
  Reinhold Publishing Corp., New York (1962), p. 288

Flat dendritic semiconductor crystals
  A. I. Bennet
  Westinghouse Electric Corp., U. S. Patent 3,031,403 (April 24, 1962), 9 pp.

Growth structure difference on opposite (111) faces of a gallium-arsenide dendrite
  G. R. Booker
  J. Appl. Phys., 33:750 (1962)

Reactions of gallium with quartz and with water vapor, with implications in the synthesis of gallium arsenide
  C. N. Cochran and L. M. Foster
  J. Electrochem. Soc., 109:149 (1962)

Crystal habits of GaAs and GaP grown from the vapor phase
  A. J. Crocker
  J. Appl. Phys., 33:2840 (1962)

Floating zone refining of gallium arsenide
  F. A. Cunnell, W. R. Harding, and R. Wickham
  Conf. on the Ultrapurification of Semiconductor Materials, Boston, Proc. (1962), pp. 513–520

Sources of contamination in GaAs crystal growth
  L. Ekstrom and L. R. Weisberg
  J. Electrochem. Soc., 109:322–327 (1962)

Preparation and properties of gallium-arsenide single crystals
  K. Harada, S. Yamada, M. Kitao, O. Ohtsuki, and S. Narita
  Kogyo Kagaku Zasshi, 65:1726–1729 (1962)
  Chem. Abstr., 58: 8467

Open-tube epitaxial synthesis of GaAs and GaP
  S. W. Ing, Jr., and H. T. Minden
  J. Electrochem. Soc., 109: 995 (1962)

Production of single-crystal plates of gallium arsenide
  G. A. Kataev, A. G. Gregor'eva, and L. N. Rozanova
  Tr. Tomskogo Gos. Univ. Ser. Khim., 154:193–194 (1962)

Leaves of gallium arsenide (grown accidentally from vapor phase)
  H. T. Minden
  J. Appl. Phys., 33: 243–244 (1962)

Preparation of epitaxial GaAs and GaP films by vapor-phase reaction
  R. R. Moest and B. R. Shupp
  J. Electrochem. Soc., 109:1061 (1962)

X-ray studies of twinned GaAs blades grown from the vapor phase
  R. R. Monchamp, W. J. McAleer, and P. I. Pollak
  J. Electrochem. Soc., 109:1108 (1962)

Preparation of solid solutions of GaP and GaAs by a gas-phase reaction
  F. A. Pizzarello
  J. Electrochem. Soc., 109:226 (1962)

On the crystallinity of GaAs grown horizontally in quartz boats
  L. R. Weisberg, J. Blanc, and E. J. Stofko
  J. Electrochem. Soc., 109:642–643 (1962)

Investigation of a method of growing crystals of GaP and GaAs from the vapour phase
  G. R. Antell
  Brit. J. Appl. Phys., 12:687–690 (1961)

Halogen-vapor transport and growth of epitaxial layers of intermetallic compounds and compound mixtures
  N. Holonyak, Jr., D. C. Jillson, and S. F. Bevacqua
  Metallurgy of Semiconductors Materials, Interscience Publishers, Inc., New York (1961), p. 49

Crystal growth of gallium arsenide using carbon boats
  J. R. Knight
  Nature, 190:1001 (June 10, 1961)

Vapor growth of GaAs
  V. J. Lyons and V. J. Silvestri
  J. Electrochem. Soc., 108:117C (1961)

Vapor-phase growth of gallium arsenide crystals
  W. J. McAleer, H. R. Barkemeyer, and P. I. Pollak
  J. Electrochem. Soc., 108:1168 (1961)

Vapor growth of gallium arsenide
  R. L. Newman and N. Goldsmith
  J. Electrochem. Soc., 108:1127 (1961)

Epitaxial vapor growth of gallium arsenide on germanium single crystal
T. Okada, T. Kano, and Y. Sasaki
J. Phys. Soc. Japan, 16:2591 (1961)

Modified rf (radio frequency) coil to facilitate floating zone techniques (use with gallium arsenide)
S. J. Silverman
J. Electrochem. Soc., 108:585-588 (1961)

Growth of gallium-arsenide crystals from dilute solutions (Bridgman)
L. J. Vieland and S. Skalski
Conf. on the Metallurgy of Elemental and Compound Semiconductors, Boston, 1960, Proc. (1961), pp. 303-315

Preparation of InAs, InP, GaAs, and GaP by chemical methods
D. Effer and G. R. Antell
J. Electrochem. Soc., 107:252 (1960)

Gallium-arsenide dendrites
R. H. Moss and H. C. Nicholson
Fall Meeting, Electrochemical Society, Houston, 1960, abstract in: J. Electrochem. Soc., 107:198C (1960)

Growth of gallium arsenide by horizontal zone melting
J. L. Richards
J. Appl. Phys., 31:600 (1960)

Properties of p-type GaAs prepared by copper diffusion
F. D. Rosi, D. Meyerhofer, and R. V. Jensen
J. Appl. Phys., 31:1105-1108 (1960)

Synthesis and crystal pulling of gallium arsenide in a magnetic crystal puller
K. Weiser and S. Blum
Fall Meeting, Electrochemical Society, Houston, 1960, abstract in: J. Electrochem. Soc., 107:189C (1960)

Distribution coefficients of various impurities in gallium arsenide (including crystal growth)
J. M. Whelan, J. D. Struthers, and J. A. Ditzenberger
Properties of Elemental and Compound Semiconductors, Metals Society Conferences, Vol. 5, 1959, Interscience, New York (1960), pp. 41-54

Materials research on GaAs and InP
L. R. Weisberg, F. D. Rosi, and P. G. Herkart
Properties of Elemental and Compound Semiconductors, Metallurgy Society Conferences, Vol. 5, 1959, Interscience, 1959, New York (1960), pp. 25-67

Preparation of crystals of InAs, InP, GaAs, and GaP by a vapor-phase reaction
G. R. Antell and D. Effer
J. Electrochem. Soc., 106:509 (1959)

On the growth of gallium-arsenide crystals from the melt
S. G. Ellis
J. Appl. Phys., 30:947-948 (1959)

The preparation and properties of gallium-arsenide single crystals
J. M. Whelan and G. H. Wheatley
Phys. Chem. Solids, 6:169-172 (1958)

Inorganic stoichiometric crystalline compounds from melts or by remelting
R. Gremmelmaier
Siemens-Schuckertwerke Akt. Ges., German Patent 1,002,741 (Feb. 21, 1957)

Preparation of GaAs single crystals
J. M. Whelan and G. H. Wheatley
Bull. Am. Phys. Soc., 2:120 (1957)

Herstellung von InAs- und GaAs-Einkristallen
R. Gremmelmaier
Z. Naturforsch., 11a:511-513 (1956)

Mixed-crystal formation in $A^{III}B^{V}$ compounds
O. G. Folberth
Z. Naturforsch., 10a:502 (1955)

## 10.d.4. GaAs Ternaries and Quaternaries

Isothermal substitutional growth of single crystals
W. Albers and J. Verberkt
Philips Res. Rept., 25(1):17-20 (1970)
GaAs—GaP

The Ga — GaP — GaAs ternary phase diagram
G. A. Antypas
J. Electrochem. Soc., 117:700-703 (1970)

Liquid epitaxial growth of GaAsSb and its use as a high-efficiency, long-wavelength threshold photoemitter
G. A. Antypas and L. W. James
J. Appl. Phys., 41:2165-2171 (1970)

Perfection of vapor-grown $GaAs_{1-x}P_x$ superlattices
A. E. Blakeslee
IBM Thomas J. Watson Research Center, Yorktown Heights, New York, RC 3055 (Sept. 1970)

Chemical inhomogeneity of epitaxial layers in solid solutions GaP — GaAs
F. A. Gimelfarb, E. M. Kistova, V. N. Maslov, B. A. Sakharov, and V. I. Fistul'
Izv. Akad. Nauk SSSR, Neorg. Mater., 6(3):461-467 (1970)
Inorg. Mater., 6(3):405-411 (1970)

Epitaxial growth of $GaAs_xP_{1-x}$ from the liquid phase
K. K. Shih
J. Electrochem. Soc., 117:387-389 (1970)

Perfection of materials technology for producing improved Gunn-effect devices, interim scientific report
R. D. Baxter, J. F. Miller, H. L. Leonard, and A. C. Beer
Battelle Memorial Institute, Columbus, Ohio, Contract NAS12-632, ISR-2; NASA-CR-86277 (Dec. 1969), 26 pp.
Vapor deposition of GaAs and the preparation of $Ga_xIn_{1-x}Sb$ ingots

Vapor-phase growth of epitaxial $GaAs_{1-x}Sb_x$ alloys using arsine and stibine
R. B. Clough and J. J. Tietjen
Trans. AIME, 245:583 (1969)

Growth and characterization of $Ga_xIn_{1-x}Sb$ solid solutions using temperature-gradient zone melting
R. W. Hamaker and W. B. White
J. Electrochem. Soc., 116:478–482 (1969)
On {111} InSb seed substrates

Ga — In — As system
I. S. Kovaleva, N. P. Luzhnaya, and S. B. Martikyan
Zh. Neorg. Khim., 14:2860–2863 (1969)

Deviation from homogeneity in epitaxial single crystals of GaP — GaAs solid solutions
A. V. Lishina, V. N. Maslov, R. L. Petrusevich, and N. V. Troneva
Rost Kristallov, Izd. Akad. Nauk, SSSR (1968), pp. 273–277
Growth of Crystals, Vol. 8 (A. V. Shubnikov and N. N. Sheftal', eds.) Consultants Bureau, New York (1969), pp. 224–227

The use of metal-organics in the preparation of semiconductor materials. I. Epitaxial gallium-V compounds
H. M. Manasevit and W. I. Simpson
J. Electrochem. Soc., 116:1725–1732 (1969)

The Ga — GaAs — GaP system: phase chemistry and solution growth of $GaAs_xP_{1-x}$
M. B. Panish
J. Phys. Chem. Solids, 30:1083–1090 (1969)
Liquid epitaxial growth of $GaAs_xP_{1-x}$ layers on GaP

Phase and thermodynamic properties of the Ga — Al — P system: solution epitaxy of $Ga_xAl_{1-x}P$ and AlP
M. B. Panish, R. T. Lynch, and S. Sumski
Trans. AIME, 245:559 (1969)

Elaboration et propriétés optiques d'alliages $Ga_xIn_{1-x}P$
Huguette Rodot, Jaromir Horak, Georges Rouy, and Jacques Bourneix
Compt. Rend., 269B:381–384 (1969)

Vapor-phase growth techniques and system for several III — V compound semiconductors, interim scientific report
J. J. Tietjen, R. Clough, A. B. Dreeben, R. Enstrom, and D. Richman
Radio Corp. of America, Princeton, N. J., Contract NAS12-538, NASA-CR-86192; ISR-2 (March 1969), 29 pp.
GaN, $In_{1-x}Ga_xP$, and $Ga_{1-x}Al_xAs$

Liquid-phase epitaxial growth of $Ga_{1-x}Al_xAs$
J. M. Woodall, H. Rupprecht, and W. Reuter
J. Electrochem. Soc., 116(6):899–903 (1969)

Close-spaced epitaxial growth of $GaAs_xP_{1-x}$ from powder GaAs and GaP
R. K. Purohit
J. Mater. Sci., 3:330–332 (1968)
Water vapor as transporting agent

GaAs and GaP for room-temperature gamma-ray counters
L. R. Weisberg and B. Goldstein
Nucleonics in Aerospace, Proceedings of the Second International Symposium, Columbus, Ohio, July 12–14, 1967, Instrument Society of America, New York (1968), pp. 182–186
Vapor-phase growth, GaAs–GaP

Solution growth of (Zn, Hg)Te and Ga(P, As) crystals
G. A. Wolff, H. E. LaBelle, Jr., and B. N. Das
Trans. AIME, 242:436 (1968)
Traveling heater method

Preparation of epitaxial $Ga_xIn_{1-x}As$
R. W. Conrad, P. L. Hoyt, and D. D. Martin
J. Electrochem. Soc., 114:164 (1967)
On (100), semi-insulating GaAs substrates

The GaAs — InSb graded gap heterojunction
E. D. Hinkley and R. H. Rediker
Lincoln Lab., Massachusetts Institute of Technology, Lexington, Contract AF 19(628)-5167, ESD-TR-67-468; TN-1967-40; AD-65777 (August 1967), 54 pp.

Preparation of GaAs — Ge and InAs — GaAs heterojunctions in a closed-tube system using iodine process
H. Kasano and S. Iida
Japan. J. Appl. Phys., 6:1038 (1967)

$GaAs_{1-x}P_x$ injection lasers
J. I. Pankova, H. Nelson, J. J. Tietjen, I. J. Hegyi, and H. P. Maruska
RCA Review, 28:560 (1967)
Vapor-transport epitaxial growth

Fundamental studies of the metallurgical, electrical, and optical properties of gallium phosphide, quarterly progress report, 1 April –30 June 1967
G. L. Pearson
Grant NsG-555, Proj. 5109, NASA-CR-87418 (June 1967), 17 pp.
Preparations and properties of rectifying junctions in GaP and $GaAs_xP_{1-x}$

Vapor-phase growth of $GaAs_{1-x}P_x$ room-temperature injection lasers
J. J. Tietjen, J. I. Pankove, I. J. Hegyi, and H. Nelson
Trans. AIME, 239:385 (1967)

Structural defects in epitaxial $GaAs_{1-x}P_x$
F. V. Williams
Trans. AIME, 239:702 (1967)

Lattice vibration spectra of $GaAs_xP_{1-x}$ single crystals
Y. S. Chen, W. Shockley, and G. L. Pearson
Phys. Rev., 151:648–656 (1966)
Open-tube epitaxial vapor growth

The crystalline perfection of melt-grown GaAs substrates and Ga(As, P) epitaxial deposits
J. K. Howard and R. H. Cox
Adv. X-Ray Anal., 9:35–50 (1966)

Gallium arsenide-phosphide: crystal, diffusion and laser properties
C. J. Nuese, G. E. Stillman, M. D. Sirkis, and N. Holonyak, Jr.
Solid-State Electron., 9:735–749 (1966)
Crystal growth, diffusion, and fabrication

The preparation and properties of vapor-deposited epitaxial $GaAs_{1-x}P_x$ using arsine and phosphine
J. J. Tietjen and J. A. Amick
J. Electrochem. Soc., 113:724 (1966)

The preparation of single crystals of multicomponent alloys
N. A. Goryunova, G. K. Averkieva, and A. A. Vaipolin
Fiz. Dokl. k XXIII-ei Nauchn. Konf. Leningr. Inzh.-Stroit. Inst., Leningrad (1965), pp. 52-53

Production and study of solid solutions GaP — GaAs and GaAs — InAs
N. A. Goryunova et al.
Symposium - Protsessy Sinteza i Rosta Kristallov i Plenok Poluprovodnik. Materialov, Tesizy Dokl., Sb., Novosibirsk (1965), pp. 7-8

Vapor-phase transport and epitaxial growth of GaAs$_{1-x}$P$_x$ using water vapor
G. E. Gottlieb
J. Electrochem. Soc., 112:192 (1965)

The preparation and properties of GaAs — InAs mixed crystals
H. T. Minden
J. Electrochem. Soc., 112:300 (1965)
Halogen-vapor transport

The preparation of homogeneous and reproducible solid solutions of GaP — GaAs
M. Rubenstein
J. Electrochem. Soc., 112:426 (1965)
Sealed-tube iodine transport

The preparation of single-crystal Ga(As$_{1-x}$P$_x$) for semiconductor-device research
C. M. Wolfe
Dissert. Abstr., 26: 2834 (1965)
Halide-vapor transport

Growth and dislocation structure of single-crystal Ga(As$_{1-x}$P$_x$)
C. M. Wolfe, C. J. Nuese, and N. Holonyak, Jr.
J. Appl. Phys., 36:3790-3801 (1965)
Halide-vapor transport

The vapor-solid interface of Ga(As$_{1-x}$P$_x$) single crystals grown by halogen-vapor transport
Nich Holonyah, Jr., and C. M. Wolfe
Appl. Phys. Letters, 5:19 (1964)

Solid solution in the gallium arsenide — indium arsenide system
N. C. Tombs, J. F. Fitzgerald, and W. J. Croft
Inorg. Chem., 2:1073-1074 (1963)
Large single crystals by gradient freezing

Preparation of double crystals of indium and gallium antimonides
V. I. Ivanov-omsky, N. K. Kiseleva, and B. T. Kolomiets
Fiz. Tverd. Tela, 3(5):1621-1622 (1961)
Sov. Phys. — Solid State, 3(5):1175-1176 (1961)

Zur Isomorphie der Verbindungen des Typs A$^{III}$B$^V$
Werner Koster and Werner Ulrich
Z. Metallk., 49:365 (1958)

## 10.d.5. Zn Doping of GaAs

Anneal behavior of defects in ion-implanted GaAs diodes
R. G. Hunsperger and O. J. Marsh
Met. Trans., 1:603-607 (1970)

Zn doping of GaAs films using coevaporation of the elements, final report
W. A. Schmidt and J. E. Davey
Naval Research Lab., Washington, D. C., NRL-7100 (April 23, 1970), 21 pp. (Also AD 705 486)

Use of a thermoelectric probe in an investigation of the diffused impurity distribution in layers
T. I. Voronina, Yu. A. Gol'dberg, O. V. Emel'yanenko, and F. P. Kesamanly
Fiz. Tekh. Poluprovod., 3(10):1591-1593 (1969)
Sov. Phys.—Semicond., 3(10):1338-1339 (1970)

Luminescence in Zn-doped GaAs
G. W. Arnold and D. K. Brice
Phys. Rev., 178:1399-1403 (1969)

Voltage dependence of capacitance of Zn-diffused GaAs diodes
G. Horak
Solid-State Electron., 12:743-746 (1969)
Anomalous diffusion; influence of lattice defects

Electrical properties of zinc- and cadmium-ion implanted layers in gallium arsenide
R. G. Hunsperger and O. J. Marsh
J. Electrochem. Soc., 116:488-492 (1969)

Distribution of a volatile impurity ahead of the crystallization front in the melt
B. D. Lainer, V. V. Voronkov, M. G. Mil'vidskii, and V. V. Rakov
Kristallografiya, 14(3) 537-539 (1969)
Sov. Phys. — Cryst., 14(3) 451-453 (1969)

Effects of heat cleaning on the photoemission properties of GaAs surfaces
Y.-Z. Liu, J. L. Moll, and W. E. Spicer
Appl. Phys. Letters, 14:275 (1969)
Out-diffusion of Zn at about 650°C

Approximative method for determining the concentration dependence of diffusion coefficients from measured concentration profiles
E. Nebauer
Phys. Stat. Sol., 33(1):K51-K54 (1969)
GaAs:Zn and AlSb:Zn

Transport of zinc in the gaseous phase during the growth of alloyed gallium-arsenide crystals
O. V. Pelevin and M. G. Mil'vidskii
Inorg. Mater., 5:1417-1440 (1969)

Properties of ion-implanted GaAs diodes
P. E. Roughan and K. E. Manchester
J. Electrochem. Soc., 116:278-279 (1969)

The diffusion of Zn in the III — V semiconducting compounds
D. Shaw and S. R. Showan
Phys. Stat. Sol., 32:109-118 (1969)

Diffusion of zinc in GaAs
Brian Tuck
J. Phys. Chem. Solids, 30:253-260 (1969)

Polishing damage and luminescence in p-type GaAs
B. Tuck
Phys. Stat. Sol., 36:285 (1969)
Diffusing a large concentration of zinc into n-type GaAs introduces dislocations, and it is suggested that this is

the reason for the low photoluminescence efficiency of zinc-doped GaAs for concentrations in excess of $10^{19}$ cm$^{-3}$

**Abrupt junctions formed in gallium arsenide by solid-to-solid diffusion**
G. R. Antell
Brit. J. Appl. Phys., 1:113 (1968)

**Reproducible diffusion of zinc into GaAs: application of the ternary phase diagram and the diffusion and solubility analyses**
H. C. Casey, Jr., and M. B. Panish
Trans. AIME, 242:406 (1968)

**Diffusion of zinc into gallium arsenide**
B. Darek
Darek Arch. Elektrotech., 17:871–887 (1968)

**Several remarks about diffusion of zinc in gallium arsenide**
S. P. Fedotov, V. A. Presnov, and V. K. Bashenov
Japan. J. Appl. Phys., 7:436 (1968)

**The close-spaced growth of degenerate p-type GaAs, GaP, and Ga(As$_x$, P$_{1-x}$) by ZnCl$_2$ transport for tunnel diodes**
P.-A. Hoss, L. A. Murray, and J. J. Rivera
J. Electrochem. Soc., 115:553–556 (1968)

**Development of ion implantation techniques for microelectronics**
R. G. Hunsperger, H. L. Dunlap, and O. J. Marsh
Hughes Research Labs., Malibu, Calif., Contract NAS12-124, Report NASA–CR-86142 (Oct. 1968), 61 pp.

**The presence of deep levels in ion implanted junctions**
R. G. Hunsperger, O. J. Marsh, and C. A. Mead
Appl. Phys. Letters, 13:295 (1968)
$\sim 10^{15}$ – $10^{16}$/cm$^2$ singly ionized ions at 70 kV into n-type GaAs substrates held at 400°C

**Effect of heat treatment on the 1.370-eV photoluminescence emission band in Zn-doped GaAs**
C. J. Hwang
J. Appl. Phys., 49:4307–4312 (1968)

**Diffusion of zinc in gallium arsenide containing a high donor concentration**
G. K. Malysk
Fiz. Tverd. Tela, 10:310 (1968)
Soviet Phys. — Solid State, 10:246–247 (1968)

**Diffusion of zinc into GaAs**
Mitsuhiro Maruyama
Japan. J. Appl. Phys., 7:476–484 (1968)

**Evaluation of bulk and epitaxial GaAs by means of x-ray topography (diffusion-induced defects)**
Eugene S. Meieran
Trans. AIME, 242:413 (1968)

**Diffusion in zinc in GaAs arsenic pressure**
S. N. Mukerjee, P. B. Parikh, and B. L. Sharma
J. Inst. Telecommun. Eng., 14:236–238 (1968)

**Behavior of zinc during growth of doped single crystals of gallium arsenide**
O. V. Pelevin and M. G. Mil'vidskii
Izv. Akad. Nauk SSSR, Neorg. Mater., 4(11):1864–1868 (1968)
Inorg. Mater., 4(11):1625–1628 (1968)

**Einkristallines Galliumarsenid ohne Versetzungen**
A. Steinemann, H. R. Winteler, and U. Zimmerli
Helv. Phys. Acta, 41:1210–1212 (1968)
Doping limits are n, p < $5 \cdot 10^{19}$ cm$^{-3}$

**Temperature dependence of the effective diffusion coefficient for zinc in GaAs as determined by isoconcentration diffusions**
C. H. Ting and G. L. Pearson
Bull. Am. Phys. Soc., 13:375 (1968)

**Some properties of semiconductor lasers based on indium phosphide**
N. G. Basov, P. G. Eliseev, I. Ismailov, A. Ya, Nashel'skii, I. Z. Pinsker, and S. V. Yakobson
Fiz. Tverd. Tela, 8(9):2610–2615 (1966)
Sov. Phys. — Solid State, 8(9):2087–2092 (1967)

**Properties of gallium arsenide crystals produced by liquid encapsulation pulling**
S. J. Bass and P. E. Oliver
Proc. International Symposium on Gallium Arsenide, Reading, Sept. 1966 (1967), pp. 41–45

**The occurrence and identification of precipitates in zinc-diffused GaAs**
J. F. Black
J. Electrochem. Soc., 114:1292 (1967)

**Precipitates induced in GaAs by the in-diffusion of zinc**
J. F. Black and E. D. Jungbluth
J. Electrochem. Soc., 114:181 (1967)

**Decorated dislocations and sub-surface defects induced in GaAs by the in-diffusion of zinc**
J. F. Black and E. D. Jungbluth
J. Electrochem. Soc., 114:188 (1967)

**Addendum to precipitates induced in GaAs by the in-diffusion of zinc**
J. F. Black and E. D. Jungbluth
J. Electrochem. Soc., 114:297 (1967)

**The effect of applied electric field on diffusion of impurities in gallium arsenide**
B. I. Boltaks and T. D. Dzhafarov
Phys. Stat. Sol., 19:705 (1967)

**Microwave oscillations in bulk semiconductors**
N. Braslau, J. M. Woodall, and C. Lanza
IBM, T. J. Watson Research Ctr., Yorktown Heights, N. Y., 5th qtr. report, 1 July to 30 Sept. 1966, Contract DA 28-043 AMC-01550(E), ECOM-01550-5 (Jan. 1967)

**Dependence of the diffusion coefficient on the Fermi level: zinc in gallium arsenide**
H. C. Casey, M. B. Panish, and L. L. Chang
Phys. Rev., 162:660 (1967)

**Gallium-arsenide lasers: powerful light sources**
J. L. Fertin, J. Lebailly, J. Thillays, and J. C. Dubois
Proc. International Symposium on Gallium Arsenide, Reading, Sept. 1966 (1967), pp. 72–77
Zn diffusion into a Te-doped crystal

**Measurement of diffusion profile of Zn in n-type GaAs by a spreading resistance technique**
H. Frank and S. A. Azim
Solid-State Electron., 10:727 (1967)

Diffusion of zinc in gallium arsenide under arsenic-vapor pressure
  M. Fujimoto, Y. Sato, and K. Kudo
  Japan. J. Appl. Phys., 6:848 (1967)

Effect of heat treatment of photoluminescence of Zn-doped GaAs
  C. J. Hwang
  J. Appl. Phys., 38:4811 (1967)

Radioactivation analysis of zinc based on the quantitative isotope dilution principle
  K. Kudo
  Radioisotopes (Tokyo), 16:199-203 (1967)

Effects of doping on the growth rate and morphology of epitaxial gallium arsenide
  Kh. A. Magomedov, Yu. N. Yarmukhamedov, and N. N. Sheftal'
  Kristallografiya, 11(4):673-680 (1966)
  Sov. Phys. — Cryst., 11(4):578-582 (1967)

Temperature dependence of the isoconcentration diffusion of zinc in gallium arsenide
  R. Sh. Malkovich and G. K. Malysh
  Fiz. Tverd. Tela, 9(2):553-558 (1967)
  Sov. Phys. — Solid State, 9(2):423-427 (1967)

Development of ion-implantation techniques for microelectronics
  O. J. Marsh, R. G. Hunsperger, H. L. Dunlap, and J. W. Mayer
  Hughes Research Labs., Malibu, Calif., Contract NAS12-124, NASA-CR-86014 (Oct. 1967), 84 pp.

Zn and Te implantations into GaAs
  J. W. Mayer, O. J. Marsh, R. Mankarious, and R. Bower
  J. Appl. Phys., 38:1975 (1967)

Crucible-fire preparation of gallium-phosphide samples and gallium-phosphide — gallium-arsenide mixed crystals
  Walter Miederer and Richard Doetzer
  Siemens A.-G., German Patent, 1,251,288 (Oct. 5, 1967; appl. May 2, 1964), 4 pp.
  By introduction of Zn, Cd, Se, or Te-alkyl in the reactor, coating can be produced with a definite foreign metal content

The concentration gradient of Zn near a p–n junction in III — V compounds
  C. van Opdorp
  J. Appl. Phys., 38:5411 (1967)

Effect of dislocations on the structure of diffused p – n junctions in gallium arsenide and on recombination radiation parameters
  V. B. Osvenskii, G. P. Proshko, and M. G. Mil'vidskii
  Fiz. Tekhn. Poluprovod., 1(6):911-917 (1967)
  Sov. Phys. — Semicond., 1(6):755-760 (1967)

The solid solubility limits of zinc in GaAs at 1000°
  M. B. Panish and H. C. Casey, Jr.
  J. Phys. Chem. Solids, 28:1673 (1967)

Large-area contacts to semiconductor devices
  Manfred H. Pilkuhn and Hans S. Rupprecht
  (To International Business Machines Corp.), U. S. Patent 3,349,476 (Cl. 29-590), Oct. 31, 1967 (appl. Nov. 26, 1963), 4 pp.

Luminescence of zinc-doped solution-grown gallium arsenide
  H. J. Queisser and M. B. Panish
  J. Phys. Chem. Solids, 28:1177 (1967)

Implantation of zinc into gallium arsenide
  J. B. Schroeder and H. D. Dieselman
  Proc. IEEE, 55:125 (1967)

Iron-doped gallium-arsenide transistors
  H. Strack
  Proc. International Symposium on Gallium Arsenide, Reading, Sept. 1966 (1967), pp. 206-212

Planar p – n junctions made by zinc diffusion into gallium arsenide
  Andrzej Uszynski
  Przegl. Elektron., 8(12):597-599 (1967)

Final report U.S.A.F. Contract AF19(628)-4970, Report AFCRL-67-0123, Jan. 1967
  Ion implantation

Tin and zinc diffusion into gallium arsenide from doped silicon dioxide layers
  W. Von Muench
  Solid-State Electron., 9:619 (1966)

Gallium arsenide-phosphide: Crystal, diffusion and laser properties
  C. J. Nuese, G. E. Stillman, M. D. Sirkis, and N. Holonyak, Jr.
  Solid-State Electron., 9:735 (1966)

Some effects of Zn diffusion on Mn-doped GaAs
  R. F. Peart, K. Weiser, J. Woodall, and R. Fern
  Thomas J. Watson Research Center, Yorktown Heights, NC-624 (June 1966)

Influence of doping on the plastic deformation of gallium-arsenide single crystals
  N. P. Sazhin, M. G. Mil'vidskii, V. B. Osvenskii, and O. G. Stolyarov
  Fiz. Tverd. Tela, 8(5):1539-1544 (1966)
  Sov. Phys. — Solid State, 8(5):1223-1227 (1966)

Crystal mosaic structures and the lasing properties of GaAs laser diodes
  D. A. Shaw, K. A. Hughes, N. F. B. Neve, D. V. Sulway, P. R. Thornton, and C. Gooch
  Solid-State Electron., 9:664 (1966)

A study of the Ga — As — Zn system with applications to the diffusion of Zn in GaAs
  Kwang Kuo Shih
  Stanford Univ., Calif., Ph. D. thesis (1966), 80 pp.
  University Microfilms, Ann Arbor, Mich.

Frequency of microwave oscillation in GaAs p – n junctions
  S. Yamashita, T. Kajiwara, T. Abe, and T. Nakano
  Japan. J. Appl. Phys., 5:843 (1966)

Mobility in heavily doped gallium arsenide
  G. H. Gooch
  Phys. Letters, 14:183 (1965)

Semiconductor materials
  HP Associates
  Palo Alto, California, interim engineering rept. No. 6, 15 July - 15 Oct. 1965, Contract NObsr-89489, AD-620267 (Oct. 1965), 46 pp.

GaAs laser diodes
  R. A. Sehr
  Korad Corporation, Santa Monica, California, third quar-
    terly progress report - 1 Jan. through 31 March 1965,
    (July 1965), Contract DA 28-043 AMC-00235 (E)

The diffusion of zinc into gallium arsenide to
achieve low surface concentrations
  H. Becks, D. Flatley, W. Kern, and D. Stolnitz
  Trans. AIME, 230:307 (1964)

Correlation of electrical measurements with
chemical analysis in zinc- and cadmium-dif-
fused GaAs
  J. Black
  J. Electrochem. Soc., 111:924 (1964)

Diffusion and electrical transport of zinc in
gallium arsenide
  B. I. Boltaks, T. D. Dzhafarov, V. I. Sokolov, and F. S.
    Shishiyanu
  TT-65-12230
  Fiz. Tverd. Tela, 6(5):1511-1519 (1964)
  Sov. Phys. — Solid State, 6(5):1181-1188 (1964)

Diffusion of zinc and tin in GaAs
  P. Gansauge
  Phys. Verh. Dtsch., 15:91 (1964)

Diffusion, solubility, and distribution coefficient of zinc in gallium arsenide and gallium
phosphide
  L. Li-Gong Chang
  Stanford Univ., California, Dissertation Abstr., 25:1094
    (1964)

Analytical study of zinc-diffusion in gallium
arsenide and the electrical properties of the
resulting diffused layers
  Rajendra Mehta
  SEL-64-062; AROD-2895-15; AD-610262 (June 1964). 71 pp.
  Contract DA-31-124-ARO(D)-155; AF 33(616)-7726

Diffusion of Zn into GaAs under the presence
of excess arsenic vapor
  H. Rupprecht and C. Z. LeMay
  J. Appl. Phys., 35:1970 (1964)

Rapid impurity diffusion in GaAs Esaki diodes
  A. Shirata
  Solid-State Electron., 7:215 (1964)

Zinc diffusion in GaAs through $SiO_2$ films
  S. R. Shortes, J. A. Kanz, and E. C. Wurst, Jr.
  Trans. AIME, 230:300 (1964)

Incorporation of zinc in vapor-grown gallium
arsenide
  V. J. Silverstri and F. Fang
  J. Electrochem. Soc., 111:1164 (1964)

Diffusion in GaAs
  L. R. Weisberg
  Trans. AIME, 230:291 (1964)

The diffusion of tin and selenium in gallium
arsenide
  R. W. Fane and A. J. Goss
  Solid-State Electron., 6:383 (1963)

Solubility of zinc in gallium arsenide
  J. O. McCaldin
  J. Appl. Phys., 34:1748 (1963)

Diffusion of $Zn^{65}$ in GaAs at constant acceptor
concentrations
  G. L. Pearson, M. G. Buehler, and N. C. Berglund
  Bull. Am. Phys. Soc., 8:229 (1963)

Diffusion with interstitial-substitutional
equilibrium. Zinc in GaAs
  L. R. Weisberg and J. Blanc
  Phys. Rev., 131:1548 (1963)

Ratio of interstitial to substitutional zinc in
GaAs and its relation to zinc diffusion
  K. Weiser
  J. Appl. Phys., 34:3387 (1963)

Surface masking in gallium arsenide during
diffusion
  T. H. Yeh
  J. Electrochem. Soc., 110:341 (1963)

Rapid zinc diffusion in gallium arsenide
  R. L. Longini
  Solid-State Electron., 5:127 (1962)

Self and impurity diffusion in GaAs
  B. Goldstein
  Compound Semiconductors, Vol. 1 (R. K. Willardson and
    H. L. Goering, eds.), Reinhold Publishing Corp., New
    York (1962), p. 345

Correlation of electrical measurements and
chemical analyses of zinc- and cadmium-dif-
fused GaAs
  J. F. Black
  J. Electrochem. Soc., 108:178C (1961)

Diffusion in compound semiconductors
  B. Goldstein
  Phys. Rev., 121:1305 (1961)

The effect of arsenic pressure on impurity dif-
fusion in gallium arsenide
  L. J. Vieland
  J. Phys. Chem. Solids, 21:318 (1961)

The diffusion of ionized impurities in semi-
conductors
  J. W. Allen
  J. Phys. Chem. Solids, 15:134 (1960)

Diffusion of zinc in gallium arsenide
  F. A. Cunnell and C. H. Gooch
  J. Phys. Chem. Solids, 15:127 (1960)

Diffusion of Cd and Zn in GaAs
  B. Goldstein
  Phys. Rev., 118:1024 (1960)

[Title Not Given]
  D. L. Kendall and M. E. Jones
  Electrochemical Society Meeting, May 1960; Solid State
    Device Research Conference, June 1960

Gallium-arsenide diffused diodes
  J. Lowen and R. H. Rediker
  J. Electrochem. Soc., 107:26 (1960)

Distribution coefficient of zinc in gallium
arsenide
  A. Shirata
  J. Phys. Soc. Japan, 15:2107 (1960)

Diffusion of Zn in GaAs
J. W. Allen and F. A. Cunnell
Nature, 182:1158 (1958)

## 10.d.6. GaSb; GaSb—InSb

Structure and properties of gallium-antimonide
dendrites
M. Ya. Dashevskii, G. V. Kukuladze, and M. S. Mirgalovskaya
Izv. Akad. Nauk SSSR, Neorg. Mater., 6(7):1239-1241 (1970)
In Russian

Crystallization of gallium antimonide from the
solutions containing >50 atomic per cent of
gallium
A. Halak and Z. Olempska
Elektronika, 4:176-178 (1970)
In Polish

Epitaxial vapor growth of gallium antimonide
M. Kakehi, R. Shimokawa, and T. Arizumi
Japan. J. Appl. Phys., 9(9):1039-1044 (1970)

Relation of the surface properties of a gallium
antimonide melt to crystal growth
M. Ya. Dashevskii, G. V. Kukuladze, V. B. Lazarev, and
M. S. Mirgalovskaya
Rost Kristallov, Izd. Akad. Nauk, SSSR (1968), pp. 104-107
Growth of Crystals, Vol. 8 (A. V. Shubnikov and N. N.
Sheftal', eds.) Consultants Bureau, New York (1969), pp.
85-90

Growth of highly doped gallium-antimonide
single crystals in polar ⟨111⟩ directions
M. Ya. Dashevskii, G. V. Kukuladze, and M. S. Mirgalovskaya
Izv. Akad. Nauk SSSR, Neorg. Mater., 5(6): 1141-1142 (1969)
Inorg. Mater., 5(6): 974-975 (1969)
Czochralski technique in purified He

Distribution of selenium during the directed
crystallization of gallium antimonide
N. L. Gryazeva, Sh. M. Mavlonov, and V. N. Vigdorovich
Dokl. Akad. Nauk Tadzh. SSR, 12:19-22 (1969)

Vapor phase growth technique and system for
several III — V compound semiconductors
J. J. Tietjen, R. Clough, R. E. Enstrom, and M. Ettenberg
Contract NAS12-538, QTR-8; NASA-CR-86306 (Sept. 1969),
14 pp.

Elaboration de monocristaux d'antimoniure de
gallium par zone fondue flottante
A. Nguyen van Mau, C. Ance and G. Bougnot
Mat. Res. Bull., 3: 901-910 (1968)

The crystal face effect in gallium-antimonide
single crystals grown by the Czochralski
method
M. S. Mirgalovskaya, G. V. Kukuladze, and A. V. Kokoshkin
Izv. Akad. Nauk SSSR, Neorg. Mater., 4(5): 694-700 (1968)
Inorg. Mater., 4(5): 606-612 (1968)

Effects of back-melting on the dislocation
density in single crystals: GaSb
R. S. Mroczkowski, A. F. Witt, and H. C. Gatos
J. Electrochem. Soc., 118:545-547 (1968)

Semiconducting thin films, an annotated bibli-
ography, 1967 supplement
W. R. Turnbull
NOLC Rept. 745 (Naval Weapons Center Corona Laboratories,
Corona, Cal. 91720), March 1, 1968, Contract AIRTASK
No. A31533212/2111/R008-03-02
Supplement to NOLC Rept. 712

Surface phenomena and crystallization pro-
cesses in gallium antimonide melts
M. Ya. Dashevskii and others
Dokl. Akad. Nauk SSSR, 172: 403-406 (1967)

Growth properties of GaSb: The structure
of the residual acceptor centres
Y. J. van der Meulen
J. Phys. Chem. Solids, 28: 25-32 (1967)

Heterogeneous distribution of impurities in
gallium antimonide
V. S. Vekshina, L. G. Elanskaya, and L. Ya. Krol
Zavod. Lab., 33(2): 200 (1967)

Epitaxial growth of composite semiconductors
by evaporation-diffusion in an isothermal
system
G. Cohen-Solal, Y. Marfaing, and F. Bailly
Rev. Phys. Appl., 1: 11-17 (1966)

Kinetics of dissolving of the gallium and indium
antimonides in hydrochloric solutions of iodine
and iron chloride
G. M. Orlova and V. P. Bruenkova
Zh. Prikl. Khim., 39: 2644-2649 (1966)

GaSb prepared from nonstoichiometric melts
F. J. Reid, R. D. Baxter, and S. E. Miller
J. Electrochem. Soc., 113: 713 (1966)

Optical properties of III — V semiconductor
compounds
E. J. Johnson
Dissertation Abstr., 26:1122 (1965)
University Microfilms, Ann Arbor, Michigan 48103

Electrical properties and resonance scattering
in heavily doped n-type GaSb and related semiconductors
Donald Long and Robert J. Hager
J. Appl. Phys., 36: 3436 (1965)

The behaviour of lithium in gallium-antimonide
single crystals
Martinus Henricus van Maaren
Thesis, March, 1965, Rijksuniversiteit of Utrecht

Deviations from stoichiometry in gallium anti-
monide
M. S. Mirgalovskaya, V. V. Sakharov, and O. G. Karpinskii
Simp. Protsessy Sint. Rosta Krist. Plenok Poluprov. Mater.,
Tezisy Dokl. (Novosibirsk, 22, 1965)

Preparation of gallium antimonide by solution-
growth method
Takeshi Miyauchi and Hazimu Sonomura
Japan. J. Appl. Phys., 4: 317-318 (1965)

Statistical approach to growth of single crystals
of GaSb by horizontal growing techniques
J. R. Peloke, R. R. Stone, and L. R. Yetter
Solid-State Electron., G. B., 8: 861-867 (1965)

Pulling n — p single crystals of gallium antimonide
>   George Petit-Le Du
>   J. Phys. (Paris), Suppl., 26: 185–189A (1965)

A fundamental study of epitaxy by flash evaporation, final report, Oct. 1, 1964–March 30, 1965
>   John L. Richards
>   AFCRL–65–412, April 30, 1965, Contract AF 19(628)–2937

The growth of crystals by solvent zone techniques, a review
>   G. A. Wolff and A. I. Mlavsky
>   Colloq. Intern. C.N.R.S., No. 152, Adsorption et croissance cristalline, Nancy, June 6–12, 1965
>   Centre National de la Recherche Scientifique, 15, quai Anatole–France–Paris 7c (1965)

Experimental investigation of conduction band of GaSb
>   A. Sagar
>   Phys. Rev., 117: 93–100 (1960)

Optical properties of gallium antimonide
>   David F. Edwards and George S. Hayne
>   J. Opt. Soc. Am., 49: 414–415 (1959)

Preparation and properties of III — V compounds
>   G. Wolff, P. H. Keck, and J. D. Broder
>   Phys. Rev., 94: 753 (1954)

## 10.e. Indium Compounds

### 10.e.1. InP

The preparation of high-purity epitaxial InP
>   R. C. Clarke, B. D. Joyce, and W. H. E. Wilgoss
>   Solid State Commun., 8: 1125–1128 (1970)

Electron diffraction study of epitaxial layers of indium phosphide
>   N. R. Aigina, M. A. Gurevich, N. M. Demenkov, L. A. Zhukova, V. N. Maslov, and B. A. Sakharov
>   Izv. Akad. Nauk SSSR, Neorg. Mater., 3(1): 175–176 (1967)
>   Inorg. Mater., 3(1): 144–146 (1967)
>   On single-crystal GaAs

On the crystallization of indium phosphide from a melt
>   Ya. A. Ugai, L. A. Bityutskaya, and A. D. Popova
>   Izv. Akad. Nauk SSSR, Neorg. Mater., 3(11):1988–1993 (1967)
>   Inorg. Mater., 3(11): 1730–1734 (1967)

InP, growth from In and $PCl_3$ at 700°C
>   V. Ya. Chernzkh and N. D. Talanov
>   Zh. Neorg. Khim., 11: 971–976 (1966)

Methods of preparing indium phosphide
>   R. J. Guire and K. Weiser
>   RCA, U. S. Patent 2,871,100, Jan. 1959

### 10.e.2. InAs

Preparation of indium-arsenide single crystals
>   J.-C. Transhart
>   Acta Electronica, 13: 13–20 (1970)
>   Liquid encapsulation

Effect of growth factors on improving InAs single crystals
>   B. P. Pyregov, I. S. Aver'yanov, S. V. Mashin, and A. S. Krasnov
>   Izv. Akad. Nauk SSSR, Neorg. Mater., 4(9): 1474–1477 (1968)
>   Inorg. Mater., 4(9): 1289–1291 (1968)

Epitaxial indium-arsenide lasers
>   M. A. C. S. Brown and P. Porteous
>   Solid-State Electron., 10: 76–77 (1967)
>   Liquid-phase epitaxy

Epitaxial InAs on InAs substrates
>   G. R. Cronin and S. R. Borrello
>   J. Electrochem. Soc., 114: 1079 (1967)

Interaction of indium arsenide with certain metals
>   Kh. D. Koppel, Z. S. Medvedeva, and N. P. Luzhnaya
>   Izv. Akad. Nauk SSSR, Neorg. Mater., 3(2): 300–310 (1967)
>   Inorg. Mater., 3(2): 267–274 (1967)
>   Crystals by spontaneous growth, directional crystallization, and by temperature-gradient

Preparation and properties of epitaxial InAs
>   J. P. McCarthy
>   Solid-State Electron., 10: 649–655 (1967)
>   Open-tube vapor-phase transport system

Epitaxial InAs on semi-insulating GaAs substrates
>   G. R. Cronin, R. W. Conrad, and S. R. Borrello
>   J. Electrochem. Soc., 113: 1336 (1966)

Experiences from the building and testing of apparatus for pulling indium-arsenide single crystals from the melt
>   H. Blumberg
>   Exper. Tech. Phys., 13: 414–419 (1965)
>   Czochralski-type crystals to 10 mm diameter and 50 mm long

Distribution coefficients of Se and Zn in InAs crystals
>   Masami Yokozawa, Morio Inoue, Rikusei Kohara, and Yoichi Okabayashi
>   Japan. J. Appl. Phys., 4: 546 (1965)
>   Horizontal Bridgman

Preparation of indium arsenide
>   R. H. Harada and A. J. Strauss
>   J. Appl. Phys., 30: 121 (1959)
>   Horizontal Bridgman

### 10.e.3. InSb

Kinetics of growth of indium-antimonide crystals from the melt
>   E. A. Dem'yanov
>   Sov. Phys. — Cryst., 15(4):690–692 (1971)

Getter evaporation of pure thin films of indium antimonides
>   R. P. Howson and V. Malina
>   J. Phys. D, Appl. Phys., 3:871–876 (1970)

Electron mobilities and photoluminescence of solution-grown indium-phosphide single crystals
>   O. Roder, U. Heim, and M. H. Pilkuhn
>   J. Phys. Chem. Solids, 31:2625–2634 (1970)

Czochralski-type crystal in transverse magnetic fields
A. F. Witt, C. J. Herman, and H. C. Gatos
J. Materials Sci., 5:822-824 (1970)

The effect of growth conditions on the density of $\alpha$-dislocations in InSb monocrystals
M. I. Dashevski, V. V. Khasikov, and L. I. Gutman
Izv. Akad. Nauk SSSR, Neorg. Mater., 5(9):1491-1495 (1969)
Inorg. Mater., 5(9):1267-1270 (1969)

A method for growing single crystals of metallic indium antimonide under pressure (from the melt at 26 kbars)
Gary L. Dorer and Hans E. Bommel
J. Appl. Phys., 40:670 (1969)

Effects of anisotropy in properties on impurity segregation for growing semiconductor crystals
U. M. Kulish and A. P. Vyatkin
Rost Kristallov, Izd. Akad. Nauk SSSR (1968), pp. 59-62
Growth of Crystals, Vol. 8 (A. V. Shubnikov and N. N. Sheftal', eds.) Consultants Bureau, New York (1969) pp. 72-77

Distribution coefficient studies in indium antimonide (Ga, Zn, Cu, Cd, Se, Zn — Cu, Zn — Cd and Zn — Se)
U. Merten, K. D. Vos, and A. P. Hatcher
J. Phys. Chem. Solids, 30:627-641 (1969)

Producing indium-antimonide single-crystal wafers (Stepanov method)
Yu. G. Nosov, P. I. Antonov, and A. V. Stepanov
Izv. Akad. Nauk SSSR, Ser. Fiz., 33:2008-2009 (1969)

Homogeneous impurity incorporation during crystal growth from the melt
August F. Witt and Harry C. Gatos
J. Electrochem. Soc., 116:511-513 (1969)

Indium-antimonide single crystals
M. Ya. Dashevskii, L. S. Okun, and G. Z. Plotkina
Moscow Institute of Steel and Alloys, U.S.S.R. 228,792 (Oct. 17, 1968); Izobret. Prom. Obraztsy, Tovarnye Znaki, 45:40 (1968)
Drawn from a melt

Growth mechanism of doped indium-antimonide dendrites and distribution of their impurities
M. Ya. Dashevskii and A. N. Poterukhin
Izv. Akad. Nauk SSSR, Neorg. Mater., 4(9):1478-1482 (1968)
Inorg. Mater., 4(9):1292-1295 (1968)
Determination of maximum growth rate on melt supercooling

Production of single-crystal plates of indium antimonide by crystallization of droplets of the melt in the gap between quartz plates
N. M. Demenkov and V. N. Maslov
Kristallografiya, 12(4):737-738 (1967)
Sov. Phys. — Cryst., 12(4):650-651 (1968)

Mechanism of single-crystal growth in InSb using temperature-gradient zone melting
Raymond W. Hamaker and William B. White
J. Appl. Phys., 39:1758 (1968)

Growth and properties of ribbonlike dendrites of semiconductor materials (A review of the foreign literature for 1962-1963)
L. P. Kalnach
Rost Kristallov, Izd. Akad. Nauk SSSR (1965), pp. 365-373

Growth of Crystals, Vol. 6B (A. V. Shubnikov and N. N. Sheftal', eds.) Consultants Bureau, New York (1968), pp. 170-179

Impurity distribution in single crystals - IV. Growth characteristics and impurity incorporation during facet growth
K. Morizone, A. F. Witt, and H. C. Gatos
J. Electrochem. Soc., 115:747 (1968)
InSb

Preparation, purification and single-crystal growth of indium antimonide
W. V. Ramana and P. R. Dastidar
J. Inst. Telecomm. Engrs., 14:287-295 (1968)

The preparation of InSb films
S. K. Sharma and V. K. Jain
Solid-State Electron., 11:423-428 (1968)

Microscopic rates of growth in single crystals pulled from the melt: indium antimonide
August F. Witt and Harry C. Gatos
J. Electrochem. Soc., 115:70 (1968)

A study on the vapour-phase deposits of InSb on single-crystal substrate
K. C. Barua and A. Goswami
Indian J. Pure Appl. Phys., 5:480-481 (1967)

Croissance des cristaux non-centrosymétriques suivant une direction polaire — rôle de la polarization superficielle
R. Cadoret and J. C. Monier
J. Crystal Growth, 1:59-66 (1967)

Preparation of homogeneous n-type InSb by thermal-neutron irradiation
W. Gilbert Clark and R. A. Isaacson
J. Appl. Phys., 38:2284 (1967)

Impurity distribution in single crystals. III. Impurity heterogeneities in single crystals rotated during pulling from the melt
K. Morizane, A. Witt, and H. C. Gatos
J. Electrochem. Soc., 114:738 (1967)

Crystallization of vacuum-deposited indium-antimonide films by the electron-beam zone-melting process
Susumu Namba, Akira Kawazu, Norio Kanekama and Toshiaki Akimoto
Japan. J. Appl. Phys., 6:1464-1465 (1967)

Structural and electrical properties of electron-beam zone recrystallized indium-antimonide thin films
N. F. Teede
Proc. Inst. Radio Electron. Engrs. Australia, 28:115-117 (1967)

Single-crystal InSb thin films by electron-beam recrystallization
N. F. Teede
Solid State Electron., 10:1069-1076 (1967)

Priprava cisteho monokrystalu InSb
Viliam Benc and Marian Morvic
Elektrotech. Casopis XVII, Cislo, 6:458-460 (1966)

A systematic study of zone refining of single-crystal indium antimonide
A. R. Murray, J. A. Baldrey, J. B. Mullin, and O. Jones
J. Mater. Sci., 1:14-28 (1966)

Elimination of solute banding in indium-antimonide crystals by growth in a magnetic field
  H. P. Utech and M. C. Flemings
  J. Appl. Phys., 37:2021 (1966)

Nuclear doping and optical properties of indium antimonide
  L. K. Vodop'yanov and N. I. Kurdiani
  Fiz. Tverd. Tela, 8(1):72-76 (1966)
  Sov. Phys. — Solid State, 8(1):55-58 (1966)

Impurity distribution in single crystals, II. Impurity striations in InSb as revealed by interference contrast microscopy
  A. F. Witt and H. C. Gatos
  J. Electrochem. Soc., 113:808 (1966)

Problems of crystal growth
  A. A. Chernov
  Vestn. Akad. Nauk SSSR (Moscow), 3:122-123 (1964),
    FTD-TT-65-538/1 + 4; AD-619470 (July 1965), 8 pp.

Indium antimonide
  K. F. Hulme
  (Royal Radar Establishment, Malvern, Worcs., England)
  Materials Used in Semiconductor Devices (C. A. Hogarth, ed.),
    Interscience Publishers, New York, London, Sydney (1965),
    p. 115

Indium antimonide of high perfection
  S. G. Parker, O. W. Wilson, and B. H. Barbee
  J. Electrochem. Soc., 112:80 (1965)

A fundamental study of epitaxy by flash evaporation
  J. L. Richards
  Philco Applied Research Lab., Blue Bell, Penn., final report covering period 1 Oct. 1964 to 30 March 1965, Contract AF 19(628)-2937. AFCRL-65-412 (April 1965)

Recent advances in crystal-growing techniques
  E. A. D. White
  Brit. J. Appl. Phys., 16:1415 (1965)

Electrical properties of flash-evaporated epitaxial InSb films
  C. Juhasz and J. C. Anderson
  Phys. Letters, 12:163 (1964)

De la possibilité d'obtenir des cristaux compensés de haute résistance d'antimoniure d'indium à l'aide d'un traitement thermique
  I. A. Mirckhulava, Z. N. Chigogidze, N. I. Kurdiani, L. V. Khvedelidze, and R. B. Dzhanelidze
  Soobshch. Akad. Nauk Gruz. SSR, 35:299 (1964)

Periodische Temperaturschwankungen in flüssigem InSb als Ursache schichtweisen Einbaus von Te in kristallisierendes InSb
  A. Muller and M. Wilhelm
  Z. Naturforsch., 19:254 (1964)

Electrical and thermal properties of indium antimonide in magnetic fields
  G. Veiss
  Izv. Akad. Nauk SSSR, Ser. Fiz., 28:969 (1964)

The influence of twin structure on growth directions in dendritic ribbons of materials having the diamond or zinc-blende structures
  N. Albon and A. E. Owen
  J. Phys. Chem. Solids, 24:899 (1963)

Dendritic growth of InSb
  H. Nicholson and J. W. Faust, Jr.
  J. Electrochem. Soc., 110:940 (1963)

Epitaxy of compound semiconductors by flash evaporation
  J. L. Richards, P. B. Hart, and L. M. Gallone
  J. Appl. Phys., 34:3418 (1963)

Dendritic growth of indium antimonide
  R. G. Seidensticker and D. R. Hamilton
  J. Phys. Chem. Solids, 24:1587 (1963)

Obtaining ultrapure InSb crystals by zone melting
  K. I. Vinogradova, V. V. Golovanov, and D. N. Nasledov
  Fiz. Metal. i Metalloved., 16:385-393 (1963)

Crystal growth and crystallography, a literature survey
  T. Cheron
  Aerospace Corp., El Segundo, California, Bibliography (1950-61, Contract AF 04(647)930, AD-274642 (Jan. 1962)

Indium antimonide — a review of its preparation, properties and device applications
  K. F. Hulme and J. B. Mullin
  Solid-State Electron., 5:211 (1962)

Review of crystal growth methods
  F. D. Loomis, C. W. Graver, and I. Mockrin
  Penn-Salt Chemicals Corp., Wyndmoor, Pa., Technical Report No. 1, Contract Nonr-3142(00), NR No. 356-423 (1962)

(100) facets in pulled crystals of InSb
  A. J. Strauss
  Solid-State Electron., 5:97 (1962)

Preparation and properties of grown p − n junctions of InSb
  H. C. Gorton, A. R. Zacaroli, F. J. Reid, and C. S. Peet
  J. Electrochem. Soc., 108:354 (1961)

Growth twins in indium antimonide
  R. K. Mueller and R. L. Jacobson
  J. Appl. Phys., 33:550 (1961)

Anisotropic segregation in InSb
  W. P. Allred and R. K. Willardson
  Battelle Memorial Inst., Columbus, Ohio, 1960 Spring Meeting Electrochem. Soc.

Growth of InSb crystals in the (111) polar direction
  H. C. Gatos, P. L. Moody, and M. C. Lavine
  J. Appl. Phys., 31:212 (1960)

Obtaining high-purity indium antimonide
  M. S. Mirgalovskaya and L. I. Matkova
  Zh. Neorgan. Khim., 5:1551-1554 (1960)

Orientation-dependent distribution coefficients of tellurium in indium-antimonide crystals
  J. B. Mullin
  Royal Radar Establishment, Gt. Malvern, Worcs., England, 1960 Spring Meeting Electrochem. Soc.

Orientation-dependent distribution coefficients in melt-grown InSb crystals
  J. B. Mullin and K. F. Hulme
  J. Phys. Chem. Solids, 17:1 (1960)

Facets and anomalous solute distribution in indium antimonide
K. F. Hulme and J. B. Mullin
Phil. Mag., 4:1286 (1959)

Intermetallic semiconductors
H. T. Minden
Semicond. Prod., 2:30 (1959)

Production of high-purity single crystals of InSb by zone melting
K. I. Vinogradova, V. V. Talavanova, D. N. Nasledov, and L. I. Solov'eva
Fiz. Tverd. Tela, 1(3):403–406 (1959)
Sov. Phys. – Solid State, 1(3):364–367 (1959)

Preparation of Single Crystals
W. D. Lawson and S. Nielsen
Butterworths Publ. and Acad. Press, New York (1958), p. 255

Improvements in or relating to processes for fusing powdered semi-conductor materials
Siemens and Halske
Brit. Pat. 806,697 (issued Dec. 31, 1958)

Melting patterns appearing on single crystals of InSb
M. F. Millea and C. T. Tomizuka
J. Appl. Phys., 27:96 (1956)

## 10.e.4. In Ternaries

Deposition of epitaxial $InAs_xP_{(1-x)}$ on GaAs and GaP substrates
H. A. Allen and E. W. Mehal
J. Electrochem. Soc., 117:1081–1082 (1970)

Electroluminescent light sources, final technical report
R. O. Bell, S. C. Foote, C. B. Lamport, A. A. Menna, and G. A. Wolff (Tyco Labs., Inc., Waltham, Mass.)
NASA-CR-113907 (July 1970), 60 pp.
$Ga_xIn_{1-x}P$, traveling heater method

Stimulated emission in $In_{1-x}Ga_xP$
R. D. Burnham, N. Holonyak, Jr., D. L. Keune, D. R. Scifres, and P. D. Dapkus
Appl. Phys. Letters, 17:430–432 (1970)
Modified Bridgman solution-growth

Crystallization of $InAs_{1-x}P_x$ solid solutions from the gas phase
L. A. Egorov and O. D. Torbova
Izv. Akad. Nauk SSSR, Neorg. Mater., 5:173–174 (1969)
Inorg. Mater., 5:144–145 (1969)

Indium gallium phosphide, a new electroluminescent semiconductor
C. Hilsum
Royal Radar Establishment Newsletter and Research Review, No. 8 (1969), pp. 5/1–5/2

The preparation and properties of vapor-deposited epitaxial $InAs_{1-x}P_x$ using arsine and phosphine
J. J. Tietjan, H. P. Maruska, and R. B. Clough
J. Electrochem. Soc., 114:492–494 (1969)

State diagram of the InAs – InP system
Ya. A. Ugai, E. G. Goncharov, Z. V. Kitina, and T. N. Shvyreva
Izv. Akad. Nauk SSSR, Neorg. Mater., 4(3):348–351 (1968)
Inorg. Mater., 4(3):291–293 (1968)

Properties of $InAs_{1-x}P_x$ solid-solution single crystals
L. A. Egorov, P. V. Shakhanov, O. D. Torbova, and I. V. Gordeev
Izv. Akad. Nauk SSSR, Neorg. Mater., 3(5):881–883 (1967)
Inorg. Mater., 3(5):787–789 (1967)

Narrow-band self-filtering junction detectors
J. W. Burns, S. Kave, D. B. Medved, M. B. Prince, and G. P. Rolik
J. Appl. Phys., 38:5388 (1967)
InAs – InP single-crystal growth

Semiconductor p – n junction lasers in the $InAs_{1-x}Sb_x$ system
N. G. Basov, A. V. Dudenkova, A. I. Krasil'nikov, V. V. Nikitin, and K. P. Fedoseev
Fiz. Tverd. Tela, 8(4):1060–1063 (1966)
Sov. Phys. – Solid State, 8(4):847–849 (1966)

Epitaxial growth of III – V materials
E. W. Mehal and G. R. Cronin
Electrochem. Tech., 4:540 (1966)

Monocrystals of indium-antimonide/gallium-antimonide alloys and their electric properties
I. S. Baukin, V. I. Ivanov-Omskii, and B. T. Kolomiets
News of the Acad. of Sci. of the USSR (June 5, 1965), pp. 23–26

Preparation of double crystals of indium and gallium antimonides
V. I. Ivanov-Omskii, N. K. Kiseleva, and B. T. Kolomiets
Fiz. Tverd. Tela, 3(5):1621 (1961)
Sov. Phys.–Solid State, 3(5):1175–1176 (1961)

# 11. III–VI Compounds

## 11.a. General, Reviews, and Bibliographies

New single-crystalline phases in the system
$Ga_2S_3 - In_2S_3$
V. Kramer, R. Nitsche, and J. Ottemann
J. Cryst. Growth, 7:285-289 (1970)

Thermal expansion of substances with diamond-like structure and volume changes during their melting
V. M. Glazov, S. N. Chizhevskaya, and S. B. Evgen'ev
Zh. Fiz. Khim., 43(2):373-379 (1969)

Thermographic study of the nature of formation of gallium and indium chalcogenides
Z. Sh. Karaev and M. Yu. Abdullaev
Issled. Obl. Neorg. Fiz. Khim. Ikh Rol Khim. Prom., Mater. Nauch. Konf. (1967), pp. 14-18 (M. S. Sogomonyan, ed.) (AzINTI, Baku, USSR, 1969)

Mechanism of growing single crystals of $A^{III}B^{VI}$- and $A_2^{III}B_3^{VI}$-type compounds by a chemical transport reaction method
P. G. Rustamov and B. A. Geidarov
Azerb. Khim. Zh., 2:143-146 (1969)

Growing semiconducting single crystals of sulfides and selenides of groups IIA and IIIA
L. A. Sysoev, L. V. Konvisar, and E. K. Raiskin
All-Union Scientific Research Institute of Monocrystals, U.S.S.R. Patent 177,844 (June 20, 1969); from Otkrytiya, Izobret., Prom. Obraztsy, Tovarnye Znaki, 46:169 (1969)

Semiconductor $A^{III}B^{VI}$ compounds
Z. S. Medvedeva
Izv. Akad. Nauk SSSR, Neorg. Mater., 4(12):2078-2084 (1968)
Inorg. Mater., 4(12):1808-1813 (1968)

Chalcogenides of Elements of Subgroup IIIB of the Periodic System
Z. S. Medvedeva
Nauka, Moscow, 1968, 216 pp.

Process for growing single crystals of sulfides, selenides and tellurides of metals of groups II and III of periodic system
Leonid Andreevich Sysoev, Leonid Viktorovich Konvisar, and Emmanuil Kelmanovich Raiskin

U. S. Patent 3,414,387 (Dec. 3, 1968; filed Jan. 1966, Ser. No. 518,887)
Holders of patent right all are subjects of USSR

Infrared absorption of $A^{III}B^{VI}$ monocrystals
G. A. Akhundov and T. G. Kerimova
Optika i Spektrosk., 22:654-655 (1967)
Optics and Spectrosc., 22:355-356 (1967)

Preparation and investigation of $A^{III}B^{VI}$ single crystals
G. A. Akhundov, G. B. Abdullayev, G. D. Guseinov, R. F. Meckhtiev, and M. K. Aliyeva
International Congress of Phys. Semiconducteurs, Paris (1964), pp. 90-91

Defect diamond-like semiconductors
N. A. Goryunova and S. I. Radautsan
Investigations on Semiconductors. New Semiconductor Materials, Izd. Kartya Moldovenyaske, Kishinev (1964), pp. 3-43

Zone leveling and crystal growth of peritectic compounds
D. R. Mason and J. S. Cook
J. Appl. Phys., 32:475-477 (1961)

Crystal growth by chemical transport reactions. I. Binary, ternary, and mixed-crystal chalcogenides
R. Nitsche, H. U. Bolsterli, and M. Lichtensteiger
J. Phys. Chem. Solids, 21:199-205 (1961)

Some properties of $In_2Te_3$ and $Ga_2Te_3$
J. C. Woolley and B. R. Pamplin
J. Electrochem. Soc., 108:874 (1961)

Semiconductors of the type $A^{III}B^{VI}$
P. Fielding, G. Fischer, and E. Mooser
J. Phys. Chem. Solids, 8:434-437, 442-443 (1959)
Advances in Semi-Conductor Science. Proceedings of the Third International Conference on Semi-Conductors held at the Univ. of Rochester, August 18-22, 1958
Pergamon Press, New York (1959), pp. 434-447

Equilibrium diagram of the $Ga_2Te_3 - In_2Te_3$ system
J. C. Woolley, D. G. Lees, and B. A. Smith
J. Less-Common Metals, 1:199 (1959)

New semiconductors with chalcopyrite structure
V. M. Glazov, M. S. Mirgalovskaya, and L. A. Petrakova
Izv. Akad. Nauk SSSR, Otd. Tekhn. Nauk, 10:68-70 (1957)

New semiconducting compounds
E. Mooser and W. B. Pearson
Phys. Rev., 101:492-493 (1956)

## 11.b. Aluminum and Boron Compounds

Liquid encapsulation important to crystal growth of semiconductor compounds ($B_2O_3$)
R. Davies
Electron. Eng., 42:58 (1970)

Phase diagram of the boron selenide — selenium system
V. A. Boryakova, Ya. Kh. Grinberg, E. G. Zhukov, V. A. Koryazhkin, and Z. S. Medvedeva
Izv. Akad. Nauk SSSR, Neorg. Mater., 5(3):477-480 (1969)
Inorg. Mater., 5(3):397-399 (1969)

Thermal emission spectrum of a new diatom, AlSe
J. Singh, D. P. Tewari, and H. Mohan
Proc. Phys. Soc. London At. Mol. Phys., 2(5):627-628 (1969)
Electronic band system in the spectral range $\lambda$ 3900-6410

Production of $B_2S_3$ monocrystals
E. G. Zhukov and I. K. Grinberg
Izv. Akad. Nauk SSSR, Neorg. Mater., 5(9):1646-1647 (1969)
Inorg. Mater., 5(9):1394-1395 (1969)

The spectrum of AlS
Mona Kronekvist and Albin Lagerqvist
Arkiv for Fysik, 39:133-137 (1968)

High-pressure, high-temperature synthesis of BS
D. C. Carlson
Thesis, Brigham Young Univer., 1966
Dissertation Abstr., 27(8):2664-2665B (1966-67)

Sulfur and selenium compounds of boron
Earl L. Muetterties
(E. I. duPont de Nemours and Co., Wilmington, Delaware)
Chem. Boron Its Compounds (1967), pp. 647-667

The crystal structure of $Al_2Se_3$
G. A. Steigmann and J. Goodyear
Acta Cryst., 20:617 (1966)

Electrical properties and optical absorptions of thin $Al_2Se_3$ films
V. P. Mushinskiy
Referativnyy Zh., Fizika, 4:213 (1960)

Study of the structure and properties of aluminum telluride
M. S. Mirgalovskaya and Ye. V. Skudnova
Izv. Akad. Nauk SSSR, Otd. Tekhn. Nauk, Met. i Toplivo, 4:148-152 (1959)

Semiconductor properties of aluminum selenide
V. P. Mushinskii
Fiz. Tverd. Tela, 1(3):515-517 (1959)
Sov. Phys. — Solid State, 1(3):463-465 (1959)

## 11.c. Gallium Compounds

Gallium, Bulletin d'information et de bibliographie 1970, fascicule No. 9
Pierre de la Breteque
Alusuisse-France S.A., Marseille, France (1971)
Pages A-13 to A-96 survey the binary gallium chalcogenides in depth; synthesis, crystal growth, crystal structures, physical properties, electrical and electronic properties, optical properties, and magnetic properties are included. Graphic or tabulated values are given for most properties

Phonon structures in optical spectra of layer compounds GaSe and GaS
Noritaka Kuroda, Yuichiro Nishina, and Tadao Fukuroi
J. Phys. Soc. Japan, 28:981-992 (1970)
49 refs.

Some optical properties of single crystals of the $Ga_2Te_3$ — $In_2Te_3$ system
K. M. Mushinskaya and V. G. Tyrziu
Fiz. Tekhn. Poluprovod., 3(7):978-981 (1969)
Sov. Phys. — Semicond., 3(7):825-827 (1970)

An open-tube technique to grow $Ga_2S_3$ crystals
C. Paorici and G. Zuccalli
J. Crystal Growth, 7:265-266 (1970)

Photoluminescence of gallium sulfide, gallium selenide and their solid solutions
G. A. Akhundov, G. M. Gasumov, and F. I. Ismailov
Opt. Spektrosk., 26:351 (1969)

Photoluminescence of $GaS_xSe_{1-x}$
G. A. Akhundov, G. M. Gasumov, and F. I. Ismailov
Opt. Spektrosk., 26(4):642-643 (1969)

Determination of the crystal structures of several three-component semiconductors with the general formula $ABX_2$
A. S. Avilov, K. A. Agaev, G. G. Guseinov, and R. M. Imamov
Kristallografiya, 14(3):443-446 (1969)
Soviet Phys. — Cryst., 14(3):364-366 (1969)

Preparation of n- and p-gallium selenide single crystals by a gas-transport method and a study of some of their physical properties
T. Kh. Azizov, M. Kh. Alieva, A. Z. Mamedova, and K. A. Sharifov
Izv. Akad. Nauk Azerb. SSR, Ser. Fiz.-Tekh. Mat. Nauk, 3:111-115 (1969)

Quelques propriétés électriques de couches minces de GaSe
Paul Benalloul and Jacques Benoit
Compt. Rend., 269B:723-726 (1969)
Electron-beam vaporization

Anomalie dans l'effet électro-optique du GaSe
J. A. Deverin
Helv. Phys. Acta, 42:597 (1969)

Reaction of $A_2^{III}B_3^{VI}$-type gallium chalcogenides with gallium antimonide
M. G. Guseinova and P. G. Rustamov
Issled. Obl. Neorg. Fiz. Khim. Ikh Rol Khim. Prom., Mater. Nauch. Konf. 1967 (M.S. Sogomonyan, ed.) (AzINTI, Baku, USSR, 1969), pp. 11-14

Thermoelectric-power measurements on gallium-sulphide single crystals. Effective density of states
A. H. M. Kipperman and T. B. A. M. Sliepenbeek
II Nuovo Cimento, Serie X, 63B:36–40 (1969)

Hall-effect measurements on gallium-sulphide single crystals
A. H. M. Kipperman and C. J. Vermij
II Nuovo Cimento, Serie X, 63B:29–35 (1969)

Negative resistance and spontaneous current oscillations in tin-doped gallium selenide
Andrea Levialdi and N. Romeo
Nuovo Cimento, 63B:41–44 (1969)

Physicochemical investigations and electrical monocrystalline gallium sulphide
Ronald M. A. Lieth
Ph. D. thesis, Technische Hogeschool, Eindhoven, Netherlands (1969), 116 pp.
Phase equilibria and crystal growth

Preparation, purity and electrical conductivity of gallium-sulphide single crystals
R. M. A. Lieth, C. W. M. Van der Heijden, and J. W. M. Van Kessel
J. Crystal Growth, 5:215–258 (1969)

Gallium-selenide – indium-selenide system
V. P. Mushinskii and N. M. Pavlenko
Krist. Tech., 4(2):K5–K7 (1969)
Single crystals grown by Bridgman method

Negative resistance in GaS single crystals
N. Romeo
Phys. Stat. Sol., 36(1):153–156 (1969)

Binary phase diagrams of chalcogen – gallium (indium) and sections of chalcogen – chalcogen – gallium (indium) ternary systems
P. G. Rustamov and M. I. Zargarova
Issled. Obl. Neorg. Fiz. Khim. Ikh Rol Khim. Prom., Mater. Nauch. Konf. 1967 (M. S. Sogomonyan, ed.) (AzINTI: Baku, USSR, 1969), pp. 9–11

Preparing gallium selenide single crystals by chemical transport reactions (CTR)
P. G. Rustamov, B. N. Mardakhaev, B. A. Geidarov, and T. N. Kuliev
Issled. Obl. Neorg. Fiz. Khim. Ikh Rol Khim. Prom., Mater. Nauch. Konf., 1967, (M. S. Sogomonyan, ed.) (AzINTI, Baku, USSR, 1969), pp. 18–21

Preparation of single crystals from gallium-sulfide – gallium-selenide solid solutions and their physical properties
P. G. Rustamov, Z. D. Melikova, Ya. N. Nasirov, and M. A. Alidzhanov
Izv. Akad. Nauk, SSSR, Neorg. Mater., 5(5):881–884 (1969)
Inorg. Mater., 5(5):750–752 (1969)

X-ray diffraction study of gallium-sulfide – gallium-selenide systems
M. G. Safarov
Azerb. Khim. Zh., 2:138–142 (1969)

Hopping conduction in gallium-selenide single crystals
R. H. Tredgold and A. Clark
Solid State Commun., 7:1519–1520 (1969)

Supralinear photoconductivity in gallium sulphide
A. T. Vink
II Nuovo Cimento, Serie X, 63B:70–79 (1969)

Growing gallium-selenide single crystals by gas-transport reactions
T. Kh. Azizov
Izv. Akad. Nauk Azerb. SSR, Ser. Fiz.-Tekh. Mat. Nauk, 3:26–28 (1968)

An open-tube reactor for preparing compounds with volatile constituents
T. B. Reed and W. J. LaFleur
Lincoln Laboratories, Massachusetts Institute of Technology, Lexington, Mass., ESD-TR-68-17 (April 1968), pp. 19–20
GaS

Preparation of single crystals from $Ga_2Te_3$ – GaSe system solid solutions
P. G. Rustamov, V. B. Cherstvova, and M. A. Alidzhanov
Izv. Akad. Nauk SSSR, Neorg. Mater., 4(8):1351–1352 (1968)
Inorg. Mater., 4(8):1186–1187 (1968)

Wurtzite modification of gallium telluride and solid solutions based on it
M. G. Safarov, R. S. Gamidov, P. G. Rustamov, and V. B. Cherstvova
Izv. Akad. Nauk SSSR, Neorg. Mater., 4(1):138–139 (1968)
Inorg. Mater., 4(1):113–114 (1968)

Reactions in the system gallium and tin tellurides
M. A. Alidzhanov, P. G. Rustamov, and M. G. Safarov
Azerb. Khim. Zh., 1:103–108 (1967)

Deviation from stoichiometry in semiconducting gallium telluride ($Ga_2Te_3$)
L. V. Atroshchenko, L. P. Gal'chinetsky, and V. M. Koshkin
Izv. Akad. Nauk SSSR, Neorg. Mater., 3(5):777–782 (1967)
Inorg. Mater., 3(5):695–698 (1967)

X-ray study of the ternary compounds $Ga_2SeTe$
B. K. Babaeva, R. S. Gamidov, and P. G. Rustamov
Izv. Akad. Nauk SSSR, Neorg. Mater., 3(5):919 (1967)
Inorg. Mater., 3(5):826 (1967)

Band structure and optical properties of graphite and of the layer compounds gallium sulfide and gallium selenide
F. Bassani and G. P. Parravicini
Nuovo Cimento, B50:95–128 (1967)

Magneto-optical absorption in GaSe
J. L. Brebner, J. Halpern, and E. Mooser
Helv. Phys. Acta, 40:385–386 (1967)

Preparation of a GaSe single crystal and determination of its elastic parameters
Kh. M. Khalilov and K. I. Rzaev
Kristallografiya, 11(6):927–930 (1966)
Sov. Phys. – Cryst., 11(6):786–787 (1967)

Vapor pressure and thermal stability of gallium sulfide
R. M. A. Lieth, H. J. M. Heijligers, and C. W. M. Heijden
Mater. Sci. Eng., 2:193–200 (1967)

Specific heat of gallium selenide and thallium selenide
K. K. Mamedov, I. G. Kerimov, V. N. Kostryukov, and M. I. Mekhtiev
Fiz. Tekhn. Poluprov., 1:441–442 (1967)
Sov. Phys. – Semicond., 1:363 (1967)

Optical and magneto-optical properties of some semiconductors
Y. Nishina, K. Tanaka, S. Kurita, M. Yamamoto, T. Jimbo, N. Kuroda, and T. Fukuroi
Sci. Rept. Res. Inst. Tohoku Univ., 18:536 (1967)

The reflection spectrum of GaSe and GaS single crystals near the fundamental absorption edge
M. A. Nizametdinova
Phys. Stat. Sol., 19:K111 (1967)

The ternary condensed phase diagram of the Ga — As — Te system
M. B. Panish
J. Electrochem. Soc., 114:91 (1967)

Liquidus of the gallium — selenium — tellurium system
P. G. Rustamov and V. B. Cherstvova
Azerb. Khim. Zh., 2:98–103 (1967)

Growing of some gallium-sulfide single crystals from gaseous phase
P. G. Rustamov and B. N. Mardakhaev
Izv. Akad. Nauk SSSR, Neorg. Mater., 3(3):575–577 (1967)
Inorg. Mater., 3(3):511–513 (1967)

Phase diagram of the gallium — sulfur system
P. G. Rustamov, B. N. Mardakhaev, and M. G. Safarov
Izv. Akad. Nauk SSSR, Neorg. Mater., 3(3):479–483 (1967)
Inorg. Mater., 3(3):429–434 (1967)

Standard enthalpy of formation of gallium selenide and indium selenide
K. A. Sharifov and T. K. Azizov
Zh. Fiz. Khim., 41:1208–1209 (1967)

Synthesis and study of the thermal decomposition of gallium diselenate
I. V. Tananaev, N. K. Bol'shakova, and V. K. Gorokhov
Z. Neorg. Khim., 12:2337–2339 (1967)

Investigation of photoconductive relaxation in p-GaSe single crystals
G. B. Abdullaev, R. F. Mekhtiev, A. Z. Mamedova, and E. S. Guseinova
Phys. Stat. Sol., 14:K127 (1966)

Infrared absorption of $A^{III}B^{VI}$ single crystals
G. A. Akhundov and T. G. Kerimova
Phys. Stat. Sol., 16:K15 (1966)

Electroluminescence of GaSe single crystals
G. A. Akhundov, I. G. Aksyanov, and A. G. Bagirov
Optika i Spektroskopiya, 1:120–121 (1966)

Optical absorption, reflection, and dispersion of GaS and GaSe layer crystals
G. A. Akhundov, N. A. Gasanova, and M. A. Nizametdinova
Phys. Stat. Sol., 15:K109 (1966)

High-frequency electroluminescence of GaAs and GaSe
A. S. Borshchevskii, Ya. A. Oksman, and V. N. Smirnov
Fiz. Tverd. Tela, 8(5):1428–1433 (1966)
Sov. Phys. — Solid State, 8(5):1139–1143 (1966)

Optical properties of a laminar GaSe crystal
M. S. Brodin and Yu. P. Gnatenko
Ukr. Fiz. Zh., 11:759–765 (1966)

Hydrothermal processes. IX. Synthesis of sulfides and thiosalts of trivalent metals
L. Cambi and M. Elli
Chim. Ind. (Milan), 48:944 (1966)

Optical reflection of the GaS and GaSe single crystals
N. A. Gasanova and G. A. Akhundov
Opt. Spektr., 20:353 (1966)

Optical reflection and absorption of $GaS_xSe_{1-x}$ single crystals
N. A. Gasanova, G. A. Akhundov, and M. A. Nizametdinova
Phys. Stat. Sol., 17:K131 (1966)

Polarization effects in the ultraviolet reflection from GaS and GaSe single crystals
N. A. Gasanova, G. A. Akhundov, and M. A. Nizametdinova
Phys. Stat. Sol., 17:K115 (1966)

Structure and growth mechanism of gallium-selenide single crystals
G. D. Guseinov and K. I. Rzaev
Slozhnye Poluprov., Akad. Nauk Azerb. SSR, Inst. Fiz. (1966), pp. 112–118

Spiral growth mechanism of gallium-selenide single crystals
G. D. Guseinov, A. M. Ramazanzade, and K. I. Rzaev
Slozhnye Poluprov., Akad. Nauk Azerb. SSR, Inst. Fiz. (1966), pp. 106–111

Dielectric constants and infrared absorption of GaSe
P. C. Leung, G. Andermann, and W. G. Spitzer
J. Phys. Chem. Solids, 27:849 (1966)

The P — T — X phase diagram of the system Ga — S
R. M. A. Lieth, H. J. M. Heijligers, and C. W. M. v.d. Heijden
J. Electrochem. Soc., 113:789 (1966)

Synthesis of gallium and indium selenides by reduction of their selenites
L. Y. Markovskii and M. S. Soboleva
Zh. Prikl. Khim., 39:2820–2821 (1966)

Investigation of the constitution diagram of the Ga — Se system
L. S. Palatnik and E. K. Belova
Izv. Akad. Nauk SSSR, Neorg. Mater., 2(4):770–771 (1966)
Inorg. Mater., 657–669 (1966)

Crystal-chemical peculiarities of sulfides and chalcogenides
E. A. Pobedimskaya and N. V. Belov
Geokhimiya (1966), pp. 152–160

Liquidus of the gallium — sulfur — tellurium ternary system
P. G. Rustamov and T. A. Dzhalilzade
Azerb. Khim. Zh., 4:93–97 (1966)

Reaction of selenium and tellurium with gallium telluride and selenide of the $A_2^{III}B_3^{VI}$ type
P. G. Rustamov, B. K. Babaeva, and V. B. Cherstvova
Azerb. Khim. Zh., 3:113–116 (1966)

Growth of crystals of sulfides, selenides, and tellurides of certain metals
All-Union Scientific-Research Institute of Monocrystals, Scintillating Materials, and Ultrahigh Purity Chemicals, Belg. 672,094 (March 1, 1966), 16 pp.
GaSe

Ternary chalcogenides of the type $A^{II}B_2^{III}C_1^{VI}$
B. V. Baranov et al.
Fiz. Dokl. k XXIII-ei Nauchn. Konf. Leningr. Inzh.-Stroit. Inst., Sb., Leningrad (1965), pp. 48–49

Les indices de refraction dans le GaS et le GaSe
J. L. Brebner and J. A. Deverin
Helv. Phys. Acta, 38:650 (1965)

Electroabsorption in GaSe
J. L. Brebner and E. Mooser
Helv. Phys. Acta, 38:656 (1965)

Optical absorption of GaTe single crystals
N. A. Gasanova and G. A. Akhundov
Opt. i Spektroskopiya, 18:731-733 (1965)

Phase diagram of the $Ga_2Se_3 - Ga_2Te_3$ section of the Ga − S − Te system
P. G. Gustamov and T. A. Dzhalilzade
Azerb. Khim. Zh., 5:90-93 (1965)

Crystal structures
Ralph W. G. Wyckoff
(Univ. Arizona, Tucson, Arizona), Second Edition, Vol. 1, Interscience Publishers, New York (1965)

Solid solutions in pseudo binary system gallium selenide − gallium telluride
Dokl. Akad. Nauk Azerb SSR, 21:8-10 (1965)

Semiconductors $A^{III}B^{VI}$
G. A. Akhundov, G. B. Abdullaev, G. D. Guseinov, R. F. Mekhtiev, M. Kh. Alieva, E. S. Guseinova, and I. A. Gasanova
Izv. Akad. Nauk Azerb. SSR, Ser. Fiz. -Mat. i Tekhn. Nauk, 3:107-114 (1964)
Single crystals of GaS, GaSe, GaTe, and InSe, 12 mm in diam. and 50 mm long, were grown by slow cooling at a constant temperature gradient

Preparation and investigation of $A^{III}B^{VI}$ single crystals
G. A. Akhundov, G. B. Abdullayev, G. D. Guseinov, R. F. Meckhtiev, and M. K. Aliyeva
Congr. International Phys. Semiconducteurs, Paris (1964), pp. 90-91

The optical absorption edge in layer structures
J. L. Brebner
J. Phys. Chem. Solids, 25:1427 (1964)

Reactions of $A_2^{III}B_3^{VI}$. Samarium and gallium selenides
G. Kh. Efendiev, Z. Sh. Karaev, and I. O. Nasibov
Azerb. Khim. Zh., 1:125-131 (1964)

Defect diamond-like semiconductors
N. A. Goryunova and S. I. Radautsan
Investigations on Semiconductors. New Semiconductor Materials, Izd. Kartya Moldovenyaske, Kishinev (1964), pp. 3-43

Optical properties of $Ga_2Te_3$ single crystals
V. I. Gramatsky and V. P. Mushinsky
Fiz. Tverd. Tela, 6(11):3478 (1964)
Sov. Phys. − Solid State, 6(11):2784 (1964)

Optical absorption edge of GaS and GaSe single crystals
F. I. Ismailov, E. S. Guseinova, and G. A. Akhundov
Fiz. Tverd. Tela, 5(12):3620 (1963)
Sov. Phys. − Solid State, 5(12):2656-2660 (1964)

Preparative methods for alloys using sulfur
P. G. Rustamov and B. N. Mardakhayev
Dokl. Akad. Nauk Azerb. SSR (Baku), 20:13-15 (1964), NASA-TT-F-9487
GaS and $Ga_2S_3$

Electron-diffraction investigation of phases in the system gallium − tellurium
S. A. Semiletov and V. A. Vlasov
Kristallografiya, 8(6):877-883 (1963)
Sov. Phys. − Cryst., 8(6):704-708 (1964)

Reflection spectra of crystals of groups II-V and III-VI
V. V. Sobolev, N. N. Syrbu, A. M. Andriesk, and S. D. Shutov
Fiz. Tverd. Tela, 6(8):2539-2540 (1964)
Sov. Phys. − Solid State, 6(8):2020-2021 (1965)

Solid solutions in the system $Ga_2Se_3 - In_2Se_3$
A. A. Vaipolin and V. S. Grigor'eva
Investigations on Semiconductors. New Semiconductor Materials, Izd. Kartya Moldovenyaske, Kishinev (1964), pp. 77-81

Some properties of $GaSb - Ga_2Se_3$ and $GaSb - Ga_2Te_3$ alloys
J. C. Woolley and K. W. Blazey
J. Electrochem. Soc., 111:951 (1964)

The optical absorption edge of GaS
J. L. Brebner and G. Fischer
Can. J. Phys., 41:561 (1963)

Metallographic examination of the phase diagram of the gallium − tellurium system
J. R. Dale
Nature, 197:242-247 (1963)

Speculation on the band structure of the layer compounds gallium sulfide and gallium selenide
G. Fischer
Helv. Phys. Acta, 36:317 (1963)

The crystal structure of $\alpha$-$Ga_2S_3$
J. Goodyear and G. A. Steigmann
Acta Cryst., 16:946 (1963)

Metastable amorphous phases in tellurium-base alloys
H. L. Luo and P. Duwez
Appl. Phys. Letters, 2:21 (1963)

Kinetics of impurity photoconductivity in gallium selenide crystals
R. F. Mekhtiev, L. G. Paritsky, and S. M. Ryvkin
Fiz. Tverd. Tela, 5(6):1649-1656 (1963)
Sov. Phys. − Solid State, 5(6):1198-1203 (1963)

Optical and photoelectric properties of thin layers of gallium telluride
V. P. Mushinskii and G. N. Manushevich
Izv. Vysshikh Uchebn. Zavedenii. Fiz., (1963), pp. 172-178

Heat capacity of gallium telluride
Yu. B. Nadzhafov and K. A. Sharifov
Tr. Inst. Fiz., Akad. Nauk Azerb. SSR, 11:31-35 (1963)

Reactions of $A_2^{III}B_3^{VI}$. Praseodymium and gallium selenides
I. O. Nasibov and Z. Sh. Karaev
Azerb. Khim. Zh., 5:105-111 (1963)

Excitation spectra of absorption, of reflection and of emission of crystalline gallium selenide
S. Nikitine et al.
J. Chim. Phys., 60:667 (1963)

The luminescence characteristics of some
group III — VI compounds
  M. Springford
  Proc. Phys. Soc. (London), 82:1020 (1963)

The luminescence of some ternary chalcogenides and mixed binary systems of group III — VI
compounds: The nature of luminescence centers
in group III — VI compounds
  M. Springford
  Proc. Phys. Soc. (London), 82:1029 (1963)

Growth spirals on gallium sulfide and gallium
selenide single crystals
  H. U. Bolsterli and E. Mooser
  Helv. Phys. Acta, 35:538 (1962)

Structure électronique des composés en couches
GaTe, GaSe et GaS
  J. L. Brebner and G. Fischer
  Can. J. Phys., 18:26 (1962)

Optical properties of the layer structures
GaTe, GaSe, and GaS
  J. L. Brebner and G. Fischer
  International Conf. on Phys. Semiconductors, Exeter (July
    1962), pp. 760

A method of growing single crystals of GaSe
and an analysis of certain of their properties
  R. F. Mekhtiev, G. B. Abdullaev, and G. A. Akhundov
  TT-66-11212, Nauk Azerb. SSR, Baku. Doklady, 18:11-15
    (1962)

Relationship between structures and disloca-
tions in GaS and GaSe
  Z. S. Basinski, D. B. Dove, and E. Mooser
  Helv. Phys. Acta, 34:373 (1961)

Crystal growth by chemical transport reactions.
I. Binary, ternary, and mixed-crystal chal-
cogenides
  R. Nitsche, H. U. Bolsterli, and M. Lichtensteiger
  J. Phys. Chem. Solids, 21:199 (1961)

Solid solutions of phosphido-selenides of gal-
lium
  S. I. Radautsan, I. A. Madan, and R. A. Ivanova
  Izv. AN Mold. SSR, 10:98-101 (1961)

Some properties of $In_2Te_3$ and $Ga_2Te_3$
  J. C. Woolley and B. R. Pamplin
  J. Electrochem. Soc., 108:874 (1961)

Photoconductivity in gallium sulfo-selenide
solid solutions
  R. H. Bube and E. L. Lind
  Phys. Rev., 119:1535 (1960)

Effects of solid solution of $Ga_2Te_3$ with $A^{II}B^{VI}$
tellurides
  J. C. Woolley and B. Ray
  J. Phys. Chem. Solids, 16:102 (1960)

Photoconductivity of gallium selenide crystals
  R. H. Bube and E. L. Lind
  Phys. Rev., 115:1159 (1959)

Semiconductors of the type $A^{III}B^{VI}$
  P. Fielding, G. Fischer, and E. Mooser
  Advances in Semi-Conductor Science: Proceedings of the
    Third International Conf. on Semi-Conductors held at
    Univ. of Rochester, August 18-22, 1958, Pergamon Press,
    New York (1959)

The sensitivity of the structure of gallium tel-
luride to small amounts of added copper
  G. Harbeke and G. Lautz
  Z. Naturforschung, 13a:771-775 (1958)

Standard x-ray diagrams of several selenides,
tellurides, arsenides, and sulfides of copper,
silver, zinc, cadmium, gallium, and indium
  V. A. Kotovich and V. A. Frank-Kamenetskii
  Uchenye Zapiski Leningradskogo Gosudarstvennogo Ordena
    Lenina Universiteta imeni A. A. Zhdanova, Seriya Geo-
    logicheskikh Nauk, 8:135-156 (1957)

Temperature dependence of the spectral distribution of photoconductivity in gallium selenide
and gallium telluride
  S. M. Ryvkin and R. Yu. Khansevarov
  Sov. Phys. — Tech. Phys., 1:2688-2691 (1957)

Crystal structures of $Ga_2S_3$, $Ga_2Se_3$, and $Ga_2Te_3$
  H. Hahn and W. Klingler
  TT-65-13054. Z. Anorg. Allgem. Chem., 259:135-142 (1949)

## 11.d. Indium Compounds

Die Dampfdruckkurve von $In_2S_3$
  G. Bollmann and H. Nelkowski
  Z. Naturforsch., 25a:301-302 (1970)

Refractive index of thin monocrystal films of
InSe
  B. Celustka, A. Persin, and D. Bidjin
  J. Appl. Phys., 41(2):813-814 (1970)

A method for the analysis of the optical absorp-
tion edge of semiconductors and application to
the absorption in $In_2Te_3$
  V. M. Koshkin, V. R. Karas', and L. P. Gal'chinetskii
  Fiz. Tekhn. Poluprovod., 3(9):1417 (1970)
  Sov. Phys. — Semicond., 3(9):1186-1188 (1970)

Preparation and electrical properties of InSe
  A. K. Sreedhar, B. L. Sharma, and R. K. Purohit
  Radiat. Effect (July 1970), pp. 121-122
  International Conference on Non-Metallic Crystals, New Delhi,
    India, Jan. 1969
  Vapor growth

Photoluminescence ($\lambda_m$ = 1.1 $\mu m$) in InSe single
crystals and its relation to photoconductivity
  G. A. Akhundov, I. B. Ermolovich, F. N. Kaziev, et al.
  Phys. Stat. Sol., 35:1065-1068 (1969)

Variation in microhardness during doping and
the solubility of impurities in indium telluride
  L. V. Atroshchenko and V. M. Koshkin
  Izv. Akad. Nauk SSSR, Neorg. Mater., 5(2):265-269 (1969)
  Inorg. Mater., 5(2):221-224 (1969)

Einige Tieftemperatur—Umwandlungen in metal-
lischen Phasen
  K. Burkhardt and K. Schubert
  Z. Metallkunde, 12:929-932 (1969)

Semiconductor properties of tellurides. 13.
Growth and investigations on indium telluride
single crystals
  D. Janowski
  Ann. Phys. (Germany), 23:71-79 (1969)

Mischkristallbildung der festen Lösung
$In_2Te_{3-x}Se_x (0 \leq x \leq 3)$
    P. Myohl and H. A. Ullner
    Ann. Phys., 23:113-128 (1969)

Negative resistivity of $In_2Se$ crystals
    A. D. Ogorodnik, I. M. Stakhira, and K. D. Tovstyuk
    Fiz. Tekhn. Poluprovod., 3(6):717-719 (1969)
    Sov. Phys. – Semicond., 3(6):847-850 (1969)

Thermal and electric properties in the indium
sulfide – indium selenide system
    P. G. Rustamov, M. A. Alidzhanov, and Z. D. Melikova
    Izv. Akad. Nauk SSSR, Neorg. Mater., 5(5):964-965 (1969)
    Inorg. Mater., 5(5):820-821 (1969)

Halleffekt und Leitfahigkeit von Kristallen des
Systems $In_2Te_3 - Sb_2Te_3$
    W. Schulz and H. Nieke
    Ann. Phys., 7 Folge, Bd. 23:129-138 (1969)

Heat conductivity of indium selenide
    S. A. Atakishiev, D. Sh. Abdinov, and G. A. Akhundov
    Phys. Stat. Sol., 28:K47 (1968)

Recrystallization of $In_3Te_4$ in thin indium-tel-
lurium films consisting of a mixture of $In_2Te_3$
and $In_3Te_4$
    A. Deksnis and V. Tolutis
    Lietuvos Fiz. Rinkinys, 8(5/6):911-915 (1968)

Quantitative phase analysis of the indium – tellurium system in the 50 to 57 atom percent
tellurium region
    A. Deksnis and V. Tolutis
    Tonkie Plenki Ikh Primen. (1968), pp. 17-20

$In_3Te_4$ formation process in a polyphase indium-
tellurium layer
    A. Deksnis, D. Sakalauskaite, and V. Tolutis
    Lietuvos Fiz. Rinkinys, 8(5/6):917-931 (1968)

Electrical properties of indium telluride – indium
selenide system solid solutions
    O. P. Derid, E. I. Gavrilitsa, and S. I. Radautsan
    Issled. Poluprov. (1968), pp. 22-27

Electron-microscope observations of twinning
and phase transformations in indium sulfide
crystals
    A. G. Fitzgerald and G. Thomas
    Phys. Stat. Sol., 25:263-271 (1968)

Some physical properties of the indium telluride – indium selenide system
    P. G. Rustamov, M. A. Alidzhanov, and F. D. Mamedaliev
    Izv. Akad. Nauk SSSR, Neorg. Mater., 4(2):297-298 (1968)
    Inorg. Mater., 4(2):247-248 (1968)

Production of monocrystals of solid solutions
in the $Ga_2Te_3 - GaSe$ system
    P. G. Rustamov, V. B. Cherstvova, and M. A. Alidzhanov
    Izv. Akad. Nauk SSSR, Neorg. Mater., 4(8):1351-1352 (1968)
    Inorg. Mater., 4(8):1186-1187 (1968)

Growth of $In_2Se$ monocrystals by Czochralski's
method
    I. M. Stakhira
    Rost Kristallov, Izd. Akad. Nauk SSSR (1965), pp. 284-287
    Growth of Crystals, Vol. 6B (A. V. Shubnikov and N. N.
       Sheftal', eds.) Consultants Bureau, New York (1968),
       pp. 93-95

A combined study of the physical properties of
thin films of the In – Te system in the range
from 57 to 60 at.% tellurium. I. Dependence
of phase composition on proportions of the components
    V. B. Tolutis and A. P. Deksnis
    Litov. Fiz. Sbornik, 8:261-271 (1968)

A combined study of the physical properties of
thin films of the In – Te system in the range
from 57 to 60 at.% tellurium. II. Investigation
of the dependence of the electrical and photoelectric properties of films on proportions of
the components
    V. B. Tolutis, A. P. Deksnis, I. Yu. Verkelis, and D. A.
       Sakalauskaite
    Litov. Fiz. Sbornik, 8:273-284 (1968)

Phase composition and physical properties of
thin indium – tellurium layers containing 57-60
percent tellurium
    I. Verkelis, A. Deksnis, D. Sakalauskaite, and V. Tolutis
    Tonkie Plenki Ikh Primen. (1968), pp. 7-11

Infrared absorption of $A^{III} B^{VI}$ monocrystals
    G. A. Akhundov and T. G. Kerimova
    Optika i Spektrosk., 22:654-655 (1967)
    Optics and Spectrosc., 22:355-356 (1967)

Low-temperature photochemical reactions in
$In_4S_5$ single crystals
N. E. Korsunskaya, N. N. Lebedeva, and M. K. Sheinkman
    Fiz. Tverd. Tela, 8:3196-3200 (1966)
    Sov. Phys. – Solid State, 8:2558 (1967)

A study of the spectral distribution of photoelectric response of $In_4S_5$ single crystals
    V. A. Lyubchenko et al.
    Ukrain. Phys. J., 12(3):497-500 (1967)

Positron annihilation in semi-conducting indium
telluride
    T. A. Murphy and M. K. Ramaswamy
    Phys. Letters, 25A:379 (1967)

Photoelectric properties of indium selenide
single crystals
    M. Kh. Alieva, F. N. Kaziev, and G. A. Akhundov
    Slozhnye Poluprov., Akad, Nauk Azerb. SSR, Inst. Fiz.,
       (1966), pp. 43-46

Thermodynamic aspects of the temperature-
pressure phase diagram of InTe
    M. D. Banus and P. M. Robinson
    J. Appl. Phys., 37:3771-3774 (1966)

Specific heat and entropy of indium monoselenide at low temperatures
    K. K. Mamedov, I. G. Kerimov, V. N. Kostryukov, and G. D.
       Guseinov
    Chemical Bonds in Semiconductors and Thermodynamics,
       Nauka i Tekhnika Minsk, (1966), pp. 179-182; Consultants
       Bureau, New York (1968), pp. 132-134

Crystal structures of $\alpha$- and $\beta$-indium selenide,
$In_2Se_3$
    K. Osamura, Y. Murakami, and Y. Tomiie
    J. Phys. Soc. Japan, 21: 1848 (1966)

On the thermodynamic properties of the tellurides of cadmium, indium, tin, and lead
P. M. Robinson
Trans. AIME, 236:814-817 (1966)

Energy structure of the bands of certain compounds of the $A^{II}B^V$, $A^V B^{VI}$, and $A^{III} B^{VI}$ types
V. V. Sobolev, N. N. Syrbu, and S. D. Shutov
Chemical Bonds in Semiconductors and Thermodynamics, Nauka i Tekhnika, Minsk (1966), pp. 221-228; Consultants Bureau, New York (1968), pp. 165-170

Thermal conductivity of indium selenide
S. M. Atakishiev, B. T. Mirzoev, and M. G. Aliev
Uch. Zap. Azerb. Gos. Univ., Ser. Fiz.-Mat., 6:59-61 (1965)

The growth of indium (2) selenide single crystals
T. N. Guliev and Z. S. Medvedeva
Izv. Akad. Nauk SSSR, Neorg. Mater., 1(6):845-846 (1965)
Inorg. Mater., 1(6):777-778 (1965)

The compound $In_5Se_6$
T. N. Guliev and Z. S. Medvedeva
Zh. Neorg. Khim. SSSR, 10:1520-1523 (1965)

Use of a method of chemical transport reactions for obtaining indium selenide single crystals
Z. S. Medvedeva and T. N. Guliev
Materialy Dokl. Nauchn.-Tekhn. Konf. Kishinevsk. Politekhn. Inst. 1st, Sb., Kishinev (1965), pp. 70-71

Growing single crystals of indium selenides from the vapor phase
Z. S. Medvedeva and T. N. Guliev
Izv. Akad. Nauk SSSR, Neorg. Mater., 1(6):848-852 (1965)
Inorg. Mater., 1(6):779-782 (1965)

Infrared absorption of indium selenide single crystals
E. L. Zorina, V. B. Velichkova, and T. N. Guliev
Izv. Akad. Nauk SSSR, Neorg. Mater., 1(5):690-691 (1965)
Inorg. Mater., 1(5):633-634 (1965)

Semiconductors $A^{III} B^{VI}$
G. A. Akhundov, G. B. Abdullaev, G. D. Guseinov, R. F. Mekhtiev, M. Kh. Alieva, E. S. Guseinova, and I. A. Gasanova
Izv. Akad. Nauk Azerb. SSR, Ser. Fiz.-Mat. Tekhn. Nauk, 3:107-114 (1964)

Preparation and investigation of $A^{III} B^{VI}$ single crystals
G. A. Akhundov, G. B. Abdullayev, G. D. Guseinov, R. F. Meckhtiev, and M. K. Aliyeva
Intern. Congr. Phys. Semiconducteurs, Paris (1964), pp. 90-91

Artificial metals: InSb, the Sn alloys with InSb, and metallic InTe
A. J. Darnell and W. F. Libby
Phys. Rev., 135:A1453 (1964)

Elektrische Eigenschaften von $In_2Te_3$ und festen $In_2Te_3 - In_2Se_3$-Lösungen
H.-G. Forner
Wiss. Z. Martin-Luther-Univ. Halle-Wittenberg, math.-naturwissensch. Reihe, 13:731-743 (1964)

Defect diamond-like semiconductors
N. A. Goryunova and S. I. Radautsan
Investigations on Semiconductors. New Semiconductor Materials, Izd. Kartya Moldovenyaske, Kishinev (1964), pp. 3-43

The phase diagram for the binary system indium — tellurium and electrical properties of $In_3Te_5$
E. G. Grochowski, D. R. Mason, G. A. Schmitt, and P. H. Smith
J. Phys. Chem. Solids, 25:551-558 (1964)

Superconductivity in the artificial metals: metallic indium antimonide, the indium — antimonide — tin alloys, and metallic indium telluride
B. R. Tittmann, A. J. Darnell, H. E. Bommel, and W. F. Libby
Phys. Rev., 135:A1460 (1964)

High-pressure transitions in $A^{(III)}B^{(VI)}$ compounds: indium telluride
M. D. Banus, R. E. Hanneman, M. Strongin, and K. Gooen
Science, 142:662-663 (1963)

Superconductivity of metallic indium telluride
H. E. Bommel, A. J. Darnell, W. F. Libby, B. R. Tittmann, and A. J. Yensha
Science, 141:714 (1963)

Indium telluride metal
A. J. Darnell, A. J. Yensha, and W. F. Libby
Science, 141:713-714 (1963)

The vapor pressure of indium sulfides as functions of composition and temperature
A. R. Miller
(Ph.D. thesis, Univ. California, Berkeley), UCRL-10857 (Oct. 1963)

Some physical properties of indium — selenium and selenium — arsenic alloy systems
G. K. Slavnova
Izv. Akad. Nauk, Moldavsk. SSR, 7:43-52 (1963)

Constitution diagram of In — Se alloys
G. K. Slavnova, N. P. Luzhnaya, and Z. S. Medvedeva
Zh. Neorg. Khim., 8:153-159 (1963)

On the order — disorder transformation in $In_2S_3$
M. Huber
Compt. Rend., 253:471-473 (1961)

The crystal structure of indium selenide $In_2Se_3$
S. A. Semiletov
Dokl. Akad. Nauk SSSR, 137:584-587 (1961)

Concerning the effects of the selection of suitable constituents on the structure of ternary and quaternary phases. III. The $In_2Se_3 - InP$, $In_2Se_3 - InAs$, $In_2Te_3 - InAs$, $InSe - InAs$ and $InTe - InAs$ systems
Harry Hahn and Dietrich Thiele
Z. Anorg. Allgem. Chem., 303:147-154 (1960)

Some electrical and optical properties of In-Se alloys of variable composition
V. P. Mushinskii
Izv. Vysshikh Uchebn. Zaveden., Fiz., 6:130-134 (1960)

Polymorphism of $In_2Te_3$
A. I. Zaslavskii and V. M. Sergeeva
Fiz. Tverd. Tela, 2(11):2872-2880 (1960)
Sov. Phys.—Solid State, 2(11):2556-2561 (1960)

Semiconductors of the type $A^{III} B^{VI}$
P. Fielding, G. Fischer, and E. Mooser
J. Phys. Chem. Solids, 8:434-437, 442-443 (1959)

Also in: Advances in Semi-Conductor Science. Proceedings of Third International Conf. on Semi-Conductors held at Univ. Rochester, August 18-22, 1958, Pergamon Press, New York (1959), pp. 434-447

Homogenization of alloys of the InAs — In$_2$Se$_3$ system by annealing under pressure
N. A. Goryunova, S. I. Radautsan, and V. I. Deryabina
Fiz. Tverd. Tela, 1(3):460-462 (1959)
Sov. Phys.—Solid State, 1(3):512-514 (1959)

Properties of various semiconductors
A. Joffe
Advances in Semi-Conductor Science. Proceedings of the Third International Conf. on Semi-Conductors held at Univ. Rochester, August 18-22, 1958, Pergamon Press, N. Y. (1959), pp. 6-14

Investigation of the section of the InAs — In$_2$Se$_3$ in the system In — As — Se
S. I. Radautsan
Zh. Neorg. Khim., 4:1121-1124 (1959)

The indium — selenium system
J. C. Brice, P. C. Newman, and H. C. Wright
Brit. J. Phys., 9:110-111 (1958)

Heat conductivity of indium and gallium arsenides and indium selenide and tellurides
N. N. Sirota and L. I. Berger
Inzhenerno-Fizicheskii Zhurnal, 11:117-120 (1958)

Standard x-ray diagrams of several selenides, tellurides, arsenides, and sulfides of copper, silver, zinc, cadmium, gallium, and indium
V. A. Kotovich and V. A. Frank-Kamenetskii
Uch. Zap. Leningradskogo Gosudarstvennogo Ordena Lenina Univ. imeni A. A. Zhdanova, Ser. Geologicheskikh Nauk, 8:135-156 (1957)

Phase transition in In$_2$Se$_3$
Hisao Miyazawa and Suezo Sugaike
J. Phys. Soc. Japan, 12:312 (1957)

## 11.e. Thallium Compounds

Kristallstruktur von Tl$_5$Te$_3$ und Tl$_2$Te$_3$
S. Bhan and K. Schubert
J. Less-Common Metals, 20:229-235 (1970)

Low-temperature electronic transport properties of Tl$_5$Te$_3$
E. Cruceanu, R. Luck, and H. Schwarz
J. Appl. Phys., 41:5223-5226 (1970)

Superconductivity in TlBiTe$_2$: A low carrier density (III-V) VI$_2$ compound
R. A. Hein and E. M. Swiggard
Phys. Rev. Letters, 24:53-55 (1970)

Far infrared reflectivity of Tl$_2$Se—As$_2$(Se$_x$Te$_{1-x}$)$_3$ glasses
P. C. Taylor and S. G. Bishop
Bull. Am. Phys. Soc., 15:290 (1970)

Über die $\gamma$-Phase im System Thallium-Tellur
Eugen Cruceanu
Z. Metallkunde, 60:852 (1969)

Some electrical transport studies on compounds of the Tl-Te system
E. Cruceanu and St. Sladaru
J. Mater. Sci., 4:410-415 (1969)

On some properties of TlInS$_2$(Se$_2$, Te$_2$) single crystals
G. D. Guseinov, E. Mooser, E. M. Kerimova, R. S. Gamidov, I. V. Alekseev, and M. Z. Ismailov
Phys. Stat. Sol., 34:33-44 (1969)
The structure, electric conductivity, Hall effect, photoconductivity

Electrical and optical properties of TlS, TlS$_{0.5}$Se$_{0.5}$, and TlSe
C. R. Kannewurf and R. S. Itoga
Ph. D. thesis research (completed), Northwestern University, Materials Research Center, 1969

Superconductivity in the $\gamma$-phase of the Tl-Te system
C. R. Kannewurf and A. G. Juodakis
Ph. D. thesis research (in progress), Northwestern University, Materials Research Center, 1969

Studies on thin polycrystalline layers of thallium selenide
Mary Juliana Mangalam, K. Nagaraja Rao, N. Rangarajan, M. I. A. Siddiqi, and C. V. Suryanarayana
Japan. J. Appl. Phys., 8:1258 (1969)

Segregation of Se, S, Tl, and Fe during the growth of single crystals CdSb and Tl$_2$Te$_3$ from the melt
S. Mavlonov
Acta Cryst., 21, Pt. 7, Suppl., A289 (1966)
Rost Kristallov, Izd. Akad. Nauk SSSR (1967), pp. 165-170
Growth of Crystals, Vol. 7 (A. V. Shubnikov and N. N. Shelftal, eds.) Consultants Bureau, New York (1969), pp. 139-144

Anisotropy of electric properties of thallium telluride monocrystals
S. Mavlonov, S. Karimov, and V. M. Glazov
Izv. Akad. Nauk SSSR, Neorg. Mater., 5(9):1648-1649 (1969)
Inorg. Mater., 5(9):1396-1397 (1969)

Thermodynamic properties of the lower selenide of thallium (Tl$_2$Se)
V. P. Vasil'ev, A. V. Nikol'skaya, and Ya. I. Gerasimov
Dokl. Akad. Nauk SSSR, 188(6):1318-1320 (1969)

(Vapor-condensed) semiconducting Tl$_2$O and Tl$_4$O$_3$ single crystals
V. P. Witte, K. Langer, F. Seifert, et al.
Naturwissenschaften, 56:414 (1969)

Optical energy gap of Tl$_2$Te$_3$
E. Cruceanu, St. Sladaru, and T. Botila
Phys. Stat. Sol., 30:K149 (1968)

Electrical conductivity and superconductivity in the $\gamma$ phase of the Tl-Te system
A. Juodakis and C. R. Kannewurf
J. Appl. Phys., 39:3003 (1968)

Metallographic and x-ray diffraction study of thallium telluride single crystals
S. Karimov, Kh. Kurbanov, and Sh. Mavlonov
Izv. Akad. Nauk Tadzh. SSR, Otd. Fiz.-Mat. Geol.-Khim. Nauk, 2:12-17 (1968)

Segregation of iron, silver, and tin in the
growth of single crystals of $Tl_2Te_3$ from a melt
by the Czochralski method
   Sh. Mavlonov and S. Karimov
   Izv. Akad. Nauk SSSR, Neorg. Mater., 4(8):1216-1219 (1968)
   Inorg. Mater., 4(8):1069-1071 (1968)

Electrophysical properties of thallium telluride
single crystals
   Sh. Mavlonov, S. Karimov, and V. M. Glazov
   Dokl. Akad. Nauk Tadzh. SSR, 11(4):24-28 (1968)

Optical energy gap in TlSe
   P. B. Pickar and H. D. Tiller
   Phys. Stat. Sol., 29:153 (1968)

Preparation of selenium samples in an ultra-
sonic field and the growing of thallium selenide
single crystals
   K. I. Rzaev and Kh. M. Khalilov
   Akust. Zh., 13(3):427-431 (1967)
   Sov. Phys. — Acoustics, 13(3):360-364 (1968)

Effect of etching thallium selenide crystals in
a field of ultrasonic waves
   G. D. Guseinov, K. I. Rzaev, and M. Z. Ismailov
   Slozhnye Poluprov., Akad. Nauk Azerb. SSR, Inst. Fiz. (1966),
      pp. 122-127

Some properties of single crystals of thallium
selenide
   G. A. Akhundov, G. B. Abdullaev, and G. D. Guseinov
   Fiz. Tverd. Tela, 2(7):1518-1521 (1960)
   Sov. Phys. — Solid State, 2(7):1378-1380 (1961)

Investigations in the tellurium — thallium sys-
tem
   A. Rabenau, A. Stegherr, and P. Eckerlin
   Z. Metallkunde, 51:295-299 (1960)

Vitreous semiconductors
   N. A. Goryunova and B. T. Kolomiets
   Voprosy Met. i Fiz. Poluprovodnikov, Akad. Nauk, SSSR,
      (1957), pp. 110-120

# 12. Germanium and Silicon

## 12.a. Germanium and Silicon

### 12.a.1. Purification

Large diameter germanium crystals for gamma-ray spectrometry
  G. Dearnaley, R. Ellis, and P. E. Gibbons
  Atomic Energy Research Establishment, Harwell, England,
    AERE-R 6345 (March 1970)

Compensation and purification of Si and Ge
  Ramesh Chaudhry
  Bhabha Atomic Research Centre (Bombay, India), BARC-443
    (1969), 24 pp.

Si – Ge isotype heterojunctions
  C. J. M. van Opdorp
  Ph. D. thesis, Technische Hogeschool, Eindhoven, Nether-
    lands, Oct. 1969, 107 pp.
  Includes critical survey of models for energy-band diagrams
    and transport mechanisms for iso- and aniso-types;
    preparation and measurement methods

Redistribution of impurities in the solid phase
during zone melting with a temperature gradient
  V. N. Lozovskii and E. A. Nikolaeva
  Izv. Akad. Nauk SSSR, Neorg. Mater., 4(7):1021-1026 (1968)
  Inorg. Mater., 4(7): 899-903 (1968)
  Includes Ge and Si

Use of transport reactions in the preparation
of high-purity materials
  A. V. Novoselova
  Zh. Vses. Khim. Obshchestva im. D. I. Mendeleeva, 13(5):
    539-542 (1968)
  Van Arkel

Zone refining of semiconductor crystals
  L. R. Crosby and H. M. Stewart
  British Patent 995,399, June 16, 1965
  Si and Ge

Materials Used in Semiconductor Devices
  C. A. Hogarth, ed.
  Interscience Publishers, New York, London, 1965
  Chapter 2: Germanium, pp. 3-28; chapter 3: Silicon, pp.
    29-48

Advances in elemental semiconductors
  C. G. Currin and E. Earleywine
  Semicond. Prod., 7:20-25 (1964)
  Ge and Si; table of properties; crystal growth

Production and properties of very pure silicon
and germanium
  G. R. Davies
  Met. Rev. 10(38):173-221 (1964)
  Review of purification and crystal growth; 341 refs.

A literature survey on the purification of elec-
tronic materials
  A. F. Armington, G. F. Dillon, and R. F. Mitchell
  Air Force Cambridge Research Labs., Hanscom Field,
    Mass., AFCRL-63-160; AD-412458 (June 1963)

The Metallurgy of Semiconductors
  Yu. V. Shashkov
  Metallurgizdat, Moscow (1960), 212 pp.
  Consultants Bureau, New York (1961), 183 pp.
  Physicochemical, electrical, and optical parameters;
    purifying; growing monocrystals; diffusion of impurities;
    doping; preparation of ohmic contacts; and etching

### 12.a.2. Melt Growth

Producing germanium and silicon crystal strips
by Stepanov method using various shapes
  L. M. Zatulovskii, P. M. Chaikin, and L. F. Nikolskii
  Izv. Akad. Nauk SSSR, Ser. Fiz., 33:1998-2000 (1969)

Inhomogeneities in doped germanium and silicon
crystals
  J. A. M. Dikhoff
  Philips Tech. Rev., 8:195-206 (1963/64)
  Czochralski growth

Constitutional supercooling during the crystal
growth of germanium and silicon
  H. Kodera
  Japan. J. Appl. Phys., 2:527-534 (1963)

Problems associated with distribution coefficient and solid solubility determinations using
crystal growth techniques
  W. Bardsley, D. T. J. Hurle, and J. B. Mullin
  J. Electrochem. Soc., 109:64-65 (1962)

Excess impurity trapping during crystal growth
A. A. Chernov
Rost Kristallov, Vol. 3, Izd. Akad. Nauk (1961), pp. 52-58
Growth of Crystals, Vol. 3 (A. V. Shubnikov and N. N. Sheftal',
eds) Consultants Bureau, New York (1962), pp. 35-39
Ge and Si

Growth of silicon bicrystals by the dash pedes-
tal-method
R. Gereth
J. Electrochem. Soc., 109:1068-1070 (1962)

The determinative part played by supercooling
of the melt in the formation of helical macroin-
homogeneities in crystals grown by the Czochralski method
D. A. Petrov and A. A. Bukhanova
Kristallografiya, 7(3):442-445 (1962)
Sov. Phys.—Cryst., 7(3):349-353 (1962)

Origin of radial heterogeneity in Ge and Si crys-
tals
D. A. Petrov, T. A. Rusakov, and S. K. Yacheva
Russ. Met. Fuels, 5:112 (1962)
Trans. of Izv. Akad. Nauk SSSR, Met. i Toplivo 5, 187-190
(1962)
Crucible rotation

The formation of faces on germanium and silicon
crystals grown by the Czochralski method
D. A. Petrov, T. A. Rusakov, and S. K. Yacheva
Dokl. Akad. Nauk SSSR, Tech. Fiz., 146(3):588-591 (1962)
Sov. Phys.- Doklady, 7(9):841-843 (1963)

The travelling solvent method of crystal growth
M. Weinstein, S. D. Axelrod, and A. I. Mlavsky
(Tyco Laboratories, Inc.) Qtr. Rept., No. 5 (1962)

Metallurgy of Elemental and Compound Semicon-
ductors
Vol. 12, Proceedings of Technical Conference, Boston,
Mass., Aug. 29-31, 1960
Ralph O. Grubel, ed.
Interscience Publishers, New York, London (1961)
Ge and Si growth, pp. 12-298

Apparatus for the growing of Ge and Si single
crystals
B. Kofoed and B. V. Jensen
Ingenioren, 70:145-150 (1961)
Resistance heated crucible; Czochralski

A survey of semiconductor materials technology
J. Myer
IRE Trans. Component Parts, CP-8, pp. 65-69 (1961)

The use of thermoelectric effects during crystal
growth
J. R. O'Connor
J. Electrochem. Soc., 108:713-715 (1961)

Supercooling of the melt and the growth of
crystals by the Czochralski method
D. A. Petrov and A. A. Bukhanova
Dokl. Akad. Nauk SSSR, Fiz.-Khim., 139:593-596 (1961)
Growth rate, pulling rate and thermal gradients; Ge, Si, Cu,
Fe, and Al

A modified rf coil to facilitate floating zone
techniques
S. J. Silverman
J. Electrochem. Soc., 108:585-588 (1961)

Crystal growth
W. A. Tiller
Thermoelectricity—Science and Engineering, Interscience
Publishers, New York (1961) pp. 181-231

Drawing of single crystals in vacuo
H. Bumen
Advances in Vacuum Science and Technology, Vol. 2, Per-
gamon Press, (1960), pp. 698-700
In German

Improvements on the pedestal method of growing
silicon and germanium crystals
W. C. Dash
J. Appl. Phys., 31:736-737 (1960)
Pedestal method; without a crucible

Methods of growing single crystals of germanium
and silicon
D. A. Petrov
Zh. Vses. Khim. Obshchestva, 5:544-552 (1960)
Czochralski

Resistivity calculations for semiconductor crys-
tals grown by the Czochralski technique
B. Pratt and M. Y. Ben-Sira
Res. Council of Israel Bull., 8C:103-108 (1960)

The Metallurgy of Semiconductors
Yu. M. Shashkov
Metallurgizdat, Moscow (1960), 212 pp.
Consultants Bureau, New York (1961), 183 pp.
Physicochemical, electrical, and optical parameters;
purifying; growing monocrystals; diffusion of impurities;
doping; preparation of ohmic contacts; and etching

Growth of large diameter silicon and germanium
crystals by the Teal — Little method
W. R. Runyan
Rev. Sci. Instr., 30(7):535-540 (1959)
6-in. diam.; modification of the Czochralski process

Temperature-gradient zone melting
W. G. Pfann
Zone Melting, John Wiley and Sons, New York, (1958), pp.
198-208

### 12.a.3. Epitaxy, Films, Vapor Deposition

Selective epitaxy using silane and germane
D. J. Dumin
J. Cryst. Growth, 8:33-36 (1971)

Gallium-doped epitaxial silicon
P. Rai-Choudhury
J. Cryst. Growth, 8:165-171 (1971)

Study of hetero-epitaxial films of germanium. I.
Mechanism of growth and crystal structure
A. P. Klimenko et al.
Ukr. Fiz. Zh., 15(5):804-811 (1970)

The mechanism of transport reactions and the
growth of Si and Ge on one another in the Ge —
Si — I system
V. F. Dorfman and I. D. Khan
Izv. Akad. Nauk SSSR, Neorg. Mater., 5(10):1670-1673
(1969)
Inorg. Mater., 5(10):1414-1416 (1969)

Single-crystal layers of germanium and silicon, prepared by pyrolysis of hydrides
   T. A. Zeveke, L. N. Kornev, and V. A. Tolornasov
   Kristallografiya, 12(6):1058-1061 (1967)
   Soviet Phys. – Cryst. 12(6):919 (1968)

A survey of epitaxial growth processes and equipment
   V. Y. Doo and E. O. Ernst
   IBM TR 22.431 (July 14, 1967)

Epitaxial growth of silicon and germanium (I)
   C. H. Li
   Phys. Stat. Sol., 15:3-56 (1966)
   Review, 121 refs

Epitaxial growth of silicon and germanium (II)
   C. H. Li
   Phys. Stat. Sol., 15:419-450 (1966)
   Review, 246 refs.

Vacuum preparation of epitaxial silicon (or germanium) films
   V. V. Postnikov, R. G. Loginova, and M. I. Ovsyannikov
   Kristallografiya, 10(4):585-586 (1966)
   Sov. Phys.–Cryst., 10(4):585-586 (1966)
   Effect of O contamination; single crystals

Growth forms of crystals of germanium and silicon grown from gaseous solution
   A. V. Sandulova, A. I. Andrievskii, and M. I. Dronyuk
   Rost Kristallov, Vol. 4, Nauka (1964), pp. 122-124
   Growth of Crystals, Vol. 4 (A. V. Shubnikov and N. N. Sheftal', eds.) Consultants Bureau, New York (1966), pp. 98-100

Single-crystal growth on $\alpha$-$Al_2O_3$ substrate
   Y. Tsunoda
   J. Phys. Soc. Japan, 21:2416-2417 (1966)
   Si and Ge

Recrystallization of thin films of germanium and silicon
   J. D. Filby and S. Nielsen
   J. Electrochem. Soc., 112:534-535 (1965)

A review of the growth and structure of thin films of germanium and silicon
   R. C. Newman
   Microelectron. Reliabil., 3:121-138 (1964)
   Vacuum sublimation, iodide disproportionation reactions, and hydrogen reduction

Infrared monitoring of semiconductor epitaxial process gas streams
   M. J. Rand
   Anal. Chem., 36:1112-1114 (1964)

Preparation of germanium and silicon single crystals from the gaseous phase by the use of a second element
   A. V. Sandulova, P. S. Bogoyavlenskii, and M. I. Dronyuk
   Dokl. Akad. Nauk SSSR, 153(1):82-85 (1963)
   Sov. Phys.–Doklady, 8(11):1112-1114 (1964)

Formula for epitaxial growth of single crystal semi-conductors
   Hidao Kaneko
   Bull. Japan Institute of Metals, 2(1):1-6 (1963)

The effects of an electric field on epitaxial vapor growth
   Y. Tarui, H. Teshima, K. Okura, and A. Minamiya
   J. Electrochem. Soc., 110:1167-1169 (1963)

Reaction kinetics of epitaxial growth
   J. J. Grossman
   Paper No. 105, Electrochemical Society Meeting, Los Angeles, May 6-10 1962
   Kinetics of vapor growth of Si and Ge epitaxial layers

Evaporation of silicon and germanium by rf levitation,
   E. A. Roth, E. A. Margerum, and J. A. Amick
   Rev. Sci. Instr., 33(6):686-687 (1962)

Mutual overgrowth of crystals of silicon and germanium
   N. N. Sheftal' and N. P. Kokorish
   Rost Kristallov, Vol. 3, Izd. Akad. Nauk (1961), pp. 363-370
   Growth of Crystals, Vol. 3 (A. V. Shubnikov and N. N. Sheftal', eds.) Consultants Bureau, New York (1962), pp. 259-263

Preparation of thin films of germanium and silicon
   B. A. Irving
   British J. Appl. Phys., 12:92 (1961)
   Combination of mechanical polishing and chemical etching

Epitaxial techniques in semiconductor devices,
   John Sigler and S. B. Watelski
   Solid State Journal, March, 1961, pp. 33-37

Reduction of chlorides of silicon and germanium and formation of crystalline films
   A. I. Melnikov
   GDA-AE60-0003-20; Ad-677092 (Oct. 1960), 12 pp.
   Transl. of Zh. Neorgan. Khim., 2:233-237 (1957), by. W. R. Eichler

### 12.a.4. Dendrites, Needles, and Whiskers

Threadlike Crystals
   G. V. Berezhkova
   Izd. Nauka, Moscow (1969), 158 pp.
   718 refs.

On dendrite filaments pulled from the melt
   G. F. Bolling and W. A. Tiller
   Can. Met. Quarterly, 8:115-118 (1969)

Growth and properties of ribbonlike dendrites of semiconductor materials
   L. P. Kalnach
   Rost Kristallov, Vol. 6, Nauka (1965), pp. 365-373
   Growth of Crystals, Vol. 6B (N. N. Sheftal', ed.) Consultants Bureau, New York (1968), pp. 170-178

Preparation and some properties of whisker and needle-shaped single crystals of germanium, silicon, and their solid solutions
   A. V. Sandulova, P. S. Bogoyavlenskii, and M. I. Dronyuk
   Fiz. Tverd. Tela, 5(9):2580-2586 (1964)
   Sov. Phys.–Solid State, 5(9):1883-1888 (1964)
   Crystallization from gas phase with aid of solvent component

The influence of twin structure on growth directions in dendritic ribbons of materials having the diamond or zinc-blende structures
   N. Albon and A. E. Owen
   J. Phys. Chem. Solid, 24:899-907 (1963)

### 12.a.5. Doping, Diffusion, and Precipitation (see also 18b)

12th Scintillation and Semiconductor Counter Symposium, Washington, D. C., March 11-13, 1970
>   IEEE Trans. Nucl. Sci., NS-17, No. 3 (June 1970)
>   Pages 125 to 309 deal almost exclusively with Ge and Si Li-doped detectors, growth and properties

Purification of phosphorus trichloride
>   Masanori Nakane, Yoshizo Miyake, Hiroyasu Ichiyanagi, and Kyoji Fujiwara
>   Kogyo Kagaku Zasshi, 73:682-686 (1970)
>   For use as a dopant

Purification of phosphorus oxychloride
>   Masanori Nakane, Yoshizo Miyake, Hiroyasu Ichiyanagi and Kyoji Fugiwara
>   Kogyo Kagaku Zasshi, 73:883-886 (1970)
>   For use as doping agent

Comportement du lithium dans le germanium: conséquences sur la réalisation et les caracteristiques des détecteurs Ge(Li)
>   G. Lopes da Silva
>   Ph. D. thesis, Univ. Strasbourg (1969), 108 pp.

Doping of semiconductors and semiconducting film, Vol.II
>   Defense Documentation Center, Alexandria, Va., AD-853000 (May 1969)
>   Comulative; 271 refs.

Solubility of carbon in silicon and germanium
>   R. I. Scace and G. A. Slack
>   J. Chem. Phys., 30:1551-1555 (1969)

Methods of fabrication for germanium and silicon detectors with compensation by lithium-ion drifting
>   Bernard Bornand
>   Commissariat à l'Energie Atomique, CEA-Bib-90 (March 1968), 74 pp.

Thermodynamic properties of copper and gold in silicon and germanium
>   R. C. Dorward and J. S. Kirkaldy
>   Trans. AIME 242:2055-2061 (1968)

Physicochemical Principles of Semiconductor Doping
>   V. M. Glazov and V. S. Zemskov
>   (Ch. Nisenbaum and B. Benny, D. Slutzkin, transl. eds.) Israel Program for Scientific Translations, Jeresalem; Davey, Hartford, Conn. (1968), 380 pp.
>   Translation from the Russian

Diffusion of As and Bi in Ge and Si semiconductors
>   V. A. Panteleev and E. I. Akinkina
>   Zh. Fiz. Khim., 42:992-993 (1968)

Diffusion in Metallen und Halbleitern
>   Alfred Seeger
>   Festkörper Probleme VIII in Referaten des Fachausschusses "Halbleiter" der Deutschen Physikalischen Gesellschaft, Berlin (1968), pp. 264-267
>   (O. Madelung, ed.) Friedr. Vieweg and Sohn. Braunschweig, Germany; Pergamon Press, New York, London (1968)

Vacancies and diffusion mechanisms in diamond-structure semiconductors
>   A. Seeger and M. L. Swanson
>   Lattice Defects in Semiconductors (R. R. Hasiguti, ed.) University of Tokyo Press, Tokyo, Japan; The Pennsylvania State University Press, University Park and London (1968), pp. 93-130

The alkali metal doping of semiconductors
>   S. D. James
>   Mat. Res. Bull., 2:773-774 (1967)
>   Introducing appropriate cations into the surrounding vacuum system by controlled electrolysis of a fused salt

Semiconductor detectors for nuclear spectrometry, I
>   F. S. Goulding
>   Nucl. Instr. Methods, 43:1-54 (1966)
>   Li drifting of Si and Ge

Diffusion of lithium into Ge and Si
>   B. Pratt and F. Friedman
>   J. Appl. Phys., 37:1893 (1966)

Doping methods for the epitaxial growth of silicon and germanium layers
>   J. Goorissen and H. G. Bruijning
>   Philips Tech. Rev., 26:194-201 (1965)

Inhomogeneities in doped germanium and silicon crystals
>   J. A. M. Dikhoff
>   Philips Tech. Rev., 25(8):195-206 (1963-64)
>   Causes of formation and suggestions for prevention

Diffusion and solubility of copper in extrinsic and intrinsic germanium, silicon, and gallium arsenide
>   R. N. Hall and J. H. Racette
>   J. Appl. Phys., 35(2):379-397 (1964)

Imperfections and Active Centres in Semiconductors
>   R. G. Rhodes
>   Pergamon Press, New York, London (1964), 373 pp.
>   Includes crystal growth, diffusion, precipitation; characterization; solubilities of active impurities in Si and Ge; 557 refs.

The interaction of impurities with dislocations in silicon and germanium
>   R. Bullough and R. C. Newman
>   Progress in Semiconductors, Vol. 7 (A. F. Gibson and R. E. Burgess, eds.) John Wiley and Sons, New York, (1963)
>   Includes precipitation in Si with O, Al, C, Cu, Au, and other impurities; Li, Ni, and Cu in Co

Constitutional supercooling during the growth of germanium and silicon
>   H. Kodera
>   Japan. J. Appl. Phys., 2(9):527-534 (1963)
>   Limiting concentrations of Ga, In, P, As, Sb, Sn, and B; Czochralski or zone leveling

Diffusion of impurities in the semiconductor melt, II. Dynamical analysis of impurity redistribution in the melting process
>   H. Kodera and S. Tauchi
>   Japan. J. Appl. Phys., 2(4):220-226 (1963)

Diffusion of impurities in the semiconductor melt, III. Experimental determination of thickness of the solute diffusion layer in the melting process
>   H. Kodera, S. Iida, and S. Tauchi
>   Japan. J. Appl. Phys., 2(4):227-232 (1963)

Vapor doping of furnace atmospheres
  J. W. Savery
    Metallurgy of Advanced Electronic Materials, Proceedings
      Meeting, Philadelphia, Aug. 27-29, 1962, pp. 283-291
    Interscience, New York (1963)
    Si and Ge

Correlation between maximum solid solubility
and distribution coefficient for impurities in Ge
and Si
  Sidney Fischler
    J. Appl. Phys., 33:1615 (1962)

Thermodynamics of binary semiconductor-metal
alloys
  K. Lehovec
    J. Phys. Chem. Solids, 23:695-709 (1962)
    Ge and Si; impurity diffusion during preparation; doping

Thermal-neutron-induced recoil and transmutation effects semiconductors
  J. W. Cleland, R. F. Bass, and J. H. Crawford, Jr.
    (Oak Ridge National Lab., Oak Ridge, Tenn.), Solid State
      Division Annual Prog. Rept., ORNL-3213 (Aug. 1961),
      pp. 66-72

Segregation and distribution of impurities in the
preparation of germanium and silicon
  J. Goorissin
    Philips Tech. Rev., 21:185-196 (1960)
    Czochralski crystals; zone refining

Compilation of calculated data useful in predicting metallurgical behavior of the elements in binary alloy systems
  E. Teatum, K. Gschneidner, Jr., and J. Waber
    Los Alamos Scientific Lab., Los Alamos, N. M., LA-2345
    Physicometallurgical data have been computed for all the bi-
      nary combinations of elements in the periodic table with
      the exception of the halogens, rare gases, and those ele-
      ments having higher atomic numbers than that of amer-
      icium; radius ratio, sublimation energy ratio, Mott bond-
      ing number, Hildebrand or heat of mixing factor, and
      electronegativity difference; helpful in predicting whether
      or not solid solubility, liquid miscibility, or compound
      formation will occur for a given pair of elements

## 12.b. Germanium Compounds

### 12.b.1. General and Review

The Chemistry of Germanium
  Frank Glockling
    Academic Press, New York (1969), 236 pp.

Negative conductivity effects and related
phenomena in germanium. I.
  J. C. McGroddy, M. I. Nathan, and J. E. Smith, Jr.
    IBM J. Res. Dev., 13(5):543-553 (1969)

Negative conductivity effects and related phenomena in germanium. II.
  J. E. Smith, Jr., M. I. Nathan, and J. C. McGroddy
    IBM J. Res. Dev., 13(5):554-551 (1969)

Germanium
  A. F. Witt and H. C. Gatos
    The Encyclopedia of the Chemical Elements (C. A. Hampel,
      ed.) Reinhold Book Corp., New York, (1968), pp. 237-244

Optical constants of germanium
  R. E. LaVilla and H. Mendlowitz
    Appl. Optics, 6:61-68 (1967)

Germanium: Semiconductor properties
  Richard Dalven
    Infrared Phys., 6:129-143 (1966)

Optical constants of germanium
  W. L. Wolfe, J. A. Jenney, W. C. Levengood, T. Limperis,
    and D. M. Szeles
    Michigan University, AD-435012 (1964)

Band parameters of semiconductors with zinc-
blende, wurtzite, and germanium structure
  Manuel Cardona
    J. Phys. Chem. Solids, 24:1543-1555 (1963)

Optical constants of germanium in the extreme
ultraviolet region
  T. Sasaki
    J. Phys. Soc. Japan, 18(5):701-703 (1963)

### 12.b.2. Purification

Optical constants of germanium
  J. S. Nodvik
    J. Appl. Phys., 33(12):3568 (1962)

Zone refining with removal of ingot ends, de-
monstrated for germanium
  Klaus Hein, Eberhard Buhrig, and Klaus-Peter Becker
    Neue Hütte, 15(4):214-216 (1970)

High-purity germanium for gamma detectors
  R. N. Hall, R. D. Baertsch, and T. J. Soltys
    General Electric Research and Development Center, Schen-
      ectady, New York
    Contract AT(30-1) 3870: June 1967 through April 1968, An-
      nual report No. 1; S-68-1088 (May 1968)

Réduction de GeO$_2$, purification par fusion de
zone et croissance de monocristaux de germanium
  I. L. Muller
    Metalurgia, Bras., 23(120):825-828 (1967)

Reaction of germanium with germanium tetrachloride
  A. N. Kochubeev and L. A. Firsanova
    Dokl. Akad. Nauk SSSR, 171(6):1337-1340 (1966)
    Dokl. Chem., Proc. Acad. Sci. USSR, 171(6):1202-1205 (1966)

Zone refining of germanium
  B. Durkovic, D. Durkovic, and D. Nikolic
    Takhnika (Yugoslavia) 2:351-355 (1965)

The removal of copper and nickel from germanium by vacuum annealing with tin and chlorinated
hydrocarbons
  G. Conrad, J. Friedman, H. Meriwether, and C. Rosenblum
    J. Electrochem. Soc., 111:1260-1263 (1964)

High-purity germanium — surface conduction and
carrier life
  F. Aurich, H. P. Kleinknecht, A. Renz, K. Schuegraf, and
    K. Seiler
    ORNL-TR-120. Ann. Physik, Ser., 7(11):83-100 (1963)
    Purification and crystal growth

Purifying crystallizable semiconductor materials by zone melting
  Heinz Henker
    U. S. Patent 3,099,550 (1963)

Improvements in the purification of germanium by zone melting
P. Gondi and G. Scacciati
Nuovo Cimento, 21:829–833 (1961)

Extraction of copper and nickel from germanium
K. P. Tissen
Fiz. Tverd. Tela, 2(2):1001–1003 (1960)
Sov. Phys.—Solid State, 2(2):916–918 (1960)

The preparation of a radiochemically pure radioactive germanium
A. N. Baraboshkin
AEC-tr-4061, pp. 307–309; Zh. Neorg. Khim., 2:2680–2681 (1957)
Purification; distillation

A simpler method for removing copper from germanium
M. Kikuchi and S. Iizima
J. Phys. Soc. Japan, 12:824 (1957)

### 12.b.3. Melt Growth

Theory of the stability of a solid–liquid interface during growth from stirred melts. II
R. T. Delves
J. Cryst. Growth, 8:13–25 (1971)

Effect of rotation of the face paths during growth of germanium single crystals
A. N. Kirgintsev and Yu. A. Rybin
Kristallografiya, 14(5):954 (1969)
Sov. Phys.—Cryst., 14(5):831–832 (1970)

Interface instability in single crystals pulled from the melt
R. Singh, A. F. Witt, and H. C. Gatos
J. Appl. Phys., 41:2730–2732 (1970)

Ge crystal growth and evaluation as Ge(Li) detector material
R. Wichner, G. A. Armantrout, and T. G. Brown
(Lawrence Radiation Lab., Univ. Calif., Livermore, Calif.),
CONF-700301-7; UCRL-72108 (March 1970), 13 pp.
From 12th Scintillation and Semiconductor Counter Symposium, Washington, D. C.

Morphology of Ge solidified in undercooled melt
T. Arizumi and N. Kobayashi
Synthetic Crystal Res. Lab., Nagoya Univ., Collection of Reports No. 6, pp. 42–50 (1969)

Impurity distribution coefficients in germanium single crystals produced by Czochralski and by Stepanov methods
N. V. Bessonova
Izv. Akad. Nauk SSSR, Ser. Fiz., 33:2022–2024 (1969)

Structure and impurity distribution in germanium single crystals of cirular cross section produced by Stepanov method
N. V. Bessonova, D. I. Levinson, V. V. Peller, et al.
Izv. Akad. Nauk SSSR, Ser. Fiz., 33:2013–2015 (1969)

Morphology of the surface of separation of germanium crystals grown by the Czochralski method
G. I. Dolivo-Dobrovol'skaya, V. A. Mokievskii, I. Yu. Litvinova, and M. D. Lyubalin
Izv. Akad. Nauk SSSR, Neorg. Mater., 5(5):853–857 (1969)
Inorg. Mater., 5(5):726–730 (1969)

Germanium single-crystal production in a rotating container
A. N. Kirgintsev and Yu. A. Rybin
Fiz. Tverd. Tela, 10:2361–2364 (1968)
Sov. Phys.—Solid State, 10(8):1857 (1969)

Electrical and structural properties of planar germanium single crystals produced by Stepanov method
A. S. Kukui, D. I. Levinson, and G. V. Sachkov
Izv. Akad. Nauk SSSR, Ser. Fiz., 33:2010–2012 (1969)

Shape of germanium crystals grown by the Czochralski method
M. D. Lyubalin and V. A. Mokievskii
Kristallografiya, 13(4):735–740 (1968)
Sov. Phys.—Cryst., 13(4):635 (1969)

Producing germanium single crystals of various profiles by Stepanov method
G. V. Sachkov, P. I. Antonov, L. M. Zatulovskii, et al.
Izv. Akad. Nauk SSSR, Ser. Fiz., 33:1996–1997 (1969)

X-ray study of structural perfection of planar germanium single crystals (Stepanov method)
E. G. Sheikhet and V. Ya. Shkot
Izv. Akad. Nauk SSSR, Ser. Fiz., 33:2016–2019 (1969)

Producing germanium with reduced oxygen content
V. A. Shershel, D. I. Levinson, G. G. Makarov, et al.
Izv. Akad. Nauk SSSR, Ser. Fiz., 33:2025–2026 (1969)
Comparison of Czochralski and Stepanov methods

Conditions for producing dislocation-free germanium single crystals by Stepanov method
Yu. M. Smirnov
Izv. Akad. Nauk SSSR, Ser. Fiz., 33:2001–2002 (1969)

Growth of germanium for lithium drift detectors
L. P. Adda, K. E. Benson, R. C. DeWit, and J. M. Kenzie
IEEE Trans. Nucl. Sci., NS-15(3):347–351 (1968)
Eleventh Scintillation and Semiconductor Counter Symposium, Washington, D. C., 1968
Hydrogen atmosphere

Convection-induced impurity distribution during germanium crystal pulling
G. Baralis and M. C. Perosino
J. Crystal Growth, 3(4):651–655 (1968)

Factors affecting the uniformity of the specific resistance in germanium crystals grown from melts
P. I. Baranskii and K. G. Marin
Rost Kristallov, Vol. 6, Nauka (1965), pp. 186–192
Growth of Crystals, Vol. 6B (N. N. Sheftal', ed.), Consultants Bureau, New York (1968), pp. 3–8

The growth of highly doped germanium monocrystals free of defects
L. V. Golubev, V. M. Tuchkevich, and Yu. V. Shmartsev
Rost Kristallov, Vol. 6, Nauka (1965), pp. 193–198
Growth of Crystals, Vol. 6B (N. N. Sheftal', ed.), Consultants Bureau, New York (1968), pp. 9–13

Croissance monocristalline du germanium à partir de solutions dans l'étain et dans les mélanges plomb-étain
A. Laugier, M. Gavand, and G. Mesnard
Bull. Soc. Franc. Mineral. Crist., 90(2):176–180 (1967)

Distribution of impurities in molten germanium and the effect of agitating the melt on the growth of germanium single crystals
    M. D. Lyubalin, K. G. Marin, E. I. Panasenko, and I. Yu. Litvinova
    Tsvetnaya Met., 1:84–86 (1967)

Preparazione di monocristalli di germanio per uso elettronico
    R. Taricco and A. Vaschetti
    Convegno metalli non ferrosi, Milano, 1967, 6 pp.

Growing germanium tubes
    S. V. Tsivinskii, Yu. I. Koptev, and A. V. Stepanov
    Fiz. Tverd. Tela, 8:2461–2462 (1966)
    Sov. Phys.–Solid State, 8(8):1961–1962 (1967)

Production of germanium crystals by the Stepanov method
    S. V. Tsivinsky
    Rost Kristallov (Growth of Crystals)
    Naukova Dumka, Kiev (1967), pp. 134–138

Production of tubular-shaped germanium crystals
    P. I. Antonov and A. V. Stepanov
    Izv. Akad. Nauk SSSR, Neorg. Mater., 2(5):950 (1966)
    Inorg. Mater., 2(5):810 (1966)

Investigations of the temperature field of a germanium melt in the growing of single crystals by the Czochralski method
    V. N. Chaknunashvili
    Progress in Heat Transfer (P. K. Konakov, ed.), Consultants Bureau, New York (1966), pp. 151–157

Growth of germanium crystals from a melt containing a considerable amount of impurity
    A. Ya. Gubenko
    Solid State Transformations (N. N. Sirota, F. K. Gorskii, and V. M. Varikash, eds.), Consultants Bureau, New York (1966), pp. 69–73
    Czochralski; Aú, As, Ga dopants

Variation of the segregation coefficient of impurities in germanium with concentration
    V. I. Korol'kov and V. N. Romanenko
    Solid State Transformations (N. N. Sirota, F. K. Gorskii, and V. M. Varikash, eds.), Consultants Bureau, New York (1966), pp. 57–60
    Zone-refining; Sb and Ga impurities

Change in mosaic structure associated with the distribution of impurities in growing single crystals and bicrystals from the melt
    A. A. Kralina and V. O. Esin
    Solid State Transformations (N. N. Sirota, F. K. Gorskii, and V. M. Varikash, eds.), Consultants Bureau, New York (1966), pp. 155–163
    Ge and Al

Growing of crystalline germanium plates
    S. V. Tsivinskii, Yu. I. Koptev, and A. V. Stepanov
    Fiz. Tverd. Tela, 8, 569–570 (1966)
    Sov. Phys.–State, 8(2):449–450 (1966)
    From melt in graphite crucible

Temperature field in germanium single crystals prepared by the Czochralski method
    A. A. Uglov
    Progress in Heat Transfer (P. K. Konakov, ed.), Consultants Bureau, New York (1966), pp. 111–117

Investigation of the temperature field on germanium single crystals
    G. E. Verevochkin and V. A. Smirnov
    Progress in Heat Transfer (P. K. Konakov, ed.), Consultants Bureau, New York (1966), pp. 117–127

Single crystal pulling without temperature programming
    J. J. Barlic
    J. Sci. Instr., 42:360–361 (1965)
    Variation of the "melted layer" method

Production of crystals by Stepanov's method
    S. V. Tsivinskii
    Rost Kristallov, Vol. 6, Nauka (1965), pp. 355–359
    Growth of Crystals, Vol. 6B (N. N. Sheftal', ed.), Consultants Bureau, New York (1968), pp. 161–164
    Ge crystals of predetermined form, dimensions, and orientation

A method of revealing the shape of the crystallization front in the growing of single crystals by the Czochralski zone refining method
    V. V. Dobrovenskii and Yu. P. Tomson
    Kristallografiya, 10(4):583–585 (1965)
    Sov. Phys.–Cryst., 10(4):492–494 (1966)

Lattice defects in single crystal of heavily doped germanium
    Y. Furukawa and N. Kakuda
    Rev. Elec. Commun. Lab. (Tokyo), 13(5–6):403–419 (1965)
    Czochralski; pulling rate, seed rotation

Floating crucible techniques for growing non-uniformly doped crystals
    C. W. Leung
    Solid-State Electron., 8:571–580 (1965)

The temperature distribution in pulled germanium crystals during growth
    J. C. Brice and P. A. C. Whiffin
    Solid-State Electron., 7:183–187 (1964)

Regrowth of germanium single crystal from indium melt
    M. Tomono
    Hitachi Rev., 9:35–44 (1964)

Resistivity striations on germanium single crystals
    G. Baralis and M. C. Perosino
    Met. Ital., 57(7):315–322 (1963)
    Impurity concentration fluctuations during crystal growth

Micro-inhomogeneities in Ge single crystals
    A. Lorinczy, T. Nemeth, and P. Szebeni
    Acta Phys. Acad. Sci. Hung., 16(1):63–67 (1963)
    Horizontal zone melting

Production of germanium monocrystals by the Czochralski method
    M. Marone
    Met. Ital., 55(4):145–151 (1963)

Some problems of horizontal-zone single-crystal growing in swinging-coil equipment
    T. Salanki
    Elektrotechnika, 56(3):107–111 (1963)
    20-kg Ge single crystals

Vertical concentration distribution of impurities in horizontal zone melting
    T. Abe, M. Kanazawa, and M. Iida
    Oyo Buturi, 31:58–69 (1962)
    Ge doped with Li, As, Sb, Bi, Zn, Ga, In, and Tl

Constitutional supercooling during crystal growth from stirred melts. Part 3. The morphology of the germanium cellular structure
W. Bardsley, J. S. Boulton, and D. T. J. Hurle
Solid—State Electron., 5:395–403 (1962)
Czochralski

Factors affecting the dislocations in germanium monocrystals
A. D. Belyaev, V. N. Vasilevskaya, and E. G. Miselyuk
Rost Kristallov, Vol. 3, Izd. Akad. Nauk (1961), pp. 380–387
Growth of Crystals, Vol. 3 (A. V. Shubnikov and N. N. Sheftal', eds.), Consultants Bureau, New York (1962), pp. 270–274
Czochralski

Growth of single crystals
Frank Halden
SRI Journal, 6(3):64–71 (1962)

Mechanisms of growth of metal single crystals from the melt
D. T. J. Hurle
Progr. Mater. Sci., 10(2):81–147 (1962)
Supercooling of Ge

Growth of dislocation-free germanium monocrystals
E. Yu. Kokorish and N. N. Sheftal'
Rost Kristallov, Vol. 3, Izd. Akad. Nauk (1961), pp. 388–394
Growth of Crystals, Vol. 3 (A. V. Shubnikov and N. N. Sheftal', eds.), Consultants Bureau, New York (1962), pp. 275–279
Czochralski, Ga and Sb doped

Production of uniform high impurity concentration semiconductor material
J. C. Marinace
(IBM Corp.)
U. S. Patent 3,031,404 (April 24, 1962), 4 pp.
Gradient freeze technique

A technique for pulling single crystals of volatile materials
E. P. A. Metz, R. C. Miller, and R. Mazelsky
J. Appl. Phys., 33:2016–2017 (1962)
Modified Czochralski method

A process for obtaining single crystals with uniform solute concentration
A. V. Valcic
Solid—State Electron., 5:131–134 (1962)

Constitutional supercooling during crystal growth from stirred melts: Part 2. Experimental — gallium-doped germanium
W. Bardsley, J. M. Callan, H. A. Chedzey, and D. T. J. Hurle
Solid—State Electron., 3:142–154 (1961)

Evolution of the horizontal crystal grower
W. D. Eisenhower
Engineer, 5:18–25 (1961)

The role of surface tension in pulling single crystals of controlled dimensions
G. K. Gaule and J. R. Pastore
Metallurgy of Elemental and Compound Semiconductors, Vol. 12
Interscience Publishers, New York (1961), pp. 201–226

Constitutional supercooling during crystal growth from stirred melts: Part 1. Theoretical
D. T. J. Hurle
Solid—State Electron., 3:37–44 (1961)

Influence of the Peltier effect on the perfection of germanium single crystals grown by the method of pulling from the melt
E. Yu. Kokorish
Kristallografiya, 5(5):815–816 (1960)
Sov. Phys.—Cryst., 5(5):778–789 (1961)

Mechanism of growth of germanium single crystals from a molecular beam
G. A. Kurow
Physikalische Abhandlungen der Sowjetunion, 5(5):565–570 (1961)

Germanium saturated with gallium antimonide
J. O. McCaldin and D. B. Wittry
J. Appl. Phys., 32:65–69 (1961)
Temperature gradient zone melting

The technology of pulling single crystals: Part 1
Henry T. Minden
Semicond. Prod., 4:25–28 (1961)

The use of thermoelectric effects during crystal growth
J. R. O'Connor
J. Electrochem. Soc., 108:713–715 (1961)
Steady-state Peltier cooling to increase the rate of crystalization

Regrowth of germanium from molten indium or lead
M. Tomano
J. Phys. Soc., Japan, 16:439–453 (1961)

An investigation of the manufacture of germanium single-crystal ingots by the levelling process
N. G. Anderson and D. Gray
Proc. IEE, Part B Suppl., 106:871–878 (1960)

Germanium crystal grown from hollow cylindrical seeds
R. C. Frank and J. E. Thomas, Jr.
J. Appl. Phys., 31:1689–1690 (1960)

La fabrication de cristaux de germanium exempts de dislocations
B. Okkerse
Rev. Tech. Philips, 21(11):342–347 (1959–60)
Czochralski

A modified rf coil to facilitate floating zone techniques
S. J. Silverman
J. Electrochem. Soc., 107:268C (1960)
Permits growth of larger diameter than floating zone procedure

Peltier effects for crystal growing
L. L. Thomas
Research Chemicals, Div. of Nuclear Corp. of America, WADD TR 60-16; AD-242707 (May 1960), 45 pp.

On the growth conditions of germanium
T. Sadaki
J. Phys. Soc., Japan, 14:381–382 (1959)
Czochralski

The growing of 5-kg single crystals of germanium
J. G. Wilkes
Proc. IEE 106, Part B Suppl. No. 17:866–870 (1959)
Czochralski

Apparatus for growing very pure single germanium crystals
    V. B. Dik and V. V. Ostroborodova
    Instr. Exp. Tech. (1958), pp. 158–159
    Czochralski

Temperature-gradient zone melting
    W. G. Pfann
    Zone Melting, Chapt. 9, John Wiley and Sons, New York
        (1958), pp. 198–208
    Pb saturated Ge

Effect of crystal growth variables on electrical and structural properties of germanium
    F. D. Rosi
    RCA Rev., XIX(3):349–387 (Sept. 1958)
    Czochralski

Production of oriented germanium crystals by the contact method
    E. Rubes
    Zh. Tekhn. Fiz., 27(8):1655–1660 (1957)
    Sov. Phys. — Tech. Phys., 2(8):1538–1543 (1958)
    Czochralski

Germanium single crystal growth by the floating crucible technique
    Jan Goorissen and Fritjof Kartensen
    Z. Metalk., 50:46–50 (1956)
    ORNL-tr-1441 (Eng. trans.), 13 pp.

## 12.b.4. Epitaxy and Films

Growth and properties of thin germanium films
    D. J. Dumin
    J. Electrochem. Soc., 117:95–100 (1970)

Control of epitaxial layers by pseudo-Kikuchi lines; Ge on Si
    P. Durupt, A. Laugier, and M. Pitaval
    Compt. Rend. B, Sci. Phys., 270:941–943 (1970)

Electrical and structural properties of vacuum deposited germanium
    J. S. Johannessen
    (Electronics Research Lab., affiliated with SINTEF, The
        Norwegian Institute of Tech., Trondheim, Norway),
        ELAB Report AE-140 (April 1970)
    Crystallite growth in polycrystalline hetero-epitaxial films
        is found to be thermally activated, and characterized by
        surface self-diffusion and grain boundary migration.
        Increasing the deposition temperature improves struc-
        tural and electrical properties, in the sense that crystal-
        lite size increases, the carrier density decreases and the
        carrier mobility increases

The preparation and characterization of electron-beam, vapor-deposited, germanium films
    Charles S. Portwood, III
    Thesis, Univ. California, Berdeley, UCRL-19164 (April
        1970), 62 pp.

Preparation of germanium films on sapphire by pyrolysis of hydrides
    T. A. Zeveke, T. I. Medvedeva, and N. N. Sheftal'
    Kristallografiya, 14(4):748–750 (1969)
    Sov. Phys.—Cryst., 14(4):649–650 (1970)

Formation of epitaxial films by the disproportionation of germanium diiodide
    G. A. Abduragimov
    Izv. Akad. Nauk SSSR, Neorg. Mater., 5(8):1332–1336 (1969)
    Inorg. Mater., 5(8):1137–1140 (1969)

Effect of oxygen on the formation of germanium films
    R. F. Adamsky, K. H. Behrndt, and W. T. Brogan
    J. Vacuum Sci., Tech., 6:542–545 (1969)

Kinetics of epitaxial crystallization under rapid cooling of the melt
    L. N. Aleksandrov, E. A. Klimenko, and A. G. Klimenko
    Kristallografiya, 13(6):1098–1100 (1968)
    Sov. Phys.—Cryst., 13(6):962 (1969)

Review: Vapor Growth of Ge
    T. Arizumi, T. Nishinaga, and M. Kakehi
    (Synthetic Crystal Research Lab., Faculty of Engineering,
        Nagoya University)
    Collection of Reports No. 6 (April 1968–March 1969)

Thermal factors in the growth of epitaxial semiconductor layers from the melt
    Yu. B. Bolkhovityanov
    Kristallografiya, 13(5):918–919 (1968)
    Sov. Phys.—Cryst., 13(5):801 (1969)

Specific heat and heat of crystallization of amorphous germanium
    H. S. Chen and D. Turnbull
    (Harvard Univ., Cambridge, Mass.), Report No. TR-22;
        Ad-691388 (May 1969), 13 pp.

Some characteristics of the growth of epitaxial germanium layers by the sandwich method
    Petko Kamadzhiev and Simeon Simeonov
    Izv. Fiz. Inst. ANEB, Bulgaria Akad. Nauk, 19:49–65 (1969)

Preparation of $0.5$-$\mu$-thick single-crystal germanium specimens
    S. V. Kopylova, E. A. Klimenko, and A. G. Klimenko
    Pribory Tekh. Eksper., No. 5:174 (1969)
    Instr. Exper. Tech., No. 5:1290 (1969)

Effect of deposited metals on the crystallization temperature of amorphous germanium film
    F. Oki, Y. Ogawa, and Y. Fujiki
    Japan. J. Appl. Phys., 8:1056 (1969)

Obtaining thin germanium foils and their morphology
    N. Pashov
    Izv. Fiz. Inst. ANEB (Bulgaria), 18:29–35 (1969)

Effects of details of the gas flow on the morphology of epitaxial germanium films
    T. A. Smorodina, A. I. Georgiev, and V. A. Rodevich
    Rost Kristallov, Vol. 8, Nauka (1968), pp. 198–203
    Growth of Crystals, Vol. 8 (N. N. Sheftal', ed.) Consultants
        Bureau, New York—London (1969), pp. 161–164

Epitaxial growth of germanium on $II$-$IV$-$V_2$ substrates
    A. J. Springthorpe, R. J. Harvey, and B. R. Pamplin
    J. Crystal Growth, 6:104–106 (1969)

The properties of an epitaxial germanium film
on a silicon crystal
   T. Stubb, H. K. J. Kanerva, and H. Stubb
   Z. Naturforsch., 24a(9):1343-1346 (1969)

Epitaxial films of germanium deposited on sap-
phire via chemical vapor transport
   R. F. Tramposch
   J. Electrochem. Soc., 116, 654-658 (1969)

Structure defects of germanium single-crystal
films
   V. D. Vasil'ev and A. A. Tikhonova
   Rost Kristallov, Vol. 8, Nauka (1968), pp. 285-295
   Growth of Crystals, Vol. 8 (N. N. Sheftal', ed.), Consultants
      Bureau, New York—London (1959), pp. 234-240

On the diffusion processes during growth of epi-
taxial films
   L. N. Aleksandrov and Y. G. Sidorov
   Proceedings of the Second Colloquium on Thin Films,
      Budapest (1967), pp. 194-200 (E. Hahn, ed.), Vandenhoeck
      and Rupprecht, Göttingen (1968), 641 pp.

Epitaxial growth of mirror-smooth Ge on GaAs
and Ge by the low-temperature GeI$_2$ disproportionation reaction
   M. Berkenblit, A. Reisman, and T. B. Light
   J. Electrochem. Soc., 115:966-969 (1968)

Conditions for growing non-dislocational germa-
nium layers from gas phase
   V. F. Dorfman and L. S. Sibirtsev
   Dokl. Akad. Nauk SSSR, 181(4):874-876 (1968)

Epitaxial growth of Ge layers on Si substrates
by vacuum evaporation
   K. Ito and K. Takahashi
   Japan. J. Appl. Phys., 7:821-826 (1968)

Über heterogene Keimbildung von Germanium
   W. Kleber and I. Mietz
   Krist. Tech., 3(4):509-512 (1968)

Single-crystal germanium films on gallium ar-
senide produced by the thermal evaporation in
vacuum
   A. P. Klimenko, V. P. Klochkov, N. N. Soldatenko, N. M.
      Torchun, and Yu. A. Tkhorik
   Kristallografiya, 13(2):367-370 (1968)
   Sov. Phys.—Cryst., 13(2):303-305 (1968)

Epitaxial growth of germanium in the open io-
dide process
   V. N. Kolesnikov, E. A. Dem'yanov, and L. I. Dereza
   Kristallografiya, 12(6):1105-1107 (1967)
   Sov. Phys.—Cryst., 12(6):975 (1968)

Device for making thin germanium foils
for the electron diffraction microscope
   Yu. G. Kostyuk
   Zavodsk. Lab., 34(4):489-491 (1968)
   Ind. Lab., 34(4):590-592 (1968)

Equipment for obtaining epitaxial layers of ger-
manium from molecular beam in vacuum
   A. E. Likhtman, G. F. Lymar', L. N. Nemirovskii, and Yu.
      D. Chistyakov
   Ind. Lab., 34:1879 (1968)
   Exchange of experience

Electron gun for floating evaporation of germa-
nium
   L. N. Nemirovskii
   Instr. Exper. Tech., No. 6:1482 (1968)

Epitaxial deposition of germanium onto semi-
insulating GaAs
   S. A. Papazian and A. Reisman
   J. Electrochem. Soc., 115:961-965 (1968)

Crystal growth phenomena of Ge in Ge-I$_2$ closed-
tube system
   T. Arizumi and T. Nishinaga
   Crystal Growth (Suppl. to J. Phys. Chem. Solids), Proc.
      Intern. Conf. on Crystal Growth, Boston, June 20-24,
      1966, pp. 295-299 (H. Steffen Peiser, ed.) Pergamon
      Press, Oxford and New York (1967)

Structural properties of vacuum-deposited ger-
manium layers
   A. Barna, P. B. Barna, E. F. Pocza, N. Croitoru, A.
      Devenyi, and R. Grigorovici
   Proc. of the Colloquium on Thin Films, Budapest, April 20-
      23, 1965, pp. 49-58 (E. Hahn, P. B. Barna, and J. Peisner,
      eds.), Hungarian Society for Optics, Acoustics, and Film
      Technics, Budapest (1967)

Propriétés optiques et photoélectriques des
couches minces de germanium
   P. B. Barna and J. Peisner
   Ibid., pp. 281-287

Epitaxial growth of Ge on GaAs by the low-tem-
perature GeI$_2$ disproportionation reaction
   M. Berkenblit, A. Reisman, and T. B. Light
   (IBM Watson Research Center, Yorktown Heights, N. Y.),
      Electrochemical Society 1967 Fall Meeting, Chicago, Ill.,
      Oct. 15-20, 1967, pp. J4, 4
   Electrochemical Society, Inc., New York (1967)

Epitaxial growth of germanium on single crystal
spinel
   D. J. Dumin
   J. Electrochem. Soc., 114:749 (1967)

Epitaxial growth of germanium
   D. M. Jackson, Jr., and R. W. Howard
   (Motorola, Inc.), U. S. Patent 3,473,978, April 24, 1967
   Uniform monocrystalline layer from GeH$_4$ on Si layer

Growth of epitaxial germanium films by the sand-
wich method
   B. S. Kurbatov, E. V. Rakova, and G. A. Kurov
   Kristallografiya, 12(1):160-162 (1967)
   Sov. Phys.—Cryst., 12(1):140-141 (1967)

Epitaxial deposition of Ge from its solution in
Sn and Pb-Sn mixtures
   A. Laugier, M. Gavand, and G. Mesnard
   Solid State Electron., 10:77-78 (1967)

The growth of epitaxial germanium films by tri-
ode sputtering
   C. K. Layton and K. B. Cross
   Thin Solid Films, 1(2):169-172 (1967)

Cristallisation des couches de germanium obtenues par évaporation sous vide sur des monocristaux de silicium
   E. G. Leon Lopez
   Vide, 129:157-160 (1967)

Recrystallization of germanium thin films on insulating substrates by electron-beam zone melting
   C. T. Naber
   J. Electrochem. Soc., 114:406-408 (1967)

Epitaxial deposition of germanium on semi-insulating GaAs
S. A. Papazian and A. Reisman
Electrochemical Society 1967 Fall Meeting, Chicago, Ill.,
Oct. 15-20, 1967, pp. J4, 1-3
Electrochemical Society, Inc., New York (1967)
$GeH_4$ pyrolysis

Wachstum und Struktur dünner, aus schmelzflüssigen Filmen auskristallisierter Germanium-Kristalle
N. Paschoff
Phys. Stat. Sol., 22(7):83-91 (1967)

Study of the germanium surface layer by mass-spectroscopic methods
G. F. Romanova and I. I. Stepko
Elektronnye processy na poverkhnosti i v monokristallicheskikh sloyakh poluprovodnikov, Simpozium, pp. 106-113 Izdat. Nauka, Novosibirsk (1967)

Growth mechanism of vapor deposited germanium films
M. S. Seltzer, N. Albon, B. Paris, and R. C. Himes
J. Electrochem. Soc., 114:102-107 (1967)

Growth of epitaxial germanium films from super-cooled drops
N. N. Sheftal', Ye. I. Givargizov, B. V. Spitsyn, and A. M. Kevorkov
Akad. Nauk SSSR. Inst. Kryst. (Moscow), 4:15-21 (1964)
FTD-HT-66-435; AD-658650 (Feb. 1967)

Growth and perfection of films of germanium deposited on sapphire via chemical vapor transport
R. F. Tramposch
Electrochemical Society 1967 Fall Meeting, Chicago, Ill.,
Oct. 15-20, 1967, pp. J4, 12-13
Electrochemical Society Inc., New York (1967)

Some aspects of the growth of epitaxial germanium films by the sandwich method
I. D. Voronova and G. V. Chaplygin
Kristallografiya, 12(1):125-129 (1967)
Sov. Phys.– Cryst., 12(1):99-102 (1967)

Optical properties of germanium films in the 1-5 $\mu$ range
J. Wales, G. J. Lovitt, and R. A. Hill
Thin Solid Films, 1:137-150 (1967)
Epitaxy on sapphire

Thin semiconductor film crystallization by zone melting
L. Beiziters
Latv. PSR Zinat. Akad. Vestis Fiz. Tehn. Ser. (USSR), 6, 34-39 (1966)

Investigation of growth defects in vacuum-deposited single-crystal germanium films
A. Catlin and R. R. Humphris
Proceedings of International Symposium: Grundprobleme der Physik dünner Schichten, Göttingen, 1965 (1966), pp. 175-180
Mixed crystals above 350°C on $CaF_2$

Electrical characteristics of epitaxial germanium films vacuum-deposited on semi-insulating GaAs up to thicknesses of $10^6$ Å
J. E. Davey
Appl. Phys. Letters, 8:164-166 (1966)

Deposition at 475 and 500°C; relation of carrier concentration to thickness

The epitaxial growth of Ge on Si by solution growth techniques
J. P. Donnelly and A. G. Milnes
J. Electrochem. Soc., 113:297-298 (1966)

Epitaxial growth of germanium by the hydrogen reduction of $GeI_4$
S. Iida
Japan. J. Appl. Phys., 5:138-144 (1966)

Epitaxy of germanium films of gallium arsenide by vacuum evaporation
R. F. Lever and E. J. Huminski
J. Appl. Phys., 37:3638-3639 (1966)

Crystallization of vacuum-evaporated germanium films by the electron-beam zone-melting process
S. Namba
J. Appl. Phys., 37:1929-1930 (1966)
On Mo or W substrates

Epitaxial growth of germanium on germanium substrates cleaned and heated by electron bombardment
J. Pfeifer
Phys. Stat. Sol., 17:K15-K18 (1966)

Nucleation and growth in vacuum-deposited germanium films
E. F. Pocza, A. Barna, and P. Barna
Proceedings of International Symposium: Grundprobleme der Physik dünner Schichten, Göttingen, 1965, pp. 153-156; discussion 155-156 (1966)

Microstructure of epitaxial Ge films deposited on (111) $CaF_2$ substrates
B. W. Sloope and C. O. Tiller
J. Appl. Phys., 37:887-893 (1966)

Source contamination effects on the epitaxy of Ge films on Ge
J. E. Davey
J. Vacuum Sci. Tech., 2:12-17 (1965)

Effect of growth conditions on certain properties of epitaxial germanium layers
V. F. Dorfman, M. S. Belokon', G. F. Krasnova, and G. N. Tolkacheva
Izv. Akad. Nauk SSSR, Neorg. Mater., 1(7):1016-1020 (1965)
Inorg. Mater., 1(7):933-937 (1965)

Vapor deposition of germanium on molybdenum
I. L. Kalnin and J. Rosenstock
J. Electrochem. Soc., 112:329-333 (1965)

Sur la croissance épitaxique sélective du germanium
L. Laukmanis and I. A. Faltyn'
Latvijas PSR Zinatnu Akad. Vestis, Fiz. Teh. Zinatnu Ser., 3:57-59 (1965)

Liquid epitaxial growth of germanium
B. S. Murthy
J. Inst. Telecommunic. Engrs., India, 11(5):172-179 (1965)

Resistivity control of Ge grown by $GeI_2$ disproportionation
A. R. Riben, D. L. Feucht, and W. G. Oldham
J. Appl. Phys., 36:3685-3686 (1965)

Formation conditions and structure of Ge films deposited on polished (111) $CaF_2$ substrates in an ultrahigh-vacuum system
B. W. Sloope and C. O. Tiller
J. Appl. Phys., 36:3174–3181 (1965)

Chemical crystallization of germanium in the open iodide process
É. A. Dem'yanov, V. N. Kolesnikov, and G. V. Sleptsov
Kristallografiya, 9(6):910–915 (1964)
Sov. Phys.—Cryst., 9(6):761–764 (1965)

Epitaxial growth of gallium arsenide on germanium substrates. Part 3. Deposition on high index planes and on curved surfaces
T. Gabor
J. Electrochem. Soc., 111:825–827 (1964)

Epitaxial growth of germanium using water vapor
R. F. Lever and F. Jona
J. Appl. Phys., 34:3139–3140 (1964)

Web growth of semiconductors
S. O'Hara and A. I. Bennett
J. Appl. Phys., 35:No. 3 (Part 1), 686–693 (1964)

Unsupported single-crystal films of germanium
C. W. Skaggs and J. R. Jones
J. Appl. Phys., 35:3013–3015 (1964)

The effect of condensation rate on the textural properties of vacuum-deposited germanium films on heated amorphous substrates
L. E. Terry and J. D. Williams
(Sandia Lab., Albuquerque, N. M.), SC-TM-64-1208 (Oct. 1964), 18 pp.

Investigation of ultra high vacuum sputtered thin films
S. P. Wolsky
(Scientific Report No. 2, Mallory and Co., Inc., Burlington, Mass.), AFCRL-64-977; AD-454939 (Nov. 1964), 43 pp.
Ge epitaxy on Ge at 150°C

Epitaxial deposition of silicon by thermal decomposition of silane
S. R. Bhola and A. Mayer
RCA Rev., 24:511–522 (1963)

Epitaxial temperature of germanium deposited on calcium fluoride
A. Catlin, A. J. Bellemore, Jr., and R. R. Humphris
J. Appl. Phys., 34:251–252 (1963)

Epitaxial deposition of silicon and germanium layers by chloride reduction
E. F. Cave and B. R. Czorny
RCA Rev., 24:523–545 (1963)

Evaporation-condensation method for making germanium layers for transistor purposes
J. C. Courvoisier, W. Haidinger, P. J. W. Jochems, and L. J. Tummers
Solid—State Electron., 6:265 (1963)

The effect of vacuum evaporation parameters on the structural, electrical and optical properties of thin germanium films
J. E. Davey, R. J. Tiernan, T. Pankey, and M. D. Montgomery
Solid—State Electron., 6:205 (1963)

Mechanism underlying the epitaxial growth of germanium films from the gaseous phase
E. I. Givargizov
Fiz. Tverd. Tela, 5(4):1150–1157 (1963)
Sov. Phys.—Solid State, 5(4):840–845 (1963)

Single-crystal germanium films by micro-zone melting
J. Maserjian
Solid—State Electron. 6(5):477–484 (1963)
On sapphire

Direct observation of Ge epitaxial growth
M. Nonura
Japan. J. Appl. Phys., 2(9):591–592 (1963)

Analysis of thin-film germanium epitaxially deposited onto calcium fluoride
A. L. Pundsack
J. Appl. Phys., 34:2306 (1963)
Amorphous to crystalline transition region between 320° and 400°C, best crystal orientation between 550° and 575°C

The growth of germanium epitaxial layers by the pyrolysis of germane
E. A. Roth, H. Gossenberger, and J. A. Amich
RCA Rev., 24:499–510 (1963)

Surface roughness of epitaxial germanium films
B. W. Sloope and C. O. Tiller
Japan. J. Appl. Phys., 2:308–309 (1963)
Deposition rate and substrate temperature

Concerning the effect of doping on the growth rate of epitaxial layers of germanium
A. N. Stepanova and E. I. Givargizov
Fiz. Tverd. Tela, 5(10):3034–3035 (1963)
Sov. Phys.—Solid State, 5(10):2220–2221 (1964)

Preparation of crystalline germanium films on metals
O. A. Weinreich and G. Dermit
J. Appl. Phys., 34:225–227 (1963)

Epitaxy of germanium films on germanium by vacuum evaporation
J. E. Davey
J. Appl. Phys., 33:1015–1016 (1962)

Growth of germanium layers from the gaseous phase
G. A. Kurov
Fiz. Tverd. Tela, 3(7):2080–2088 (1961)
Sov. Phys.—Solid State, 3(7):1512–1517 (1962)

On growth conditions of germanium crystals in thin films
G. A. Kurov, V. D. Vasil'ev, and M. G. Kosaganova
Fiz. Tverd. Tela, 3(11):3541–3542 (1961)
Sov. Phys.—Solid State, 3(11):2571–2572 (1962)

Preparation of single-crystal films
N. G. Nifontov
Bull. Acad. Sci. USSR, 25:663 (1962)
Ge on $CaF_2$

Vapor growth of twinned germanium platelets
C. Pritchard
J. Electrochem. Soc., 109, 993 (1962)

Effect of the temperature of formation on the crystallinity and electrical properties of germanium films on fluorite
R. L. Schalla, L. H. Thaller, and A. E. Potter, Jr.
J. Appl. Phys., 33:2554 (1962)

Formation conditions and structure of thin epitaxial germanium films on single-crystal substrates
B. W. Sloope and C. O. Tiller
J. Appl. Phys., 33:3458-3463 (1962)

Epitaxial vapor growth of single-crystal Ge
M. Takabayashi
Japan J. Appl. Phys., 1:22-29 (1962)

Epitaxial growth of germanium on GaP
L. J. Van Ruiven and W. Dekker
Physica, 28:307 (1962)

Radiotracer studies of the incorporation of iodine into vapor-grown Ge
W. E. Baker and D. M. J. Compton
IBM J. Res. Dev., 4:269-274 (1960)

Epitaxial vapor growth of Ge single crystals in a closed-cycle process
J. C. Marinace
IBM J. Res. Dev. 4:248-255 (1960)

Electrical properties of vapor-grown Ge junctions
M. J. O'Rourke, J. C. Marinace, R. L. Anderson, and W. H. White
IBM J. Res. Dev., 4:256-263 (1960)
Closed-cycle iodide vapor growth

Vapor-deposited single-crystal germanium
R. P. Ruth, J. C. Marinace, and W. C. Dunlap
J. Appl. Phys., 31:995 (1960)

High-vacuum evaporator for radioactive materials
H. Widmer and J. Kirsch
Rev. Sci. Instr., 31:791 (1960)
Deposition of radioactive Ge on Ge substrates

### 12.b.5. Dendrites, Needles, and Whiskers

A new method for growing crystal ribbons
C. E. Bleil
J. Crystal Growth, 5:99-104 (1969)

Obtaining germanium monocrystal whiskers
P. Kamadzhiev and L. Mladzhov
Izv. Fiz. Inst. ANEB (Bulgaria), 18:37-47 (1969)
Chemical transport reaction with iodine carrier

Growth and properties of acicular gold-doped germanium single crystals
N. I. Makarova, S. G. Yudin, and V. I. Revyakina
Izv. Akad. Nauk SSSR, Neorg. Mater., 5(8):1328-1331 (1969)
Inorg. Mater., 5(8):1133-1136 (1969)

Lateral growth of dendritic germanium crystals
I. M. Skvortsov
Rost Kristallov, Vol. 7, Nauka (1967), pp. 160-164
Growth of Crystals, Vol. 7 (N. N. Sheftal', ed.) Consultants Bureau, New York-London (1969), pp. 135-138

Dislocation distribution in germanium crystals grown in the form of thin strips
S. V. Tsivinskii and A. V. Stepanov
Fiz. Tverd. Tela, 10:2519-2521 (1968)
Sov. Phys.—Solid State, 10(8):1977 (1969)

Thermoelectric effects during the growth of dendritic germanium crystals from melts
A. A. Davydov and V. S. Maslov
Izv. Akad. Nauk SSSR, Neorg. Mater, 4(9):1449-1452 (1968)
Inorg. Mater., 4(9):1269-1271 (1968)

Mechanism of dendritic growth in germanium crystals with two and three twinning planes
D. A. Petrov and A. A. Bukhanova
Izv. Akad. Nauk SSSR, Neorg. Mater, 4(9):1439-1444 (1968)
Inorg. Mater., 4(9):1262-1265 (1968)

Growth of nondendritic single-crystal ribbons of Ge
D. E. Swets
Electrochem. Tech., 5:385-389 (1967)

The preparation of "perfect" dendritic germanium crystals
C. Andrle
Czech. J. Phys., B16:342-352 (1966)

The structure of "perfect" dendritic germanium crystals
C. Andrle
Czech. J. Phys., B16:855-863 (1966)
Pulling from an undercooled melt

Device for pulling dendritic crystals
S. Ionescu-Bujor and E. Gruciany
Instr. Exp. Tech., 1:229-230 (1966)
Pribory Tekh. Eksper. USSR, 1:213-214 (1966)

Filamentary growth of germanium crystals
P. R. Kamadjiev, L. K. Mladjov, and N. B. Velchev
Compt. Rend. Acad. Bulgare Sci., 19(9):779-781 (1966)

On lateral growth of germanium dendritic crystals
I. M. Skvorcov
Acta Cryst. 21, Pt. 7, Suppl., A292 (1966)
Seventh International Congress and Symposium International Union of Crystallography, Moscow (1966)

Germanium dendritic growth in the so-called "difficult" directions
A. A. Bukhanova and D. A. Petrov
Fiz. Tverd. Tela, 6(8):2518-2519 (1964)
Sov. Phys.—Solid State, 6(8):1999-2000 (1965)

Dendritic seed crystals having a critical spacing between three interior twin planes
J. W. Faust, Jr., and H. F. John
U. S. Patent 3,130,040 (1964)

Epitaxis of germanium on germanium dendrites
V. N. Maslov, L. J. D'yakonov, A. A. Davydov, and M. P. Shaforostov
Kristallografiya, 9(6):938-939 (1964)
Sov. Phys.—Cryst., 9(6):789-790 (1965)

Germanium "fiber" crystal
T. Nakagawa
Electron. Eng. (Japan), 84(10):50-55 (1964)

Method of continuously growing thin strip crystals
F. L. Vogel, Jr., Glen Gardner, and E. F. Cave
U. S. Patent 3,124,489 (1964)

The influence of twin structure on growth directions in dendritic ribbons of materials having the diamond or zinc-blende structures
N. Albon and A. E. Owen
J. Phys. Chem. Solids, 24(7):899–907 (1963)

Growing of single-crystal germanium in strips
J.-J. Brissot and H. Raynaud
Electrochem. Tech., 1:304–307 (1963)

Growth mechanisms of germanium dendrites: Kinetics and the nonisothermal interface
D. R. Hamilton and R. G. Seindensticker
J. Appl. Phys., 34:1450–1460 (1963)

Association in melted semiconductors
D. R. Hamilton and R. G. Seidensticker
J. Appl. Phys., 34(9):2697–2699 (1963)

Grain boundary states in silicon and germanium
Y. Matukura
Japan. J. Appl. Phys., 2(2):91–98 (1963)

Growth mechanisms in germanium dendrites: Three twin dendrites: Experiments on and models for the entire interface
R. G. Seidensticker and D. R. Hamilton
J. Appl. Phys., 34(10):3113–3119 (1963)

Some features of the growth and twin structure of germanium dendrites and of the anomalous segregation of impurities in the process of dendritic crystallization
M. A. Medvedev, B. G. Anokhin, I. M. Skvortsov, A. S. Korotkov, and E. V. Myakinenkova
Fiz. Tverd. Tela, 4(1):36–43 (1962)
Sov. Phys.—Solid State, 4(1):25–30 (1962)

Diffusion of vacancies during quenching of Ge and Si
J. Melngailis and S. O'Hara
J. Appl. Phys., 33(8):2596–2601 (1962)

Vapor growth of twinned germanium platelets
C. Pritchart
J. Electrochem. Soc., 109(10):993–995 (1962)

Growth processes of three twin germanium dendrites
R. G. Seidensticker and D. R. Hamilton
J. Electrochem. Soc., 109:202C (1962)

Note on the electrochemical preparation of germanium dendrites
J. O'M. Bockris, J. Diaz, and M. Green
Electrochim. Acta, 4:362–363 (1961)

Growth steps on germanium dendrites
G. R. Booker
J. Electrochem. Soc., 108:574–578 (1961)

Dendritic growth of Ge and Si
C. H. Church
Nucl. Sci. Abstr., 15(14):2382 (Abstract No. 18424) (1961)

Germanium dendrite studies. Part 1. Studies of twin structures and the seeding mechanism
J. W. Faust, Jr., and H. F. John
J. Electrochem. Soc., 108:855–860 (1961)

Germanium dendrite studies. Part 2. Lateral growth processes
J. W. Faust, Jr., and H. F. John
J. Electrochem. Soc., 108:860–863 (1961)

Germanium dendrite studies. Part 3. Dislocations
J. W. Faust, Jr., and H. F. John
J. Electrochem. Soc., 108:864–868 (1961)

Dendritic growth of germanium
Metallurgy of Elemental and Compound Semiconductors, Vol. 12
R. O. Grubel, ed.
Interscience Publishers, New York and London (1961), pp. 97–200
Proceedings of Technical Conference held at Boston, Mass., August 29–31 (1960)

A study of growth processes in germanium dendrites using pulse electroplating techniques
R. C. Smith
J. Electrochem. Soc., 108:238–241 (1961)

Liquid-solid interface studies in germanium dendrites
A. I. Bennett and S. O'Hara
Bull. Am. Phys. Soc., Ser. 2, 5:165 (1960)

Dislocations in dendrites
J. W. Faust, Jr., and H. F. John
Bull. Am. Phys. Soc., Ser. 2, 5:165 (1960)

Propagation mechanism of germanium dendrites
D. R. Hamilton and R. G. Seidensticker
J. Appl. Phys., 31:1165–1168 (1960)

The role of the twin plane in the dendritic propagation of germanium crystals
D. R. Hamilton, R. G. Seidensticker, J. W. Faust, Jr., and H. F. John
Bull. Am. Phys. Soc., Ser. 2, 5:165 (1960)

Etch pits on dendritic germanium
P. J. Holmes
Phys. Rev., 119:131–132 (1960)

Simple apparatus for the growth of germanium dendrites
R. F. Lever, J. K. Powers, J. L. Richards, and H. V. Sirgo
Rev. Sci. Instr., 31:1334–1335 (1960)

Growth of atomically flat surfaces on germanium dendrites
R. L. Longini, A. I. Bennett, and W. J. Smith
J. Appl. Phys., 31:1204–1207 (1960)

The propagation of germanium dendrites
R. G. Seidensticker and D. R. Hamilton
Electrochemical Society, Inc., Electronics Division Abstracts, 9(1):159–162 (1960)

Dendritic growth studies by plating techniques
R. C. Smith
Electrochemical Society, Inc., Electronics Division Abstracts, 9(1):159–162 (1960)

Dendritic growth of germanium crystals
A. I. Bennett and R. L. Longini
Phys. Rev., 116:53–61 (1959)

### 12.b.6. Doping, Diffusion, and Precipitation (see also 18b)

A method for the synthesis of $BI_3$ for use as a dopant source
M. Berkenblit and A. Reisman
J. Electrochem. Soc., 117:1100–1101 (1970)

Large-diameter germanium crystals for gamma-ray spectrometry
G. Dearnaley, R. Ellis, and P. E. Gibbons
Nuclear Physics, Div., Electronics and Applied Physics Div., UKEA Research Group, Atomic Energy Research Establishment, Harwell, England, AERE-R 6345 (March 1970)
Vacuum-pulled; Li-drifted; oxygen-free; area 40 cm$^2$; depletion thickness 8 mm; resolution >3.7 keV at 1.33 MeV

The distribution coefficient of boron in germanium
W. D. Edwards
J. Electrochem. Soc., 117:1062-1065 (1970)

Effect of impurity precipitations on the anomalous x-ray transmission in heavily arsenic-doped germanium
O. N. Efimov, E. G. Sheikhet, and L. I. Datsenko
Phys. Stat. Sol., 38:489-498 (1970)

Polytropy of dopants in semiconductors
V. I. Fistul', P. M. Grinshtein, and N. S. Rytova
Fiz. Tekhn. Poluprovodnikov 4:84-88 (1970)
Sov. Phys.-Semicond., 4:67-69 (1970)
Ge heavily doped with As; precipitation of a supersaturated solid solution

An x-ray multiple diffraction study of crystals of arsenic-doped germanium
B. J. Isherwood and C. A. Wallace
J. Appl. Cryst., 3:Pt. 2, 66-71 (1970)
Diffusion during and after Czochralski growth

Materials problems in lithium-drifted germanium radiation detectors
G. R. Jindal and J. W. Faust, Jr.
J. Appl. Phys., 41:2106-2109 (1970)

Structure of the ground state of the interstitial lithium donor in germanium
Hisashi Nara and Haruhiko Yamazaki
J. Phys. Soc. Japan, 28:1485-1488 (1970)

Lithium driftability in detector-grade germanium
A. H. Sher and J. A. Coleman
IEEE Trans. Nucl. Sci., NS-17 (3):125-129 (1970)

Development of lithium-activated germanium semiconductor nuclear detectors
Raul Gallardo Villegas
Thesis, Universidad Nacional Autonoma de Mexico, Mexico City, NP-18180 (1970), 72 pp.

Effect of intrinsic radiation from the Sb$^{124}$ isotope on the surface processes accompanying the diffusion of antimony into germanium
V. S. Arakelyan, Vikt. I. Spitsyn, and V. B. Lazerev
Dokl. Akad. Nauk SSSR, 187(1):116-119 (1969)
Dokl. Phys. Chem. USSR, 187(1):429-433

Equilibrium distribution of the micro-impurities arsenic and phosphorus in germanium tetraiodide between liquid and vapor
A. A. Efremov, Ya. D. Zelvenskii, and I. L. Kostandova
Russ. J. Inorg. Chem., 14:263-266 (1969)

An experimental study of the accuracy of compensation in lithium drifted germanium detectors
A. Lauber and B. Malmsten
Aktiebolaget Atomenergi, Nykoping, Sweden, AE-373 (Oct. 1969)

Some evidence for a strong decrease of mobile carrier recombination lifetime with increasing drifted depth

A method for copper-staining germanium crystals
E. J. River
Nucl. Instr. Method, 67:349-351 (1969)
To show nonuniformities in germanium crystals prior to Li drifting, and as a tool for studying the diffusion and drifting of Li in germanium

Lithium-ion drift mobility in germanium
A. H. Sher
J. Appl. Phys., 40:2600-2607 (1969)

The effect of crystallographic orientation on the geometrical arrangement of dislocations formed in germanium by the diffusion of arsenic
I. M. Sukhodreva and L. D. Cheryukanova
Fiz. Tverd. Tela, 11:760-762 (1969)
Sov. Phys.-Solid State, 11(3):608-609 (1969)

On the diffusion processes during growth of epitaxial films
L. N. Aleksandrov and Y. G. Sidorov
Proceedings of the Second Colloquium on Thin Films, Budapest (1967), pp. 194-200 (E. Hahn, ed.) Vandenhoeck and Rupprecht in Göttingen, 1968, 641 pp.
In vapor transport of Ge

The interaction between oxygen and boron in liquid germanium
W. D. Edwards
J. Appl. Phys., 39:1784-1790 (1968)

The effects of heat treatments of germanium on resistivity, photoconductive decay, and on lithium drift properties
E. Fischer-Colbrie, T. G. Brown, and A. V. Friensehner
Lawrence Radiation Lab., University of California, Livermore, UCRL-71082 (Rev. 1); CONF-681017-10 (Oct. 1968), 13 pp.
From 15th Nuclear Science Symposium, Montreal, Canada

The solubility of calcium in germanium
V. F. Kalabukhova, L. V. Kurtinova, N. M. Morozova, A. F. Prokofeva, and E. B. Sokolov
Izv. Akad. Nauk SSSR, Neorg. Mater., 4(11):1845-1848 (1968)
Inorg. Mater., 4(11):1609-1611 (1968)

Experiments concerning the diffusion of indium in germanium
M. Kuisl
BMwF-FB-W-68-72 (Oct. 1968), 54 pp.

Production of p-doped Ge-epitaxial layers
H. J. Schnabel
Wiss. Z. Elektrotech., 11(4):193-201 (1968)
Boron, spark doping

Formation of dislocations in germanium during diffusion of arsenic
I. M. Sukhodreva and L. D. Cheryukanova
Fiz. Tverd. Tela, 10:932-935 (1968)
Sov. Phys.-Solid State, 10(3):737-738 (1968)

Control of lithium migration and precipitation in germanium single crystals
W. E. Wheeler
Nucleonics in Aerospace – Proceedings of the Second International Symposium, Columbus, Ohio (July 12-14, 1967), pp. 176-181
Instrument Society of America (1968)

Electrical and x-ray measurements of slowly diffusing elements in single-crystal germanium
 Adelbert Zielasek
 Ph. D. thesis, Technische Hochschule, Hannover, West
  Germany (1968), 130 pp.
 Sb, Al, Ga, P, As, Bi, In, Tl

Fabrication, properties and applications of Ge(Li) gamma detectors
 F. Adams
 Atomic Energy Rev, 5:31-92 (1967)
 213 refs.

Metallurgy and physical properties of mercury-doped germanium related to the performances of the infrared detector
 Y. Darviot, A. Sorrentino, B. Joly, and B. Pajot
 Infrared Phys., 7:1-10 (1967)
 Zone leveling in inert gas with Hg vapor

Production and fundamental properties of germanium single crystals with respect to their application in $\gamma$-ray detectors
 L. De Laet
 BLG-425, Paper 2 (1967), 16 pp.

Oxygen-free, aluminum-doped germanium for lithium drifting
 W. D. Edwards and C. Wilburn
 Nucl. Instr. Methods, 48:357-358 (1967)

A technique for drifting lithium into germanium
 D. A. Jenkins
 Lawrence Radiation Lab., Univ. of California, Berdeley,
  UCRL-17317 (Jan. 1967)

Distribution of impurities in crystals as a function of condition of their growth
 M. G. Kekua
 FTD-MT-24-341-67 (Dec. 1967), 13 pp.
 Akad. Nauk Gruzinskoi SSR, Tiflis. Institut Metallurgii.
  Trudy 14:171-178 (1964)

Diffusion of antimony in epitaxial germanium layers
 G. S. Kulikov and E. I. Givargizov
 Fiz. Tverd. Tela, 8(11):3344-3349 (1966)
 Sov. Phys.—Solid State, 8(11):2670 (1967)

Diffusion von Aluminium und Bor in Germanium
 W. Meer and D. Pommerrenig
 Z. Angew. Physik, 23:369-372 (1967)

Investigation of the diffusion of antimony and indium in germanium, including effect of internal electric field
 P. V. Pavlov and V. A. Uskov
 Fiz. Tverd. Tela, 8(10):2977-2981 (1966)
 Sov. Phys.—Solid State, 8(10):2377 (1967)

A bibliography on methods for the measurement of inhomogeneities in semiconductors, Final rept. 1966-Dec. 1967
 Harry A. Schafft and Susan Needham
 (National Bureau of Standards, Wash., D.C.), RADC-TR-
  68-96; AD-671524 (June 1968), 58 pp.

Preparation and use of lithium-drifted germanium detectors
 D. Srnka
 Ceskoslovenska Akademie Ved, Rez, Ustav Jaderneho
  Vyzkumu
 UJV-1873 (Aug. 1967), 23 pp.

Solid state diffusion of antimony in germanium, from the vapour phase, in a vacuum furnace
 G. N. Willis
 Solid-State Electron., 10:1-8 (1967)

Lithium-Drifted Germanium Detectors
 Proceedings of a Panel on the Use of Lithium-Drifted Germanium Gamma-Ray Detectors for Research in Nuclear
  Physics, Vienna (6-10 June, 1966)
 International Atomic Energy Agency, Vienna (1966)

Investigation of the influence of the specific radioactivity of the isotope $Sb^{124}$ upon the diffusion of antimony into germanium
 V. S. Arakelyan and Vikt. I. Spitsyn
 Dokl. Akad. Nauk SSSR, 170(6):1352-1355 (1966)
 Dokl. Phys. Chem., Proc. Acad. Sci. USSR, 170(6):678-681
  (1966)

Study of heterogeneous equilibrium during the crystallization of germanium from melts containing elements of the donor and acceptor types
 A. D. Belaya and V. S. Zemskov
 Solid State Transformations (N. N. Sirota, F. K. Gorskii,
  and V. M. Varikash, eds.) Consultants Bureau, New York
  (1966), pp. 61-67
 Dopants Al, Sb, In, Ga

A technique for adding volatile dopants to the melt during crystal pulling
 H. A. Chedzey and D. T. J. Hurle
 Brit. J. Appl. Phys., 17:699-700 (1966)
 Zn doping of Ge

Lithium drift rates and oxygen contamination in germanium
 R. J. Fox
 IEEE Trans. Nucl. Sci., NS-13:367-368 (1966)

Diffusion of arsenic in germanium from a germanium arsenide source. Prediffusion and diffusion
 G. F. Foxhall and L. E. Miller
 J. Electrochem. Soc., 113:698-701 (1966)

Controlled doping of germanium layers made by the evaporation-condensation method
 W. Haidinger, J. C. Courvoisier, P. J. W. Jochems, and
  L. J. Tummers
 Solid-State Electron., 9:689-693 (1966)
 Ga, In, As, and Sb

Drift rate and precipitation of lithium in germanium
 R. Henck, L. Stab, G. Lopes da Silva, P. Siffert, and
  A. Coche
 IEEE Trans. Nucl. Sci., NS-13:245-251 (1966)

The effect of doping on gold diffusion in germanium
 M. F. Millea
 J. Phys. Chem. Solids, 27:309 (1966)

Diffusion of antimony in germanium during plastic straining
 C. D. Calhoun and L. A. Heldt
 Acta Met., 13:932-933 (1965)

The influence of counterdoping on the distribution of manganese over substitutional and interstitial sites in germanium
 F. N. Hooge and T. Vrijheid-Lammers
 Philips Res. Repts., 20:292-305 (1965)

# Germanium

**Growth and characterization of uranium-doped germanium single crystals**
H. R. Killias, S. N. Dermatis, and H. C. Gatos
J. Electrochem. Soc., 112:362–363 (1965)
Czochralski method

**Solubility interactions in compensated heavily-doped germanium**
J. O. McCaldin
J. Appl. Phys., 36:211–213 (1965)
As, Sb, Ga

**Properties of heavily doped germanium**
J. I. Pankove
Progress in Semiconductors, Vol. 9, pp. 47–86 (A. F. Gibson and R. E. Burgess, eds.) A Heywood Book, Temple Press Books Led., London (1965)

**A supplement to "thermodynamics of impurity doping reactions in vapor growth of Ge"**
T. Arizumi and I. Akasaki
Japan. J. Appl. Phys., 3:87 (1964)

**Techniques for the fabrication of lithium drifted germanium gamma detectors**
W. L. Hansen and B. V. Jarrett
Nucl. Instr. Methods, 31:301–306 (1964)
Lithium drifting

**Influence of counterdoping on the distribution of Mn over substitutional and interstitial sites in Ge**
F. N. Hooge
Congres International Physics Semiconducteurs, Paris (1964), pp. 63–64
Impr. Jouve, Paris (1964)
As or Sb

**A mechanism causing super-saturation of impurities during semiconductor crystal growth**
K. Lehovec
Surface Sci., 1:165–70 (1964)
Ge(In) or Ge(Sb)

**Impurity content of germanium crystallized from the liquid ternary alloy Ge-In-Sb**
K. Lehovec and A. Slobodskoy
J. Electrochem. Soc., 111:65–73 (1964)

**Infrared radiation detector made of gold-doped germanium**
Z. Majewski, A. Ambroziak, and J. Swiderski
Przeglad Elektroniki, 4, 251–254 (1963)
FTD-TT-64-80/1 + 2; AD-452368 (Nov. 1964), 7 pp.
Gold diffusion into pure germanium monocrystals

**Electric-field-enhanced precipitation of Li in Ge**
Joseph Blanc and M. S. Abrahams
J. Appl. Phys., 34:3638–3639 (1963)

**Effect of local electric fields on the diffusion of antimony in germanium**
B. I. Boltaks and T. D. Dzhafarov
Fiz. Tverd. Tela, 5(10):2818–2824 (1963)
Sov. Phys.—Solid State, 5(10):2061–2065 (1964)

**Fabrication of highly doped germanium single crystals**
W. Heinke
Z. Angew. Physik., 15:128–130 (1963)
Drawing As-doped Ge single crystals from Ge—As solutions

**The influence of an electric field on the diffusion of thallium in germanium single crystals**
I. I. Ibragimov, M. G. Shakhmakhtinski, and A. A. Kuliev
Fiz. Tverd. Tela, 5(3):862–864 (1963)
Sov. Phys.—Solid State, 5(3):632–634 (1963)

**Constitutional supercooling during the crystal growth of germanium and silicon**
H. Kodera
Japan. J. Appl. Phys., 2:527–534 (1963)
Limiting concentrations of Ga, In, P, As, Sb, Sn, and B in Ge and Si grown by the Czochralski or zone-leveling methods

**Heavily doped materials for Esaki-tunnel diodes**
F. A. Trumbore
Met. Adv. Electron. Mater., 19:15–34 (1963)
Crystal growth; doping; solvent evaporation; Czochralski

**Effect of indium and gallium on the crystallization of germanium from melts containing these elements**
V. S. Zemskov, A. D. Belaya, and T. E. Puris
Fiz. Tverd. Tela, 5(4):1100–1103 (1963)
Sov. Phys.—Solid State, 5(4):802–804 (1963)

**Preparation and properties of mercury-doped germanium**
S. R. Borrello and H. Levinstein
J. Appl. Phys., 33:2947–2950 (1962)
Modified zone-leveling technique

**Transmutation doping and recoil effects in semiconductors exposed to thermal neutrons**
J. H. Crawford, Jr., and J. W. Cleland
Radioisotopes in the Physical Sciences and Industry, Vol. 1
International Atomic Energy Agency (1962), pp. 269–286

**Diffusion of tantalum in single-crystal germanium**
V. I. Tagirov and A. A. Kuliyev
Izv. Akad. Nauk Azerb. SSR, Ser. Fiz.-Mat. Tekhn. Nauk, 1:65–68 (1962), AD-428546 (1963)

**Diffusion of arsenic in germanium from the vapour phase**
W. Albers
Solid—State Electron., 2:85–95 (1961)

**The diffusion of beryllium in germanium**
Yu. I. Belyaev and V. A. Zhidkov
Fiz. Tverd. Tela, 3(1):182–184 (1961)
Sov. Phys.—Solid State, 3(1):133–134 (1961)

**Donor equilibria in the germanium-oxygen system**
C. S. Fuller, W. Kaiser, and C. D. Thurmond
Phys. Chem. Solids, 17:301–307 (1961)

**Solubility of oxygen in germanium**
W. Kaiser and C. D. Thurmond
J. Appl. Phys., 32:115–118 (1961)

**Growth of germanium films from the gas phase**
G. A. Kurow
Physik. Abhand. Sowjetunion, 5:446 (1961)

**Growing heavily compensated germanium crystals of known impurity concentrations**
L. M. Lambert
Solid—State Electron., 3:316–317 (1961)
Floating crucible, Czochralski

Gaseous diffusion of arsenic and phosphorus into germanium
K. Lehovec and C. Pihl
J. Electrochem. Soc., 108:552–560 (1961)

Epitaxial vapor growth and doping of germanium by a germanium-iodine reaction
J. C. Marinace, W. E. Baker, and D. M. J. Compton
Metallurgy of Elemental and Compound Semiconductors, Vol. 12
Interscience Publishers, Inc., New York (1961), p. 271

Growth and properties of heavily doped germanium single crystals
J. R. Patel, R. F. Tramposch, and A. R. Chaudhuri
Metallurgy of Elemental and Compound Semiconductors, Vol. 12 (Ralph O. Grubel, ed.), Interscience Publishers, New York, London (1961), pp. 45–63
Proceedings of Technical Conference sponsored by the Semiconductors Committee of the Inst. of Metals Div., The Metallurgical Soc., and Boston Section of AIME, held in Boston, Mass., Aug. 29–31, 1960
Ga and As; >$10^{20}$/cm$^3$

Diffusion of silver, cobalt and iron in germanium
L. Y. Wei
J. Phys. Chem. Solids, 18:162 (1961)

Germanium films on Ge obtained by thermal evaporation in vacuum
O. Weinreich, G. Dermit, and C. Tufts
J. Appl. Phys., 32:1170 (1961)

Diffusion of antimony in germanium alloyed with aluminum
I. P. Akimchenko and L. S. Milevskii
Fiz. Tverd. Tela, 2(9):2109–2116 (1960)
Sov. Phys.—Solid State, 2(9):1891–1896 (1961)

Incorporation of As into vapor-grown Ge
W. E. Baker and D. M. J. Compton
IBM J. Res. Dev., 4:275–279 (1960)

On the kinetics and mechanism of the precipitation of lithium from germanium
J. R. Carter, Jr., and R. A. Swalin
J. Appl. Phys., 31:1191–1200 (1960)

The diffusion of hydrogen in single-crystal germanium
R. C. Frank and J. E. Thomas, Jr.
Phys. Chem. Solids, 16:144–151 (1960)

Solvent evaporation technique for the growth of arsenic-doped germanium single crystals for Esaki diodes
F. A. Trumbore and F. M. Porbansky
J. Appl. Phys., 31:2068 (1960)
Ge – 5 to 10% As melt

Evaluation and control of diffused impurity layers in germanium
H. S. Veloric and W. J. Greig
RCA Rev., XXI:437–456 (1960)
Powder diffusion method for carefully controlled impurity layers; As, Sb

On the kinetics and mechanism of precipitation of lithium from germanium
J. R. Carter, Jr., and R. A. Swalin
Univ. of Minnesota, Minneapolis
Tech. Rept., No. 1:NP-8452 (Nov. 1959), 31 pp.

Über die Diffusion in Germaniumkristallen, die eine Korngrenze enthalten
F. Karstensen
Z. Naturforsch., 14a:1031–1039 (1959)
In, As, Sb, Ga, P

The effect of an electric field on the diffusion of silver in germanium
V. E. Kosenko, E. G. Miselyuk, and L. A. Khomenko
Fiz. Tverd. Tela, 1:100 (1959)

The diffusion of boron in germanium
M. D. Sturge
Proc. Phys. Soc., 73:320–322 (1959)

Solid solubilities of aluminum and gallium in germanium
F. A. Trumbore, E. M. Porbansky, and A. A. Tartaglia
J. Phys. Chem. Solids, 11:239–245 (1959)

Determination of the limiting segregation of gallium in zone-refined germanium
L. W. Davies
Trans. AIME, 212:799–781 (1958)

The effect on ion-pair and ion-triplet formation on the solubility of lithium in germanium — effect of gallium and zinc
Howard Reiss and C. S. Fuller
J. Phys. Chem. Solids, 4:58–67 (1958)

Diffusion of silver, copper, cobalt and iron in germanium
Ling Yun Wei
(Univ. of Illinois, Urbana), Tech. Note, No. 4:AFOSR-TN-58-853; AD-203496 (Aug. 1958), 71 pp.
58 refs., 24 figs.

Diffusion and solubility of iron in germanium
A. A. Bugai, V. E. Kosenko, and E. G. Miseliuk
Zh. Tekh. Fiz., 27(1):210–211 (1957)
Sov. Phys.—Tech. Phys., 2(1):183–184 (1957)

Mechanism of diffusion of copper in germanium
F. C. Frank and D. Turnbull
Phys. Rev., 104:617–618 (1956)

Solid solubilities and electrical properties of tin in germanium single crystals
F. A. Trumbore
J. Electrochem. Soc., 103:597–600 (1956)
Crystal pulling and crystal growth from the melt in a thermal gradient

## 12.c. Silicon

### 12.c.1. General and Reviews

Impurity limits for growth of silicon crystals
L. D. Dyer and H. J. Moltzan
Materials Engineering and Sciences Division Biennial Conference (1970), pp. 129–134

A re-evaluation of bonding features in diamond and silicon
J. F. McConnell and P. L. Sanger
Acta Cryst., A 26:83–93 (1970)

Precipitation in high-purity silicon single crystals
Aasmund Erik Nes
Thesis, University of California, Berkeley, Calif., UCRL-19608 (Aug. 1970), 96 pp.

A new method for measuring the ionized-impurity
concentration in high-purity materials
A. Alberigi Quaranta, A. Canali, G. Ottaviani, and A. Taroni
Phys. Stat. Sol., A1(2):315-322 (1970)

Silicon device process
NBS Special Publ. 337 (Nov. 1970).

Assessment of defects in silicon by IR microscopy
D. M. Connah
Automat. Inspect. Defects and Dimensions. Conf. Eastbourne
(1969), pp. 184-193

Semiconductor Silicon. First International Symposium on Silicon Materials Science and Technology, New York, May 1969
R. R. Haberecht and E. L. Kern, eds.
Electronics Division and Electrothermics and Metallurgy
Division, Electrochemical Society, New York (1969),
766 pp.

Silicon
Meta Neuberger and S. J. Welles
(Hughes Aircradt Co., Culver City, Calif.), DS-162 (Oct.
1969), 264 pp., 734 refs.
Compilation of a wide range of electronic properties; data
table which includes a wide range of mechanical, physical,
and thermal properties is included as well as a summary
of crystal structure and phase transitions

Origine, symétrie et largeur des transitions
électroniques et vibrationnelles des impuretés
dans les semiconducteurs. Application au phos-
phore et à l'oxygene dans le silicium
B. Pajot
Ph. D. thesis, Paris, CNRS No. 3366 (May 1969), 141 pp.

Radial resistivity profile in high-purity silicon
A. Alberigi Quaranta, C. Canali, G. Ottaviani, A. Taroni,
and G. Zanarini
(Istituto Nazionale di Fisica Nucleare, Sezione di Bologna),
INFN/TC-69/11 (Dec. 1969), 10 pp.

Silicon Semiconductor Data
Helmut F. Wolf
Pergamon Press, New York (1969), 648 pp.
Emphasis on graphical and tabular data; summary of most
important properties; properties of Si; impurities in Si;
silicon surfaces and surface structures; general diffusion
characteristics; epitaxial growth; P-N junctions; characteristics of $SiO_2$

An x-ray study of the lattice parameter of sili-
con
N. E. Moyer
Ph. D. thesis, Purdue University (1968)

Silicon technology for semiconductors. II.
Growing single crystals
H.-F. Hadamovsky
Neue Hütte, 11(1):28-33 (1966)

Table 24 – Silicon, pp. 1-7 in Selected Values of
Chemical Thermodynamic Properties. Part 2.
Tables for the elements twenty-three through
thirty-two in the standard order of arrangement
D. D. Wagman, W. H. Evans, I. Halow, V. B. Parker,
S. M. Bailey, and R. H. Schumm
(National Bureau of Standards, Wash., D. C.), NBS Tech.
Note 270-272 (1966)

Silicon
C. A. Hogarth
Materials Used in Semiconductor Devices
Interscience Publishers, New York, London, Sydney (1965),
pp. 29-49

Silicon Semiconductor Technology
W. R. Runyan
McGraw-Hill Book Co., New York, London, Sydney (1965),
277 pp.
Purification, crystal growth, doping, diffusion, electrical
properties, physical and optical properties

Silicon: Semiconductor properties
M. L. Schultz
Infrared Phys., 4:93-112 (1964)

Properties of high-purity silicon
Walter Heywang and Manfred Zerbst
Siemens Rev., 25:44-50 (1958)

## 12.c.2. Purification

Method of crucible-free zone melting of semi-
conductor material, particularly silicon
Konrad Reuschel
(Siemens Akt.-Ges.), U. S. Patent 3,454,367 (July 8, 1969)

Preparation of high-purity silicon
Lallan Prasad Pandey Kartarsingh
J. Sci. Ind. Res. (India), 27:386-395 (1968)
Review; 174 refs.

Formation de carbone au cours de la fabrication
du silicium ultrapur
J. Martin and E. Haas
Solid-State Electron., 11:993-996 (1968)

Preparation of silicon of high purity
E. Bonnier, H. Pastor, and J. Driole
Metallurgie, 5(7):299-317 (1965/66)
Reduction of fluorosilicate; treatment with Sb or Sn; electron-
beam zone refining

Studies on the preparation of high-purity silicon
by the iodide process
N. J. Prakash, I. D. Varma, and N. S. K. Prasad
(Atomic Energy Establishment Trombay, Bombay, India),
A. E. E. T.-231 (1965), 22 pp.

Preparation of semiconductor-grade silicon by
the iodide process
H. Baba and H. Araki
Rev. Elec. Commun. Lab. (Tokyo), 12:430-446 (1964)

The fundamental experiments of thermal decomposition and hydrogen reduction of trichlorosilane. The preparation of high-purity silicon (II)
T. Yagihashi and T. Wada
Trans. Natl. Res. Inst. Metals (Tokyo), 5:49-56 (1963)

Production experiment by hydrogen reduction of
silicon tetraiodide and its behaviours. Preparation of pure silicon by the iodide process (IV)
T. Kurosawa and T. Yagihashi
Trans. Res. Inst. Metals (Tokyo), 5:22-27 (1963)

Preparation of pure silicon by the iodide process
(1) Preparation of silicon tetraiodide
T. Kurosawa and T. Yagihashi
Trans. Natl. Res. Inst. Metals (Tokyo), 5:28-33 (1963)

Purification of silicon tetraiodide by recrystallization and distillation methods, preparation of pure silicon by the iodide process (II)
  T. Kurosawa, T. Ishikawa, and T. Yagihashi
  Trans. Natl. Res. Inst. Metals (Tokyo), 5:34-38 (1963)

Fundamental experiment of hydrogen reduction of silicon tetraiodide. Preparation of pure silicon by the iodide process (III)
  T. Kurosawa, T. Ishikawa, and T. Yagihashi
  Trans. Res. Inst. Metals (Tokyo), 5:39-42 (1963)

The experiments on fractional distillation and infrared absorption spectrum of trichlorosilane. The preparation of high-purity silicon (III)
  T. Yagihashi, T. Kurosawa, and T. Wada
  Trans. Res. Inst. Metals (Tokyo), 5:17-24 (1963)

Relation between conditions of hydrogen reduction of trichlorosilane and pulled single crystals. The preparation of high-purity silicon (V)
  T. Yagihashi, T. Kurosawa, and T. Wada
  Trans. Natl. Res. Inst. Metals (Tokyo), 5:32-37 (1963)

The large-scale experiment on the preparation of high-purity silicon by hydrogen reduction of trichlorisilane. The preparation of high-purity silicon (IV)
  T. Yagihashi, T. Kurosawa, and T. Wada
  Trans. Natl. Res. Inst. Metals (Tokyo), 5:65-71 (1963)

Application of radiochemistry to the preparation of high-purity silicon
  T. Ichimiya, H. Bara, and T. Nozaki
  Radioisotopes in the Physical Sciences and Inudstry, Vol. 1, International Atomic Energy Agency (1962), pp. 533-542
  Recrystallization, zone refining

The reduction of $SiCl_4$ with hydrogen in the electric field
  T. Kurosawa and J. Minamiya
  Trans. Natl. Res. Inst. Metals (Tokyo), 4:28-35 (1962)

Die Reinigung von Silicium durch Transportreaktionen
  R. Lesser and E. Erben
  60 Jahre Quarzglas—25 Jahre Hochvakkuumtechnik, W. W. Heraeus GmbH, Hanau (1961), pp. 3-13

Über die Reinigung von Silicium durch Transportreaktionen
  R. Lesser and E. Erben
  Z. Anorg. Allgem. Chem., 309:297-303 (1961)

Preparation of high-purity silicon from silane
  C. H. Lewis, H. C. Kelly, M. B. Giusto, and S. Johnson
  J. Electrochem. Soc., 108:1114-1118 (1961)
  Preparation, purification, and thermal decomposition of silane

Obtaining high-purity silicon
  S. S. Alikberov and L. P. Shklover
  Zh. Neorgan. Khim., 513-518 (1960)
  $SiHCl_3 + H_2$

An apparatus for the preparation of semiconductor-grade silicon by film boiling
  R. C. Ellis, Jr.
  J. Electrochem. Soc., 107:222-225 (1960)

High-purity silicon from an iodide-process pilot plant
  C. S. Herrick and J. G. Krieble
  J. Electrochem. Soc., 107:111-117 (1960)

Preparation of high-purity single crystal silicon. The floating-zone technique
  J. L. Parmee
  Engineer, pp. 2-8 (June 10, 1960)
  Floating zone refiner

Zone melting of silicon with an electron beam
  V. Gusa, I. Krzhizh, and I. Ladnar
  Fiz. Tverd. Tela, 1(2):290-293 (1959)
  Sov. Phys.—Solid State, 1(2):261-263 (1960)

Extraction of high-purity silicon by thermal decomposition of silanes
  Izv. Vysshikh Uchebn. Zaveden., Tsvetnaya Metal., 1:99-105 (1959)

Electrical properties of high-purity silicon made from silicon tetraiodide
  L. V. McCarty
  J. Electrochem. Soc., 106:1036-1042 (1959)
  Distillation; recrystallization; zone refining

Production of pure elemental silicon
  D. R. Stern and Z. H. McKenna
  (American Potash and Chemical Corp.)
  U. S. Patent 2,892,763 (June, 30, 1959)
  Electrolysis in a fused salt bath

Method of purification of silicon compounds
  G. A. Wolff
  U. S. Patent 2,877,097 (March 10, 1959)
  Removing $BCl_3$ from $SiCl_4$

Zone purification of silicon
  E. A. Taft and F. H. Horn
  J. Electrochem. Soc., 105:81-83 (1958)

Semiconductor purification process
  W. E. Taylor
  (Motorola)
  U. S. Patent 2,835,612 (May 20, 1958)
  A metal that reduces the distribution coefficient of the impurity is added to the molten zone

### 12.c.3. Melt Growth

Silicon crystals almost free of dislocations
  W. N. Borle, S. Tata, and S. K. Varma
  J. Cryst. Growth, 8:223-225 (1971)

Distribution coefficient of boron and phosphorus in silicon
  H. R. Huff, T. G. Digges, Jr., and O. B. Cecil
  J. Appl. Phys., 43:1235-1236 (1971)

The stability of the growth front in crystallization by the moving-solvent method
  V. N. Lozovskii and V. P. Popov
  Kristallografiya, 15:149-155 (1970)
  Sov. Phys. — Cryst., 15:116-121 (1970)

Effect of the thermal conditions of crystallization on the impurity distribution in single crystals grown from the melt
  B. N. Savel'ev, V. V. Dobrovenskii, and V. S. Chudakov
  Kristallografiya, 15:473-476 (1970)
  Sov. Phys. — Cryst., 15(3):473-476 (1970)

Growth of silicon crystals
  K. E. Benson
  Semiconductor Silicon (R. R. Haverecht and E. L. Kern, eds.), The ECS, Inc., New York (1969), pp. 97-123
  Vapor, liquid phase, Czochralski, float-zone; 71 refs.

Monocristal de silicium sans dislocations, et étude de ses propriétés physiques
  T. Kawamura and J. Matsui
  Oyo Buturi, 38(4):365–373 (1969)
  69 refs.

Particularités de la structure et caractère de la disposition des sous-joints de dislocations dans les monocristaux de Si préparés par la méthode Czochralski
  L. V. Layner and B. M. Turovskii
  Fiz. Khim. Obrabot. Mater. SSSR, 4:63–66 (1969)

Oscillations of crystallization front in producing silicon by Czochralski method
  Yu. M. Shaskov, G. M. Stepanov, and V. M. Nikitin
  Izv. Akad. Nauk SSSR, Ser. Fiz., 33:2027–2030 (1969)

Types of impurity bands in silicon single crystals
  N. Ya. Shushlebina, Yu. M. Shashkov, and B. A. Sakharov
  Izv. Akad. Nauk SSSR, Neorg. Mater., 5(4):712–716 (1969)
  Inorg. Mater., 5(4):605–608 (1969)
  Influence of growth rate

Producing silicon crystals of desired profiles
  Yu. M. Smirnov, A. A. Mashnitskii, V. A. Kuznetsov, et al.
  Izv. Akad. Nauk SSSR, Ser. Fiz. 33:2003–2004 (1969)

Spontaneous generation of dislocations during growth of silicon single crystals
  M. Y. Ben–Sira and S. Bukshpan
  J. Crystal Growth, 2:248–250 (1968)
  Czochralski

Report on the work on growth of semiconductor-grade silicon single crystals
  W. N. Borle, Sudhakar Tata, and Vishwa Prakash
  J. Inst. Telecommun. Engrs. (India), 14(8):370–373 (1968)
  Purification and Czochralski growth

Moving mask growth of single-crystal silicon films on amorphous quartz substrates
  M. Braunstein, R. R. Henderson, and A. I. Braunstein
  Appl. Phys. Letters, 12:66–67 (1968)
  Vapor–liquid–solid mechanism + moving mask; crystals to $50 \times 300 \ \mu$

A study of the effect of growth parameters on some properties of Czochralski-grown silicon crystals
  R. O. DeNicola
  M. S. thesis, Materials Research Center, Lehigh Univ., Pa., June 1968

Influence de la vitesse de cristallisation sur les gradients de température lors de la croissance des cristaux par la méthode de Czochralski
  V. P. Grishin, L. A. Timofeeva, B. M. Turovskii, and and K. D. Cheremin
  Fiz. Khim. Obrabot. Mater., SSSR, 4:155–160 (1958)

Sur la détermination des gradients de température dans les cristaux préparés par la méthode de Czochralski
  V. P. Grishin and B. M. Turovskii
  Fiz. Khim. Obrabot. Mater., SSSR, 6:130–133 (1968)

Effet de l'écran sur le champ thermique du lingot lors de la croissance de monocristaux de Si par la méthode de Czochralski
  V. M. Gurevich, V. B. Silkin, and Yu. M. Shashkov
  Fiz. Khim. Obrabot. Mater., SSSR, 4:149–155 (1968)

Production of dislocation-free silicon single crystals
  P. H. Hunt
  (Texas Instruments, Inc.), U. S. Patent 3,397,042 (Aug. 13, 1968)
  Pulled from HF heated molten zone

Apparition de "l'effet de frontière" dans les monocristaux de silicium à haute valeur ohmique
  R. Kh. Karimov, A. N. Suvorov, A. S. Lyutovich, and E. I. Morkovkin
  Izv. Akad. Nauk USSR, Fiz.-Mat. Nauk, 12(6):30–33 (1968)
  Czochralski

Application de la méthode vapeur – liquide pour la croissance de cristaux de Si
  E. Komatsu, Y. Higuchi, and T. Niina
  Suiyokwai-Shi (Japan), 16(7):476–479 (1968)

Axial temperature gradients in the vertical-zone melting of silicon without a crucible
  G. I. Kononov and Yu. M. Shashkov
  Dokl. Akad. Nauk SSSR, 179(3):656–659 (1968)
  Dokl. Chem. Technol., Proc. Acad. Sci. USSR, 179(3):55–57 (1968)

Mechanism of the crystallization of silicon from an aluminum-silicon melt by zone melting with a temperature gradient
  V. N. Lozovskii and A. I. Udyanskaya
  Kristallografiya, 13(3):565–566 (1968)
  Sov. Phys.–Cryst., 13(3)477–478 (1968)

Axial temperature gradients in growing single crystals of silicon by the Czochralski method
  Yu. M. Shashkov and V. P. Grishin
  Dokl. Akad. Nauk SSSR, 179(2):404–406 (1968)
  Dokl. Chem. Technol., Proc. Acad. Sci. USSR, 179(1):45–47 (1968)

Causes of the appearance of impurity bands during growth of Si single crystals by the Czochralski method
  Yu. M. Shashkov and N. Ya. Shushlebina
  Dokl. Akad. Nauk SSSR, 178(1):160–163 (1968)
  Dokl. Chem. Technol., Proc. Acad. Sci. USSR, 178(1):6–9 (1968)

Fluctuation of crystallization front during growth of silicon by the Czochralski technique
  Yu. M. Shashkov and G. M. Stepanova
  Dokl. Akad. Nauk SSSR, 179(4):840–844 (1968)
  Sov. Phys.–Doklady, 13(4):288–290 (1968)

Transfert de masse lors de la croissance des monocristaux de silicium par la méthode de Czochralski
  V. P. Silkin, G. M. Stepanova, and Yu. M. Shashkov
  Fiz. Khim. Obrabot. Mater., SSSR, 6:75–79 (1968)

Radial solute segregation in Czochralski growth
  J. R. Carruthers
  J. Electrochem. Soc., 114:959 (1967)
  Greatly influence by the effects of secondary liquid flaws not associated with crystal rotation

The part played by oxygen in the structural changes of silicon single crystals during thermal treatment
  M. I. Iglitsyn, G. P. Kekelidze, L. V. Layner, and M. G. Mil'vidskii
  Soobshch. Akad. Nauk Gruz. (Tiflis), 45:77–84 (1967)
  NASA-TT-F-11083 (July 1967), 8 pp.

Metallography of silicon single crystals
H. Nagorsen and H. Schreiner
Prakt. Metallog., 4(5):221-235 (1967)
Producing Si single crystals and preparing for metallographic
examination; 14 refs.

Growing two-inch silicon crystals
E. M. Reithard
Semicond. Prod. Solid State Tech., 10:46-47 (1967)
Czochralski

Effect of a superposed axial direct current on
the properties of silicon rods during zone refin-
ing
J. Dorner
TT-66-11618; ATS-56S86G (1966), 15 pp. Transl. of
Elektrowaerme, 22:331-339 (1964)

Technology of semiconducting silicon.  Part 2.
Single-crystal growing
H. F. Hadamovsky
Neue Heutte, 11:28-33 (1966)

Graphit als Schmelztiegel für Halbleiter-Silicium
A. Mühlbauer and H. H. Kocher
Z. Naturforsch., 21a:2116 (1966)

Automatic diameter control in the float-zone re-
fining of silicon
E. E. Rowton
IEEE Trans., EC1-13:66-71 (1966)

Thermal conductivity of silicon in the solid and
liquid states near the melting point
Yu. M. Shashkov and V. P. Grishin
Fiz. Tverd. Tela, 8:567-569 (1966)
Sov. Phys.—Solid State, 8:447-448 (1966)
From the heat balance at the crystallization front in the
Czochralski method

Growth of dislocation-free silicon web crystals
Tom Tucker and G. H. Schwuttke
Appl. Phys. Letters, 9:219 (1966)

Radial solute distribution in Czochralski-grown
silicon crystals
K. E. Benson
Electrochem. Tech., 3:332-335 (1965)

Single-crystal semiconductor growth without
seeding by using the anisotropy of the Peltier
and Thomson effects in zone melting or crystal
pulling
A. Lebek and S. Raab
E. Germany Patent 42,117 (Appl. Nov. 4, 1964; publ. Oct.
25, 1965)

Distribution of impurities in single crystals of
silicon during growth according to the Czochralski
method
M. G. Mil'vidskii, S. P. Grishina, and V. V. Eremeev
Izv. Akad. Nauk SSSR, Neorg. Mater., 1(11):1864-1872
(1965)
Inorg. Mater., 1(11):1686-1692 (1965)

Striations in Czochralski-grown Si crystals
A. Mühlbauer, R. K. Kappelmeyer, and F. Keiner
Z. Naturforsch., 20a:1089-1092 (1965)

The growth of silicon and other materials with
an adjustable close-spaced RF furnace
P. A. Hoss and L. A. Murray
J. Electrochem. Soc., 111:196C (1964)

Rate of growth of silicon from a melt
Yu. M. Shashkov and V. P. Grishin
Zh. Fiz. Khim., 38(12):2992-2994 (1964)

Solute striations in Czochralski-grown silicon
crystals:  Effect of crystal rotation and growth
rates
J. R. Carruthers and K. E. Benson
Appl. Phys. Letters, 3:100-102 (1963)
P-doped

The development of a "faceting effect" in sili-
con single crystals during growth by the
Czochralski method
M. G. Mil'vidskii and A. V. Berkova
Fiz. Tverd. Tela, 5(2):513-517 (1963)
Sov. Phys.—Solid State, 5(2):374-377 (1963)

Growth of silicon bicrystals by the Dash pedes-
tal method
R. Gereth
J. Electrochem. Soc., 109:1068-1070 (1962)

An improvement to the floating-zone method of
growing single crystals
G. W. Green
J. Sci. Instr., 38:167(L) (1961)

On the method of determining the form of the
crystallization front of silicon single crystals
grown from the melt by the Czochralski method
M. G. Mil'vidskii and B. M. Turovskii
Kristallografiya, 6(1):143-145 (1961)
Sov. Phys.—Cryst., 6(1):118-119 (1961)

Surface phenomena in semiconductors and growth
of semiconductor crystals
W. A. Albers, Jr., V. E. Noble, R. P. Poplawsky, and
J. E. Thomas
(Wayne State Univ.), AD-239132 (March 1960), pp. 14-26
Arc image furnace

The shape of melt-crystal interfaces during float
zoning of silicon
J. H. Braun and R. A. Pellin
J. Electrochem. Soc., 107:268C (A) (Dec. 1960)

Improvements on the pedestal method of growing
silicon and germanium crystals
W. C. Dash
J. Appl. Phys., 31:736-737 (L) (1960)

Liquid-solid interface shape observed in silicon
crystals grown by the Czochralski method
W. D. Edwards
Can. J. Phys., 38:439-443 (1960)

Purification of silicon and the production of ho-
mogeneous single crystals
J. Goorissen and B. Okkerse
Acta Electron., 4:479-481 (1960)
Doped crystals by zone melting and pulling, review 123 refs.

The effects of seed rotation of silicon crystals
A. J. Goss and R. E. Adlington
Solid State Physics in Electronics and Telecommunications,
Vol. 1. Semiconductors (M. Desirant and J. L. Michiels,
eds.), Academic Press, London (1960), pp. 28-43
Proceedings of International Conference held at Brussels,
June 2-7, 1958

Floating-zone crystals using an arc image fur-
nace
R. P. Poplawsky and J. E. Thomas, Jr.
Rev. Sci. Instr., 31:1303-1308 (1960)

The preparation of single-crystal silicon for the production of voltage-reference diodes
G. Ashton and M. H. Issott
Proc. IEE, 106B Suppl.:273-276 (1959)
Czochralski; P-doped

Growth of silicon crystals free from dislocations
W. C. Dash
J. Appl. Phys., 30:459-474 (1959)

A silicon-ingot-growing furnace using electron-bombardment heating
D. B. Gasson
Proc. IEE, 106B Suppl.:854-857 (1959)
Czochralski

The effects of seed rotation on silicon crystals
A. J. Goss and R. E. Adlington
Marconi Rev., 22(132):18-36 (First Quarter 1959)

Zone melting of silicon with an electron beam
V. Gusa, I. Krzhizh, and I. Ladnar
Fiz. Tverd. Tela, 1(2):290-293 (1959)
Sov. Phys.—Solid State, 1(2):261-263 (1960)

Uniform resistivity p-type silicon by zone leveling
E. D. Kolb and M. Tanenbaum
J. Electrochem. Soc., 106:597-599 (1959)
Al-doped

The control of resistivity in pulled silicon crystals
A. Trainor and P. T. Harris
Proc. Phys. Soc., 74:669-670(L) (Nov. 1959)

Silicon crystal perfection study
H. J. Yearlian
(Purdue Res. Found.), Final Report, AD-246269 (Aug. 1959), 78 pp.
Modified Czochralski method

The preparation of single-crystal ingots of silicon by the pulling technique
E. Billig and D. B. Gasson
J. Sci. Instr., 35:360-365 (1958)

The contamination of silicon by the crucible
S. E. Bradshaw
Solid State Physics in Electronics and Telecommunications, Academic Press, London (1960), pp. 44-60
Proceedings of an International Conference held in Brussels, June 2-7, 1958

## 12.c.4. Epitaxy, Films, Vapor Deposition

Epitaxy of silicon in cathode sputtering
L. N. Aleksandrov, R. N. Lovyagin, E. A. Krivorotov, and N. E. Dozhdikova
Kristallografiya, 15:203-204 (1970)
Sov. Phys.—Cryst., 15:171-173 (1970)

Heteroepitaxial growth of Si on $ZnSiP_2$
I. Bertoti
J. Mater. Sci., 5:1073-1077 (1970)

Interfacial reactions in an epitaxial Si layer on an $Al_2O_3$ substrate
G. Blet and J. Mercier
Compt. Rend. B, Sci. Phys., 270:964-966 (1970)

Silicon epitaxy from mixtures of $SiH_4$ and HCl
J. Bloem
J. Electrochem. Soc., 117:1397-1401 (1970)

The epitaxial growth of silicon on sapphire and spinel substrates: suppression of changes in the film properties during device processing
G. W. Cullen, G. E. Gottlieb, and C. C. Wang
RCA Review, 31:355-371 (1970), 36 refs.

A new technique for liquid-phase epitaxy
J. A. Donahue and H. T. Minden
J. Crystal Growth, 7:221-226 (1970)

Anisotropy of macrostep motion and pattern edge-displacements during growth of epitaxial silicon on silicon near {100}
C. M. Drum and C. A. Clark
J. Electrochem. Soc., 117:1401-1405 (1970)

Stagnant layer model for the epitaxial growth of silicon from silane in a horizontal reactor
F. C. Eversteyn, P. J. W. Severin, C. H. J. v. d. Brekel, et al.
J. Electrochem. Soc., 117:925-931 (1970)

Stress in silicon films deposited heteroepitaxially on insulating substrates with particular reference to corundum
D. M. Jefkins
J. Phys. D. Appl. Phys., 3:770-777 (1970)

Silicon layers produced by vacuum sublimation at 430-600°C
V. P. Kuznetsov, V. V. Postnikov, and V. A. Tolomasov
Kristallografiya, 15:391-392 (1970)
Sov. Phys. — Cryst., 15:335-336 (1970)

The optimum conditions for the vacuum deposition of silicon on sapphire
R. W. Lawson and Diana M. Jefkins
J. Phys. D: Appl. Phys., 3:1627-1640 (1970)

Dopant transfer in heteroepitaxial Si layers on sapphire substrates
J. Mercier
J. Electrochem. Soc., 117:812-814 (1970)

Thin-film silicon: preparation, properties, and device applications
J. F. Allison, D. J. Dumin, F. P. Heiman, C. W. Mueller, and P. H. Robinson
Proc. IEEE, 57:1490-1498 (1969)

Effect of reactor geometry on growth of epitaxial silicon
R. W. Andrews, D. M. Rynne, and E. G. Wright
Solid State Tech., 12(10):61-66 (1969)

A new method for the epitaxial growth of silicon and its thermodynamical analysis
T. Arizumi, T. Nishinaga, M. Kasuga, and H. Ogawa
Japan. J. Appl. Phys., 8:32 (1969)

Condensation and nucleation processes of single-crystal thin films, Final Report, Oct. 15, 1967, through April 30, 1969
M. Braunstein and R. Kikuchi
(Hughes Research Labs., Malibu, Calif.), AFML-TR-69-149 (April 1969)

Epitaxial growth and properties of silicon on alumina-rich single-crystal spinel
G. W. Cullen, G. E. Gottliev, C. C. Wang, and K. H. Zaininger
J. Electrochem. Soc., 116:1444-1449 (1969)

Localized epitaxy by iodine transfer, final contract report
R. Dubois and H. Valdman
(Société Européenne de Semiconducteurs et de Microélectronique, Paris, France), Contract DGRST-67-90-859-00-212-75-01, RD/AD-600 (Sept. 1969), 65 pp.

Deposition and evaluation of silicon films formed by pyrolytic decomposition of silane on oxidized silicon single crystals
A. L. Fripp, Jr.
M. S. thesis, Univ. of Virginia, NASA-TM-X-61916 (June 1969), 68 pp.

Étude de l'épitaxie à profil controlé
P. Gibeau
(Thomson-CSF, Corbeville, Orsay, France), Rapp. Final D. G. R. S. T., Action concertée: Electron., Contrat No. 6700762 (July 1969), 31 pp.
Si doped with As, P, or B

Silicon epitaxial layers with abrupt interface impurity profiles
D. C. Gupta and R. Yee
J. Electrochem. Soc., 116:1561 (1969)

Epitaxial films of silicon on spinel by vacuum evaporation
T. Itoh, S. Hasegawa, and N. Kaminaka
J. Appl. Phys., 40:2597-2600 (1969)

The epitaxial growth of semiconductors
Morton E. Jones
Reactivity of Solids (1969), pp. 433-451

Heteroepitaxy. 1 (Si on sapphire)
H. F. Matare
Sci. Elec., 15(3):95-109 (1969)
28 refs.

Evaporation of silicon by vacuum-arc discharge
M. Naoe and S. Yamanaka
Japan. J. Appl. Phys., 8:287-288 (1969)

Preparation and properties of homoepitaxial silicon grown at low temperatures from silane
D. Richman and R. H. Arlett
Semiconductor Silicon
(R. R. Haberecht and E. L. Kern, eds.), The ECS, Inc., New York (1969), pp. 200-207

Low-temperature epitaxial growth of single crystalline silicon from silane
D. Richman and R. H. Arlett
J. Electrochem. Soc., 116:872-873 (1969)

The status of silicon epitaxy
W. R. Runyan
Semiconductor Silicon
(R. R. Haberecht and E. L. Kern, eds.), The ECS, Inc., New York (1969), pp. 169-188
101 refs.

Role of surface migration in growth of Si films on Si substrate and on oxide layer
M. Samokhvalov, S. Tajibaeva, and E. Levshin
Phys. Letters, 29A:38 (1969)

Apparatus for producing vapor growth of silicon crystals
T. L. Chu
(Westinghouse Electric Corp.), U. S. Patent 3,372,671 (March 1, 1968)

Surface imperfections arising during thermal treatment of silicon
W. A. FitzGibbons and K. M. Busen
Electrochem. Tech., 6:52 (1968)

Low-temperature silicon epitaxy
R. G. Frieser
J. Electrochem. Soc., 115:401-405 (1968)
Si <111> substrates

Diffusion masking of silicon nitride and silicon oxynitride films on Si
F. K. Heumann, D. M. Brown, and E. Mets
J. Electrochem. Soc., 115:99 (1968)

Eigenschaften epitaxialer Siliziumschichten auf Spinell-Einkristallen
W. Heywang
Mat. Res. Bull., 3:315-328 (1968)

Selection of orientation of sapphire substrates for silicon films
V. D. Ignatkov and V. E. Kosenko
Ukr. Fiz. Zh., 13:1921-1912 (1968)

Growth and perfection of chemically-deposited epitaxial layers of Si and GaAs
B. A. Joyce
J. Crystal Growth, 3:43-59 (1968)

Perfect single-crystal silicon films on sapphire
V. P. Klochkov, V. E. Kosenko, and A. V. Stadnik
Ukr. Fiz. Zh., 13:664-669 (1968)

Epitaxial growth with light irradiation
M. Kumagawa, H. Sunami, T. Terasaki, and Jun-ichi Nishizawa
Japan. J. Appl. Phys., 7:1332 (1968)

Epitaxial silicon and gallium arsenide thin films on insulating ceramic substrates
J. T. Milek
(Hughes Aircraft Co., Culver City, Calif.), Report S-9 (Aug. 1969), 146 pp.
A state-of-the-art literature survey; 265 refs.

Silicon vapour epitaxy
B. S. Murthy, R. Kesavan, and K. V. Ratnavati
J. Inst. Telecommun. Engrs. (India), 14:486-495 (1968)

The deposition of silicon on sapphire in ultra-high vacuum
C. T. Naber and J. E. O'Neal
Trans. AIME, 242:470-479 (1968)

Vacuum evaporated silicon single crystal films on sapphire
S. Namba, A. Kawazu, and T. Maruyama
Finommechanika (Hungary), 7:45-49 (1968)

The deposition of silicon on single-crystal spinel substrates
P. H. Robinson and D. J. Dumin
J. Electrochem. Soc., 115:75 (1968)

Method of forming a crystalline semiconductor layer on an alumina substrate
P. H. Robinson and D. J. Dumin

(Radio Corp. of America), Patent U. S. 3,413,145 (Nov. 29, 1965; publ. Nov. 26, 1968)
Single-crystal silicon

The epitaxial growth of silicon in horizontal reactors
P. C. Rundle
Intern. J. Electron., 24:405-413 (1968)

Doping of epitaxial silicon films
W. H. Shepherd
J. Electrochem. Soc., 115:541 (1968)
P, As, Sb, as tetrachlorides

Epitaxial growth of Si on sapphire by hydrogen reduction of SiCl₄
M. Tamura and M. Nomura
Oyo Buturi, 37(6):496-504 (1968)

Low-temperature epitaxial growth of pn junctions by ultra-high vacuum sublimation
R. N. Thomas and M. H. Francombe
Appl. Phys. Letters, 13:270-272 (1968)

Vacuum deposition of single-crystalline silicon on sapphire
L. R. Weisberg and E. A. Miller
Trans. AIME, 242:479-484 (1968)

The preparation of sapphire for silicon epitaxy
T. A. Zeveke, L. N. Kornev, and V. A. Tolomasov
Kristallografiya, 13(3):579-581 (1968)
Sov. Phys.—Crystal., 13(3):493-495 (1968)

Large-area silicon junctions by epitaxial growth techniques
J. Electrochem. Soc., 114:522-525 (1967)

Apparatus for vapor deposition of silicon
F. Bischoff
(Siemens und Halske Akt.-Ges.), U. S. Patent 3,335,697 (1967)

The effects of gas pressure and velocity on epitaxial silicon deposition by the hydrogen reduction of chlorosilanes
S. E. Bradshaw
Intern. J. Electron., 23:381-391 (1967)

Isoépitaxie du silicium par évaporation thermique sous ultra-vide
C. Constantin, R. Cinti, and R. Montmory
Vide, 22:105 (1967)

Electrical properties of silicon films grown epitaxially on sapphire
D. J. Dumin
J. Appl. Phys., 38:1909 (1967)

Induction heating in the epitaxial process
W. A. Emerson
Semicond. Prod.-Solid State Tech., 10:50 (1967)
Three epitaxial reactor designs—the horizontal boat, the vertical tube, and the "pancake"—are outlined

A study of nucleation in chemically grown epitaxial silicon films using molecular beam techniques. III. Nucleation rate measurements and the effect of oxygen on initial growth behaviour
B. A. Joyce, R. R. Bradley, and G. R. Booker
Phil. Mag., 15:1167-1187 (1967)
On (111) silicon substrates

Condensation and nucleation processes of single-crystal thin films
R. Kikuchi and M. Braunstein
(Hughes Research Labs., Malibu, Calif.), AFML-TR-67-410 (Dec. 1967), 157 pp.

Silicon heteroepitaxy on oxides by chemical vapor deposition
T. J. La Chapelle, Arnold Miller, and F. L. Morritz
Progress in Solid State Chemistry, Vol. 3 (H. Reiss, ed.), Pergamon Press, New York, London (1967), pp. 1-44

Vacuum-deposited single-crystal silicon film on sapphire
S. Namba, A. Kawazu, and T. Maruyama
Sci. Papers Inst. Phys. Chem. Res. (Tokyo), 61:45-54 (1967)

Epitaxial deposition of silicon by pyrolysis of SiH₄
Y. Nishi and M. Watanabe
Japan. J. Appl. Phys., 6:550-551 (1967)

Growth from vapor of thin single-crystal silicon on spinel-structured substrates for thin-film electronics
North American Aviation
(U. S. appl. corresponding to) Neth. appl. 6,609,160; appl. U. S. June 30, 1965; publ. Jan. 2, 1967

The growth and etching of Si through windows in SiO₂
W. G. Oldham and R. Holmstrom
J. Electrochem. Soc., 114:381-388 (1967)

A cause and cure of stacking faults in silicon epitaxial layers
D. Pomerantz
J. Appl. Phys., 38:5020 (1967)

Epitaxial silicon layers obtained by sublimation in vacuum
V. V. Postnikov et al.
Dokl. Akad. Nauk SSSR, 175(4):817-818 (1967)
Sov. Physics—Doklady, 12(8):822-823 (1968)

Vacuum deposition of silicon on corundum
F. H. Reynolds and A. B. M. Elliot
Solid State Electron., 10:1093 (1967)

Behavior of large-scale surface perturbations during silicon epitaxial growth
W. R. Runyan, E. G. Alexander, and S. E. Craig, Jr.
J. Electrochem. Soc., 114:1154 (1967)

Characteristics of the growth of silicon single crystals from the gas phase
I. V. Salli, E. S. Fal'kevich, Yu. S. Dement'ev, E. N. Chukal'skii, V. V. Belousova, and V. G. Khrebtishchev
Kristallografiya, 12(3):499-507 (1967)
Sov. Phys.—Cryst., 12(3):425-431 (1967)

A low-energy electron diffraction study of the epitaxial silicon layers on a Ge (111) surface
Y. Takeishi, I. Sasaki, and K. Hirabayashi
Appl. Phys. Letters, 11:330 (1967)

Observation of chemical-vapor-deposited silicon on sapphire by transmission electron microscopy
M. Tamura and M. Nomura
Appl. Phys. Letters, 11:196 (1967)

Low-temperature vacuum deposition of homoepitaxial silicon
    L. R. Weisberg
    J. Appl. Phys., 38(11):4537 (1967)

A study of factors affecting silicon growth on
amorphous $SiO_2$ surfaces
    E. A. Alexander and W. R. Runyan
    Trans. AIME, 236:284–290 (1966)

Thin single-crystal silicon layers by vaporization onto a substrate, then recrystallization,
both by use of an electron beam
    M. Balkanski
    French Patent 1,449,900 (Appl. July 9, 1965; publ. Aug. 19,
    1966)

Nucleation and crystal growth of silicon on sapphire
    J. M. Blank and V. A. Russell
    Trans. AIME, 236:291–294 (1966)

A study of nucleation in chemically grown epitaxial silicon films using molecular beam techniques.  II. Initial growth behaviour on clean
and carbon-contaminated silicon substrates
    G. R. Booker and B. A. Joyce
    Phil. Mag., 14:301–315 (1966)
    On (111) silicon

The kinetics of epitaxial silicon deposition by
the hydrogen reduction of chlorosilanes
    S. E. Bradshaw
    Intern. J. Electron., 21:205–227 (1966)

Epitaxial growth of Si on Si webs and Si slices
    T. L. Chu
    J. Electrochem. Soc., 113:717–720 (1966)

In situ etching of silicon substrates prior to
epitaxial growth
    T. L. Chu, G. A. Gruber, and R. Stickler
    J. Electrochem. Soc., 113:156–158 (1966)

Autodoping of silicon films grown epitaxially on
sapphire
    D. J. Dumin and P. H. Robinson
    J. Electrochem. Soc., 113:469–472 (1966)
    Al doping

Growth of epitaxial layers of silicon by sublimation through thin alloy zones
    J. D. Filby and S. Nielsen
    Brit. J. Appl. Phys., 17:81–86 (1966)
    Vapour–liquid–solid mechanism

Selected area deposition of single-crystal silicon on amorphous quartz
    J. D. Filby and S. Nielsen
    J. Electrochem. Soc., 113:1091–1092 (1966)

Epitaxial deposition of silicon using ultra-thin
alloy zone crystallization
    J. D. Filby and S. Nielsen
    Microelectron. Reliabil., 5:11–14 (1966)

Mechanism of growth of epitaxial layers of silicon
    J. D. Filby, S. Nielsen, and G. J. Rich
    The Use of Thin Films in Physical Investigations, pp. 233–
    243; discussion, pp. 253–257 (J. C. Anderson, ed.), Aca-
    demic Press, London and New York (1966)
    A Nato Advance Study Inst. held at Imperial College of Sci-
    ence and Technology, Univ. London, July 19–24, 1965

Épitaxie du silicium sur corindon.  Mesures des
caractéristiques électroniques des dépôts épitax-
iaux
    J. L. Fraimbault, I. Gyomlai, R. Montmory, and J. Vuillod
    Basic Probl. Thin Film Phys. Proc. Intern. Symp. Clausthal-
    Göttingen (1965), Vandenhoeck and Ruprecht (1966), pp.
    638–645

Study of the early stages of the epitaxy of silicon on silicon
    F. Jona
    Appl. Phys. Letters, 9:235 (1966)
    Low–energy electron diffraction

Condensation and nucleation processes of single-
crystal thin films
    R. Kikuchi, A. F. Kaspaul, M. Braunstein, E. E. Kaspaul,
    and W. E. McKee
    (Hughes Research Labs., Malibu, Calif.), Summary Techni-
    cal Report July 1, 1965, to June 10, 1966, AFML-TR-
    66–326 (Oct. 1966), 211 pp.
    Crystalline and amorphous dielectric substrates

Single-crystal silicon on spinel
    H. Manasevit and D. H. Forbes
    J. Appl. Phys., 37:734–739 (1966)

Growth of single-crystal silicon on beryllium
oxide
    H. M. Manasevit, D. H. Forbes, and I. B. Cadoff
    Trans. AIME, 236:275–279 (1966)

Single-crystal silicon epitaxy on foreign substrates
    A. Miller and H. M. Manasevit
    J. Vacuum Sci. Tech., 3:68–78 (1966)
    $\alpha$-$Al_2O_3$, spinel, BeO

The preparation of the surface of single-crystal
substrates for epitaxial growth of silicon
    Pavel Polivka, Zdenek Lhotak, and Juraj Eckstein
    Cesk. Casopis Fys., 16:108 (1966)
    Substrate plates of silicon

The deposition of silicon upon sapphire substrates
    P. H. Robinson and C. W. Mueller
    Trans. AIME, 236:268–274 (1966)

Die Abtragung von Silicium im System Si–Cl–H
    H. Seiter and E. Sirtl
    Z. Naturforsch., A21:1696–1702 (1966)
    Growth; surface purity

Vapor-deposited microcrystalline silicon
    E. Sirtl and H. Seiter
    J. Electrochem. Soc., 113:506–507 (1966)

Epitaxial growth of silicon on hexagonal silicon
carbide
    R. L. Tallman, T. L. Chu, G. A. Gruber, J. J. Oberly, and
    E. D. Wolley
    J. Appl. Phys., 37:1588–1593 (1966)

Directional growth of single-crystal silicon films
across silicon carbide by a moving deposition-
zone technique
    R. L. Tallman, T. L. Chu, and J. J. Oberly
    Solid–State Electron., 9:327–330 (1966)

Rheotaxial growth of silicon with germanium —
silicon alloys
    Y. Tarui, H. Teshima, and N. Gomyo
    Bull. Electrotech. Lab. (Japan), 30:109–116 (1966)
    Single crystalline films of polycrystalline W substrates

Thermochemistry of Si halides. 4. Equilibrium
reaction $SiCl_4(g) + Si(f) = 2SiCl_2(g)$ studies by
the flow method
R. Teichmann and E. Wolf
Z. Anorg. Allgem. Chem., 347:145–155 (1966)

Growth and structure of evaporated silicon layers
D. J. D. Thomas
Phys. Stat. Sol., 13:359–372 (1966)

Low-resistivity epitaxial layers of silicon
H. Thomas and W. G. Townsend
Solid-State Electronics, 9:1137–1139 (1966)

Controlled vapor — liquid — solid growth of sili-
con crystals
R. S. Wagner and C. J. Doherty
J. Electrochem. Soc., 113:1300–1305 (1966)

Thermochemistry of Si halides. 3. Equilibrium
reaction $SiBr_4(g) + Si(f) = 2SiBr_2(g)$ in the 1200–
1550°K range
E. Wolf and C. Herbst
Z. Anorg. Allgem. Chem., 347:113–122 (1966)

Anomalous photovoltaic films produced by subli-
mation of silicon atoms from the surface of a
current-carrying silicon plate
E. I. Adirovich and L. M. Gol'dshtein
Dokl. Akad. Nauk SSSR, 158(2):309–312 (1964)
Sov. Phys.–Doklady, 9(9):795–797 (1965)

Kinetics of the silicon — silicontetrachloride re-
action in a flow system
Ole Alstrup and C. O. Thomas
J. Electrochem. Soc., 112:319–323 (1965)

Évaporation par bombardement électronique de
métaux à hauts points de fusion
M. Aubecz, M. Brabers, M. Henuset, and M. Meulemans
Mem. Sci. Rev. Met., 62:373–378 (1965)
V, Nb, Cr, Si, and B

Investigation of growth of single-crystal films
on dielectric substrates
J. M. Blank, E. C. Henry, K. K. Reinhartz, and V. A.
Russell
(General Electric Co., Syracuse, N. Y.), Tech. Documentary
Report, Contract AF 33(615)-1539, Project No. 7371
and 737102 (Aug. 25, 1965), 70 pp.
Nucleation and growth of silicon on sapphire

Low-temperature epitaxial growth of Si (inverted
transport in close-spaced technique)
J. Bloem and J. W. A. Scholte
J. Electrochem. Soc., 112:1211 (1965)

Growth of epitaxial silicon layers by vacuum
evaporation. II. Initial nucleation and growth
G. R. Booker and B. A. Unvala
Phil. Mag., 11–30 (1965)
On (111) Si substrates

Low-temperature epitaxy of silicon by sublima-
tion onto thin alloy layers
J. D. Filby and S. Nielsen
J. Electrochem. Soc., 112:535–536 (1965)

Progress toward single-crystal silicon films on
amorphous substrates
J. D. Filby and S. Nielsen
J. Electrochem. Soc., 112:957–958 (1965)
Ultrathin alloy-zone recrystallization

Single-crystal silicon on sapphire: a new dimen-
sion in microelectronics flexibility
D. H. Forbes and H. M. Manasevit
Wescon Tech. Papers, Part 2, 9: art. 16A. 2 (1965)

Control of stacking faults in epitaxial silicon
Y. Haneta
Japan. J. Appl. Phys., 4:69–70 (1965)

The epitaxial deposition of silicon on quartz and
alumina
B. A. Joyce, R. J. Bennett, R. W. Bicknell, and P. J. Etter
Trans. AIME, 233:556–562 (1965)

Kinetics of epitaxial silicon deposition by a low-
pressure iodide process
J. E. May
J. Electrochem. Soc., 112:710–713 (1965)

Evaporation of silicon to obtain epitaxial films
A. I. Petrin and G. A. Kurov
Kristallografiya, 10(5):754–756 (1965)
Sov. Phys.—Cryst., 10(5):634–635 (1966)
On Si substrates

Integrated silicon device technology
Research Triangle Institute, Durham, N. C., ASD-TDR-63-
316, Vol. IX (Aug. 1965) and Vol. X (Nov. 1965)
Vol. IX— Methods and practical applications of silicon epi-
taxy; Vol. X— Chemical/metallurgical properties of silicon
Complete set of silicon binary phase diagrams

Epitaktische Siliziumschichten auf Mg — Al- Spi-
nell
H. Seiter and C. Zaminer
Z. Angew. Phys., 20:158–161 (1965)

Vapor-phase deposition and etching of silicon
W. H. Shepherd
J. Electrochem. Soc., 112:988–994 (1965)

Chemical vapour deposition promoted by r.f. dis-
charge
H. F. Sterling and R. C. G. Swann
Solid-State Electron., 8:653–654 (1965)

The effects of substrate orientation on epitaxial
growth
S. K. Tung
J. Electrochem. Soc., 112:436–438 (1965)

Epitaxiales Wachstum von Silizium im Ultrahoch-
vakuum
H. Widmer
Phys. Verh. D. P. G., 5:130 (1965)
Si on Si

Research on epitaxial growth by the mass trans-
fer method
M. Yasufuki, K. Maeda, and E. Yamagami
Fujitsu Sci. Tech. J. (Japan), 1(2):51–71 (1965)

The epitaxial deposition of silicon on quartz
R. W. Bicknell, J. M. Charig, B. A. Joyce, and D. J.
Stirland
Phil. Mag., 9:965–978 (1964)

Si epitaxial growth by iodide transport between
piled up pellets
R. M. A. Lieth and A. G. M. Eggels
J. Appl. Phys., 35:3015–3016 (1964)

Multiple slice epitaxial deposition of silicon in resistance heated furnace
B. A. Lombos and T. R. Somogyi
J. Electrochem. Soc., 111:1097-1098 (1964)

Polycrystalline silicon films on foreign substrates
W. J. McAleer, M. A. Kozlowski, and P. I. Pollak
J. Electrochem. Soc., 111:877-878 (1964)
Zone melting; vapor deposition

A kinetic study of the system Si-SiCl
R. R. Monchamp, W. J. McAleer, and P. I. Pollak
J. Electrochem. Soc., 111:879-881 (1964)

Preparation of epitaxial layers of silicon. II. Effect of impurities
S. Nielsen and G. J. Rich
Microelectron. Reliabil., 3:171-173 (1964)

Analysis of the hydrogen reduction of silicon tetrachloride process on the basis of a quasi-equilibrium model
T. O. Sedgwick
J. Electrochem. Soc., 111:1381-1383 (1964)

The kinetics of epitaxial growth of silicon from the trichlorosilane-hydrogen reaction
A. M. Stein
J. Electrochem. Soc., 111:483-484 (1964)

Epitaxial growth of silicon by vacuum sublimation
E. Tannenbaum Handelman and E. I. Povilonis
J. Electrochem. Soc., 111:201-206 (1964)

Growth of epitaxial silicon layers by vacuum evaporation. I. Experimental procedure and initial assessment
B. A. Unvala and G. R. Booker
Phil. Mag., 9:691-701 (1964)

The use of electron beams in the preparation of silicon "epitaxial" films
J. Wales
Symposium Electron Beam Technology. Microelectron. Malvern, Worcs., 1964, pp. 21-22

Epitaxial growth of Si on Si in ultra-high vacuum
H. Widmer
Appl. Phys. Letters, 5:108-110 (1964)

Epitaxial deposition of silicon on quartz
B. A. Joyce, R. W. Bicknell, J. M. Charig, and D. J. Stirland
Solid State Commun., 1:107-108 (1963)

Epitaxial growth of silicon from the pyrolysis of monosilane on silicon substrates
B. A. Joyce and R. R. Bradley
J. Electrochem. Soc., 110:1235-1240 (1963)

Preparation of evaporated silicon films
B. F. Kilgore and R. W. Roberts
Rev. Sci. Instr., 34:11-12 (1963)
From an etched silicon filament

Epitaxial growth of silicon by vacuum sublimation
Y. Nannichi
Nature, 200:1087-1088 (1963)
Single-crystal silicon substrate

Die "Sandwich-Methode" – ein neues Verfahren zur Herstellung epitaktisch gewachsener Halbleiterschichten
E. Sirtl
J. Phys. Chem. Solids, 24:1285-1289 (1963)

Thermodynamical approach to the growth rate of epitaxial silicon from SiCl$_4$
W. Steinmaier
Philips Res. Repts., 18:75-81 (1963)

Epitaxial growth of silicon by hydrogen reduction of SiHCl$_3$ onto silicon substrates
J. M. Charig and B. A. Joyce
J. Electrochem. Soc. 109, 957-962 (1962)

Growth mechanism and defect structures in epitaxial silicon
J. M. Charig, B. A. Joyce, D. J. Stirland, and R. W. Bicknell
Phil. Mag., 7:1847-1860 (1962)

Epitaxial growth of silicon
C. H. Li
J. Electrochem. Soc., 109:952 (1962)

Single-crystal silicon overgrowths
Albert Mark
J. Electrochem. Soc., 108:880-885 (1961)
Pyrolytic vapor phase deposition on parent substrates

Epitaxial silicon films by the hydrogen reduction of SiCl$_4$
H. C. Theuerer
J. Electrochem. Soc., 108:649-653 (1961)

Epitaxial growth of silicon
E. S. Wajda, B. W. Kippenhan, and W. H. White
IBM J. Res. Dev., 4:288-295 (1960)

Research directed toward the growth of silicon crystals from silane
T. J. Carroll, R. F. Lever, and J. K. Powers (Philco)
U. S. Gov. Res. Rept. 32, 385(A), Sept. 11, 1959, PB151841

Method of producing semiconductor crystal bodies
R. C. Sangster
(Hughes Aircraft), U. S. Patent 2,895,858 (July 21, 1959)
Si on Si seed, thermal decomposition of silicon halides

### 12.c.5. Dendrites, Needles, and Whiskers

Growth features of silicon filamentary crystals grown from solution
W. N. Borle
J. Appl. Phys., 41:3184-3186 (1970)

Production of fine silicon crystals using photolithography for electron microscopic investigation
Yu. P. Boitsov, V. I. Prokhorov, and L. M. Sorokin
Pribory Tekh. Eksper., No. 5:171 (1969)

Growth of silicon filamentary crystals from solution
W. N. Borle, S. Tata, and V. Prakash
Indian J. Pure Appl. Phys., 6(10):576-577 (1968)

Effective impurity-distribution coefficients during growth of silicon needles from a supercooled melt
Yu. M. Shashkov and V. P. Grishin

Izv. Akad. Nauk SSSR, Neorg. Mater., 4(9):1429–1433 (1968)
Inorg. Mater., 4(9):1255–1257 (1968)

**Growth of thread-like silicon crystals**
S. V. Starodubtsev, D. A. Agievskii, and S. A. Bakaev
Izv. Akad. Nauk Uzb. SSR, Ser. Fiz.-Mat. Nauk, 12(4):76–78 (1968)

**Process for growth of single-crystal silicon ribbon**
J. C. Boatman and P. C. Goundry
Electrochem. Tech., 5:98–101 (1967)

**A p-n junction in silicon whiskers grown by VLS methods**
E. Komatsu, Y. Higuchi, and T. Niina
Appl. Phys. Letters, 10:42 (1967)
From gold-silicon alloys

**Morphology of needle, whisker, and strip crystals of silicon**
A. V. Sandulova, A. I. Andrievskii, and M. I. Dronyuk
Rost Kristallov, Vol. 4, Nauka (1964), pp. 125–128
Growth of Crystals, Vol. 4 (A. V. Shubnikov and N. N. Sheftal', eds.), Consultants Bureau, New York (1966), pp. 101–103
From Br gas

**Silicon whisker growth by the vapour-liquid-solid process**
P. R. Thornton, D. W. F. James, C. Lewis, and A. Bradford
Phil. Mag., 14:165–168 (1966)

**Growth and morphology of silicon ribbons**
S. N. Dermatis, J. W. Faust, Jr., and H. F. John
J. Electrochem. Soc., 112:792–796 (1965)
From supercooled melt

**Silicon whisker growth and epitaxy by the vapour-liquid-solid mechanism**
D. W. F. James and C. Lewis
Brit. J. Appl. Phys., 16:1089–1094 (1965)

**Sur le mécanisme VLS de croissance des whiskers**
B. Mutaftschiev, R. Kern, and C. Georges
Phys. Letters, 16:32–33 (1965)

**Structure des dendrites de silicium**
Ju. M. Shashkov and V. P. Grishin
Dokl. Akad. Nauk SSSR, 162(6):1349–1351 (1965)

**The vapor-liquid-solid mechanism of crystal growth and its application to silicon**
R. S. Wagner and W. C. Ellis
Trans. AIME, 233:1053–1064 (1965)

**Vapor-liquid-solid mechanism of single crystal growth**
R. S. Wagner and W. C. Ellis
Appl. Phys. Letters, 4:89–90 (1964)
Growth "catalysis" from impurity; whiskers, epitaxial, and large crystals; Si

**Study of the filamentary growth of silicon crystals from the vapor**
R. S. Wagner, W. C. Ellis, K. A. Jackson, and S. M. Arnold
J. Appl. Phys., 35:2993–3000 (1964)

**Formation of silicon whiskers on a sublimating surface**
Y. Nannichi
Appl. Phys. Letters, 3:139–142 (1963)

**Preparation of silicon ribbons**
E. S. Greiner, J. A. Gutowski, and W. C. Ellis
J. Appl. Phys., 32:2489–2490 (1961)

**Spiral growth of silicon crystals**
N. A. Shamba and N. N. Sheftal'
Kristallografiya, 2(3):441–446 (1957)
Sov. Phys.—Cryst., 2(3):439–443 (1958)
Needles 6–7 cm long

## 12.c.6. Doping, Diffusion, and Precipitation (see also 18b)

**Precipitation of impurities into silicon**
R. Schuttler, X. Fortin, J. F. Roux, and A. Roizes
ONERA-NT-02-25 (July 1971), 19 pp.
Service de Physique Electronique des Solides, O. N. E. R. A./C. E. R. T./D. E. R. T. S., BP 4025, Toulouse, 31, France

**Diffusion of Au in n-type Si**
A. Z. Badalov and V. B. Shuman
Fiz. Tekh. Poluprovodnikov, 3:1366–1369 (1969)
Sov. Phys.—Semicond., 3:1137–1139 (1970)

**Diffusion, solubility, and electrical properties of cobalt in silicon**
M. K. Bakhadyrkhanov, B. I. Boltaks, and G. S. Kulikov
Fiz. Tverd. Tela, 12(1):181–189 (1970)
Sov. Phys.—Solid State, 12(1):144–149 (1970)

**Diffusion, solubility and electrical properties of zinc in silicon**
M. K. Bakhadyrkhanov, B. I. Boltaks, G. S. Kulikov, and E. M. Pedyash
Fiz. i Tekhn. Poluprovod., 4(5):873–878 (1970)
Sov. Phys.—Semicond., 4(5):739–743 (1970)

**Doped oxides as diffusion sources. II. Phosphorus into silicon**
M. L. Barry
J. Electrochem. Soc., 117:1405–1410 (1970)

**Precipitation kinetics of solid solutions of chromium in silicon**
N. T. Bendnik, V. S. Garnyk, and L. S. Milevskii
Fiz. Tverd. Tela, 12(1):190–195 (1970)
Sov. Phys.—Solid State, 12(1):150–154 (1970)

**Solid solubility of Zn in Si**
M. M. Blouke, N. Holonyak, Jr., B. G. Streetman, and H. R. Zwicker
J. Phys. Chem. Solids, 31:173–177 (1970)

**Electrical properties of silicon doped with platinum**
H. Carchano and C. Jund
Solid-State Electron., 13:83–90 (1970)
Diffused into n-type and p-type silicon at 900, 1100, and 1250°C

**Propriétés de diffusion du platine dans le silicium**
J. Charlot and A. Vapaille
Compt. Rend., B 270:609–611 (1970)

**Silicon defect structure induced by arsenic diffusion and subsequent steam oxidation**
S. Dash and M. L. Joshi
IBM J. Res. Develop., 14:453–460 (1970)

**Bulk diffusion of phosphorus in silicon in hydrogen atmosphere**
R. N. Ghoshtagore
Appl. Phys. Letters, 17(4):137–138 (1970)

Experimental-condition dependence of phosphorus diffusivity in silicon
R. N. Ghoshtagore
Phys. Rev. Letters, 25(13):856–858 (1970)

Diffusion of phosphor in silicon with gaseous doping substance
W. Henning, D. Exner, K. Herrmann, and Y. Yucelen
Z. Angew. Physik, 29:114–117 (1970)

Phosphorous diffusion in silicon using phosphine
Y. W. Hsueh
J. Electrochem. Soc., 117:807–811 (1970)

Boron diffusion in silicon along different crystallographic orientations
R. A. Kovalev, V. B. Bernikov, Yu. I. Pashintsev, and V. A. Marasanov
Fiz. Tverd. Tela, 11:1953–1955 (1969)
Sov. Phys.—Solid State, 11:1576–é573 (1970)

Donor behavior in indium-alloyed silicon
J. O. McCaldin and J. W. Mayer
Appl. Phys. Letters, 17(9):365–366 (1970)
Anomalous doping behavior of Si regrown from In solution

Optical properties of single-crystal silicon arsenide
L. C. E. Miller and C. R. Kannewurf
J. Phys. Chem. Solids, 31:849–855 (1970)
If arsenic is added in significant amounts to prepare a stoichiometric binary composition, new semiconductor structures are formed

Determination of diffusion profiles in silicon 1962–1969. Diffusion in silicon 1967–1969
A. R. Pierce and B. A. Stevens
Bell Telephone Laboratories Bibl. No. 151 (March 1970)

Doping of epitaxial silicon; equilibrium gas phase and doping mechanism
P. Rai-Choudhury and E. I. Salkovitz
J. Crystal Growth, 7:353–360 (1970)

Doping of epitaxial silicon; effect of dopant partial pressure
P. Rai-Choudhury and E. I. Salkovitz
J. Crystal Growth, 7:361–367 (1970)

The diffusion of iron and nickel to silicon surfaces
J. W. T. Ridgway and D. Haneman
Phys. Stat. Sol., 38:K31 (1970)

Diffusion of boron in epitaxial silicon
I. B. Sladkov, V. V. Tuchkevich, and N. M. Shmidt
Fiz. i Tekhn. Poluprovod., 4(4):793–796 (1970)
Sov. Phys. — Semicond., 4(4):673–675 (1970)

Beryllium as an acceptor in silicon
E. A. Taft and R. O. Carlson
J. Electrochem. Soc., 117:711–713 (1970)

Precipitation of Li in Si
T. Takabatake, T. Furuya, and Y. Ueda
Japan. J. Appl. Phys., 9:416–417 (1970)

Anomalous diffusion in semiconductors — a quantitative analysis
N. D. Thai
Solid State Electron., 13:165–172 (1970)

Concentration-dependent diffusion of boron and phosphorus in silicon
N. D. Thai
J. Appl. Phys., 41:2859 (1970)

Diffusion of antimony, phosphorus, and boron in silicon with various surface concentrations of the diffusant
V. A. Uskov, P. V. Pavlov, E. V. Kuril'chik, and V. I. Pashkov
Fiz. Tverd. Tela, 12(5):1504–1510 (1970)
Sov. Phys.—Solid State, 12(5):1181–1185 (1970)

Preparation of thin silicon crystals by electrochemical thinning of epitaxially grown structures
H. Ja. van Dijk and J. de Jonge
J. Electrochem. Soc., 117:553–554

Anomalous diffusion of phosphorus into silicon
K. Yagi, N. Miyamoto, and J. Nishizawa
Japan. J. Appl. Phys., 9:246–254 (1970)

Dissociative diffusion of gold in silicon
M. Yoshida and K. Saito
Japan. J. Appl. Phys., 9(10):1217–1228 (1970)

Zinc diffusion into silicon by a closed-tube, two-temperature technique
S. M. Zalar
J. Appl. Phys., 41(8):3458–3464 (1970)

Lithium-boron ion pairing in silicon
C. A. J. Ammerlaan and W. E. van der Vliet
Phys. Rev. Letters, 23:470–472 (1969)
Equilibrium constant of reaction $Li^+ + B^- \rightleftharpoons LiB$

Diffusion parameters of platium in silicon
R. F. Bailey and T. G. Mills
Semiconductor Silicon (R. R. Haberecht and E. L. Kern, eds.), Electrochemical Society, New York (1969), pp. 481–489

Oxygen and carbon content of Czochralski silicon crystals
J. A. Baker
Semiconductor Silicon (R. R. Haberecht and E. L. Kern, eds.), Electrochemical Society, New York (1969), pp. 566–573

Doped oxides as diffusion sources. I. Boron into silicon
M. L. Barry and P. Olofsen
J. Electrochem. Soc., 116:854–860 (1969)

Electron microscopic observations of $SiO_2$ precipates at dislocations in silicon
D. Bialas and J. Hesse
J. Mater. Sci., 4:779–783 (1969)
Formed preferentially at dislocations

Segregation of gold to the silicon (111) surface observed by Auger emission spectroscopy and by LEED
H. E. Bishop and J. C. Riviere
Brit. J. Appl. Phys., 2:1635–1642 (1969)

Stresses in silicon after boron diffusion. I. Determination of residual stresses in boron-diffused silicon slices using Lang's method
O. Brummer and H. R. Hoche
Kristal. Tech., 4:279–285 (1969)

Stresses in silicon after boron diffusion. II. Stresses between the glaze-like reaction phase and silicon and their detection by means of the Berg-Barrett method

O. Brummer and H. R. Hoche
Kristal. Tech, 4:287–291 (1969)

**Enhanced gold solubility effect in heavily n-type silicon**
S. F. Cagnina
J. Electrochem. Soc., 116:498 (1969)

**Experimental investigation of diffusion of dopants along dislocations introduced into silicon by electron beam heating**
G. V. Dudko, M. A. Kolegaev, and V. A. Panteleev
Tverd. Tela, 11:1356–1359 (1969)
Sov. Phys.—Solid State, 11:1097–1100 (1969)
P, Sb, B, Al

**Concentration profiles of two-step diffusions of boron into silicon**
E. P. Dudley
M. S. thesis, Univ. Maryland, AD-684907; HDL-TR-1424 (Jan. 1969), 44 pp.

**Hydride sources for diffusion of dopants into silicon-on-sapphire films**
D. J. Dumin
J. Electrochem. Soc., 116:133–137 (1969)

**Doping epitaxial silicon layers with arsenic**
T. A. Dutkina, I. M. Skvortsov, and G. B. Fedosova
Izv. Akad. Nauk SSSR, Neorg. Mater., 5(4):789–790 (1969)
Inorg. Mater., 5(4):670–671 (1969)

**Silicon nitride as a mask in phosphorus diffusion**
I. Franz and W. Langheinrich
Solid State Electron., 12:955–962 (1969)

**Distribution of copper and gold in silicon crystals grown by crucible-free zone melting**
H.-F. Hadamovsky
Rost Kristallov, Vol. 7, Nauka (1967), pp. 236–240
Growth of Crystals, Vol. 7 (N. N. Sheftal', eds.), Consultants Bureau, New York—London (1969), pp. 206–210

**A mechanism of gold diffusion into silicon**
Y. Hayashi, Y. Tarui, and T. Komuro
Bull. Electrotech. Lab. (Japan), 33(6):624–630 (1969)

**Phosphorus diffusion into silicon using phosphine**
M. S. R. Heynes and P. G. G. van Loon
J. Electrochem. Soc., 116:890–893 (1969)

**Diffusion in silicon**
D. L. Kendall and D. B. DeVries
Semiconductor Silicon (R. R. Haberecht and E. L. Kern, eds.), Electrochemical Society, New York (1969), pp. 358–421

**Ambipolar diffusion of impurities into silicon**
D. P. Kennedy
Proc. IEEE, 57:1202–1203 (1969)

**Doping of silicon epitaxial layers with arsenic at very high concentrations**
H. Krause
Krist. Tech., 4(3):359–364 (1969)

**Effects and control of surface states during lithium ion drift in silicon**
R. P. Lothrop
Lawrence Radiation Lab., Univ. California, Berkeley
UCRL-19413 (Nov. 1969), 43 pp.

**Arsenic isoconcentration diffusion studies in silicon**
B. J. Masters and J. M. Fairfield
J. Appl. Phys., 40:2390–2394 (1969)

**Boron diffusion into silicon using elemental boron**
M. Okamura
Japan. J. Appl. Phys., 8:1440–1447 (1969)

**Diffusion of antimony into silicon from a layer bombarded by ions**
V. A. Panteleev and L. N. Isavtseva
Izv. Vysshykh. Uchebn. Zavedenii SSSR, 8:27–30 (1969)
Sov. Phys. J. (USA).

**Oxidation, defects and vacancy diffusion in silicon**
I. R. Sanders and P. S. Dobson
Phil. Mag., 20:881–893 (1969)
P and B

**Entry of impurities into epitaxial silicon films during growth**
S. V. Starodubtsev, A. S. Lyutovich, V. V. Kharchenko, and V. P. Prutkin
Rost Kristallov, Vol. 8, Nauka (1968), pp. 264–270
Growth of Crystals, Vol. 8 (N. N. Sheftal', eds.), Consultants Bureau, New York—London (1969), pp. 217–221

**Phosphorus and arsenic doping of epitaxial silicon films in the 1000° to 1200°C temperature range**
T. B. Swanson and R. N. Tucker
J. Electrochem. Soc., 116:1271–1274 (1969)

**Antimony diffusion into silicon using $SbCl_5$**
Y. Tarui, Y. Hayashi, and K. Awauda
Bull. Electrotech. Lab. (Japan), 33(6):716–718 (1969)

**Phosphorous shallow diffusion in silicon using $POCl_3$**
Y. Tarui, Y. Hayashi, and K. Tanaka
Bull. Electrotech. Lab. (Japan), 33(6):708–715 (1969)

**Boron diffusion into silicon using $BBr_3$**
Y. Tarui, Y. Hayashi, and N. Yui
Bull. Electrotech. Lab. (Japan), 33(6):700–707 (1969)

**Shallow phosphorus diffusion profiles in silicon**
J. C. C. Tsai
Proc. IEEE, 57:1499–1506 (1969)

**Solid solubility and diffusion coefficients of boron in silicon**
G. L. Vick and K. M. Whittle
J. Electrochem. Soc., 116:1142–1144 (1969)

**Reaction of copper and phosphorus in molten silicon**
G. I. Voronkova, V. V. Voronkov, V. P. Grishin, and M. I. Iglitsyn
Izv. Akad. Nauk SSSR, Neorg. Mater., 5(10):1691–1694 (1969)
Inorg. Mater., 5(10):1432–1435 (1969)

**The orientation dependent diffusion of boron in silicon under oxidizing conditions**
G. N. Wills
Solid—State Electron., 12:133–134 (1969)

Defects induced by deep diffusion of phosphorus into silicon
Y. Yukimoto
Japan. J. Appl. Phys., 8:568-581 (1969)
Surface treatment before diffusion

Precipitation of boron in silicon
P. S. Dobson and J. D. Filby
J. Crystal Growth, 3:209-213 (1968)
Epitaxial, melt growth, and diffused

Effects of high phosphorus concentration on diffusion into silicon
M. C. Duffy, F. Barson, J. M. Fairfield, and G. H. Schwuttke
J. Electrochem. Soc., 115:84 (1968)

Interactions of lithium with impurities and defects in silicon
P. H. Fang
Lattice Defects in Semiconductors (R. R. Hasiguti, ed.), University of Tokyo Press, Tokyo and The Pennsylvania State University Press, University Park and London (1968), pp. 155-158

Arsenic diffusion in silicon using arsine
Y. W. Hsue
Electrochem. Tech., 6:361-365 (1968)

Precipitates formed by high-concentration phosphorus diffusion in silicon
R. J. Jaccodine
J. Appl. Phys., 39:3105 (1968)

Behavior of dislocations in silicon semiconductor devices: diffusion, electrical
J. E. Lawrence
J. Electrochem. Soc., 115:860-865 (1968)
Boron diffusion

Diffusion of phosphorus and boron into silicon through silicon monoxide layers
Chao-Chen Mai
Ph. D. thesis, Utah State State University, Logan (1968), 74 pp.
Available from University Microfilms, Inc., Ann Arbor, Mich., Order No. 68-13754

Solid-solid diffusion of boron in silicon using reactive sputtering
K. Nagano, S. Iwauchi, and T. Tanaka
Japan. J. Appl. Phys., 7:1361-1367 (1968)

Diffusion of boron in silicon
T. Nakamura
NEC Res. Dev., 11:93-105 (1968)

Diffusion of antimony into silicon epitaxial layers from buried layers
S. Nakanuma and S. Yamagishi
J. Electrochem. Soc. Japan, 36:3-10 (1968)

The retarded diffusion of gallium in silicon. I
M. Okamura
Japan. J. Appl. Phys., 7:1067 (1968)

The retarded diffusion of gallium in silicon. II
M. Okamura
Japan. J. Appl. Phys., 7:1231-1236 (1968)

Diffusion in silicon. Part 3. Generation of excess vacancies at climbing diffusion induced dislocations and dislocation enhanced diffusion in (001) crystals
T. J. Parker
J. Appl. Phys., 39:2043 (1968)

Electrophoretic process for controlled deposition of boron diffusion sources on silicon
L. C. Scala and J. C. Sandor
Electrochem. Tech., 6:434 (1968)

Impedance and other properties of gold doped silicon p-n junctions
D. K. Schroder
Ph. D. thesis, University of Illinois, Urbana (1968), 155 pp.
Available from University Microfilms, Inc., Ann Arbor, Mich., Order No. 68-12191
Diffusion times necessary to achieve gold saturation

Effective impurity-distribution coefficients during growth of silicon needles from a supercooled melt
Yu. M. Shashkov and V. P. Grishin
Izv. Akad. Nauk SSSR, Neorg. Mater., 4(9):1429-1432 (1968)
Inorg. Mater., 4(9):1255-1257 (1968)
Al and P

Autodoping of epitaxial silicon
W. H. Shepherd
J. Electrochem. Soc., 115:652-656 (1968)

Investigation of precipitates formed in silicon during heat treatment
I. L. Shul'pina, A. I. Zaslavskii, and T. T. Dedegkaev
Fiz. Tverd. Tela, 10:1347-1354 (1968)
Sov. Phys.-Solid State, 10:1070-1075 (1968)
$SiO_2$ and SiC

Radial aluminum distribution in silicon monocrystals raised by Czochralski's method
B. M. Turovskii
Izv. Akad. Nauk SSSR, Neorg. Mater., 4(3):307-311 (1968)
Inorg. Mater., 4(3):255-258 (1968)

Anomalous diffusion effects in silicon (a review)
A. F. W. Willoughby
J. Mater. Sci., 3:89-98 (1968)

Diffusion of tin into silicon
T. H. Yeh, S. M. Hu, and R. H. Kastl
J. Appl. Phys., 39:4266-4271 (1968)

Dissociative diffusion of nickel in silicon and self-diffusion of silicon
M. Yoshida and K. Saito
Lattice Defects in Semiconductors (R. R. Hasiguti, ed.), University of Tokyo Press, Tokyo and The Pennsylvania State University Press, University Park and London (1968), pp. 148-154

Contribution à l'étude de techniques de diffusion de bore dans le silicium. Diborane, tribromure de bore, anhydride borique
A. Zann
Ph. D. thesis, Univ. Grenoble, 1968, 102 pp.

Influence of oxygen in silicon on the motion of dislocation generated by diffusion
V. V. Batavin
Fiz. Tverd. Tela, 8(10):3100-3102 (1966)
Sov. Phys.-Solid State, 8(10):2478 (1967)

Diffusion of nickel in silicon
H. P. Bonzel
Phys. Stat. Sol., 20:493 (1967)

Diborane for boron diffusion into silicon
M. C. Duffy, D. W. Foy, and W. J. Armstrong
Electrochem. Tech., 5:29 (1967)

The nucleation and precipitation of lithium in silicon
J. W. Ferman
Ph. D. thesis, University of Minnesota, Duluth (1967) 120 pp.
Available from University Microfilms, Ann Arbor, Mich.,
Order No. 67-14606

Zum Mechanismus der Dotierung von Silicium mit Phosphor bei Anwesenheit einer Oberflächenschicht aus $(SiO_2)_n \cdot P_2O_5$
Helga Garski
Z. Naturforsch., 22a:66–75 (1967)

Beitrag zur Verteilung von Kupfer und Gold in tiegelfrei gezogenen Siliziumkristallen
H.-F. Hadamovsky
Krist. Tech., 2:415–418 (1967)

Boron diffusion into silicon using diborane
M. S. R. Heynew
Electrochem. Tech., 5:25–29 (1967)

On the diffusion of Al into Si
Y. C. Koo
Electrochem. Tech., 5:90–94 (1967)

Diffusion of phosphorous into silicon through an oxide film
G. S. Kulikov, B. I. Boltaks, and É. P. Savin
Izv. Akad. Nauk SSSR, Neorg. Mater., 3(1):26–28 (1967)
Inorg. Mater., 3(1):20–22 (1967)

Solute diffusion in plastically deformed silicon crystals
J. E. Lawrence
Brit. J. Appl. Phys., 18:405–410 (1967)
B, Ga, P, and Sb

Diffusion into silicon from an arsenic-doped oxide
D. B. Lee
Solid-State Electron., 10:623–624 (1967)

Diffusion of boron in silicon
T. Nakamura
Oyo Butsuri, 36(8):615–625 (1967)

Studies of anomalous diffusion of impurities in silicon
K. H. Nicholas
Philips Tech. Rev., 28: No. 5, 6, and 7, p. 149 (1967)
Group III and group V elements

Temperature dependence of the distribution coefficient of silver in silicon
E. A. Nikolaeva and V. N. Lozovskii
Fiz. Tekh. Poluprovodnikov, 1(3):458 (1967)
Sov. Phys.—Semicond., 1(3):381–382 (1967)

Diffusion of phosphorous and boron in silicon
Pravin Chandra Parekh
Ph. D. thesis, Delaware Univ., Newark (1967), 126 pp
Available from University Microfilms, Inc., Ann Arbor,
Mich., Order No. 68-15547

Deep impurities in silicon
E. Schibli and A. G. Milnes
Mater. Sci. Eng., 2:173–180 (1967)
Review; 127 refs. including In, Tl, Zn, Cu, Ag, Au, Ni, Pd, Pt, Co, Fe, Mn, Cr, W, O, S, Se, and Te

Diffusion of gold in silicon
V. B. Shuman
Fiz. Tekh. Poluprovodnikov 1(6):947–948
Sov. Phys.—Semicond., 1(6):790–791 (1967)

Diffusion and electrotransport of indium and silver along dislocations in silicon
V. A. Sterkhov, V. A. Panteleev, and P. V. Pavlov
Fiz. Tverd. Tela, 9(2):681 (1967)
Sov. Phys.—Solid State, 9(2):533–534 (1967)

Solubility and diffusion coefficient of sodium and potassium in silicon
L. Svob
Solid State Electron., 10:991–996 (1967)

Formation of precipitates in gold diffused silicon
E. D. Wolley and R. Stickler
J. Electrochem. Soc., 114:1287 (1967)

An electrophotographic method for determining the diffusion coefficient of lithium in p-type silicon
L. G. Yuskeselieva and A. S. Antonov
Fiz. Tverd. Tela, 8(9):2527–2531 (1966)
Sov. Phys.—Solid State, 8(9):2025–2028 (1967)

Integrated Silicon Device Technology. Vol. IX. Epitaxy
Research Triangle Institute, ASD-TDR-63-316, Vol. IX
(August 1965), 150 pp.; 112 refs.
Theory, methods, applications; doping; evaluation methods

A closed-tube technique for diffusing impurities into Si
W. J. Armstrong and M. C. Duffy
Electrochem. Tech., 4:475–478 (1966)

Properties of gold in silicon
W. M. Bullis
Solid-State Electron., 9:143–168 (1966)
Electrical properties, solubility and diffusion characteristics of gold in silicon reviewed

Solute incorporation during the cyclic solidification of silicon
J. R. Carruthers
Can. Met. Quarterly, 5:55–75 (1966)
P in Czochralski Si

Autodoping of silicon films grown epitaxially on sapphire
D. J. Dumin and P. H. Robinson
J. Electrochem. Soc., 113:469–472 (1966)
Al from $\alpha$-$Al_2O_3$

Dislocation-induced deviation of phosphorus-diffusion profiles in silicon
M. L. Joshi and S. Dash
IBM J. Res. Dev., 10:446–454 (1966)

Cooperative diffusion effect
J. E. Lawrence
J. Appl. Phys., 37:4106–4112 (1966)

Radiochemische Untersuchungen zur Diffusion von Gold in Silizium
J. Martin, E. Haas, and K. Raithel
Solid-State Electron., 9:83–85 (1966)

Diffusion of impurities into silicon from gaseous sources
   R. M. McLouski
   Westinghouse Defense and Space Center, Aerospace Division, Baltimore, Md., NASA CR-524 (1966), 24 pp.

The effect of heavy doping on the diffusion of impurities in silicon
   M. F. Millea
   J. Phys. Chem. Solids, 24:315 (1966)

Anomalous diffusion of impurities in silicon
   K. H. Nicholas
   Solid State Electron., 9:35-47 (1966)

Deep diffusions of phosphorus and chlorine in silicon under neutron irradiation
   P. F. Schmidt
   Appl. Phys. Letters, 8:264 (1966)

Helical dislocations and dislocation loops in silicon induced by platinum diffusion
   Y. Tokumaru and M. Kikuchi
   Japan. J. Appl. Phys., 5:847-848 (1966)

Closed-tube technique for diffusing impurities into silicon
   W. J. Armstrong and M. C. Duffy
   J. Electrochem. Soc., C 112:186 (1965)

Deep diffusion of boron into silicon single crystal
   M. Y. Ben-Sira and S. Bukshpan
   Solid State Commun., 3:15-17 (1965)

Radial solute distribution in Czochralski-grown silicon crystals
   K. E. Benson
   Electrochem. Tech., 3:331-335 (1965)
   Effects of crystal diameter and the solutes B, Ga, In, Al, P, As, and Sb

Diffusion-induced defects in thin silicon films
   M. L. Joshi, B. J. Masters, and S. Dash
   Appl. Phys. Letters, 7:306-308 (1965)
   Precipitation of P, doping method, electron transmission microscopy

Impurity distribution processes in epitaxial silicon layers
   B. A. Joyce, J. C. Weaver, and D. J. Maule
   J. Electrochem. Soc., 112:1100-1106 (1965)
   P, Ga, As, In

Surface concentration of gallium-diffused silicon as functions of gallium temperature and diffusion system
   A. N. Knopp
   Electrochem. Tech., 3:60-62 (1965)

Anomalous diffusion of boron into silicon grain boundary
   Y. Matsukura, J. Oda, and S. Koreeda
   Japan. J. Appl. Phys., 4:1022-1023 (1965)

Controlled phosphorus diffusion into silicon from $P_2O_5$ vapor using a red phosphorus source
   A. G. Nassibian and G. Whiting
   Solid State Electron., 8:843-853 (1965)

Direct dynamic observations of impurity flow patterns during gas-source boron diffusions of silicon
   A. M. Smith and R. P. Donovan
   Research Triangle Institute's Design Parameters and Pro-

cedures for Functional Electron. Struct., pp. 139-163 (Sept. 1965)
   Presented at the Electrochem. Soc. Meeting, Washington, Oct. 1964, N66-12334-03-09

Impurity diffusion in silicon
   A. M. Smith and R. P. Donovan
   Ibid., pp. 165-181
   Diffusion coefficients of 28 impurities in silicon

Interstitial-substitutional diffusion in a finite medium, gold into silicon
   G. J. Sprokel
   J. Electrochem. Soc., 112:807-812 (1965)

Diffusion of gold into silicon crystals
   G. J. Sprokel and J. M. Fairfield
   J. Electrochem. Soc., 112:200-203 (1965)

Investigation and study of rare-earth doped silicon
   C. R. Betz
   Nuclear Corp. of America, Phoenix, Arizona, AFCRL-64-406; AD-602031 (April 1964), 39 pp.
   Producing single silicon crystals doped with rare-earth elements and uranium; modifications of Czochralski

Research on the influence of surface conditions on diffusion in silicon
   R. K. Gereth et al.
   Clevite Corp., Shockley Research Lab., Palo Alto, Calif., AROD-3330-4; AD-604660 (July 1964), 163 pp.
   Ion bombardment and simultaneous doping of silicon; P

Imperfections of silicon induced by diffusion of the impurities
   H. Ino, T. Kawamura, and M. Yasufuku
   Japan. J. Appl. Phys., 3:692-697 (1964)
   B and P

The preparation of compensated silicon samples with high concentration of impurities
   J. Jurkowski and W. Nazarewicz
   Institute of Nuclear Research, Warsaw, Rept. No. 513/11 (Feb. 1964)

Formation and composition of surface layers and solubility limits of phosphorus during diffusion in silicon
   E. Kooi
   J. Electrochem. Soc., 111:1383-1387 (1964)
   Neutron activation analysis

Effect of surface imperfections on gallium diffusion in silicon
   J. G. Kren, B. J. Masters, and E. S. Wajda
   Appl. Phys. Letters, 5:49-50 (1964)

Phosphorus nitride as a diffusion source for silicon
   T. J. La Chapelle and H. B. Heller
   Trans. AIME, 230:311-314 (1964)

Diffusion of boron into silicon
   S. Maekawa and T. Oshida
   J. Phys. Soc. Japan, 19:253-257 (1964)

Incorporation of boron into epitaxially grown silicon
   Y. Matukura, K. Suzuki, and Y. Miura
   J. Electrochem. Soc., 111:491-492 (1964)

Obtention of silicon of high resistivity
   J. Messier
   Mem. Soc. Roy. Sci. Liege 10(2):115-118 (1964)

High resistivity silicon compensated by transmutation of Si into P by thermal neutron flux; purified by floating zone technique

Preparation of epitaxial layers of silicon — III. Heavy doping
S. Nielsen, G. J. Rich, and K. M. Fairhurst
Microelectron. Reliabil., 3:233–237 (1964)
B, As

Diffusion von Wismut in Silizium
D. Pommerrenig
Phys. Verh. D. P. G., 15(11):267 (1964)

Diffusion of arsenic in silicon
P. S. Raju, N. R. K. Rao, and E. V. K. Rao
Indian J. Pure Appl. Phys., 2:353–355 (1964)

Mechanism of gold diffusion into silicon
W. R. Wilcox and T. J. La Chapelle
J. Appl. Phys., 35:240–246 (1964)

Gold in silicon: effect on resistivity and diffusion in heavily-doped layers
W. R. Wilcox, T. J. La Chapelle, and D. H. Forbes
J. Electrochem. Soc., 111:1377–1380 (1964)

Behavior of nickel as an impurity in silicon
M. Yoshida and K. Furusho
Japan. J. Appl. Phys., 3:521 (1964)

Integrated Silicon Device Technology. Vol. IV, Diffusion
Research Triangle Institute, Durham, N. C., AD–603716 (Feb. 1964)

Diffusion of gallium in inhomogeneous silicon
B. I. Boltaks and T. D. Dzhafarov
Fiz. Tverd. Tela, 5(12):3611–3613 (1963)
Sov. Phys.—Solid State, 5(12):2649–2650 (1964)

Vapor growth and doping of silicon crystals with tellurium as carrier
S. Fischler
Metal. of Advanced Electronic Materials, Proc., Philadelphia (Aug. 27–29, 1962), pp. 273–281
Interscience Publishers, New York (1963)

Generation of large dislocation loops in silicon crystals
T. Iizuka, K. Kanasaki, and M. Kikuchi
Japan. J. Appl. Phys., 2:442–444 (1963)
Simultaneous diffusion of nickel and gold

Diffusion coefficients of impurities in silicon melt
H. Kodera
Japan. J. Appl. Phys., 2:212–219 (1963)
B, Al, Ga, In, P, As, and Sb

Diffusion of impurities in the semiconductor melt. III. Experimental determination of thickness of the layer in the melting process
H. Kodera, S. Iida, and S. Tauchi
Japan. J. Appl. Phys., 2:227 (1963)

Individual and joint solubilities of aluminum and phosphorus in germanium and silicon
N. K. H. Abrikosov, V. M. Glazov, and C. Y. Liu
Russ. J. Inorg. Chem., 7:429–431 (1962)

The diffusivity of arsenic in silicon
W. J. Armstrong
J. Electrochem. Soc., 109:1065 (1962)

Problems associated with distribution coefficient and solid solubility determinations using crystal-growth techniques
W. Bardsley, D. T. J. Hurle, and J. B. Mullin
J. Electrochem. Soc., 109:64–65 (1962)
Si; solvent evaporation method; Sb–doped melts; faceting effect

Study of aluminum fusion into silicon
T. –I. Chung
J. Electrochem. Soc., 109:229–235 (1962)

Anomalous impurity diffusion in epitaxial silicon near the substrate
D. Kahng, C. O. Thomas, and R. C. Manz
J. Electrochem. Soc., 109:1106–1108 (1962)

Diffusion of phosphorus in silicon
I. M. Mackintosh
J. Electrochem. Soc., 109:392–401 (1962)

Diffusion of impurities in the semiconductor melt
S. Touchi
J. Phys. Soc. Japan, 17:102–113 (1962)

Diffusion, solubility, and the effect of silver impurities on electrical properties of silicon
B. I. Boltaks and Hsüeh Shih–yin
Fiz. Tverd. Tela, 2(11):2677–2685 (1960)
Sov. Phys.—Solid State, 2(11)2383–2388 (1961)

Distribution coefficient of antimony in silicon from solvent evaporation experiments
F. A. Trumbore, P. E. Freeland, and R. A. Logan
J. Electrochem. Soc., 108:458–460 (1961)

Impurity redistribution and junction formation in silicon by thermal oxidation
M. M. Atalla and E. Tannenbaum
Bell System Tech. J., 39:933–946 (1960)
Ga and Sb

Diffusion and oxide masking in silicon by the box method
L. A. D'Asaro
Solid State Electron., 1:3–12 (1960)
B and P

Reactions of group III acceptors with oxygen in silicon crystals
C. S. Fuller, F. H. Doleiden, and Katherine Wolfstirn
Phys. Chem. Sol., 13:187–203 (1960)
B, Al, and Ga

Impurity introduction during epitaxial growth of silicon
R. Glang and B. W. Kippenhan
IBM J. Res. Dev., 4:299–301 (1960)
As, Sb, B, P, Al

Diffusion of boron into silicon
A. D. Kurtz and R. Yee
J. Appl. Phys., 31:303 (1960)

Effect of Li – B ion pairing on $Li^+$ ion drift in Si
E. M. Pell
J. Appl. Phys., 31:1675–1679 (1960)

Diffusion of Li in Si at high T and the isotope effect
E. M. Pell
Phys. Rev., 119:1014–1021 (1960)

Diffusion rate of Li in Si at low temperatures
  E. M. Pell
  Phys. Rev., 119:1222 (1960)

Sur la diffusion des éléments des colonnes III
et V dans le silicium
  R. Saintesprit
  Solid-State Electron., 1:123–130 (1960)

Sulfur in silicon
  R. O. Carlson, R. N. Hall, and E. M. Pell
  J. Phys. Chem. Solids, 8:81–83 (1959)

Diffusion of phosphorus into silicon under condi-
tions of controlled vapour pressure
  M. J. Coupland
  Proc. Phys. Soc., 73:577–584 (1959)

Effect of oxygen in silicon on phosphorus diffu-
sion
  J. L. Hartke
  J. Appl. Phys., 30:1469 (1959)

Solid solubilities of impurity elements in germa-
nium and silicon
  F. A. Trumbore
  Bell Telephone Labs. Monograph (1959)

Diffusion of gallium in silicon
  A. D. Kurtz and C. L. Bravel
  J. Appl. Phys., 29:1456–1459 (1958)

# 13. IV—VI Compounds

## 13.a. General, Reviews, and Bibliographies

Lead telluride, tin telluride and the lead tel-
luride tin telluride system
  M. Neuberger
  (Electronic Properties Information Center, Hughes Aircradt
    Company, Culver City, California), DS-164 (Jan. 1970),
    200 pp.

Principles of the growth of thin single-crystal
lead chalcogenide films
  L. S. Palatnik, V. K. Sorokin, and L. P. Zozulya
  Izv. Akad. Nauk SSSR, Neorg. Mater., 6(2):224–227 (1970)
  Inorg. Mater., 6(2):257–261 (1970)

Semiconducting Lead Chalcogenides
  Yu. I. Ravich, B. A. Efimova, and I. A. Smirnov
  L. S. Stil'bans, ed. (translated from Russian by Albin Tybule-
    wicz), Plenum Press, New York (1970), 400 pp.

Electronic structure of IV–VI semiconductors
  Yvonne Y. W. Tsang
  UCRL-19144 (Jan. 1970), 98 pp. (Thesis)

SemiconductingII-VI, IV-VI, and V-VI Compounds
  N. Kh. Abrikosov, V. F. Bankina, L. V. Poretskaya,
    L. E. Shelimova, and E. V. Skudnova
  Plenum Press, New York (1969)

A review of the semiconductor properties of
PbTe, PbSe, PbS, and PbO
  R. Dalven
  Infrared Phys., 9:141–184 (1969)
  288 refs.

IV-VI semiconducting compounds data tables
  M. Neuberger
  (Electronic Properties Information Center, Hughes Aircraft
    Company, Culver City, California)
  Contract F33615-68-C-1225, Project No. 7381 and 8975,
    Task No. 738103 and 897503, Report S-12 (1969), 102 pp.

Lead selenide — mercury selenide system
  V. G. Vanyarkho, V. P. Zleomanov, and A. V. Novoselova
  Izv. Akad. Nauk SSSR, Neorg. Mater., 5(11):2025–2026 (1969)
  Inorg. Mater., 5(11):1726–1727 (1969)

International Conference on IV-VI Semiconductors
  Gif-sur-Yvette, July 15-18, 1968, J. Phys., 29:(11-12) (Nov.
    1968)

The dielectric properties of the cubic IV-VI com-
pound semiconductors
  E. Burstein, S. Perkowitz, and M. H. Brodsky
  (Pennsylvania Univ., Philadelphia, Dept. of Phys.), Rept.
    No. TR-21; AD-679586 (1968), 22 pp.

Lead chalcogenide crystalline solutions: phase
equilibria, mechanisms of phase separation, and
micro-indentation hardness studies
  M. S. Marius
  Thesis, Penn. State University, University Park, Pa. (1968),
    171 pp.
  Available from University Microfilms, Ann Arbor, Mich.,
    Order No. 69-9751

Deviations from stoichiometry and lattice defects
in IV-VI compounds and their alloys
  A. J. Strauss and R. F. Brebrick
  J. Phys. (Suppl), 29:C4 (1968)

Crystal chemistry and band structures of the
group V semimetals and the IV-VI semiconductors
  M. H. Cohen, L. M. Falicov, and S. Golin
  IBM J. Res. Dev., 8(3):215–227 (1964)

## 13.b. Silicon Compounds

Silicon sulfide and selenide
  Hans H. Emons, Siegfried Moehlhenrich, and Lothar Theisen
  East German Patent 65,914 (March 21, 1969), 2 pp.
  Vitreous

Optical transmission in single-crystal silicon
diselenide
  E. A. Hauschild and C. R. Kannewurf
  J. Phys. Chem. Solids, 30:353 (1969)

Elektrisches Dipolmoment und Mikrowellen-Rota-
tionsspektrum von SiS
  J. Hoeft, F. J. Lovas, E. Tiemann, and T. Torring
  Z. Naturforschung, 24a:1422–1423 (1969)

## 13.c. Germanium Compounds

**High-temperature modifications of germanium monosulfide and monoselenide**
S. G. Karbanov, V. P. Zlomanov, and Yu. M. Ukrainskii
Izv. Akad. Nauk SSSR, Neorg. Mater., 6(1):125–126 (1970)
Inorg. Mater., 6(1):104–105 (1970)

**Preparation and structure of thin films of germanium telluride and selenide**
A. G. Mikolaichuk and A. N. Kogut
Kristallografiya, 15:353–357 (1970)
Soviet Phys.—Cryst., 15:294–298 (1970)

**Preparation and some general properties of germanium monoselenide thin layers**
S. M. Nikolov, H. M. Vodenicharov, S. G. Karbanov, and P. Petrov
Compt. Rend. Acad. Bulgare Sci., 23(3):257–260 (1970)

**Czochralski encapsulation growth of GeTe, SnTe and PbTe single crystals**
R. J. Baughman and R. A. Lefever
Mat. Res. Bull., 4:721–726 (1969)

**Amorphous versus crystalline GeTe films. I. Growth and structural behavior**
K. L. Chopra and S. K. Bahl
J. Appl. Phys., 40:4171–4178 (1969)

**New method of obtaining monocrystals from germanium monoselenide by means of vacuum sublimation**
S. G. Karbanov, P. Petrov, and S. Ivanov
Compt. Rend. Acad. Bulgare Sci., 22(12):1381–1384 (1969)

**Phase diagram of the germanium—tellurium system**
S. G. Karbanov, V. P. Zlomanov, and A. V. Novoselova
Izv. Akad. Nauk SSSR, Neorg. Mater., 5(7):1171–1174
Inorg. Mater., 5(7):997–1000 (1969)

**Vapor phase of germanium telluride containing additions of bismuth and antimony**
A. P. Lyubimov, I. I. Bespal'tseva, and L. D. Dudkin
Izv. Akad. Nauk SSSR, Neorg. Mater., 5(7):1287–1288 (1969)
Inorg. Mater., 5(7):1095–1096 (1969)

**Ferroelectric semiconductors**
Atomic Energy of Canada Ltd.
French Patent 1,542,778 (October 1968), 7 pp.
GeTe and Ge$_x$Sn$_{1-x}$Te

**Apparent activation energies for electrical conduction of solid and liquid germanium (II) sulfide**
M. d'Amboise, G. Handfield, and M. Bourgon
Can. J. Chem., 46:3545–3550 (1968)

**Absorption edge of GeSe$_2$**
N. P. Gavaleshko, M. V. Kurik, and A. I. Savchuk
Fiz. Tekh. Poluprov., 1(7):1099–1100 (1967)
Sov. Phys.—Semiconductor, 1(7):920–921 (1968)

**New crystal forms of germanium selenide**
Howard S. Young
E. I. duPont deNemours, Co.
U. S. Patent 3,375,071, March 26, 1968, 4 pp.

**Crystal growth by chemical transport reactions. IV. New results on the growth of binary, ternary and mixed-crystal chalcogenides**
R. Nitsche
Proc. Intern. Conf. on Crystal Growth, Boston, June 20–24, 1966, Crystal Growth (H. Steffen Peiser, ed.), Pergamon Press, New York (1967) pp. 215–220
GeS and GeS$_2$

**Epitaxial growth of composite semiconductors by evaporation-diffusion in an isothermal system**
G. Cohen-Solal, Y. Marfaing, and F. Bailly
Rev. Phys. Appl., 1:(1) 11–17 (1966)

**Single-crystal growth of SnTe and GeTe**
P. F. Weller
J. Electrochem. Soc., 113:90–92 (1966)

**Optical absorption and photoconductivity in germanium selenide**
C. R. Kannewurf and R. J. Cashman
J. Phys. Chem. Solids, 22:293–298 (1961)

## 13.d. Tin Compounds

### 13.d.1. SnS

**Structure and electrophysical properties of epitaxial films of tin chalcogenides**
A. G. Mikolaichuk and D. M. Freik
Fiz. Tverd. Tela, 11(9):2520–2525 (1969)
Sov. Phys.—Solid State, 11(9):2033–2036 (1970)
Epitaxial films of SnS, SnSe, and SnTe

**Polymorphism in some IV-VI compounds induced by high pressure and thin-film epitaxial growth**
A. N. Mariano and K. L. Chopra
Appl. Phys. Letters, 10:282 (1967)

**Crystal growth by chemical transport reactions— I. Binary, ternary, and mixed-crystal chalcogenides**
R. Nitsche, H. U. Bolsterli, and M. Lichtensteiger
J. Phys. Chem. Solids, 21:199 (1961)

**The preparation and the electrical and optical properties of SnS crystals**
W. Albers, C. Haas, and F. van der Maesen
J. Phys. Chem. Solids, 15:306 (1960)

### 13.d.2. SnS$_2$

**Electronic properties of inorganic solid solutions, final report, 16 March 1965 –30 June 1969**
L. E. Conroy
(Dept. of Chemistry, Minnesota Univ., Minneapolis), Contract DA-ARO(D)-31-124-G661, AROD-5532-3-C; AD-695697 (Oct. 1969), 5 pp.
TiS$_2$, ZrS$_2$, HfS$_2$, SnS$_2$

**Optical and electrical properties of some Group IV disulfides**
Kyu Chang Park
University Microfilms, Ann Arbor, Mich., Order No. 68-1958, 1967. Ph. D. thesis from University of Minnesota
Large single crystals of TiS$_2$, ZrS$_2$, HfS$_2$, and SnS$_2$ have been prepared by the chemical transport method, using iodine as the transporter

**Preparation and optical properties of group IV-VI$_2$ chalcogenides having the CdI$_2$ structure**
D. L. Greenaway and R. Nitsche
J. Phys. Chem. Solids, 26:1445 (1965)

Crystal growth by chemical transport reactions-
I. Binary, ternary, and mixed-crystal chalco-
genides
  R. Nitsche, H. U. Bolsterli, and M. Lichtensteiger
  J. Phys. Chem. Solids, 21:199 (1961)

The growth of single crystals of binary and ternary chalcogenides by chemical transport reactions
  R. Nitsche
  J. Phys. Chem. Solids, 17:163-165 (1960)

## 13.d.3. Sn Se

Sublimation coefficient of tin selenide
  R. C. Blair and Z. A. Munir
  J. Am. Ceram. Soc., 53(6):301-303 (1970)

Optical and electrical properties of $SnSe_2$
  B. L. Evans and R. A. Hazelwood
  British J. Appl. Phys., 2:1507-1516 (1969)
  Growth by iodine vapor transport

Crystal growth and chemical synthesis under hy-
drothermal conditions (Au, CuS, CuI, PbSe, CdS,
$GeI_2$, $Sb_2S_3$, BiSBr, $Pb_5S_2I_6$, Te, $PbI_2$, Pt, SnSe)
  A. Rabenau and H. Rau
  Philips Tech. Rev., 30(4):89-96 (1969)

Optical properties of tin di-selenide single crys-
tals
  P. A. Lee and G. Said
  Brit. J. Appl. Phys. (J. Phys. D), Ser., 2:(1)837-843 (1968)

Structure, electrical and thermoelectrical prop-
erties of $SnSe_2$
  G. Busch, C. Frohlick, and F. Hulliger
  Helv. Phys. Acta, 34:359-368 (1961)

The electrical and optical properties of the semi-
conductor tin selenide
  Dean Lewis Mitchell
  Dissertation Abstr., 20:2857-2858 (1960)

## 13.d.4. SnTe

Thermal expansion of tin telluride
  Henry S. Belson and Bland Houston
  J. Appl. Phys., 41:422-424 (1970)
  Czochralski

The capillary seed technique in crystal pulling
  S. E. R. Hiscocks
  J. Mater. Sci., 4:310-312 (1969)

Préparation, mesures de résistivité, d'éffet
Hall et étude des propriétés cristallographiques
de couches minces de SnTe, déposées sur des
substrats amorphes ou orientés
  Pierre C. Bourgeois
  Compt. Rend., Ser. B, 265:819-821 (1967)

Polymorphism in some IV-VI compounds induced
by high pressure and thin-film epitaxial growth
  A. N. Mariano and K. L. Chopra
  Appl. Phys. Letters, 10:282 (1967)

Low-temperature phase transition in tin telluride
  S. I. Novikova and L. E. Shelimova
  Fiz. Tverd. Tela, 9(5):1046-1047 (1967)
  Sov. Phys.—Solid State, 9(5):1336-1338 (1967)

Preparation of epitaxial SnTe films of control-
led carrier concentration
  H. R. Riedl, R. B. Schoolar, and Bland Houston
  Solid-State Commun., 4:399-402 (1966)

Single-crystal growth of SnTe and GeTe
  P. F. Weller
  (Thomas J. Watson Research Center, IBM Corp., Yorktown
    Heights, N. Y.), J. Electrochem. Soc., 113:90-92 (1966)

Conducting properties of SnTe
  M. Moldovanova, S. Dimitrova, and S. Decheva
  Fiz. Tverd. Tela  6(12):3717 (1964)
  Sov. Phys.—Solid State, 6(12):2979 (1965)

Evidence that SnTe is a semiconductor
  J. A. Kafalas, R. F. Brebrick, and A. J. Strauss
  Appl. Phys. Letters, 4:93 (1964)

Vapor pressures of tin selenide and tin telluride
  Chikara Hirayama, Yoshio Ichikawa, and Anthony M. DeRoo
  J. Phys. Chem., 67:1039 (1963)

## 13.d.5. Sn Ternary Chalcogenides

The $SnSe-SnSe_2$ eutectic; a p-n multilayer struc-
ture
  W. Albers and J. Verberkt
  J. Mater. Sci., 5:24-28 (1970)
  Conditions required to grow a two-phase p-n heterojunction
    from a eutectic melt

Isothermal substitutional growth of single crys-
tals
  W. Albers and J. Verberkt
  Philips Res. Rept., 25(1):17-20 (1970)
  SnS-SnSe

Crystal growth and neutron studies of large sin-
gle crystals of the alloy series SnTe-GeTe
  I. Lefkowitz, M. Shields, and G. Dolling
  J. Crystal Growth, 6:143-146 (1970)

Das Dreistoffsystem Zinn-Antimon-Tellur
  A. Stegherr
  Thesis, Technical University Aachen, Germany, Philips Res.
    Repts., Suppl. No. 6 (1969)
  Phase diagrams

Thermoelectric power and electrical conductivity
of crystals of the system SnTe-InTe
  E. Beleites and H. Nieke
  Ann. Phys. (Germany), 18:258-267 (Nov. 1966)

Electronographic investigation of the nature of
crystal orientation and formation of phases in
$SnSb_2Te_4$ and $PbBi_4Te_7$ compounds
  A. G. Talybov
  Azerb. Khim. Zh. (Baku), 30(6):111-118 (1963)
  FTD-TT 65-1399/1+2+4; AD-633672

Preparation and properties of mixed crystals
$SnS_{1-x}Se_x$
  W. Albers, C. Haas, H. Ober, G. R. Schodder, and J. D.
    Wasscher
  J. Phys. Chem. Solids, 23:215-220 (1962)

## 13.e. Lead Compounds

### 13.e.1. PbS

Lamellar dendritic growth in lead sulfide
O. C. Kopp and G. W. Clark
J. Cryst. Growth, 8:135–136 (1971)

Some structural and electrical properties of lead
sulfide films
R. V. Kudryavtseva, S. A. Semiletov, and G. L.
Perevezentseva
Kristallografiya, (12)1:109–112 (1969)
Sov. Phys.—Cryst., (12)1:86–89 (1969)

Formation of epitaxial films by vacuum deposition
S. A. Semiletov, V. A. Vlasov, and Z. A. Magomedov
Rost Kristallov, Vol. 8, Izd. Akad. Nauk (1968), pp. 184–187
Growth of Crystals, Vol. 8. (A. V. Shubnikov and N. N.
Sheftal', eds.), Consultants Bureau, New York (1969),
pp. 150–152

The growth of single crystals of lead sulphide in
silica gels at ambient temperatures —preliminary
characterization and effect of various organic
compounds as sulphide ion donors
Zvi Blank, Walter Brenner, and Yoshiyuki Okamoto
Mat. Res. Bull., 3:555–562 (1968)

An investigation of the crystal growth of the
heavy metal sulfides in supercritical hydrogen
sulfide
Leroy C. Lewis
University Microfilms, Inc., Ann Arbor, Mich., Order No.
68-11, 915
Ph. D. thesis, Oregon State Univ., 1968

Epitaxial films of PbTe, PbSe, and PbS grown on
mica substrates
R. F. Egerton and C. Juhasz
Brit. J. Appl. Phys., 18:1009 (1967)

High-mobility PbS and CdS films deposited under
ultrahigh vacuum equilibrium conditions
P. Hudock
Trans. AIME, 239:338 (1967)

Crystal syntheses and growth in strong acid solutions under hydrothermal conditions
H. Rau and A. Rabenau
Solid State Commun., 5:331 (1967)

Activation energies for high-temperature steady-
state creep in lead sulfide
M. S. Seltzer
Trans. AIME, 239:650 (1967)

The morphology of artificial lead sulphide crystals
V. V. Badikov and A. A. Godovikov
Acta Cryst., 21:Pt.7, Suppl. A257 (Dec. 1966)
Seventh International Congress and Symposium International
Union of Crystallography, Moscow, 1966

Characteristics of the morphology of galena
crystals obtained under hydrothermal conditions
V. V. Badikov and A. A. Godovikov
Zap. Vses. Mineralog. Obshchestva SSSR, 95:526–536 (1966)

Growth of single crystals of lead sulphide in
silica gels near ambient temperatures
W. Brenner, Z. Blank, and Y. Okamoto
Nature, 212:392–393 (1966)

A study on the influence of the imperfect structure of crystals on growth processes
G. I. Distler, S. A. Kobzareva, and V. S. Cudakov
Acta Cryst., 21:Pt.7, Suppl. A275 (Dec. 1966)
Seventh International Congress and Symposium International
Union of Crystallography, Moscow, 1966

Physics of Thin Films, Vol. 3
George Hass and Rudolf E. Thun, eds.
Academic Press, New York and London (1966)

Growth of lead sulfide monocrystals by Tammann's
method
B. A. Kazennov
Rost Kristallov, Vol. 4, Izd.Akad. Nauk (1964), pp. 101–112
Growth of Crystals, Vol. 4. (A. V. Shubnikov and N. N.
Sheftal', eds.), Consultants Bureau, New York (1966),
pp. 82–90

Über die epitaktische Abscheidung von Zinksulfid und Bleisulfid auf Zinksulfid mit Hilfe von
chemischen Transportreaktionen und durch Sublimation
W. Kleber and I. Meusel
Z. Physik. Chem. (Frankfurt), 231:191–202 (1966)

Development of techniques for preparing homo-
geneous single crystals of lead telluride, lead
selenide, and lead sulfide
J. F. Miller, J. W. Moody, and R. C. Himes
Lincoln Lab., Massachusetts Inst. of Tech., Lexington, Mass.,
Contract AF 19(628)-5167
Final report 15 July 1965–15 July 1966, ESD-TR-67-37
(July 1966), 12 pp.

Lead salt epitaxial films with near bulk properties
E. G. Bylander and A. S. Rodolakis
Proc. IEEE, 53:395 (1965)

Epitaxies de cristaux de sulfure de plomb et de
sulfure de cadmium sur differents supports monocristallins
Lucien Capella and Jean-Claude Heyraud
Compt. Rend., 261:4053–4054 (1965)

Crystal growth in hydrothermal systems
J. W. Moody and R. C. Himes
Battelle Tech. Rev., 14:3 (1965)

Growth of single crystals at high gas pressures
A. Ya. Preobrazhenskii and V. A. Stepanov
Pribory i Tekhn. Éksperim., 10:196–198 (1965)

Lead sulphide, selenide and telluride
E. H. Putley
(Royal Radar Establishment, Malvern, Worcs., England),
Materials Used in Semiconductor Devices (C. A. Hogarth,
ed.), Interscience Publishers, New York, London, Sydney
(1965), p. 71

Electrical and optical properties of epitaxial
films of PbS, PbSe, PbTe, and SnTe
J. N. Zemel, J. D. Jensen, and R. B. Schoolar
Phys. Rev., 140:A330 (1965)

Growth and annealing of Pb sulfide films
Hans Norden
Cavendish Lab., Cambridge, England
Proc. European Reg. Conf., Electron Microscopy, 3rd,
Prague (1964)

Dislocations in evaporated lead sulphide films
J. W. Matthews and K. Isebeck
Phil. Mag., 8:469 (1963)

Crystallization of the sulfides of lead and zinc
from aqueous solutions of the chlorides
L. V. Bryatov and I. P. Kuz'mina
Rost Kristallov, Vol. 3, Izd. Akad. Nauk (1961), pp. 416–420
Growth of Crystals, Vol. 3. (A. V. Shubnikov and N. N.
Sheftal', eds.), Consultants Bureau, New York (1962), pp.
294–296

Production of crystals of zinc, cadmium, and
lead sulfides, selenides, and tellurides
Bernard Kopelman
U. S. Patent 3,174,823 (Cl. 23–50) (March 23, 1965)
Appl. Dec. 15, 1961, 3 pp.

Study of the influence of atmosphere and annealing on crystal growth performed with electrically
sustained heat source, Appendix 3. Development
of crystal growth since 1945
ITT Laboratories, Physical Sciences and Materials Laboratory, Nutley, New Jersey, AFCRC-TR-60-149 (July
1960)
Contract AF 19(604)-2261

Controlled conductivity in lead sulphide single
crystals
J. Bloem
Philips Res. Rep., 11:273–336 (1956)

Oxygen-free single crystals of lead telluride,
selenide and sulfide
W. D. Lawson
J. Appl. Phys., 23:495 (1952)

## 13.e.2. PbSe and PbTe

Scattering of current carriers and transport
phenomena in lead chalcogenides, I. Theory
Yu. I. Ravich, B. A. Efimova, and V. I. Tamarchenko
Phys. Stat. Sol., 43 B:11–33 (1971)
Review

The sublimation pressure and sublimation coefficient of (100) oriented lead telluride single
crystals
E. E. Hansen and Z. A. Munir
J. Electrochem. Soc., 117:121–124 (1970)

Preparation of lead telluride single crystals
M. I. Karklina, M. S. Ablova, and V. M. Muzhdaba
Izv. Akad. Nauk SSSR, Neorg. Mater., 6(5):985–987 (1970)
Inorg. Mater., 6(5):860–862 (1970)

The growth mechanism of single-crystal PbTe
films on mica
L. S. Palatnik, V. M. Kosevich, L. P. Zozulya, L. F.
Zozulya, and V. K. Sorokin
Fiz. Tverd. Tela, 11(9):2586–2589 (1969)
Sov. Phys.–Solid State, 11(9):2086–2088 (1970)

Influence of formation mechanism on the sub-
structure and semiconductor properties of con-
densed films of lead chalcogenides
L. S. Palatnik, V. K. Sorokin, and L. P. Zozulya
Izv. Akad. Nauk SSSR, Neorg. Mater., 6(2):224–239 (1970)
Inorg. Mater., 6(2):198–202 (1970)

Regularities in the growth of thin monocrystal-
line lead chalcogenide films
L. S. Palatnik, V. K. Sorokin and L. P. Zozulya
Izv. Akad. Nauk SSSR, Neorg. Mater., 6(2):224–239 (1970)
Inorg. Mater., 6(2):198–202 (1970)

Coalescence in epitaxial films of lead chalco-
genide
L. S. Palatnik, V. K. Sorokin, and L. P. Zozulya
Izv. Akad. Nauk SSSR, Neorg. Mater., 6(3):441–446 (1970)
Inorg. Mater., 6(3):387–391 (1970)

Phase diagram of the PbTe–PbSe pseudobinary
system
Jacques Steininger
Met. Trans., 1:2939–2941 (1970)

Phase diagram of PbTe – PbSe system
J. M. Steininger
Lincoln Lab., Mass. Inst. of Tech.
ESD–TR–70–33 (May 1970), pp. 15–16

Growth and characterization of lead telluride
epitaxial layers
J. W. Wagner and A. G. Thompson
J. Electrochem. Soc., 117:936–940 (1970)

The capillary seed technique in crystal pulling
S. E. R. Hiscocks
J. Mater. Sci., 4:310–312 (1969)

Vapor growth conditions of lead selenide single
crystals and controlling the number of nuclei
K. Igaki and T. Suzuki
Nippon Kinzoku Gakkaishi, 33(2):190–194 (1969)

Crystal growth and chemical synthesis under hy-
drothermal conditions
A. Rabenau and H. Rau
Philips Tech. Rev., 30(4):89–96 (1969)
Au, CuS, CuI, PbSe, CdS, GeI$_2$, Sb$_2$S$_3$, BiSBr, Pb$_5$S$_2$I$_6$, Te,
PbI$_2$, Pt, SnSe

Preparation and properties of single-crystal lead
telluride films
G. G. Sumner and L. L. Reynolds
J. Vacuum Sci. Tech., 6:493–497 (1969)

Preparation and properties of Pb$_{1-x}$Ln$_x$Te (Ln = Eu,
Yb) films
R. Suryanarayanan and C. Paparoditis
J. Vac. Sci. Tech., 6:497 (1969)

Defects of vapor-grown lead selenide single
crystals
T. Suzuki, N. Ohashi, and K. Igaki
J. Japan. Inst. Metals, 33(2):194–198 (1969)

Precipitation studies of Pb and Se in PbSe single crystals
Halle Abrams
Ph. D. thesis, Materials Research Center, Lehigh University (1968)

Data on the preparation and structure of PbSe
single crystals
  I. S. Aver'yanov, N. P. Markina, F. P. Volkova, G. V.
      Pertsev, and S. P. Chashchin
  Izv. Akad. Nauk SSSR, Neorg. Mater., 4(6):825–828 (1968)
  Inorg. Mater., 4(6):723–726 (1968)

Growth of lead selenide monocrystals
  Ya. S. Budzhak
  Rost Kristallov, Vol. 6B, Izd. Akad. Nauk (1965), pp. 229–230
  Growth of Crystals, Vol. 6B (A. V. Shubnikov and N. N.
      Sheftal', eds.), Consultants Bureau, New York (1968), pp.
      43–44

Zone crystallization of lead telluride from a so-
lution in lead
  M. I. Karklina
  Izv. Akad. Nauk SSSR, Neorg. Mater., 4(8):1344–1345 (1968)
  Inorg. Mater., 4(8):1178–1179 (1968)

Variant synthesis and the growing of single crystals of solid solutions using the lead telluride-
lead sulfide system as an example
  S. Kasimov
  Dokl. Akad. Nauk Tadzh. SSR, 11(1):26–29 (1968)

The preparation and examination of PbTe by
transmission electron microscopy
  E. Levine and R. N. Tauber
  J. Electrochem. Soc., 115:107 (1968)

Growth and morphology of epitaxial lead telluride
deposits on rocksalt
  B. Lewis and D. J. Stirland
  J. Cryst. Growth, 3(4):200–205 (1968)

Single-crystal semiconductors and p-n junctions
  Michel M. Moulin
  French Patent 1,541,127 (Oct. 4, 1968; appl. Aug. 1967), 7 pp.

Growth of PbTe monocrystals by Czochralski's
method
  P. M. Starik and P. I. Voronyuk
  Rost Kristallov, Vol. 6B, Izd. Akad. Nauk (1965), pp. 281–283
  Growth of Crystals, Vol. 6B (A. V. Shubnikov and N. N.
      Sheftal', eds.), Consultants Bureau, New York (1968),
      pp. 91–92

Producing lead selenide crystals with perfect
structure
  I. S. Aver'yanov, M. P. Markina, F. P. Volkova, G. V.
      Pertsev, and S. P. Chashchin
  Izv. Akad. Nauk SSSR, Neorg. Mater., 3(5):877–878 (1967)
  Inorg. Mater., 3(5):783–784 (1968)

Synthesis of lead telluride by deposition from
solutions
  M. A. Berchenko and A. I. Beliaev
  Chalcogenides, Naukova Dumka (1967), pp. 94–100
  Lockheed Missiles and Space Co., Palo Alto, California, 7 pp.

Préparation et étude du PbSe en couches minces
évaporées sous vide, sur des substrats amorphes
ou orientés
  P. Bourgeois and P. Moch
  Compt. Rend., 264:1732–1735 (1967)

Préparation et étude des propriétés cristallographiques du PbTe en couches minces evaporées
sous vide, sur des substrats amorphes ou orientés
  P. C. Bourgeois and M. Y. Moulin
  Compt. Rend., 264:1830 (1967)

The role of sodium in lead telluride
  A. J. Crocker
  J. Phys. Chem. Solids, 28:1903 (1967)

Epitaxial films of PbTe, PbSe and PbS grown on
mica substrates
  R. F. Egerton and C. Juhasz
  Brit. J. Appl. Phys., 18:1009 (1967)

Polymorphism in some IV-VI compounds induced
by high pressure and thin-film epitaxial growth
  A. N. Mariano and K. L. Chopra
  Appl. Phys. Letters, 10:282 (1967)

Growing lead telluride crystals of stoichiometric
composition in the vapor-gas phase
  T. A. Mashaev, V. V. Krapkhim, and S. P. Pavlov
  Chalcogenides, Kiev, pp. 101–107
  PB-176441T (1967), Lockheed Missiles and Space Co., Palo
      Alto, California, 6 pp.

The preparation of PbTe crystals
  J. F. Miller, J. W. Moody, and R. C. Himes
  Trans. AIME, 239:342 (1967)

A physicochemical investigation of lead selenide
  A. V. Novoselova, V. P. Zlomanov, and O. V. Mateev
  Izv. Akad. Nauk SSSR, Neorg. Mater., 3(8):1323–1329 (1967)
  Inorg. Mater., 3(8):1154–1159 (1967)

Preparation of lead selenide using selenium sul-
fide
  G. G. Rybnikova, B. A. Popovkin, V. G. Butkevich, and A. V.
      Novoselova
  Izv. Akad. Nauk SSSR, Neorg. Mater., 3(12):1934–1937 (1967)
  Inorg. Mater., 3(12):2217–2220 (1967)

Preparation, structure, and some photoelectric
properties of single- and multilayer monocrys-
talline chalcogenide films of the types $A^{II}B^{IV}$ and
$A^{IV}B^{VI}$
  L. A. Sergeeva, I. P. Kalinkin, and V. B. Aleskovskii
  Kristallografiya, 12(1):113–118 (1967)
  Sov. Phys.—Cryst., 12(1):90–94 (1967)

Obtaining lead telluride by the hydrogen reduc-
tion of the tellurite
  O. I. Tananaeva, R. A. Sapozhnikov, and V. A. Novoselova
  Izv. Akad. Nauk SSSR, Neorg. Mater., 3(3):578 (1967)
  Inorg. Mater., 3(3):514–515 (1967)

Synthesis of lead telluride
  A. I. Belyaev, M. A. Berchenko, and A. A. Telegrin
  Sb. Kosk. Inst. Stali i Splavov, No. 41:366–374 (1966)

Tellurium and Tellurides
  D. M. Chizhikov and V. P. Shastlivii
  Izd. Nauka, Moscow (1966), 279 pp.

Mosaic-free lead telluride crystals
  A. J. Crocker
  Brit. J. Appl. Phys., 17:433 (1966)

Growth of high-alloy n-type lead telluride single
crystals and the determination of the distribu-
tion coefficients for iodine, chlorine, and bromine
  V. I. Kaidanov, R. B. Mel'nik, and E. Sh. Fedorenko
  Izv. Akad. Nauk SSSR, Neorg. Mater., 2(12):2246–2247 (1966)
  Inorg. Mater., 2(12):1939–1941 (1966)

Zone crystallization of lead telluride from a so-
lution in tellurium
  M. I. Karklina and T. L. Koval'chik
  Izv. Akad. Nauk SSSR, Neorg. Mater., 2(7):1190–1193 (1966)
  Inorg. Mater., 2(7):1015–1017 (1966)

Development of techniques for preparing homogeneous single crystals of lead telluride, lead selenide, and lead sulfide
J. F. Miller, J. W. Moody, and R. C. Himes
Mass. Inst. of Tech., Lincoln Lab., Lexington, Mass.
Final report 15 July 1965 to 15 July 1966, Contract
AF 19(628)-5167, ESD-TR-67-37 (July 1966), 12 pp.

Oriented growth of lead telluride and selenide films
L. S. Palatnik and V. K. Sorokin
Fiz. Tverd. Tela, 8(4):1088-1090 (1966)
Sov. Phys.-Solid State, 8(4):869-870 (1966)

Croissance epitaxique de PbTe par sublimation
Noboru Takahashi, Henri Martina, and Jean-Jacques Trillat
Compt. Rend., 262:824-826 (1966)

Growth of lead selenide monocrystals
Ya. S. Budzhak
Rost Kristallov, Vol. 6B, Izd. Akad. Nauk (1965), pp. 229-230
Growth of Crystals, Vol. 6B (A. V. Shubnikov and N. N. Sheftal', eds.), Consultant Bureau, New York (1968), pp. 43;44

Lead salt epitaxial films with near bulk properties
E. G. Bylander and A. S. Rodolakis
Proc. IEEE, 53:395-396 (1965)

Über Struktur und Halbleitereigenschaften epitaxialer Bleiselenid-Schichten auf Alkalihalogenideinkristallen
H. Gobrecht, K.-E. Boeters, and H.-J. Fleischer
Z. Physik, 187:232-242 (1965)

Materials Used in Semiconductor Devices
C. A. Hogarth, ed.
John Wiley and Sons, New York (1965), 243 pp.
Preparation and properties of Ge, Si, Se, PbS, PbSe, PbTe, InSb, Bi₂Te₃, CdSb, and ZnSb

Production of crystals of zinc, cadmium, and lead sulfides, selenides, and tellurides
Bernard Kopelman
U. S. 3,174,823 (Cl. 23-50), March 23, 1965

Préparation des pellicules semiconductrices de PbTe à l'aide de la méthode des échantillons de composition variable
L. S. Palatnik and V. K. Sorokin
Izv. Vyssh. Uchebn. Zaved., Fiz., SSSR 8:48-52 (1965)

Preparation of PbTe films of variable composition
L. S. Palatnik and V. K. Sorokin
Izv. Vyssh. Ucheben. Zaved., Fiz., SSSR, No. 3:48-52
Sov. Phys. J., No. 3:33-35 (May/June, 1965),

Lead sulphide, selenide and telluride
E. H. Putley
Materials Used in Semiconductor Devices, John Wiley and Sons, New York (1965), p. 71-114

Vacuum-deposited thin films using a ruby laser
H. M. Smith and A. F. Turner
Appl. Opt., 4:147 (1965)

Growth of PbTe monocrystals by Czochralski's method
P. M. Starik and P. I. Voronyuk
Rost Kristallov, Vol. 6B, Izd. Akad. Nauk (1965), pp. 281-283

Growth of Crystals, Vol. 6B (A. V. Shubnikov and N. N. Sheftal', eds.), Consultants Bureau, New York (1968), pp. 91-92

Electrical and optical properties of epitaxial films of PbS, PbSe, PbTe, and SnTe
J. N. Zemel, J. D. Jensen, and R. B. Schoolar
Phys. Rev., 140:A330 (1965)

Obtaining lead selenide single crystals from the gas phase
V. P. Zlomanov, O. V. Matveev, and A. V. Novoselova
Zh. Neorg. Khim., SSSR, 10:1753 (1965)

Diffused junction diodes of PbSe and PbTe
J. F. Butler
J. Electrochem Soc., 111:1150 (1964)

PbTe thin film prepared by vacuum evaporation on mica
Yoshimi Makino
J. Phys. Soc. Japan, 19:580 (1964)

Preparation of fine-grained PbTe by ultrasonic agitation of a solidifying melt
Martin Weinstein
Trans. AIME, 230:321 (1964)

Chemical deposition of thin films of lead selenide
R. A. Zingaro and D. O. Skovlin
J. Electrochem. Soc., 111:42 (1964)

Temperature-pressure projection of lead telluride phase diagram
Richard F. Bis
J. Phys. Chem. Solids, 24:579-581 (1963)

The growth of crystals from compounds with volatile components
G. R. Gronin, M. E. Jones, and O. Wilson
J. Electrochem. Soc., 110:582 (1963)

Controlled deviation from stoichiometry in PbSe
Kenzo Igaki and Nobumitsu Ohashi
J. Phys. Soc. Japan, 18(11):143 (1963)

Dielectric constant of PbTe
Yasuo Kanai and Katsufusa Shohno
Japan. J. Appl. Phys., 2:6 (1963)

Polishes and etches for tin telluride, lead sulfide, lead selenide and lead telluride
M. K. Norr
AD-423367 (1963), 27 pp.

Phase studies of the group IV-A tellurides
R. Mazelsky, M. S. Lubell, and W. E. Kramer
J. Chem. Phys., 37:45-47 (1962)

A technique for pulling single crystals of volatile materials
E. P. A. Metz, R. C. Miller, and R. Mazelsky
J. Appl. Phys., 33:2016 (1962)

Precipitation of Te and Pb in PbTe crystals
W. W. Scanlon
Phys. Rev., 126:509-513 (1962)

Growth from the vapor of large single crystals of lead selenide of controlled composition
A. C. Prior
J. Electrochem. Soc., 108:82-87 (1961)

Vacuum radiation furnace with precise control
of temperature gradients for crystal growth by
sublimation
    A. C. Prior
    J. Sci. Instr., 38:198–201 (1961)

The growth of a lead selenide layer
    V. A. Dorin and G. M. Filaretova
    Fiz. Metal. i Metalloved., 9:718–721 (1960)

Determination of the standard free energy of formation of lead selenide
    C. B. Finch and J. B. Wagner, Jr.
    J. Electrochem. Soc., 107:932 (1960)

Investigation of the methods of preparing lead
selenide and of its semiconductor properties
    N. I. Glistenko and A. A. Yeremina
    Zh. Neorgan. Khim., 5:1003 (1960)

A contribution to the preparation of evaporated
lead-selenide layers of high mechanical and electrical stability
    H. Gobrecht, F. Niemeck, and K.-E. Boeters
    Z. Physik, 159:533–540 (1960)

Preparation and properties of PbTe photoresis-
tors of controlled stoichiometric composition
    Karl Gurs
    Z. Physik, 158:533–552 (1960)

Influence of crystal size on the spectral response
limit of evaporated PbTe and PbSe photoconductive cells
    W. D. Lawson and A. S. Young
    J. Electrochem. Soc., 107:206–210 (1960)

Stoichiometry of lead telluride
    E. Miller, K. Komarek, and I. Cadoff
    Trans. AIME, 215:882 (1959)

Preparation, properties, and stoichiometry of
PbTe single crystals
    E. Miller, I. Cadoff, and K. Komarek
    J. Electrochem. Soc., 105:252C (1958)

The influence of several gases and vapors on the
semiconductor properties of vapor deposited lead
selenide films
    H. Gobrecht, F. Niemeck, and K. E. Boeters
    Z. Physik, 148:281–297 (1957)

Oxygen-free single crystals of lead telluride,
selenide and sulfide
    W. D. Lawson
    J. Appl. Phys., 23:495 (1952)

A method of growing single crystals of lead telluride and lead selenide
    W. D. Lawson
    J. Appl. Phys., 22:1444 (1951)

Bridgman crystals of lead telluride
    Frederick C. Abbott
    University of Delaware, Newark, Dissertation Abstr. 66-
      5530, 162 pp.
    Available from University Microfilms, Ann Arbor, Mich.

Lead sulphide, selenide and telluride
    E. H. Putley
    Materials Used in Semiconductor Devices (C. A. Hogarth,
      ed.), Interscience Publishers, New York, p. 71

## 13.f. Lead—Tin Mixed Systems

Review of the $Pb_{1-x}Sn_x$ chalcogenide semiconductors
    J. R. Butler
    J. Vacuum Sci. Tech., 7:174 (1970)
    Proceedings of the 16th National Symposium of the American
      Vacuum Soc. including the Symposium of the Thin Film
    and Surface Science Divisions, Seattle, Wash., Oct. 28–31,
    1969

Preparation and properties of $Pb_{1-x}Sn_xTe$ epitaxial
films
    T. O. Farinre and J. N. Zemel
    J. Vacuum Sci. Tech., 7:121–126 (1970)
    Single-crystalline on KCl and NaCl, polycrystalline on $CaF_2$

Semiconductor materials for electroluminescent
diodes and lasers, technical summary report
    A. G. Thompson
    Bell and Howell Co., Pasadena, Calif., AD-709 969 (May
      1970), 103 pp.
    Growth of high-quality $Pb_{1-x}Sn_xTe$ single crystals by the
      Czochralski method using encapsulation

Shubnikov − de Haas effect in $Pb_{0.83}Sn_{0.17}Te$
    G. A. Antcliffe, R. T. Bate, and J. S. Wrobel
    Bull. Am. Phys. Soc., 14:330 (1969)
    A vapor-phase transport technique has been developed which
      yields single crystals of these alloys of up to 30 g

Electrical and optical properties of lead − tin −
telluride semiconducting alloys
    Richard F. Bis and Jack R. Dixon
    AD-697647, NOLTR-69-146 (Sept. 3, 1969), 102 pp.
    Single crystals in a new-type thin-film evaporator

The capillary seed technique in crystal pulling
    S. E. R. Hiscocks
    J. Mater. Sci., 4:310–312 (1969)

Phase diagram of the ternary system Pb − Sn − Te
    K. J. Linden and C. A. Kennedy
    J. Appl. Phys., 40:2595–2597 (1969)

Growth of $Pb_{1-x}Sn_xTe$ single crystals from non-
stoichiometric melts
    John W. Wagner and Robert K. Willardson
    Trans. AIME, 245:461 (1969)

Epitaxic films of lead chalcogenides and related
compounds
    Joy Norman Zemel
    Solid State Surface Sci., 1:291–403 (1969)
    $Pb_xSn_{1-x}(Se, Te)$; 167 refs.

Long-wavelength infrared $Pb_{1-x}Sn_xTe$ diode lasers
    J. F. Butler and T. C. Harman
    Appl. Phys. Letters, 12:347–349 (1968)

Crystal growth, annealing, and diffusion of lead—
tin chalcogenides
    A. R. Calawa, T. C. Harman, M. Finn, and P. Youtz
    Trans. AIME, 242:374 (1968)

Infrared detector materials
    S. E. R. Hiscocks and J. B. Mullin
    Royal Radar Establishment Newsletter and Res. Rev. No. 7,
      1968, 3 pp.; also N69-28664 15-09
    Pb—Sn—Te single crystals; liquid encapsulation technique

Crystal pulling and constitution in $Pb_{1-x}Sn_xTe$
S. E. R. Hiscocks and P. D. West
J. Mater. Sci., 3:76–79 (1968)

Crystal growth from the vapor phase by forced convection
T. B. Reed and W. J. LaFleur
Lincoln Lab., Mass. Inst. of Tech.
Rept. ESD–TR–68–17 (April 1968), pp. 20–23

On the solid solutions of tin telluride and lead telluride
A. M. Reti, A. K. Jena, and M. B. Bever
Trans. AIME, 242:371 (1968)

Excess carrier concentrations in cation–saturated $Pb_xSn_{1-x}Te$ solid solutions
Koichi Sugiyama
Japan. J. Appl. Phys., 7:961–962 (1968)

Growth and characterization of single crystals of PbTe–SnTe
John W. Wagner and Robert K. Willardson
Trans. AIME, 242:366 (1968)

Crystal growth, composition control and junction formation in the lead–tin chalcogenides
T. C. Harman, A. R. Calawa, and Mary C. Finn
ESD–TR–67–266 (May 1967), Contract AF 19(628)–5167, p. 1

Vacuum–deposited thin films of the type $PbS_xSe_{1-x}$
B. A. Riggs
J. Electrochem. Soc., 114:708 (1967)

Inversion of conduction and valence bands in $Pb_{1-x}Sn_xSe$ alloys
A. J. Strauss
Phys. Rev., 157:608–611 (1967)

Alloy films of $PbTe_xSe_{1-x}$
R. F. Bis and J. N. Zemel
J. Appl. Phys., 37:228 (1966)

Reproducible preparation of $Sn_{1-x}Pb_xTe$ epitaxial films with moderate carrier concentrations
E. G. Bylander
Mater. Sci. Eng., Netherl., 1:190–194 (1966)

Investigation of alloys of the system PbTe–SnTe
A. A. Machonis and I. B. Cadoff
Trans. AIME, 230:333 (1964)

Mischkristallsysteme zwischen halbleitenden Chalkogeniden der vierten Hauptgruppe
H. Krebs, K. Grun, and D. Kallen
Z. Anorg. Allgem. Chem., 312:307 (1961)

Crystal growth, composition control, and junction formation in $Pb_{1-x}Sn_xTe$
T. C. Harman, Mary C. Finn, and A. E. Paladino
(Lincoln Labs., Massachusetts Inst. of Tech.)
ESD–TR–67–162 p. 6

# 14. V—VI Compounds

## 14.a. General, Reviews, and Bibliographies

Mass-spectrometry of laser-induced vaporization of compounds. 2. Bismuth with VIa elements
V. S. Ban and B. E. Knox
J. Chem. Phys., 52:243–247 (1970)

Mass-spectrometry of laser-induced vaporization of arsenic and antimony compounds with group VIa elements
V. S. Ban and B. E. Knox
J. Chem. Phys., 52:248–253 (1970)

Heats of fusion of the $A_2^V B_3^{VI}$ compounds ($As_2S_3$, $As_2Se_3$, $As_2Te_3$, and $Sb_2S_3$)
M. B. Myers and E. J. Felty
J. Electrochem. Soc., 117:808–820 (1970)

Optical properties of $As_2S_3$ and $Sb_2S_3$ single crystals
M. Ya. Valakh and T. N. Nikolaeva
Fiz. Tekhn. Poluprovod., 4:80–83 (1970)
Sov. Phys.—Semicond., 4:63–66 (1970)

Semiconducting II-VI, IV-VI, and V-VI Compounds
N. Kh. Abrikosov, V. F. Bankina, L. V. Poretskaya, L. E. Shelimova, and E. V. Skudnova
Plenum Press, New York (1969)

Synthesis and infrared spectra of some Group Va chalcogenides
George N. Chremos
Thesis, Texas A and M Univ., State College, Tex. (1969), 101 pp.
University Microfilms, Ann Arbor, Mich., Order No. 69-14, 130

Messungen der magnetischen Anisotropie von $Bi_2Te_3$- und $Sb_2Te_3$-Einkristallen und ihren Legierungen
N. van Deynse, A. van Itterbeek, and R. Dekeyser
Z. Angew. Phys., 26:174–178 (1969)

Investigations on evaporated films of bismuth oxide. II. Determination of type of conductivity and photoconductivity measurements on doped and undoped layers
Heinrich Gobrecht, Siegfried Seeck, H. E. Bergt, A. Maertens, and K. Kossmann
Phys. Stat. Sol., 34:569–576 (1969)

Crystal structure and electrical properties of bismuth selenide – antimony triselenide system alloys
V. G. Kuznetsov, K. K. Palkina, A. V. Dmitriev, and A. A. Reshchikova
Khim. Svyaz Krist. (N. N. Sirota, ed.), Izd. "Nauka i Tekhnika," Minsk, USSR (1969), pp. 380–387

Band structure and transport coefficients of V-VI compound semiconductors and their alloys
Gerald R. Miller
Utah Univ., Salt Lake City, Div. of Materials Science and Engineering, UTEC-MSE-69-113, AD-696 244 (Sept. 1969), 125 pp.

The average heat of atomization and the properties of semiconductors
Varadachari Sadagopan and Harry C. Gatos
Chemical Bonds in Crystals (N. N. Sirota, ed.), Science and Technology Publishers, Minsk, USSR (1969), pp. 220
Correlating the physical properties of known semiconductors and predicting the properties of new ones, including the vitreous semiconductors and the limits of stability of various semiconductor structures

Mass spectrometric determination of the dissociation energies of the molecules BiO, BiS, BiSe, and BiTe
O. M. Uy and J. Drowart
Trans. Faraday Soc., 65(2):3221–3230 (1969)

An investigation of the crystal growth of the heavy metal sulfides in supercritical hydrogen sulfide
Leroy Crawford Lewis
Ph. D. thesis, Oregon State University, Corvallis, Oregon (1968), 172 pp.
University Microfilms, Ann Arbor, Michigan, Order No. 68-11915

Reflectivity spectra of the rhombohedral crystals $Bi_2Te_3$, $Bi_2Se_3$, and $Sb_2Te_3$, over the range from 0.7 to 12.5 eV
V. V. Sobolev, S. D. Shutov, Yu. V. Popov, and others
Phys. Stat. Sol., 30(1):349–355 (1968)

The pseudo-binary $V_2VI_3$–IV·VI compounds systems, $Bi_2Te_3$–$PbTe$, $Bi_2Te_3$–$SnTe$, $Sb_2Te_3$–$PbTe$, $Sb_2Te_3$–$SnTe$, and $Bi_2Se_3$–$SnSe$
Tadamasa Hirai, Yutaka Takeda, and Kazuhiro Kurata
J. Less-Common Metals, 13:352–356 (1967)

Crystal structure of bismuth selenides and bismuth and antimony tellurides
M. M. Stasova
Zh. Strukt. Khim., 8(4):655–661 (1967)

Energy structure of the bands of certain compounds of the $A^{II}B^V$, $A^VB^{VI}$, and $A^{III}B^{VI}$ types
V. V. Sobolev, N. N. Syrbu, and S. D. Shutov
Chemical Bonds in Semiconductors and Thermodynamics (in Russian), Nauka i Tekhnika, Minsk (1966), pp. 221–228

Energy gaps in bismuth trioxide
D. M. Mattox and L. Gildart
J. Phys. Chem. Solids, 18:215–217 (1961)
Unique method of preparing polycrystalline films of highly reactive oxide

Stoichiometry of bismuth telluride and related compounds
G. Offergeld and J. van Cakenberghe
Nature (Suppl. No. 4), 184:185–186 (1959)
Bismuth telluride, bismuth selenide, and antimony telluride; semiconductors for thermoelectric cooling

## 14.b. Arsenic Compounds

### 14.b.1. As—S

Comparative investigation of the optical properties of arsenic chalcogenides during transition from the crystalline to glassy state
M. L. Belle, B. T. Kolomiets, and B. V. Pavlov
Fiz. Tekhn. Poluprovod., 2(10):1448–1453 (1968)
Sov. Phys.—Semicond., 2(10):1210–1214 (1969)

Refractive indexes of amorphous $As_2S_5$
F. Kosek and J. Cermak
Cesk. Casopis Fys., 19A:271–275 (1969)

Glass transition in amorphous precipitates
N. Onodera, H. Suga, and S. Seki
J. Non-Cryst. Solids, 1(4):331–334 (1969)

Heat capacities of glassy and liquid $As_2S_3$ and $As_2Se_3$
U. E. Schnaus, C. T. Moynihan, R. W. Gammon, and P. B. Macedo
Catholic Univ. of America, Washington, D. C., Vitreous State Lab., TR-6, AD-698 842 (Dec. 1969), 18 pp.

On the structure of glasses in the system As-S. III. The properties and structure of sulfide glasses
S. Tsuchihashi and Y. Kawamoto
J. Ceram. Soc. Japan, 77:35–39 (1969)

Production and certain properties of single crystals of $As_2Se_3$
S. A. Dembovskii, Yu. A. Polyakov, and A. A. Vaipolin
Izv. Akad. Nauk SSSR, Neorg. Mater., 4(5):767–768 (1968)
Inorg. Mater., 4(5):669–671 (1968)

Measurements of electrical conductivity and optical absorption in chalcogenide glasses
J. T. Edmond
J. Non-Cryst. Solids, 1(1):39–48 (1968)
$As_2S_3$–$As_2Se_3$ and $As_2Se_3$–$As_2Te_3$

Dielectric constants of the systems As—S and As—Se
Akikazu Shibata
Joint Meeting of Phys. Soc. Japan and Soc. of Appl. Phys. (1968)
In Japanese; abstract in English

Synthetic crystals of arsenic trisulphide
C. Bowlt and B. N. Ghosh
Brit. J. Appl. Phys., 16:1762 (1965)

The absorption edge of arsenic-sulphur-selenium mixtures
C. Hilsum
Proc. Phys. Soc., 74(5):667–669 (1959)

### 14.b.2. As—Se

Frequency dependent conductivity of vitreous $As_2Se_3$
M. Kitao, F. Araki, and S. Yamada
Phys. Stat. Sol., 37: K119–21 (1970)

Interaction of glassy arsenious selenide with alkali metals
Z. U. Borisova and V. V. Bakulina
J. Appl. Chem. USSR, 42:1176–1179 (1969)

An electron microscope study of the diffusion of metals in amorphous arsenic triselenide films
L. A. Freeman, R. F. Shaw, and A. D. Yoffe
Thin Solid Films, 3:367–376 (1969)

High-frequency conductivity of arsenic selenide
E. B. Ivkin and B. T. Kolomiets
J. Non-Cryst. Solids, 3(1):41–45 (1969)

Synthesis of the third polymorphic modification of arsenic selenide at high pressure
V. A. Kirkinskii, A. P. Ryaposov, and V. G. Yakushev
Fiz. Tverd. Tela, 11(8):2382–2383 (1969)
Sov. Phys. — Solid State, 11(8):1923–1924 (1970)

Temperature dependence of the magnetic susceptibility of $As_2Se_3$ at the crystal-melt and glass-melt phase transitions
S. K. Novoselov, L. P. Strakhov, and L. A. Baidakov
Fiz. Tverd. Tela, 11(6):1564–1568 (1969)
Sov. Phys. — Solid State, 11(6):1266–1269 (1969)

The chalcogenide semiconductor system $As_2Se_3$–$Sb_2Se_3$
N. S. Platakis, H. C. Gatos, and V. Sadagopan
RC 2499 (12064), June 6 (1969)

Preparation of vitreous materials in the form of well-defined hollow cylinders
N. S. Platakis, H. C. Gatos, and A. F. Witt
J. Electrochem. Soc., 116:510–511 (1969)
$xAs_2Se_3 \cdot ySb_2Se_3$

Influence of manganese, iron, nickel, and cobalt on the electrical conductivity and chemical stability of vitreous arsenious selenide
Ya. Savan and Z. Yu. Borisova
J. Appl. Chem. USSR, 42:970–974 (1969)

Effect of manganese, iron, nickel, and cobalt on the electrical conductivity and chemical solubility of glassy arsenic selenide
   Ya. Savan and Z. U. Borisova
   Zh. Prikl. Khim. (Leningrad), 42(5):1017–1023 (1969)

Heat capacities of glassy and liquid $As_2S_3$ and $As_2Se_3$
   U. E. Schnaus, C. T. Moyhihan, R. W. Gammon, and P. B. Macedo
   Catholic Univ. of America, Washington, D. C., Vitreous State Lab., TR-6, AD-698 842 (Dec. 1969), 18 pp.

Dynamics of vibrations of selenium and arsenic sesquiselenide in crystalline and vitreous states from low-temperature heat-capacity data
   V. V. Tarasov, V. M. Zhdanov, A. K. Mal'tsev, and S. A. Dembovskii
   Zh. Fiz. Khim., 43:467–471 (1969)
   Russ. J. Phys. Chem., 43:249–251 (1969)

Saturated vapor pressure of arsenic selenide and telluride
   G. P. Ustyugov, A. A. Kuryavtsev, B. M. Kuadzhe, and E. N. Vigdorovich
   Izv. Akad. Nauk SSSR, Neorg. Mater., 5(2):378–379 (1969)
   Inorg. Mater., 5(2):315–316 (1969)

Magnetic susceptibility of vitreous arsenic triselenide
   L. M. Blinov, L. A. Baidakov, and L. P. Strakhov
   Ser. Fiz. Nauk, 345:49–51 (1968)

Study of the linear expansion of vitreous and polycrystalline selenium and $As_2Se_3$
   S. F. Chistov, A. P. Chernov, and S. A. Dembovskii
   Izv. Akad. Nauk SSSR, Neorg. Mater., 4(12):2085–2088 (1968)
   Inorg. Mater., 4(12):1814–1816 (1968)

New polymorphic modification of arsenic selenide, prepared at high pressures
   V. A. Kirkinskii and V. G. Yakushev
   Dokl. Akad. Nauk SSSR, 182(5):1083–1086 (1968)

Preparation of $As_2Se_3$ single crystals
   Michihiko Kitao, Norio Asakura, and Shoji Yamada
   Japan. J. Appl. Phys., 8:499–500 (1968)
   Absorption edge at about 1.77 eV

Drift mobility in amorphous $As_2Se_3$ films
   V. I. Kruglov, L. P. Strakhov, and N. A. Grishin
   Fiz. Khim., 23(10):62–68 (1968)

Mixed selenide-oxide semiconducting glasses
   J. E. Lang and J. D. Mackenzie
   Amer. Ceram. Soc. Bull., 47(4):398 (1968)
   Glass-forming tendencies of $As_2Se_3$ with metal oxides, particularly well with copper oxide

Photoelectric properties of $As_2Se_3$ layers
   Okio Yoshida
   Japan. J. Appl. Phys., 6:875 (1967)

Optical reflection spectra of the arsenic chalcogenides
   A. M. Andriesh and V. V. Sobolev
   Chemical Bonds in Semiconductors and Thermodynamics (in Russian), Nauka i Tekhnika (Minsk) (1966), pp. 212–216
   $As_2Se_3$ and $As_2Te_3$; crystalline and vitreous

## 14.b.3. As—Te

Filamentary conduction in semiconducting glass diodes
   A. David Pearson and C. E. Miller
   Appl. Phys. Letters, 14:280 (1969)
   Electrical switching in thin-film $As_2SeTe_2$

New polymorphic modification of arsenic telluride, produced at high pressures
   V. G. Yakushev and V. A. Kirkinskii
   Dokl. Akad. Nauk SSSR, 186(4):882–884 (1969)
   Dokl. — Phys. Chem., 186(4):369–371 (1969)

Phase diagram of the arsenic-tellurium system
   S. A. Dembovskii, I. A. Kirilenko, and A. S. Khvorostenko
   Zh. Neorg. Khim., 13(5):1462–1463 (1968)

The phase diagram of the system tellurium — arsenic
   J. R. Eifert and E. A. Peretti
   J. Mater. Sci., 3:293–296 (1968)

Hall effect in vitreous semiconductors of the system As — Ge — Te
   V. R. Panus, Ya. M. Ksendzov, and Z. U. Borisova
   Izv. Akad. Nauk SSSR, Neorg. Mater., 4(6):885–888 (1968)
   Inorg. Mater., 4(6):778–781 (1968)

Thermal conductivity of glassy semiconductors
   I. A. Rozov, A. F. Chudnovskii, and V. F. Kokorina
   Fiz. Tekhn. Poluprovod., 1(8):1159–1163 (1967)
   Sov. Phys.—Semicond., 1(8):969–972 (1968)

Effect of the nature of the chemical bond on the electrical conductivity of glassy semiconductors
   Z. U. Borisova
   Zh. Fiz. Khim., 41(8):1942–1945 (1967)

The determination of the crystal structure of arsenic telluride
   G. J. Carron
   Dissertation Abstr., 27(10):3540 (1967)

Fluctuation levels in the glassy semiconductor $Tl_2Se \cdot As_2Te_3$
   B. T. Kolomiets, T. N. Mamontova, and G. L. Stepanov
   Fiz. Tverd. Tela, 9(1):29–30 (1967)
   Sov. Phys. — Solid State, 9(1):19–21 (1967)

Hall effect measurement in semiconducting chalcogenide glasses and liquids
   J. C. Male
   Brit. J. Appl. Phys., 18:1543 (1967)

Optical reflection spectra of arsenic chalcogenides
   A. M. Andreish and V. V. Sobolev
   Khim. Svyaz Poluprov. Termodin. Inst. Fiz. Tverd. Tela Poluprov. Akad. Nauk Beloruss. SSR (1966), pp. 212–216

Electronic conduction in $As_2Se_3$, $As_2Se_2Te$ and similar materials
   J. T. Edmond
   Brit. J. Appl. Phys., 17:979 (1966)

The electrical properties of some chalcogenide glasses
   A. E. Owen, N. Clare, and S. Frank
   Conference on Electronic Processes in Low Mobility Solids, Sheffield (1966), Univ. of Sheffield, Dept. of Glass Technology (1966), pp. 109–113

Preparation and properties of sulfide, selenide, and telluride glasses
  A. David Pearson
  Glass Industry (1965), pp. 18–21

Electrical and optical properties of some $M_2^{V-B}N_3^{VI-B}$ semiconductors
  J. Black, E. M. Conwell, L. Seigle, and C. W. Spencer
  J. Phys. Chem. Solids, 2:240–251 (1957)

Preparation and some physical properties of $Bi_2Te_3$, $Sb_2Te_3$, and $As_2Te_3$
  T. C. Harman, B. Paris, S. E. Miller, and H. L. Goering
  J. Phys. Chem. Solids, 2:181–190 (1957)

X-ray crystallographic data on $As_2Te_3$
  Joseph Singer and Chester W. Spencer
  J. Metals, 7:144 (1955)

## 14.c. Antimony Compounds

### 14.c.1. Sb—S

Some physical properties of the semiconductor system antimony trisulfide-antimony triselenide
  A. F. Skubenko and P. E. Mozol
  Ukr. Fiz. Zh., 15:687–689 (1970)

Optical properties of antimony trisulfide single crystals in the region of phase-transition temperatures
  A. I. Audzionis and A. S. Karpus
  Fiz. Tverd. Tela, 11(4):1053–1055 (1969)
  Sov. Phys. — Solid State, 11(4):859–860 (1969)

Space-charge-limited currents in vitreous antimony trisulfide films
  T. Budinas, P. Mackus, A. Smilga, and J. Viscakas
  Phys. Stat. Sol., 31:375 (1969)

Density of antimony chalcogenides in the solid and liquid states as a function of temperature
  V. M. Glazov, N. N. Glagoleva, and S. B. Evgen'ev
  Izv. Akad. Nauk SSSR, Neorg. Mater., 5(7):1181–1184 (1969)
  Inorg. Mater., 5(7):1005–1007 (1969)

Special features of the measurement of pyrocurrents in ferroelectric semiconductors
  M. P. Mikalkevichyus and V. S. Rinkyavichyus
  Fiz. Tverd. Tela, 11(3):769–770 (1969)
  Sov. Phys. —Solid State, 11(3):617–618 (1969)

Glass transition in amorphous precipitates
  N. Onodera, H. Suga, and S. Seki
  J. Non-Cryst. Solids, 1(4):331–334 (1969)

The hydrothermal crystallization of $Sb_2S_3$
  V. I. Papolitov
  Kristallografiya, 14(3):545–547 (1969)
  Sov. Phys. — Cryst., 14(3):461–463 (1969)

Calorimetric study of phase change in antimony trisulfide
  T. A. Pikka and V. M. Fridkin
  Izv. Akad. Nauk SSSR, Ser. Fiz., 33(2):364–365 (1969)

Crystal growth and chemical synthesis under hydrothermal conditions
  A. Rabenau and H. Rau
  Philips Tech. Rev., 30:89–96 (1969)

Au; CuS; CuI; PbSe; CdS; GeI$_2$; Sb$_2$S$_3$; BiSBr; Pb$_5$S$_2$I$_6$; Te; PbI$_2$; Pt, SnSe

X-ray investigation of crystallization processes in antimony trisulfide films
  E. M. Smirnova, T. A. Mingazin, and M. N. Odlis
  Kristallografiya, 14:367–369 (1969)
  Sov. Phys. —Cryst., 14:303–304 (1969)

Some peculiarities of the optical properties of $Sb_2S_3$ in the near and middle infrared optical spectrum range
  A. I. Audzionis, P. P. Balyulis, and A. Karpus
  Litov. Fiz. Sbornik (USSR), 8:407–418 (1968)

Photoconductivity of $Sb_2S_3$ single crystals in infrared range
  M. Mikalkevichyus and V. Rinkyavichyus
  Litov. Fiz. Sbornik (USSR), 8(4):624–635 (1968)

Antimony semiconductor materials
  Danfoss A/S
  French Patent 1,498,954 (Oct. 20, 1967), 4 pp.

Study of the dielectric constant of $Sb_2S_3$ crystals at high frequencies
  J. Grigas and A. Karpus
  Litov. Fiz. Sbornik (USSR), No. 2, 437–445 (1967)

Crystal syntheses and growth in strong acid solutions under hydrothermal conditions
  H. Rau and A. Rabenau
  Solid State Commun., 5:331–332 (1967)
  Includes $Sb_2S_3$

### 14.c.2. Sb—Se

Dynamic growth of single crystals from the vapor phase
  B. P. Grigas and M. P. Mikalkevichyus
  Izv. Akad. Nauk SSSR, Neorg. Mater., 6(1):141–142 (1970)
  Inorg. Mater., 6(1):119–120 (1970)

Properties of acicular single crystals of antimony selenide
  N. I. Butsko and A. P. Oksenyuk
  Fiz. Tverd. Tela, 10(7):2237–2238 (1968)
  Sov. Phys. — Solid State, 10(7):1759–1761 (1969)

Density of antimony chalcogenides in the solid and liquid states as a function of temperature
  V. M. Glazov, N. N. Glagoleva, and S. B. Evgen'ev
  Izv. Akad. Nauk SSSR, Neorg. Mater., 5(7):1181–1184 (1969)
  Inorg. Mater., 5(7):1005–1007 (1969)

Photoelectric properties of alloys of the antimony selenide-bismuth sulfide system
  L. G. Gribnyak and N. M. Bondar
  Ukr. Fiz. Zh., 14:1223–1225 (1969)

Local levels in antimony selenide single crystals grown from the gas phase
  B. Grigas and M. Mikalkevichyus
  Liet. Fiz. Rink., 9(2):369–379 (1969)

Rectifying properties of an antimony — antimony selenide diffused junction
  N. N. Koren and N. N. Sirota
  Dokl. Akad. Nauk Beloruss. SSR, 13:595–596 (1969)

Calibration curves for x-ray fluorescence analysis of some binary alloy systems
Tetsuya Kariya
Kochi Daigaku Kenkyu Hokoku (Shizen Kagaku), 17:83-90 (1968)
Sb-Te, Sb-Se

Phase transformation of $As_2Se_3$ and $Sb_2Se_3$ films
M. D. Coutts and E. R. Levin
J. Appl. Phys., 38:4039 (1967)

Antimony semiconductor materials
Danfoss A/S
French Patent 1,498,954 (Oct. 20, 1967)

The effect of zone refining on some physical properties of $Sb_2Se_3$ single crystals
B. P. Grigas and M. P. Mikalkevichyus
Izv. Akad. Nauk SSSR, Neorg. Mater., 3(12):2179-2183 (1967)
Inorg. Mater., 3(12):1901-1904 (1967)

High-temperature enthalpy studies of $Bi_2S_3$ and $Sb_2Se_3$
A. C. Glatz and K. E. Cordo
J. Phys. Chem., 70:3757-3760 (Nov. 1966)

Étude des niveaux de piégeage dans des monocristaux de $Sb_2Se_3$ par la méthode de conductivité thermostimulée
M. P. Mikalkevichyus and B. P. Grigas
Liet. Fiz. Rink., 6(4):543-550 (1966-1967)

Photoconductivity of $Sb_2Se_3$
B. T. Kolomiets and A. Kh. Zeinally
Fiz. Tverd. Tela, 1(6):979-980 (1960)
Sov. Phys. — Solid State, 1(6):896-897 (1960)

Electric conductivity and photoconductivity of $Sb_2Se_2$ monocrystals
M. V. Kot and S. D. Shutov
Uch. Zap. Kishinev. Univ., 39:45-48 (1959)

Temperature dependence of optical properties of layers of $Sb_2Se_3$
A. Yu. Shileyka and P. P. Brazdzhyunas
Vil'nyus Trudy Akad. Nauk Litovskoy SSR, Ser. B, 4(20): 31-43 (1959)

Photoconductivity of monocrystals of antimony selenide
A. Kh. Zeinally and B. T. Kolomiets
Uch. Zap. Azerb. Univ., Ser. Fiz.-Mat. Khim., No. 1, 79-83 (1959)

## 14.c.3. Sb—Te

Segregation of silver during the growth of antimony telluride single crystals from a melt
P. Sherov, S. Karimov, and Sh. Mavlonov
Dokl. Akad. Nauk Tadzh. SSR, 13(2):19-21 (1970)

Vapor depositing infrared-sensitive antimony tritelluride ($Sb_2Te_3$)
Stanley V. Forgue
U. S. Patent 3,424,610 (Jan. 28, 1969)

Epitaxial growth of $Sb_2Te_3$ films
L. H. Gadgil and A. Goswami
J. Vac. Sci. Tech., 6:591-593 (1969)

Density of antimony chalcogenides in the solid and liquid states as a function of temperature
V. M. Glazov, N. N. Glagoleva, and S. B. Evgen'ev
Izv. Akad. Nauk SSSR, Neorg. Mater., 5(7):1181-1184 (1969)
Inorg. Mater., 5(7):1005-1007 (1969)

The Peltier coefficient at the solid-liquid phase boundary in antimony and some of its semiconducting compounds
N. V. Kolomoets and N. I. Strekopytova
Fiz. Tverd. Tela, 11(4):866-868 (1969)
Sov. Phys. — Solid State, 11(4):711-713 (1969)

Calibration curves for x-ray fluorescence analysis of some binary alloy systems
Tetsuya Kariya
Kochi Daigaku Kenkyu Hokoku (Shizen Kagaku), 17:83-90 (1968)
Sb-Te, Sb-Se

Reflectivity spectra of the rhombohedral crystals $Bi_2Te_3$, $Bi_2Se_3$, and $Sb_2Te_3$ over the range from 0.7 to 12.5 eV
V. V. Sobolev, S. D. Shutov, Yu. V. Porov, and S. N. Shestatskii
Phys. Stat. Sol., 30:349-355 (1968)

Phase relations and thermoelectric properties of the alloy systems $SnTe-Bi_2Te_3$ and $PbTe-Sb_2Te_3$
Richard A. Reynolds
J. Electrochem. Soc., 114:526 (1967)

Hall-Effekt und Leitfähigkeit von Antimontellurid-Einkristallen (Halbleitereigenschaften von Telluriden. II)
L. Liebe
Annal. Physik, 15:179 (1965)

A study of the Sb-Te system in the vicinity of $Sb_2Te_3$
L. V. Poretskaya, N. Kh. Abrikosov, and V. M. Glazov
Zh. Neorg. Khim., 8(5):1196-1198 (1963)

Large dislocation loops in antimony telluride
P. Delavignette and S. Amelinckx
Phil. Mag., 6:601 (1961)

The antimony—tellurium system
N. Kh. Abrikosov, L. V. Poretskaya, and I. P. Ivanova
Zh. Neorg. Khim., 4:2525-2530 (1959)
Beside the compound $Sb_2Te_3$, two intermediate phases, $\mu$- and $\gamma$-phases, exist with a broad region of homogeneity

## 14.c.4. Sb Mixed Systems

A study of the system Sb—Pb—Te
N. Kh. Abrikosov, E. V. Skudnova, L. V. Poretskaya, and T. A. Osipova
Izv. Akad. Nauk SSSR, Neorg. Mater., 5(4):741-745 (1969)
Inorg. Mater., 5(4):630-633 (1969)

Halleffekt und Leitfähigkeit von Kristallen des Systems $In_2Te_3-Sb_2Te_3$ (Halbleitereigenschaften von Telluriden. XV)
W. Schulz and H. Nieke
Ann. Physik, 23:129-138 (1969)

Polarization effects in the reflectivity spectra of orthorhombic crystals of antimony trisulfide and antimony triselenide

S. D. Shutov, V. V. Sobolev, Yu. V. Popov, and S. N.
  Shestatskii
Phys. Stat. Sol., 31:K23–K27 (1969)

Saturated vapor pressure of the antimony chal-
cogenides
  G. P. Ustyugov, E. N. Vigdorovich, B. M. Kuadzhe, and
  I. A. Timoshin
  Izv. Akad. Nauk SSSR, Neorg. Mater., 5(3):589–590 (1969)
  Inorg. Mater., 5(3):498–499 (1969)

Crystal structures of solid solutions of the sys-
tem Sb$_2$S$_3$—Sb$_2$Se$_3$
  N. Kh. Abrikosov and V. I. Ivlieva
  Izv. Akad. Nauk SSSR, Neorg. Mater., 4(6):868–872 (1968)
  Inorg. Mater., 4(6):763–766 (1968)

Optical and photoelectric properties of antimony
trisulfide—antimony triselenide system single
crystals
  A. O. Aliev, B. V. Pavlov, A. Kh. Zeinally, and B. T.
  Kolomiets
  Uch. Zap. Azerb. Gos. Univ., Ser. Fiz. Mat. Nauk, No. 1,
  72–76 (1968)

Halleffekt, Leitfähigkeit und Thermokraft von
Kristallen des Systems SnTe—Sb$_2$Te$_3$
  E. Beleites, H. Nieke, and H. Saul
  Ann. Physik 21:375–386 (1968)

Electrical conductivity of Sb$_2$S$_3$—Sb$_2$Se$_3$ single
crystals
  A. O. Aliev, I. S. Baukin, A. Kh. Zeinally, and B. T.
  Kolomiets
  Uch. Zap. Azerb. Gos. Univ., Ser. Fiz.-Mat. Nauk, No. 6
  89–91 (1967)

Antimony semiconductor materials
  Danfoss A/S
  French Patent 1,498,954 (Oct. 20, 1967)
  SbS; Sb$_2$Se$_3$

Dielectric properties of antimony selenide and
antimony sulfide
  K. Sh. Kocharli, N. M. Bezdetnyi, M. K. Iskenderov, and
  D. M. Nasrullaev
  Uch. Zap. Azerb. Gos. Univ., Ser. Fiz.-Mat. Nauk, No. 6,
  92–96 (1967)

On the thermal conductivity of the solid solution
(Sb$_2$Te$_3$)–(Bi$_2$Se$_3$)
  G. V. Kokosh, Kh. M. Kuliev, L. V. Prokof'eva, and L. S.
  Stil'bans
  AEC-tr-6587 Izv. Akad. Nauk Turkm. SSR, Ser. Fiz.- Tekhn.,
  Khim. Geol. Nauk, No. 5, 23–28 (1963), 10 pp.

## 14.d. Bismuth Compounds

### 14.d.1. Bi—S

Electrical properties of bismuth trisulfide
whiskers
  M. N. Bilyi and E. D. Stefanishin
  Ukr. Fiz. Zh., 14:1225–1226 (1969)

Electron emission, pyroelectric effect and di-
electric properties of bismuth trisulfide
  A. V. Yuodvirshis, V. V. Kedavichus, and P. A. Pipines
  Fiz. Tverd. Tela, 11(5):1420–1421 (1969)
  Sov. Phys.— Solid State, 11(5):1158–1159 (1969)

Alloys for thermoelectric devices
  Joseph V. Fisher
  U. S. Patent 3,414,405 (Dec. 3, (1968), 4 pp.

High-temperature, high-pressure synthesis of a
new bismuth sulfide
  M. S. Silverman
  Inorg. Chem., 3:1041 (1964)
  BiS$_2$

Electrical properties of single-cyrstal Bi$_2$S$_3$
  Richard C. Heckman and D. M. Mattox
  J. Phys. Chem. Solids, 24:973–975 (1963)

Some semiconducting properties of bismuth tri-
sulfide
  L. Gildart, J. M. Kline, and D. M. Mattox
  J. Phys. Chem. Solids, 18:286–289 (1961)
  Single crystals, dendrites, and crystalline films of Bi$_2$S$_3$

### 14.d.2. Bi—Se

Semiconducting properties of Bi$_2$Te$_3$ and Bi$_2$Se$_3$
films
  A. Goswami and S. S. Koli
  Indian J. Pure Appl. Phys., 7(3):166–169 (1969)

Preparation of mercury and bismuth selenides
  G. S. Klebanov, N. A. Ostapkevich, N. V. Pakhomova, and
    A. V. Morozova
  Zh. Prikl. Khim., 42(8):1715–1719 (1969)

Band structure and transport coefficients of V—
VI compound semiconductors and their alloys,
final report
  Gerald R. Miller
  Division of Materials Science, University of Utah, Salt Lake
    City
  Contract N00014-67-A-0325, AD-696 244: Rept. No. UTEC-
    MSE-69-113 (Sept. 1969), 125 pp.

The saturated vapor pressure of bismuth sele-
nide and telluride
  G. P. Ustyugov, E. N. Vigdorovich, and I. A. Timoshin
  Izv. Akad. Nauk SSSR, Neorg. Mater., 5(1):138–139 (1969)
  Inorg. Mater., 5(1):166–167 (1969)

Determination of the saturated vapor pressure
of solid bismuth selenide
  Z. Boncheva-Miladenova, A. S. Pashinkin, and A. V.
    Novoselova
  Izv. Akad. Nauk SSSR, Neorg. Mater., 4(7):1027-1031 (1968)
  Inorg. Mater., 4(7):904–907 (1968)

Alloys for thermoelectric devices
  Joseph V. Fisher (Semi-Elements, Inc.)
  U. S. Patent 3,414,405, Dec. 3, 1968

Thermal properties of vitreous Bi-Se alloys
  M. B. Myers and J. C. Schottmiller
  Am. Ceram. Soc. Bull., 47(4):402 (April 1968)
  70th Annual Meeting, American Ceramic Society, Chicago,
    April 20-25, 1968

Specific heat of n- and p-type bismuth telluride,
bismuth selenide, and platinum from 1.4 to 90°K
  Gary E. Shoemake
  Thesis, Carnegie-Mellon Univ., Pittsburgh, Penn., 1968,
    185 pp.
  University Microfilms, Ann Arbor, Mich., Order No. 69-
    6587

Reflectivity spectra of the rhombohedral crystals $Bi_2Te_3$, $Bi_2Se_3$, and $Sb_2Te_3$ over the range from 0.7 to 12.5 eV
V. V. Sobolev, S. D. Shutov, Yu. V. Porov, and S. N. Shestatskii
Phys. Stat. Sol., 30:349-355 (1968)

Über Kristallstruktur und elektrische Eigenschaften der Wismutselenide $Bi_2Se_2$ und $Bi_2Se_3$
H. Gobrecht, K.-E. Boeters, and G. Pantzer
Z. Physik, 177:68-83 (1964)
Crystal growth described

Monotectic reaction in the bismuth-selenium system
R. J. Knight, Che-Yu Li, and C. W. Spencer
Trans. AIME, 227:18 (1963)

Determination of the nonstoichiometric doping mechanism in $Bi_2Se_3$
M. J. Smith
Appl. Phys. Letters, 1:79 (1962)

Investigation of the phase diagram of the system Bi-Se
N. Kh. Abrikosov, V. F. Bankina, and K. F. Kharitonovich
Zh. Neorg. Khim., 5(9):2011-2016 (1960)

Diffusion of certain impurities in $Bi_2Se_3$ and ZnSb
A. A. Kuliev
Fiz. Tverd. Tela, 1(8):1176-1178 (1960)
Sov. Phys. — Solid State, 1(8):1076-1078 (1960)

An investigation of the diffusion of Zn and Se in $Bi_2Se_3$, BiSe and CdSb
A. A. Kuliev and G. B. Abdullaev
Fiz. Tverd. Tela, 1(4):603-605 (1959)
Sov. Phys. — Solid State, 1(4):545-547 (1959)

## 14.d.3. Bi—Te

Impurity scattering and two-band effects in n-type $Bi_2Te_3$
H. A. Ashworth and J. A. Rayne
Bull. Am. Phys. Soc., 15:31 (1970)

Thermal conductivity of bismuth telluride in the liquid phase and in the premelting region
V. I. Fedorov and V. I. Machuev
Fiz. Tverd. Tela, 11(9):2690-2691 (1969)
Sov. Phys.— Solid State, 11(9):2179-2180 (1970)

Nucleation in undercooled Bi-Te alloys
G. L. F. Powell and G. A. Colligan
Met. Trans., 1:1349-1351 (1970)

Partition coefficients of copper and halides in doping of bismuth telluride with copper monohalides
É. T. Abdukarimov, S. N. Belorukova, V. A. Kutasov, A. G. Orlov, and V. N. Romanenko
Izv. Akad. Nauk SSSR, Neorg. Mater., 5(9):1655-1656 (1969)
Inorg. Mater., 5(9):1403-1404 (1969)

Summary report of activities concerning electrical, thermal, and optical properties of semiconductors related to energy conversion, final technical report, June 15, 1958-Nov. 30, 1969
R. B. Adler and A. C. Smith
Massachusetts Institute of Technology, Cambridge, Mass.,

Contract Nonr-1841 (51), AD-693235 (August 1, 1969), 26 pp.

Magnetoresistance of n-type $Bi_2Te_3$
H. A. Ashworth, J. A. Rayne, and R. W. Ure
Phys. Letters, 30A:231-232 (1969)

Semiconducting properties of $Bi_2Te_3$ and $Bi_2Se_3$ films
A. Goswami and S. S. Koli
Indian J. Pure Appl. Phys., 7(3):166-169 (1969)

Elastic moduli of $Bi_2Te_3$ from 4.2°K to 300°K
J. O. Jenkins, J. A. Rayne, and R. W. Ure
Phys. Letters, 30 A:349-350 (1969)

The band structure of bismuth telluride
Shin-ichi Katsuki
J. Phys. Soc. Japan, 26:58 (1969)

Electromotive force investigation of the bismuth-tellurium system
Chung-Chiun Liu and John C. Angus
J. Electrochem. Soc., 116(8):1054-1060 (1969)

De Haas-van Alphen effect in n-type $Bi_2Te_3$
Richard B. Mallinson
Thesis, Carnegie-Mellon Univ., Pittsburgh, Pa., 1969, 156 pp.
University Microfilms, Ann Arbor, Mich., Order No. 69 17,506

Band structure and transport coefficients of V-VI compound semiconductors and their alloys, final report
Gerald R. Miller
Division of Materials Science, University of Utah, Salt Lake City
Contract N00014-67-A-0325, AD-696 244; Rept. No. UTEC-MSE-69-113 (Sept. 1969), 125 pp.

Thermal conductivity measurements of semiconductors, final report
R. Daniel Redin
South Dakota School of Mines and Technology, Rapid City
Contract Nonr-2964 (01), Rept. No. F2964 (Nov. 15, 1969), 13 pp.

Specific heat of n- and p-type $Bi_2Te_3$ from 1.4 to 90°K
G. E. Shoemake, J. A. Rayne, and R. W. Ure, Jr.
Phys. Rev., 185:1046 (1969)

Phase studies using the traveling solvent method of crystal growth: the bismuth-tellurium system
John Strassburger
J. Electrochem. Soc., 116:640-645 (1969)

Characterization of phases in the 50-60 at.% tellurium region of the bismuth-tellurium system by x-ray powder diffraction patterns
R. F. Brebrick
J. Appl. Cryst., Vol. I, Pt. 4, 241-246 (1968)

Superconductivity in semiconductors
Daniel Wayne Deis
Ph. D. thesis, Duke Univ., Durham, N. C., 1968, 130 pp.
University Microfilms, Ann Arbor, Michigan, Order No. 68-14301
Impurity-doped lead telluride and $Bi_2Te_3$ were investigated but superconductivity was not exhibited by either compound down to 0.0138°K and 0.0188°K, respectively

An apparatus for growing bismuth telluride mo-
monocrystals by Czochralski's method
  B. M. Gol'tman and S. D. Prokhorova
    Rost Kristallov, Vol. 6B, Izd. Akad. Nauk. (1965), pp. 231-233
    Growth of Crystals, Vol. 6B (A. V. Shunikov and N. N.
      Sheftal', eds.), Consultants Bureau, New York (1968),
      pp. 45-47

Calibration curves for x-ray fluorescence analysis of some binary alloy systems
  Tetsuya Kariya
    Kochi Daigaku Kenkyu Hokoku (Shizen Kagaku), 17:83-90
      (1968)

Phase equilibrium studies and thermodynamic
properties of the bismuth-tellurium system
  Chung-Chiun Liu
    Ph. D. thesis, Case Western Reserve University, Cleveland,
      Ohio, 1968, 127 pp.
    University Microfilms, Ann Arbor, Mich., Order No. 69-9419

Specific heat of n- and p-type bismuth telluride,
bismuth selenide, and platinum from 1.4 to 90°K
  Gary E. Shoemake
    Thesis, Carnegie-Mellon Univ., Pittsburgh, Penn., 1968,
      185 pp.
    University Microfilms, Ann Arbor, Mich., Order No. 69-
      6587

Reflectivity spectra of the rhombohedral crystals $Bi_2Te_3$, $Bi_2Se_3$, and $Sb_2Te_3$ over the range
from 0.7 to 12.5 eV
  V. V. Sobolev, S. D. Shutov, Yu. V. Porov, and S. N.
    Shestatskii
    Phys. Stat. Sol., 30(1):349-355 (1968)

Peltier cooling
  W. Lechner
    Philips Tech. Rev., 27(5):113-130 (1966)
    Preparation and properties of the best materials for Peltier
      cooling, namely compounds derived from bismuth
      telluride

Effect of added microsegregation on the thermoelectric properties of $Bi_2Te_3$
  Tsan Chi Lo, Shih Tuan Yin, Tai Wei T'ang, and Sheng
    Yang Yuan
    Wu Li Hsueh Pao, 22:515-524 (1966)

The preparation of materials for thermoelectric
coolers
  W. N. Borle, R. K. Purohit, and A. K. Sreedhar
    J. Sci. Instr., 42:55 (1965)

Bismuth telluride
  H. J. Goldsmid
    Materials Used in Semiconductor Devices, John Wiley and
      Sons, Inc., New York (1965), pp. 165-197
    Crystal growth; properties

Preparation and properties of bismuth telluride
and its alloys
  P. T. Chiang
    47th Conf. Canad. Chi., Kingston, 1964 (Ottawa, Inst. Chim.
      Canada), pp. 56-57

The physical properties of single-crystal bismuth
telluride
  J. R. Drabble
    Progress in Semiconductors, Vol. 7 (Alan F. Gibson and
      R. E. Burgess, eds.), John Wiley and Sons, Inc., New York
      (1963), pp. 47-98

Anisotropy of thermoelectric power in bismuth
telluride
  Jane Hodgson Dennis
    Thesis, Mass. Inst. of Tech. Tech. Rept. 377 (January 15,
      1961)

The electrical properties of single crystals of
bismuth and its alloys. II. Galvanomagnetic properties of bismuth-tellurium alloys (solid solutions)
  D. V. Gitsu and G. A. Ivanov
    Fiz. Tverd. Tela, 2(7):1464-1476 (1960)
    Sov. Phys. — Solid State, 2(7):1330-1340 (1961)

The thermal conductivity and thermoelectric
power of bismuth telluride at low temperatures
  P. A. Walker
    Proc. Phys. Soc., Vol. 76, Part 1, 113-126 (July 1960)
    p- and n-type between 6 and 200°K

Ternary sulfides, selenides, and tellurides of
bismuth and thallium
  Tom A. Bither, Jr.
    E. I. du Pont de Nemours and Co., U. S. Patent 2,893,831,
      July 7, 1959

n- and p-type single-crystal bismuth telluride
  A. Yang and F. D. Shepherd
    AFCRC-TR-59-360 (December 1, 1959) pp. 57-59

Galvanomagnetic effects in bismuth telluride
  J. R. Drabble
    Advances in Semi-Conductor Science. Proc. Third Intern.
      Conf. on Semi-Conductors held at University of Rochester,
      N. Y., August 18-22, 1958
    Pergamon Press, New York (1959), pp. 428-430

Preparation and thermoelectric properties of
$Bi_2Te_3$ and all alloys with $Bi_2Se_3$
  T. C. Harman, M. J. Logan, B. Paris, and E. H. Lougher
    (Battelle Memorial Inst.), Fall Meeting of Electrochemical
      Society, Sept. 1958

Preparation of Single Crystals
  W. D. Lawson and S. Nielsen
    Butterworths and Academic Press (1958)

Electrical and thermal properties of $Bi_2Te_3$
  C. B. Satteraite and R. W. Ure, Jr.,
    Phys. Rev., 108:1164 (1957)

Single-crystal bismuth telluride
  L. Ainsworth
    Proc. Phys. Soc., 69 B:606 (1956)

### 14.d.4. Bi Mixed Systems

Anisotropy of the electrical properties of low-
temperature thermoelectric materials
  L. D. Dudkin, N. P. Zykova, and S. N. Lyuskin
    Izv. Akad. Nauk SSSR, Neorg. Mater., 6(1):127-128 (1970)
    Inorg. Mater., 6(1):106-107 (1970)

Investigation of bismuth-lead-tellurium system
  N. Kh. Abrikosov, E. V. Skudnova, L. V. Poretskaya, and
    T. A. Osipova
    Izv. Akad. Nauk SSSR, Neorg. Mater., 5(10):1682-1686 (1969)
    Inorg. Mater., 5(10):1424-1427 (1969)

Thermoelectric properties of p-$Bi_{0.5}Sb_{1.5}Te_3$ and
p-$Bi_{0.5}Sb_{1.5}(TeSe)_2$ compositions
  B. M. Gol'tsman and V. Sh. Sarkisyan
    Izv. Akad. Nauk Arm. SSR, Fiz., 4(1):33-39 (1969)

Photoelectric properties of alloys of the antimony selenide-bismuth sulfide system
  L. G. Gribnyak and N. M. Bondar
  Ukr. Fiz. Zh., 14(7):1223–1225 (1969)

Thermoelectric alloy
  F. K. Heumann
  General Electric Company, German Patent 1,295,044 (May 14, 1969)
  US Appl. June 26, 1961, 4 pp.

Crystal structure and electrical properties of bismuth selenide-antimony triselenide system alloys
  V. G. Kuznetsov, K. K. Palkina, A. V. Dmitriev, and A. A. Reshchikova
  Khim. Svyaz Krist. (N. N. Sirota, ed.), Izd. Nauka i Tekhnika, Minsk, USSR (1969), pp. 380–387

Thermal conductivity measurements of semiconductors, final report
  R. Daniel Redin
  South Dakota School of Mines and Technology, Rapid City
  Contract Nonr–2964 (01), Rept. No. F2964 (Nov. 15, 1969), 13 pp.

Thickness dependence of optical constants ($Bi_8Te_7S_5$)
  H. H. Soonpaa
  J. Vacuum Sci. Tech., 6:741–743 (1969)

Semiconductor properties of tellurides. 9. Hall effect, conductivity and thermoelectric power of $SnTe-Sb_2Te_2$ system. 10. Thermal conductivity and electrical properties of $HgTe-CdTe$ system. 11. Thermal power, conductivity, and specific heat of thallium telluride. 12. Region of mixed crystal formation in $Bi_{2-x}Sb_xTe_{3-y}Se_y$
  E. Beleites, H. Nieke, H. Saul, and others
  Ann. Phys. (Germany), 21(7/8):375–413 (1968)

Semiconductor properties of tellurides. XII. Limits of mixed-crystal formation in the solid solution $Bi_{2-x}Sb_xTe_{3-y}Se_y$
  G. Fienhold, H. Strachauer, and H. A. Ullner
  Ann. Phys. (Germany), 21(7–8):411–413 (1968)

The scattering mechanism of highly doped semiconductors of the system Bi–Te–Se
  W. Heldmann and J. Klugel
  Z. Naturforsch. (Germany), 23 A(5):670–675 (May 1968)

Optical constants of $Bi_8Te_7S_5$
  W. C. Malm, A. J. D. Liu, and H. H. Soonpaa
  Surface Sci., 16(1):365–369 (1969)
  Proceedings of symposium on recent developments in ellipsometry, Lincoln, Nebraska, Aug. 7–9, 1968

Thermodynamic properties of Bi(III) telluride and selenide
  B. T. Melekh and S. A. Semenkovich
  Izv. Akad. Nauk SSSR, Neorg. Mater., 4(8):1346–1348 (1968)
  Inorg. Mater., 4(8):1180–1182 (1968)

The lattice thermal conductivity of bismuth telluride and some bismuth telluride-bismuth selenide alloys
  D. H. Damon, R. W. Ure, Jr., and J. Gersi
  NBS Spec. Publ. No. 302, 111–22 (1968)

Proceedings of the seventh conference on Thermal Conductivity, held at National Bureau of Standards, Gaithersburg, Maryland, Nov. 13-16, 1967

High-temperature enthalpy studies of $Bi_2S_3$ and $Sb_2Se_3$
  A. C. Glatz and K. E. Cordo
  J. Phys. Chem., 70:3757–3760 (Nov. 1966)

Electron transport and thermoelectric properties of n-type $Bi_{1.75}Sb_{0.25}Te_3$
  W. B. Muir, P. T. Chang, and C. H. Champness
  Can. J. Phys., 44:2797 (1966)
  Crystal growth described

New manipulable variable in crystal growing
  H. H. Soonpaa and J. W. Dunning
  J. Appl. Phys., 37:454–455 (1966)
  $Bi_8Te_7S_5$

Neuere Untersuchungen an halbleitenden Mischkristallen unter besonderer Berücksichtigung von Zustandsdiagrammen
  J. Rupprecht and R. G. Maier
  Phys. Stat. Sol., 8:3 (1965)
  $(Bi,Sb)_2(Te,Se)_3$

Energy band studies from thin films
  H. H. Soonpaa
  Research Center, Honeywell, Inc., Hopkins, Minn., Nov. 1965, 30 pp.
  $Bi_8Te_7S_5$ prepared by cleavage from a large crystal

The crystal structure of $Bi_2Te_{3-x}Se_x$
  Seizo Nakajima
  J. Phys. Chem. Solids, 24:479–485 (1963)

Doping properties of $Bi_2Te_{2.7}Se_{0.3}$
  Olof Beckman and Par Bergvall
  Arkiv. Fysik, 24:113 (1962)

The analysis of bismuth telluride and related thermoelectric materials
  H. J. Cluley and P. M. C. Proffitt
  Analyst, 85:815 (1960)

Variation with temperature of the thermoelectric properties of solid solutions of $Bi_2Te_3/Sb_2Te_3$
  H. Rodot and M. G. Weill
  J. Phys. Rad., 21(5):502–503 (May 1960)

An investigation of the $Bi_2Se_3-Bi_2S_3$ system
  M. L. Beglarian and N. Kh. Abrikosov
  Dokl. Akad. Nauk SSSR 128(2):345–347 (Sept. 11, 1959)

Transport properties of the pseudo-binary alloy system $Bi_2Te_{3-y}Se_y$
  N. Fuschillo, J. N. Bierly, and F. J. Donahoe
  Advances in Semi-conductor Science. Proc. Third Intern. Cong. on Semi-conductors held at the University of Rochester, N. Y., August 18-22, 1958, Pergamon Press, New York (1959), pp. 430–433

The magnetic susceptibility of selenides and tellurides of the heavy metals
  M. Matyas
  Cesk. Cas. Fys., No. 4, 439–443 (1958)
  $Sb_2Se_3$, $Sb_2Te_3$, $Bi_2Te_3$, $Bi_2Se_3$

# 15. Bismuth, Antimony, and Bismuth—Antimony

## 15.a Bismuth

Investigation of efficacy of zone refinement
of bismuth, tellurium and their compounds
N. M. Kinysheva and L. A. Firsanova
Physico-Chemical Principles of Crystallization Processes
Used in the Superpurification of Metals (V. N. Vigdorovich,
ed.), pp. 108–112, JPRS 52552 (March 1971)

Preparation of very large bismuth single crystals
for neutron filters
S. Bednarski
J. Cryst. Growth, 6:193–194 (1970)

Etude de la conductivité des couches minces
d'antimoine et de bismuth pendant leur formation
Roland Gerber and Jean-Louis Petit
Compt. Rend., 271B:55–57 (1970)

Spherical single crystals of bismuth
J. P. Issi and A. Moureau
J. Less-Common Metals, 20:67–69 (1970)

Recrystallisation and grain growth in vacuum-
evaporated bismuth films
S. K. Sharma and O. P. Bahl
Thin Solid Films, 6:239–248 (1970)

Separation of Sb and Bi using ion exchangers
(Dowex)
Z. Sulcek, M. Boseova, and J. Dolezal
Collect. Czech. Chem. Commun., 34:787–794 (1969)

Vacuum decanting of bismuth and bismuth alloys
J. J. Frawley, W. R. Maurer, and W. J. Childs
Trans. AIME, 242:1517–1521 (1968)
The object of this investigation was to determine the growth
habit of bismuth and bismuth alloy dendrites as a function
of supercooling

Experimental device for the study of crystal
growth from the melt
M. Gautherir
Rev. Phys. Appl. (France), 3(2):131–142 (June 1968)
Normal emissivity, thermal conductivity, and growth and
melting rates of tin, bismuth, and lead

Method for the preparation and analysis of sam-
ples of rare earth and other metals with special
properties
A. N. Gladkikh, V. A. Skudnov, and L. D. Sokolov
Zavod. Lab., 34:218–221 (1968)
Tabulated data are given on the characteristics of Tl, Ga, In,
Cd, Bi, La, Nd, Pr, Y, Sc, Ho, Er, Tb, Lu, and Co

Vacuum evaporation technology
E. H. Kobisk
Review of Isotopes Target Program, Jan. 1965–Dec. 1967
Contract No. W-7405-eng-26, ORNL-4308 (Oct. 1968), p. 9

Instruments for electroerosion cutting of single
crystals of bismuth
I. M. Pilat, V. B. Orletskii, V. G. Okhrem, and S. V. Chaika
Instr. Exper. Tech., 3:482 (1968)

Temperature dependence of growth rate of
twinned layers in bismuth single crystals
V. I. Startsev, V. P. Soldatov, and M. M. Brodskii
Fiz. Metal. i Metalloved. 25:1111–1115 (1968)

Mechanism governing the growth of certain metal
crystals from the melt
G. A. Alfintsev and D. E. Ovsienko
Rost Kristallov (Growth of Crystals), Naukova Dumka, Kiev,
(1967), 40–52

Growing orientated strain-free bismuth single
crystals of specialised shapes
C. J. Creasey and J. E. Aubrey
J. Less-Common Metals, 12:508 (1967)

Mesures des vitesses de croissance et de décrois-
sance de cristaux de bismuth
Michel Gautherie
Compt. Rend., 264B:1826 (1967)

Calculation of the optimum conditions of the
process for concentrating impurities in bismuth
by the method of zone melting
E. E. Konovalov and Sh. I. Peizulaev
Zh. Anal. Khim., 22:736–740 (1967)

Growing zinc and bismuth single crystals of a
predetermined shape and orientation
F. F. Lavrent'ev, V. P. Soldatov and Yu. G. Kazarov
Fiz. Metal. i Metalloved., 21:793 (1966)

Growing zinc and bismuth single crystals of pre-
arranged shape and crystallographic orientation
    F. F. Lavrent'ev, V. P. Soldatov, and Yu. G. Kazarov
    Rost Kristallov (Growth of Crystals), Naukova Dumka (Kiev),
        (1967), pp. 139–143

Growth of large bismuth single crystals from a
supercooled melt
    W. A. Nordland
    Trans. AIME, 239:2002–2003 (1967)

Characteristics of growth features of bismuth
single crystals
    N. S. Pandya and B. S. Shah
    Indian J. Pure Appl. Phys., 5:143–144 (1967)

Obtaining single crystals of bismuth
    H. Riveros and G. Torres
    Rev. Mexicana Fis., 16:115–122 (1967)

Preparation and properties of eutectic Bi–MnBi
single crystals
    W. M. Yim and E. J. Stofko
    J. Appl. Phys., 38:5211 (1967)
    Single crystals of Bi containing ordered MnBi filaments,
        approximately 4% by volume, grown by zone-melting

Growth of Bi whiskers from gaseous phase in ultrahigh vacuum
    A. Grohman and L. Wojda
    Acta Phys. Polon. (Poland), 29:419–422 (1966)

The method of semicontinuous distillation in vacuum and its use for the production of high-purity
bismuth and selenium for rectifiers
    S. Richter and L. Mueller
    Freiberger Forschungsh. B, 112:127–158 (1966)

Dendritic solidification in bismuth and bismuth
alloys
    E. J. Schneider
    Rensselaer Polytechnic Inst., Troy (N. Y.), Dissertation
        Abstr., 27:505 (1966)

The rate of twin layer growth in bismuth single
crystals
    V. I. Startsev, V. P. Soldatov, and M. M. Brodsky
    Phys. Stat. Sol., 18:863–871 (1966)

The use of directional freezing of the melt for
the preparation of large bismuth single crystals
    S. Bednarski
    Phys. Stat. Sol., 9:839 (1965)

Growth of deformation twins
    A. A. Bochyar, Yu. P. Pshenichnov, and I. N. Chuvilina
    Dokl. Akad. Nauk SSSR, 164(2):305–306 (1965)
    Soviet Phys.–Dokl., 10(9):866–867 (1966)

Growing monocrystals of Bi and Pb in the form
of plates through controlled crystallization
    V. Cioca
    Stud. Cercetari Fiz. (Rumania), 17:1093–1102 (1965)

Bismuth films regrown from the liquid phase
    A. R. Clawson
    Solid — State Electron., 8:967 (1965)

Growth of single crystals of Bi
    H. Riveros, G. Torres, and J. Podolsky
    Rev. Mexican Fis., 14:198 (1965)
    Conf. of the Mexican Phys. Soc., Yucatan, 1965

Single-crystal filters for attenuating epithermal
neutrons and $\gamma$-rays in reactor beams
    B. M. Rustad, J. Als-Nielsen, A. Bahnsen, C. J. Christensen,
        and A. Nielsen
    Rev. Sci. Instr., 36:48–54 (1965)

Transition metal and intermetallic compound single crystals
    A. I. Schindler, M. E. Glicksman, G. N. Kamm, B. C. La Roy,
        R. J. Schaefer, T. J. Schriempf, and C. L. Vold
    N66-15564 (August 1965)

Factors affecting the growth and the mechanical
and physical properties of bismuth single crystals
    R. E. Slonaker, M. Smutz, H. Jensen, and E. H. Olson
    J. Less-Common Metals, 8:327 (1965)

Dendrite Growth in Supercooled Metals
    Wylie J. Childs
    Rensselaer Polytechnic Inst. Annual Report, TID-20234
        (Feb. 1964). Contract AT(30-1)-3004. 7 pp.

The Properties and Preparation of Thin Bismuth
Films
    George A. Condas
    UCRL-12001 (Nov. 1, 1964).  Contract W-7405-eng-48. 15 pp.

Rapid-freeze method for growth of bismuth single crystals
    Sidney Fischler
    Trans. Met. Soc. AIME, 230:340 (1964)

Growth of bismuth single crystals in the form
of shaped tensile specimens
    R. E. Slonaker, Jr., M. Smutz, H. Jensen, and E. H. Olson
    J. Less-Common Metals, 7:165 (1964)

Melting of gallium and bismuth
    J. D. MacKenzie and R. L. Cormia
    J. Chem. Phys., 39:250 (1963)

Obtaining pure bismuth
    B. N. Aleksandrov
    Fiz. Metal. i Metalloved., 14:733–736 (1962)

Épitaxies de cristaux d'arsenic, d'antimoine et
de bismuth obtenus par réduction en phase gazeuse d'un de leurs sulfures et déposés sur différents supports monocristallins
    Lucien Capella
    Compt. Rend., 254:1309 (1962)

Mechanism of the vacuum condensation of metals
    A. A. Chernov
    Rost Kristallov, Vol. 3, Izd. Akad. Nauk (1961), pp. 126–132
    Growth of Crystals, Vol. 3 (A. V. Shubnikov and N. N.
        Sheftal', eds.), Consultants Bureau, New York (1962), pp.
        174–183

Growth of Single Crystals of Lead, Cadmium and
Bismuth Using the Bridgman Method
    James Herriott, Morton Smutz, and Edwin H. Olson
    IS-542 (Nov. 1962)

Growing of Bismuth Single Crystals
    Bruce K. Long, Ed Olson, and Morton Smutz
    IS-567 (1962)

Bismuth whisker growth
    Ludwig Mayer, Robert Rickett, and Heinrich Stenemann
    J. Appl. Phys., 33:982 (1962)

Growing of single crystals of zinc, bismuth, and white tin by the Bridgman Method
Mikio Yamamoto and Jiro Watanabe
Sci. Rept. Res. Inst., Tohoku Univ., Ser. A, 12:486 (1960)

## 15.b. Antimony

Growth features on the cleavage face of antimony single crystals
V. P. Bhatt, M. C. Talati, and H. M. Shah
Indian J. Pure Appl. Phys., 8(4):236 (1970)

Etude de la conductivité des couches minces d'antimoine et de bismuth pendant leur formation
Roland Gerber and Jean-Louis Petit
Compt. Rend., 271B:55-57 (1970)

High-resistance ratio antimony
D. A. Huntley and J. S. Shah
J. Cryst. Growth, 6:216-218 (1970)

Electroerosion etching of bismuth, tellurium, and antimony single crystals
I. M. Bagai
Elektron. Obrab. Mater., 3:18-21 (1969)

Crystallization of amorphous antimony films prepared by vacuum evaporation
Kanwar Bahadur and K. L. Chaudhary
Appl. Phys. Letters, 15(9) (Nov. 1, 1969)

Separation of Sb and Bi using ion exchangers (Dowex)
Z. Sulcek, M. Boseova, and J. Dolezal
Collect. Czech. Chem. Commun., 34:787-794 (1969)

Studies on the preparation of high-purity antimony
R. Krishna Iyer and T. R. Bhat
Trans. Indian Inst. Metals, 21:49-52 (1968)

Production of epitaxial films of antimony on disordered substrates
K. G. Robbins and J. M. Thomas
Nature, 217:1251-1252 (1968)

Measurements of the Peltier coefficient at the boundary between solid and liquid antimony
N. I. Strekopytova
Fiz. Tekhn. Poluprovod., 2(6):886-887 (1968)
Sov. Phys. — Semicond., 2(6):737-738 (1968)

Electrodeposition of Sb as influenced by the specific radioactivity of the $Sb^{124}$ isotope
V. S. Arakelyan, V. I. Spitsyn, and L. A. Uvarov
Dokl. Akad. Nauk SSSR, 173:1367 (1967)

Composition of antimony evaporating from different sources
F. Baumann, J. Kessler, and W. Roessler
J. Appl. Phys., 38:3398 (1967)

Concerning the evaporation of Sb and $Sb_2$ from a PtSb source
Gerd M. Rosenblatt
J. Appl. Phys., 38:888 (1967)

Vacuum distillation of antimony metal
A. J. Singh, P. Suryanarayana, and B. S. Mathur
Indian J. Technol., 5:162-164 (1967)

Purification of antimony by sublimation
E. Bonnier and M. Charveriat
A. T. B. Metallurgie, 5:319 (1965-66)

Effect of film area on crystallization of evaporated antimony
Ryoji Suganuma and Yoshiaki Tanaka
Japan. J. Appl. Phys., 4:232-233 (1965)

Crystallization of evaporated antimony films as revealed by optical and electron microscopy
Ryoji Suganuma and Yoshiaki Tanaka
Mem. Res. Inst. Sci. and Engineering, Ritumeikan Univ., Kyoto, Japan, No. 13, 9 (1965)

Optical properties of semiconductors
M. Cardona and G. Harbeke
Synthesis and characterization of electronically active materials, RCA Tech. Rept. No. 1, May 15, 1963-Feb. 15, 1964, Contract SD-182, ARPA order 446

Studies on the preparation of high-purity antimony
R. Krishna Iyer and T. R. Bhat
AEET/CD/34, 1964

Low-temperature antimony evaporator
G. A. Shifrin and D. D. Kelly
Rev. Sci. Instr., 35:1712 (1964)

A literature survey on the purification of electronic materials
A. F. Armington, G. F. Dillon, and R. F. Mitchell
AFCRL-63-160, June 1963

Épitaxies de cristaux d'arsenic, d'antimoine et de bismuth obtenus par réduction en phase gazeuse d'un de leurs sulfures et deposés sur différents supports monocristallins
Lucien Capella
Compt. Rend., 254:1309 (1962)

Mechanism of the vacuum condensation of metals
A. A. Chernov
Rost Kristallov, Vol. 3, Izd. Akad. Nauk (1961), pp. 126-132
Growth of Crystals, Vol. 3 (A. V. Shubnikov and N. N. Sheftal', eds.), Consultants Bureau, New York (1962), pp. 174-183

Properties and preparation of thin antimony films of high uniformity
George A. Condas
Rev. Sci. Instr., 33:987-991 (1962)

A simple technique for seeding and growing oriented, relatively unstrained, single-crystal antimony, square-sectioned rods
Seymour Epstein
J. Electrochem. Soc., 109:738 (1962)

Etching reagents for antimony single crystals
Jiro Shigeta, Shigero Kitagawa, and Makoto Hiramatsu
Rept. Res. Lab. Surface Sci., Fac. Sci., Okayama Univ., 2: 39-44 (1962)

Refining antimony by the zone-recrystallization method
V. N. Vigdorovich, V. S. Ivleva, and L. Ya. Krol'
Izv. Akad. Nauk SSSR, Metallurgiya i Toplivo, 1:44-49 (1960)

Preparation of high-purity antimony
V. S. Ivleva
Symp. Pure Metals and Semiconductors, Proc. Inter-Institute Conf. on Pure Metals, Intermetallic Compounds, and Semiconducting Materials, Oct. 15-18, 1957, publ. Metallurgizdat, Moscow, 1959

Preparation of high-purity antimony by zone melting
L. Ya. Krol' and V. S. Ivleva
Symp. Scientific Papers of Giredmet Inst., Vol. 1, Metallurgizdat, Moscow, 1959

New methods of purification of very pure antimony
D. A. Petrov, V. A. Butov, and N. G. Gil'yadova
Zh. Neorg. Khim., 4:1970-1971 (1959)
Russ. J. Inorg. Chem., 4:894-895 (1959)

Levitation melting of Ga, In, Au, and Sb
L. R. Weisberg
Rev. Sci. Instr., 30:135 (1959)

Arsenic and carbon-free antimony for intermetallic compounds
R. R. Harberecht, A. Herzog, A. E. Middleton, and P. R. Mellory
J. Electrochem. Soc., 105:72 (1958)

The production of ultrapure antimony metal
I. Szep and P. Endroedi
Mag. Kemi. Folyoi., 64:409-412 (1958)

Dislocation etch pits in antimony
J. H. Wernick, J. N. Hobstetter, L. C. Lovell, and D. Dorsi
J. Appl. Phys.. 29:1013 (1958)

[Title Not Given]
H. A. Schell
Z. Metallkunde, 46:58 (1955)

Purification of antimony and tin by a new method of zone refining
M. Tannenbaum, A. I. Gross, and W. G. Pfann
J. Met., 6:762 (1954)

[Title Not Given]
M. Tannenbaum, A. J. Gross, and W. G. Pfann
Trans. AIME, 200:962 (1954)

Investigations on antimony single crystals in a transverse magnetic field
K. Rausch
Ann. Phys. (Leipzig), 1:190 (1947)

Handling Cominco high-purity metals
Data sheet from Cominco American, Inc., Electronic Materials Division, 818 West Riverside Avenue, Spokane, Washington 99201

## 15.c. Bismuth—Antimony

Bismuth—antimony alloys
H. J. Goldsmid
Phys. Stat. Sol., A1:7-28 (1970)

A semimetal-semiconductor transformation in bismuth-antimony alloys
V. M. Grabov, G. A. Ivanov, V. L. Naletov, V. S. Ponaryadov, and T. A. Yakovleva
Fiz. Tverd. Tela, 11:3653-3655 (1969)
Sov. Phys. — Solid State, 11:3069-3071 (1970)

The Nernst and the Seebeck effects in Te-doped Bi-Sb alloys
Sheng San Li and T. A. Rabson
Solid—State Electron., 13:153-160 (1970)

Lattice parameters of bismuth-antimony alloys at 4.2°K
V. Meisalo
J. Appl. Cryst., 3(4):224-226 (1970)

Thermogalvanomagnetic effects in tellurium-doped $Bi_{95}Sb_5$
C. B. Thomas and H. J. Goldsmid
J. Phys. C (Solid State Phys.), 3:696-705 (1970)

Nonquadratic nature of the conduction band of bismuth and bismuth-antimony alloys
V. S. Voloshin, G. A. Ivanov, V. A. Kulikov, and Yu. N. Saraev
Fiz. Tverd. Tela, 11:1511-1514 (1970)
Sov. Phys.—Solid State, 11:1511-1514 (1970)

Growing single crystals of Bi — Sb solid solutions by pulling
N. K. H. Abrikosov, V. S. Zemskov, and V. V. Rozhdestvenskaya
Fiz. Khim. Obrabot. Mater. SSSR, 5:47-51 (1969)

Band structure of $Bi_{88}Sb_{12}$
G. A. Antcliffe
Phys. Letters, 28a:601-602 (1969)

Appearance of Baushinger effect during elastic broadening of wedge-shaped twins (bismuth-antimony alloy)
V. I. Bashmakov, L. A. Skalko, and N. G. Yakovenko
Fiz. Metal. i Metalloved., 28:937-939 (1969)

Thermoelectric properties of bismuth and Bi-12Sb alloy ingots and powder compacts
W. H. Bear, G. Cochrane, and W. V. Youdelis
Trans. AIME, 245:2357-2359 (1969)

Pressure-induced electron transitions in bismuth-tin, bismuth-lead, bismuth-antimony, and bismuth-antimony-lead alloys
N. B. Brandt and Ya. G. Ponomarev
Zh. Éksp. Teor. Fiz., 55(4):1215-1237 (1969)
Sov. Phys.—JETP, 28(4):635-646 (1969)

Investigation of the semiconductor-metal transition in the bismuth-antimony system in a magnetic field
N. B. Brandt, E. A. Svistova, and R. G. Valeev
Zh. Éksp. Teor. Fiz., Pis'ma, 55(2):469-485 (1969)
Sov. Phys.—JETP, 28(2):245-254 (1969)

Anomalies of the magnetoresistance in Bi-Sb semiconducting alloys in strong magnetic fields at low temperatures
N. B. Brandt, E. A. Svistova, Yu. G. Kashirskii, and L. V. Lyn'ko
Zh. Éksp. Teor. Fiz., 56(1):65-72 (1969)
Sov. Phys.—JETP, 29(1):35-39 (1969)

Single-crystal preparation
G. Burnet and M. J. Murtha
Annual Summary Research Report Ceramic and Mechanical Engineering, Chemical Engineering, Chemistry, Mathematics and Computer Science, Metallurgy, Physics, and Reactor Divisions, July, 1968-June 30, 1969, Ames Laboratory (Ames, Iowa), Contract W-7405-eng-82, IS-2100 (July 1969)
1 to 12 at.% antimony in bismuth

Ideal resistivity of bismuth-antimony alloys and
the electron-electron interaction
  E. W. Fenton, J. P. Jan, A. Karlsson, and R. Singer
  Phys. Rev., 184:663–667 (1969)

Electrical conductivity and density of the bis-
muth-antimony system in the liquid state
  F. Gaibullaev, M. Mamadaliev, A. R. Regel', and
    Kh. Khusanov
  Fiz. Tverd. Tela, 10(10):3174–3175 (1968)
  Sov. Phys. — Solid State, 10(10):2516–2517 (1969)

Ideal resistivity of bismuth-antimony alloys and
the electron-electron interaction
  A. Karlsson, E. W. Fenton, and R. Singer
  Bull. Am. Phys. Soc., 14:306 (1969)

Hot electron effects in $Bi_{1-x}Sb_x$ alloys
  J. S. Lannin
  Bull. Am. Phys. Soc., 14:1159 (1969)

Die elektrischen Transportgrössen von dotiertem
$Bi_{88}Sb_{12}$
  W. Lehnefinke and G. Schneider
  Z. Naturforsch., 24A(10):1594–1601 (1969)
  Crystal growth by zone refining

Growth of Bi-Sb alloy single crystals
  O. Leistiko, Jr., and A. L. Andersen
  J. Appl. Phys., 40:4659–4660 (1969)

High-temperature study of electrical and
galvanomagnetic properties of bismuth-antimony
alloys
  Yu. T. Levitskii and G. A. Ivanov
  Fiz. Metal. i Metalloved., 28:804–812 (1969)

Pressure-temperature phase diagrams of bismuth
and bismuth-antimony alloy
  Hisao Mii, Ikuya Fujishiro, and Yutaka Uchida
  Synthetic Crystal Research Labs., Nagoya Univ. Collection
    of Reports, No. 6, 25–31 (1969)

Microwave emission from BiSb alloys
  C. A. Nanney and E. V. George
  Bull. Am. Phys. Soc., 14:384 (1969)

Coherent microwave radiation from BiSb alloys
  C. A. Nanney and E. V. George
  Phys. Rev. Letters, 22:1062–1065 (1969)

Shubnikov-de Haas effect in the temperature
range between $T = 2°K$ and $T = 77.4°K$
  D. Schneider
  Solid State Commun., 7:1167–1171 (1969)
  Erratum: Vol. 7, p. vi (1969)

Magneto-optical investigation of Bi-Sb alloys
  E. J. Tichovolsky and J. G. Mavroides
  Bull. Am. Phys. Soc., 14:433 (1969)

Magnetoreflection studies on the band structure
of bismuth-antimony alloys
  E. J. Tichovolsky and J. G. Mavroides
  Solid State Commun., 7:927–931 (1969)

Nonohmic conductivity in $Bi_{88}Sb_{12}$
  G. A. Antcliffe
  Phys. Letters, 27A:606–607 (1968)

Anomalies of longitudinal resistance of semiconducting Bi-Sb alloys in magnetic fields to 500 kOe
at liquid-helium temperature
  N. B. Brandt, E. A. Svistova, Yu. G. Kashirskii, and L. V.
    Lyn'ko
  JETP Letters, 7:347 (1968)

Recherches sur la structure électronique de
semi-conducteurs à bande interdite étroite:
Pyrocarbones et alliages bismuth-antimoine
  E. Dupart
  Ph. D. thesis, Bordeaux, France (1968)

Electric spark method for preparing samples
from single crystals of some semimetals
  S. P. Fursov, A. E. Gitlevich, Yu. V. Goloshchapov, A. S.
    Fedorko, and V. G. Bivol
  Electron. Obrab. Mater., 2:11–16 (1968)

Band model for bismuth-antimony alloys
  Stuart Golin
  Phys. Rev., 176:830–832 (1968)

Transport properties of the Bi-Sb alloy system
  R. B. Horst, K. F. Cuff, and S. R. Hawkins
  Bull. Am. Phys. Soc., 13:384 (1968)

Study of conduction bands in bismuth-antimony
alloys by means of electrical properties of (bismuth-antimony)-tellurium alloys
  G. A. Ivanov, G. N. Kolpachnikov, and T. A. Yakovleva
  Uch Zap. Leningrad. Gos. Pedagog, Inst. im. A. I. Gertsena,
    384(4):54–59 (1968)

Double-layer Bi + Sb films as test objects for
electron microscopes
  V. M. Kosevich, L. S. Palatnik, and A. I. Fedorenko
  Instrum. Exper. Tech., No. 3, 707–709 (1968)

Energy-band parameters and relative band-edge
motions in the Bi-Sb alloy system near the semi-
metal-semiconductor transition
  L. S. Lerner, K. F. Cuff, and L. R. Williams
  Rev. Modern Phys., 40:770–775 (1968)

De Haas-Van Alphen effect in Bi-Sb-Pb alloys
  L. G. Lyubutina
  Fiz. Tverd. Tela, 10:77–80 1968
  Sov. Phys. — Solid State, 10:55–57 (1968)

Propriétés électroniques d'alliages BiSb à faible
taux d'antimoine
  A. Marchand, E. Dupart, and J. W. McClure
  Mat. Res. Bull., 3:971–982 (1968)

Large magneto-Seebeck effect in an extrinsic Bi-
Sb alloy
  C. B. Thomas and H. J. Goldsmid
  Phys. Letters A, 27:369–370 (1968)

Electronic band structure and electronic properties, 1. Magneto-optical investigation of Bi-Sb
alloys
  E. Tichovolsky, J. G. Mavroides, and D. F. Kolesar
  (Massachusetts Inst. of Technology, ESD-TR-68-353 (Dec.
    27, 1968) pp. 35–36

Die magnetische Suszeptibilität von Bi und Bi-
Sb-Legierungen
  L. Wehrli
  Phys. Kondens. Materie, 8:87–128 (1968)

The magnetic susceptibility of bismuth and bismuth-antimony alloys
Leo Wehrli
Ph. D. thesis, Eidgenössische Technische Hochschule, Zurich, Switzerland (1968), 45 pp.

Thermal conductivity of bismuth-antimony alloy single crystals
T. Yazaki
J. Phys. Soc. Japan, 25:1054–1060 (1968)

Magneto plasma wave propagation at high magnetic fields in the bismuth-antimony alloy system
M. Greenebaum
Bull. Am. Phys. Soc., 12:708 (June 1967)

Investigation of the band structure of group V semi-metals and the IV-VI semiconductors
A. L. Jain
(State Univ. of New York, Stony Brook)
Annual Summary Report 1 April 1966–31 March 1967, Contract N0016-66-C0194, AD-649989 (1967)

Galvanomagnetic properties of solid solutions of bismuth-antimony
A. A. Kuliev et al.
Izv. Akad. Nauk Azerb SSR, Ser. Fiz.-Tekhn. i Mat. Nauk (Baku), No. 3-4:17–24 (1967)
NRC-TT-1386 (1969)

The Hall and Nernst effect in Bi and Bi-Sb alloys
Sunil Narayan Shabde
Ph. D. thesis, Rice University, Houston, Texas
University Microfilms, Ann Arbor, Michigan, Order No. 67-13120 (1967), 87 pp.

Magnetic properties of metals: solid solutions of antimony in bismuth
B. I. Verkin, L. B. Kuz'micheva, and I. V. Svechkarev
JETP Letters, 6:225–227 (1967)
Zh. Éksper. Teor. Fiz., Pis'ma (USSR), 6:757–758 (1967)

Growth of homogeneous Bi-Sb alloy single crystals
W. M. Yim and J. P. Dismukes
Crystal Growth (Suppl. to J. Phys. Chem. Solids), Pergamon Press, New York (1967), pp. 187–196

Thermoelectric and galvanomagnetic properties of Bi-Sb single crystals
I. Ia. Khirch
Ukr. Fiz. Zh. (Kiev), 11:501–506 (1966)
N68-33799 (July 1966), 8 pp., Lockheed Missiles and Space Co., Palo Alto, California

Dendritic solidification in bismuth alloys
E. J. Schneider
Dissertation Abstr. 27, 505 (1966)

Superconducting transitions of amorphous bismuth alloys
J. S. Shier and D. M. Ginsberg
Phys. Rev., 147:384–391 (1966)
Bismuth with lead, thallium, or antimony

Low-temperature specific heat measurements in arsenic, antimony, bismuth, and dilute bismuth-tellurium alloys
William Alvin Taylor
(Univ. of California, Riverside), Dissertation Abstr. B 28 (10), 4259 (1968)
University Microfilms, Ann Arbor, Michigan, Order No. 68-4929 (1966), 124 pp.

Thermomagnetische und galvanomagnetische Effekte in Bi-Sb-Legierungen
H. G. Busse, E. Justi, and G. Schneider
Advan. Energy Conversion, 5:331–344 (1965)

Low-field galvanomagnetic effects on Bi-Sb alloys at 4.2°K
B. Ronnlund, L. Ericsson, and O. Beckman
Arkiv Fysik, 29:237 (1965)

Neuere Untersuchungen an halbleitenden Mischkristallen unter besonderer Berücksichtigung von Zustandsdiagrammen
J. Rupprecht and R. G. Maier
Phys. Stat. Sol., 8:3 (1965)

Preparation of homogeneous single-crystal bismuth-antimony alloys
M. A. Short and J. J. Schott
J. Appl. Phys., 36:659–660 (1965)
Bi-Sb crystals

Growth of bismuth-antimony single-crystal alloys
Dale M. Brown and Fred K. Heumann
J. Appl. Phys., 35:1947–1951 (1964)

Magnetic susceptibility of Bi-Sb alloys
E. I. Blount and J. J. Hauser
Bull. Am. Phys. Soc., Ser. II, 8:206(A) (Mar. 1963)

The thermomagnetic figure of merit and Ettingshausen cooling in Bi-Sb alloys
K. F. Cuff, R. B. Horst, J. L. Weaver, S. R. Hawkins, C. F. Kooi, and G. M. Enslow
Appl. Phys. Letters, 2:145 (1963)

Size dependence of the magneto-Seebeck effect in bismuth-antimony alloys
M. E. Ertl, G. R. Pfister, and H. J. Goldsmid
Brit. J. Appl. Phys., 14:161–162 (1963)

The Ettingshausen figure of merit of bismuth and bismuth-antimony alloys
H. J. Goldsmid
Brit. J. Appl. Phys., 14:271–274 (1963)

Diffusionless crystallization of the binary alloys
Yu. A. Krishtal
Izv. Vysshikh Uchebn. Zavoden., Chernaya Metallurgiya, pp. 110–116 (March 1960)

# 16. Organic Semiconductors

## 16.a. General, Reviews, and Bibliographies

X-ray Diffraction Methods in Polymer Science
Leroy E. Alexander
Wiley-Interscience Publishers, New York (1970), 544 pp.

Organic Semiconductors and Biopolymers
L. I. Boguslavskii and A. V. Vannikov
Monographs in Semiconductor Physics, Plenum Press
(1970), 186 pp.

Organic Solid State Chemistry
George Adler, ed.
Gordon and Breach Science Publishers, New York (1969),
524 pp.

The range of generalized crystallography
J. D. Bernal and C. H. Carlisle
Kristallografiya, 13(5):927-954 (1968)
Sov. Phys. — Cryst., 13(5):811-832 (1969)

Photoemission from metal contacts into anthra-
cene crystals: A critical review
J. M. Caywood
Molecular Crystals and Liquid Crystals, 12:1-26 (1970)

Thin Film Technology
R. W. Berry, P. M. Hall, and M. T. Harris
D. Van Nostrand Co., Inc., Princeton, N. J. (1969)

Rotational energy barriers and electron tunnel-
ing in organic solids
Felix Gutmann
Japan. J. Appl. Phys., 8:1417-1423 (1969)

Electrical conductivity — activated carrier mobil-
ity in organic solids
F. Gutmann, A. M. Hermann, and A. Rembaum
Nature, 221:1237 (1969)

Contributions to the Structural Chemistry of Organic Compounds. Vol. II. Systems for Collection and Treatment of X-Ray Diffraction Data
Peder Kierkegaard, Bjorn G. Brandt, Lars-Ove Hagman, et al.
Progress report on the work conducted during the period
Jan. 1, 1967-Dec. 31, 1968, University of Stockholm Institute of Inorganic and Physical Chemistry Data and Information Series No. 33 (1969)
Lists and describes many developed computer programs

Diffusion in glassy polymers
T. K. Kwei and H. M. Zupko
J. Polym. Sci., Part A-2, Polym. Phys., 7:867-887 (1969)

Chemical structure and crystal size in polymer
single crystals
H. K. Livingston
Macromolecules, 2:98-103 (1969)
Review; 31 refs.

Doping mechanism of organic semiconductors
H. Meier, W. Albrecht, and U. Tschirwitz
Ber. Bunsen. Physik Chem., 73(8-9):795-804 (1969)

Structural effects on photoconduction in amor-
phous films of organic solids
D. J. Morantz and H. James
J. Vac. Sci. Tech., 6:637-640 (1969)

Polymere organische Halbleiter
H. Naarman
Naturwissenschaften, 56(6):308-313 (1969)

Characteristics of the thermoresistance of
organic semiconductors
Ya. M. Paushkin, A. F. Lunin, V. A. Aleksandrova, S. S.
Oganesov, and V. B. Markovich
Izv. Vysshikh Uchebn. Zaveden., Fiz. Tomsk, 12(3):90-93
(1969)

Polymer synthesis: philosophy and approaches
E. M. Pearce
Trans. New York Acad. Sci., 2:629-636 (1969)

Yielding behavior of glassy polymers. I. Free-
volume model
K. C. Rusch and R. H. Beck, Jr.
J. Macromol. Sci. — Phys., B3(3):365-383 (1969)

Proceedings of the Second International Confer-
ence on Thermal Analysis, Worcester, Mass.,
August 1968
Robert F. Schwenker, Jr., and Paul D. Garn, eds.
Academic Press, New York (1969), Vol. 1, Instrumentation,
Organic Materials, and Polymers, 720 pp.

Proceedings of the Third International Confer-
ence on Photoconductivity, held at Stanford,
August 12-15, 1969
Solid State Commun., 7(11) (1969)

Propriétés électriques des cristaux moléculaires
A. Barraud
Rapp. C. E. A., France, No. 3548, 15 pp. (1968)

Polymers: their structure and dielectric properties
C. A. Buehler
J. Electrochem. Soc., 115:271c–276c (1968)

Conducting and semiconducting organic materials
Electricien (France), 96(2907):173 (1968)

Piezoelectricity, pyroelectricity and thermoelectricity of polymer films
T. Furukawa, Y. Uematsu, K. Asakawa, et al.
J. Appl. Polym. Sci., 12:2675–2689 (1968)

The electron conductivity of organic materials
V. Hadek
Cesk. Casopis Fys., 18A:17–46 (1968)
94 refs.

Far infrared absorption of molecular crystals
Armand Hadni
Univ. Nancy, Nancy, France, Excitons, Magnons, Phonons
Mol. Cryst., Proceedings International Symposium (A. B. Zahlan, ed.), Cambridge Univ. Press, Cambridge, England (1968), pp. 31–41
Review; 16 refs.

Organic Semiconducting Polymers
J. E. Katon, ed.
Dekker, New York (1968)

Polymer crystals
A. Keller
Reports on Progress in Physics, Vol. XXXI, Part II, C. I. Pedersen (Managing Editor) and R. A. Cook (Editorial Assistant)
The Institute of Physics and The Physical Society, London (1968), pp. 624–704

Organic semiconductor junctions
Esther Krikorian
General Dynamics, Pomona, California, Pomona Division, AD-679 601 (June 1968), 283 pp.

Electrical and photoelectric properties of organic polymeric semiconductors
V. S. Mylnikov
Russ. Chem. Rev., 37:25–38 (1968)
300 refs.

Organic semiconductors
R. Pethig
Electronics and Power, 14:271–274 (1968)

Catalytical properties of organic semiconductors
S. Z. Roginskii and M. M. Sakharov
Zh. Fiz. Khim., 42:1331–1345 (1968)
Review; 121 refs.

Pre-exponential factor in semiconducting organic substances
B. Rosenberg, B. B. Bhowmic, H. C. Harder, and E. Postow
J. Chem. Phys., 49:4108–4114 (1968)
Methods of evaluation of constants in conductivity expression

The Science and Technology of Polymer Films
Orville J. Sweeting, ed.
Interscience Publishers, New York (1968)

Organic Semiconductors
Group of authors at Topchiev Institute of Inorganic Synthesis, Nauka Publishing House, Moscow (1968)

Molecular dynamics and structure of solids, NBS Spec. Publ. 301
R. S. Carter and J. J. Rush, eds.
Presented at the Second Materials Research Symposium, held at the National Bureau of Standards, October 16–19, 1967
Includes structure of organic crystals; spectroscopy of organic crystals; dynamics of polymers

Electrical conductivity in organic semiconductors
S. C. Datt, J. K. D. Verma, and B. D. Nag
J. Sci. Ind. Res., 26:57–75 (1967)
Review; 208 refs.

Organic Semiconductors
Felix Gutmann and E. Lyons
John Wiley and Sons, Inc., New York (1967), 858 pp.

Organic metals? The electrical conductance of organic solids
The Liversidge Lecture, Sydney, 1966
L. E. Lyons
J. Proc. Roy. Soc. New S. Wales, 101:1–9 (1967)

Glass transition in polymers
M. C. Shen and Adi Eisenberg
Progress in Solid State Chemistry, Vol. 3 (H. Reiss, ed.), Pergamon Press, New York (1967), pp. 407–482

Proceedings of Symposium on Electrical Conduction Properties of Polymers
A. M. Hermann and A. Rembaum, eds.
Jet Propulsion Lab., Pasadena, California, 1966

Organic semiconductor-analytical survey
Serge Markov
ATD-66-84, AD-644 463 (June 1966), 57 pp.
(Survey of Soviet scientific and technical literature)

Physics and Chemistry of the Organic Solid State, Vol. II
D. Fox, M. M. Labes, and A. Weissberger, eds.
Interscience Publishers, Inc., New York (1965)

Intrinsic and extrinsic conductivity of organic solids
L. E. Lyons
Presented at the Third International Organic Crystal Symposium, Univ. of Chicago, May 1965

Proceedings of the International Conference on Physics of Non-Crystalline Solids, Delft, July 1964
J. A. Prins, ed.
North-Holland Publishing Co., Amsterdam; Interscience Publishers, a division of John Wiley and Sons, Inc., New York (1965)

Electrical properties of thin organic films
A. Bradley and J. P. Hammes
J. Electrochem. Soc., 110:15–22 (1963)

Transition metal complexes as potential semiconductors
A. M. Zwickel and S. Kwan
Clark Univ., Worcester, Mass., Contract Nonr-3694 (01), Task No. NR 051-440, Tech. Rept. No. 1 (Feb. 1963)

Organic Semiconductors-Proceedings of an Inter-Industry Conference, Chicago, April 1961
J. J. Brophy and J. W. Buttrey, eds.
The Macmillan Co., New York (1962)

[Title Not Given]
E. O. Forster
Organic Crystal Symposium (National Research Council, Ottawa), Oct. 1962, 125

Organic semiconductors
C. Sosnovsky, O. W. Adams, D. Laskowski, M. Salkowski, and L. U. Berman
Armour Research Foundation of Illinois Institute of Technology, Chicago, ARD 3142-12 (1962)

Further studies of some semiconducting polymers
J. A. Bornmann and H. A. Pohl
Plastics Lab., Princeton Univ. Technical Rept. 63A (1961)

Electrical Conductivity in Organic Solids
H. Kallmann and M. Silver, eds.
Interscience Publishers, Inc., New York (1961)

Organic Chemical Crystallography
A. I. Kitaigorodskii
Consultants Bureau, New York (1961), 541 pp.

Solid State Physics, Vol. 12. Advances in Research and Applications
Frederick Seitz and David Turnbull, eds.
Academic Press, New York (1961), 471 pp.
An outline is given of the preparation, purification electric conductivity, and photoelectric properties of organic semiconductors

Proceedings of the Conference on Semiconduction in Molecular Solids
H. A. Pohl, ed.
Princeton University Press, Princeton, New Jersey (1960)

Low-resistance organic semiconductor
J. Kommandeur and F. R. Hall
Bull. Am. Phys. Soc. (II), 4:421 (1959)

## 16.b. Purification

Purification of Inorganic and Organic Materials. Techiques of Fractional Solidification
Morris Zief, ed.
Marcel Dekker, Inc., New York (1969), 344 pp.

Zone refining techniques for liquid organics
J. N. Carides
Rev. Sci. Instr., 39:1811-1813 (1968)

Induction heating in zone melting of organic compounds
Henry Plancher, J. C. Morris, and W. E. Haines
Anal. Chem., 40:1592-1594 (1968)

Purification of organic semiconductors
S. C. Datt, J. K. D. Verma, and B. D. Nag
J. Sci. Ind. Res., 25:455-470 (1966)
Includes assay methods; review, 162 refs.

Zone refining as a purification tool (for organic compounds, review and assessment)
E. F. G. Herington
Purification of Materials, Vol. 137, Part 1, New York Academy of Science (Jan. 20, 1966), pp. 63-71
21 refs.

New zone-refining techniques for chemical compounds
W. G. Pfann, C. E. Miller, and J. D. Hunt
Rev. Sci. Instr., 37:649-652 (1966)

Zone melting of organic compounds
Hermann Schildknecht
Zone Melting, Express Translation Service, London. Academic Press, New York (1966), pp. 170-198

Theoretical aspects of polymer crystallization with chain folds: bulk polymers
J. D. Hoffman
SPE Trans. (Oct. 1964), 315 pp.

Pure diffusional mass transfer in zone melting
W. R. Wilcox and C. R. Wilke
A. I. Ch. E. Journal, 10:160 (1964)

Ultrapurity and ultrapurification of pharmaceuticals by zone melting
R. M. Friedenberg
Ph. D. thesis, Univ. of Connecticut (1963)

Syntheses of high polymeric semiconductors
Masaniro Hatano
J. Soc. Organic Synthetic Chem., Japan, 20:326-342 (1962)
Noburu Hiraga, Sandia Corp., Albuquerque, N. M., SCL-T-462 (Jan. 1963), 40 pp.

Zone melting of organic compounds
E. F. G. Herington
John Wiley and Sons Inc., New York (1963)

An improved Bridgman apparatus for preparation of organic mixed crystal scintillators
S. Hayakawa, T. Nakamura, and S. Kobayashi
Bull. Tokyo Inst. Tech., 49:91 (1962)

Synthesis and characterization of some highly conjugated semiconducting polymers
H. A. Pohl and E. H. Engelhardt
J. Phys. Chem., 66:2085 (1962)

Purification of some organic medicinal compounds by zone melting
W. G. Walter
Ph. D. thesis, Univ. of Connecticut (1962)

Research and investigation of materials for laser applications
K. A. Yamakawa
Electro-optical Systems, Inc., Pasadena, Calif., Contract AF 33(657)-8918, Interim Engineering Report for 1 May-31 July 1962, EOS Report 2110-Q-1; AD-282 296 (Aug. 1962)

Pressurestat for high polymer crystallization
Bernhard Wunderlich
Rev. Sci. Instr., 32:1424-1425 (1961)

The growth habits of single polymer crystals
Fraser P. Price
J. Polymer Sci., 42:49-56 (1960)

Zone melting and phase studies of some organic compounds
D. R. Makenzie
M. S. thesis, Univ. of Tennessee, Knoxville, Tenn. (1958)

Distillation: Techniques of Organic Chemistry Vol. IX
A. Weissberger, ed.
Interscience Publishers Inc., New York (1951)

## 16.c. Crystal Growth

Four pour la croissance de cristaux organiques par la méthode de Bridgman
G. Guillaud, M. Le Helley, and G. Mesnard
Bull. Soc. Fr. Mineral. Cristallogr., 93:131–132 (1970)

Solution-grown polymer crystals
A. Keller
Kolloid–Z., 231(1–2):385–421 (1969)

Charge separation by crystallization
A. M. Mel'nikova
Kristallografiya, 14(3):548–563 (1969)
Sov. Phys. — Cryst., 14(3):464–479 (1969)

Crystallization from the melt
R. T. Southin and G. A. Chadwick
Sci. Prog., Oxford, 57:353–370 (1969)

The growth of polymer crystals
D. C. Bassett
J. Crystal Growth, 3(4):92–96 (1968)

Crystal growth of organic solids and mechanism of single-crystal growth from melt
S. C. Datt and J. K. D. Verma
J. Sci. Instr. Res., India, 27:11–27 (1968)
Review of methods and mechanisms

Influence of some crystal defects on the morphology of its growing surface. II. Formation of defects in the growing over of inclusions, and the part played by block boundaries at high supersaturation
E. D. Dukova
Kristallografiya, 12(3):483–487 (1967)
Sov. Phys. — Cryst., 12(3):413–416 (1967)

On the nature of crystal growth from the melt
K. A. Jackson, D. R. Uhlmann, and J. D. Hunt
J. Crystal Growth, 1:1–36 (1967)

Spherulitic crystallization from the melt. I. Fractionation and impurity segregation and their influence on crystalline morphology
H. D. Keith and F. J. Padden, Jr.
J. Appl. Phys., 35:1270 (1964)

Spherulitic crystallization from the melt. II. Influence of fractionation and impurity segregation on the kinetics of crystallization
H. D. Keith and F. J. Padden, Jr.
J. Appl. Phys., 35:1286 (1964)

Cristaux unitaires de polymères
A. Keller
Bull. Inst. Textile France, 17:301 (1963)

Equipment for single-crystal growth from aqueous solution
J. L. Torgesen, A. T. Horton, and C. P. Saylor
J. Res. Natl. Bur. Stand. –C, 67C:25 (1963)

On the habit of polyethylene crystals
D. C. Bassett and A. Keller
Phil. Mag., 7:1553 (1962)

The epitaxial growth of organic semiconductor materials on mica
N. Uyeda, M. Ashida, and E. Suito
Fifth International Congress for Electron Microscopy, Academic Press, Inc., New York (1962)
Co–, Ni–, Cu–, Zn–, Sn–, and Pt-phthalocyanines, flavanthrone, pyranthrone, violanthrone, and iso-violanthrone

Organic semiconductors
E. R. Biehl, G. F. Deebel, R. E. Dolle, and B. S. Wildi
Monsanto Research Corp., Dayton, Ohio, Contract AF19(604)–8497, Semiannual Sci. Rept. No. 1; AFCRL 62–245 (Dec. 1961)
Coronene, perylene/iodine complex, chloranil/durenediamine complex, 2,4,6-tricyano-s-triazine

Electrical properties of organic solids. I. Kinetics and mechanism of photoconductivity of metal-free phthalocyanine. II. Effects of added electron acceptors and donors
David R. Keams
Thesis, Lawrence Radiation Lab., Univ. California, Berkeley, Contract W-7405-eng-48, UCRL-9120 (March 1960)

The effect of surface-active agents on crystal growth rate and crystal habit
A. S. Michaels and A. R. Colville, Jr.
J. Phys. Chem., 64:13 (1960)

Growth features on crystals of long-chain compounds. III
S. Amelinckx
Acta Cryst., 9:217 (1956)
$C_{34}H_{70}$ and $C_{34}H_{47}OH$

# 17. Amorphous Semiconductors

## 17.a. General, Reviews, and Theory

Theories and models of the amorphous state
R. S. Allgaier
J. Vac. Sci. Technol., 8:113-124 (1971)
Mainly those developed for amorphous semiconductors by
Mott, Cohen, and Boer

Electro-thermal effects in ovonics
K. W. Boer
Phys. Stat. Sol., 4A:571-596 (1971)
Review; switching effect not a "thermal runaway," 58 refs.

Glasses in the Ge − As − Te system and devices
made from them
G. A. Yurlova, I. D. Gudkov, and B. T. Kolomiets
Soviet Phys. − Semicond., 4(9):1401-1404 (1971)

Switching and breakdown in films
N. Klein
Thin Solid Films, 7:149-177 (1971)
Crystalline and amorphous; the main classes of events,
thermal, electronic and electronic modified thermal are
discussed

Electrical Conductivity of Vitreous Substances
Rudol'f L. Myuller
Consultants Bureau, New York (1971), 240 pp.

On the frequency dependence of conductivity in
amorphous solids
M. Pollak
Phil. Mag., 23:519-542 (1971)

Switching effects in glassy semiconductors
D. Armitage, D. Brodie, and P. Eastman
Bull. Am. Phys. Soc., 15(6):765 (1970)
Breakdown is generally agreed to be thermal in the case of
films thicker than about 1 $\mu$m, but for thinner films other
effects may be important

A simple test for double-injection initiation of
switching
I. Balberg
Appl. Phys. Letters, 16(12) (June 15, 1970)
"Thermal microheating" model supported; amorphous semi-
conductors

Formation of amorphous films
Klaus H. Behrndt
J. Vac. Sci. Tech., 7:385-398 (1970)

Deflecting of light beams using a gradient of the
index of refraction in semiconducting glasses
K. W. Boer
Appl. Phys. Letters, 16(6) (March 15, 1970)
Due to a temperature gradient in a current channel in a semiconducting glass or near a heated wire in a glass or liquid
or liquid

Ideal-real semiconducting glass and low-high
conductivity transition
K. W. Boer
Phys. Stat. Sol., 3A:1007-1018 (1970)
Conduction mechanism in real glasses is suggested to be
carrier drift at a conductivity edge over barriers provided
by Coulomb-repulsive defects

Semiconductivity of glasses
K. W. Boer and R. Haislip
Phys. Rev. Letters, 24(3):230-233 (1970)

Behaviour of the charge carriers in slowly varying random fields and semiphenomenological approach to the theory of electronic processes in
disordered systems
V. L. Bonch-Bruevich
J. Non-Cryst. Solids (Netherlands), 4:410-416 (1970)

On the theory of absorption of low frequency
electromagnetic waves in disordered semiconductors
V. L. Bonch-Bruevich
Mater. Res. Bull., 5:555-566 (1970)

Electronic dielectric constant of amorphous
semiconductors
M. H. Brodsky and P. J. Stiles
Phys. Rev. Letters, 25:798-800 (1970)

Some experiments on switching-type amorphous
semiconductors
P. Calella, S. Defeo, W. Doremus, J. Hall, and R. Nicolaides
PA − TR − 3935; AD − 703 858 (Jan. 1970), 40 pp.

Optical absorption of semiconductors from 15 to 170 eV
 M. Cardona, W. Gudat, B. Sonntag, and P. Y. Yu
 Interner Bericht DESY F41-70/6 (July 1970), paper to be presented at the 10th Intern. Conf. Physics of Semiconductors, Aug. 17-21, 1970, Cambridge, Mass.
 Amorphous and crystalline Ge, Se, and III-V compounds

Rapid quenching technique for preparation of thin uniform films of amorphous solids
 H. S. Chen and C. E. Miller
 Rev. Sci. Instr., 41:1237-1238 (1970)

On the Frenkel exciton theory of amorphous solids. I. General theory
 W. Christiaens and P. Phariseau
 Physica, 46:569-576 (1970)

Exciton theory in amorphous materials
 W. Christiaens and P. Phariseau
 Z. Physik, 234:268-280 (1970)

Review of the theory of amorphous semiconductors
 M. H. Cohen
 J. Non-Cryst. Solids (Netherlands), 4:391-409 (1970)

Model for the resistive-conductive transition in reversible resistance-switching solids
 E. L. Cook
 J. Appl. Phys., 41(2):551-554 (1970)

Thermal mechanism of the switching phenomenon
 N. Croitoru and C. Popescu
 Phys. Stat. Sol., 3A:1047-1055 (1970)

Conduction of non-crystalline systems, V. Conductivity, optical absorption and photoconductivity in amorphous semiconductors
 E. A. Davis and N. F. Mott
 Phil. Mag., 22(179):903-922 (1970)

A model approach to glasses
 P. Dean and J. Bell
 New Sci., 45:104-111 (1970)

On the switching initiation in ovonics
 G. Dohler
 Phys. Stat. Sol., 1A:125-134 (1970)
 Thermally forced "double tunneling"

Proceedings of the Symposium on Semiconductor Effects in Amorphous Solids, New York, May 14-17, 1969
 W. Doremus, ed.
 North-Holland Publishing Co., Amsterdam (1970)

Anderson's theory of localization and the Mott-CFO model
 E. N. Economou and M. H. Cohen
 Mater. Res. Bull., 5:577-590 (1970)
 Electronic theory of disordered materials

Switching in elemental amorphous semiconductors
 C. Feldman and K. Moorjani
 J. Non-Cryst. Solids, 2:82-90 (1970)
 B, Si, Ge

Elastoresistivity of amorphous semiconductors
 W. Fuhs and J. Stuke
 Mater. Res. Bull., 5:611-620 (1970)

Bistable switching and the Mott transition
 L. Gildart
 J. Non-Cryst. Solids, 2:240-249 (1970)
 Proceedings of the Symposium on Semiconductor Effects in Amorphous Solids, New York, May 14-17 (1969)

Switching mechanism in amorphous semiconductors
 A. K. Goswami
 J. Non-Cryst. Solids, 2:205-209 (1970)
 Proceedings of the Symposium on Semiconductor Effects in Amorphous Solids, New York, May 14-17 (1969)

Switching cycle and reversible switching effect in glass semiconductors
 D. R. Haberland
 Frequenz, 24:185-190 (1970)

Charge-determined switching mechanism in glass semiconductors
 D. R. Haberland
 Solid State Electron., 13:207-217 (1970)

Microscopical study of glassy semiconductor solid-state switches
 D. R. Haberland and H. P. Kehrer
 Solid State Electron., 13:451-455 (1970)

Semiconducting glass-ceramics
 L. L. Hench
 J. Non-Cryst. Solids, 2:250-277 (1970)
 Proceedings of the Symposium on Semiconductor Effects in Amorphous Solids, New York, May 14-17, 1969

The investigation of local states in vitreous semiconductors by photoconductivity and thermally stimulated depolarization methods
 B. T. Kolomiets, V. M. Ljubin, and V. L. Averjanov
 Mater. Res. Bull., 5:655-664 (1970)

Mössbauer spectroscopy in inorganic glasses
 C. R. Kurkjian
 J. Non-Cryst. Solids, 3:157-194 (1970)
 85 refs.

Semiconducting glass switches (state of the art report)
 H. F. Matare
 Int. Elek. Rundsch., 24:171-176 (1970)

International Conference on Amorphous and Liquid Semiconductors, Cambridge University, 24-27 Sept. 1969
 Robert G. Morris
 ONRL-C-1-70; AD-701192 (Jan. 1970), 24 pp.

Optical properties of amorphous InSb, GaSb, and GaAs
 J. Niklas, J. Stuke, and G. Zimmerer
 UCRL-Trans-10480, from Spring Meeting of the German Physical Soc., Freundenstadt, Germany (April 6-11, 1970), 9 pp.
 CONF-700423-1

Semiconducting glasses. I. Glass as an electronic conductor
 A. E. Owen
 Contemporary Phys., 11(3):227-255 (May 1970)

Semiconducting glasses. II. Properties and interpretation
 A. E. Owen
 Contemporary Phys., 11(3):257-286 (May 1970)

Classification of non-crystalline solids
R. Roy
J. Non-Cryst. Solids, 3:33-40 (1970)

Bistable switching and memory devices
P. O. Sliva, G. Dir, and C. Griffiths
J. Non-Cryst. Solids, 2:316-333 (1970)
Proceedings of the Symposium on Semiconductor Effects in
Amorphous Solids, New York, May 14-17, 1969

Applications of amorphous semiconductors in
electronic devices
J. G. Simmons
Contemp. Phys., 11(1):21-41 (1970)

Absorption edge and internal electric fields in
amorphous semiconductors
J. Tauc
Mater. Res. Bull., 5:721-730 (1970)

Experimental results in amorphous semiconduc-
tor switching behavior
P. J. Walsh, J. E. Hall, R. Nicolaides, S. Defeo, P. Calella,
J. Kuchmas, and W. Doremus
J. Non-Cryst. Solids, 2:107-124 (1970)
Proceedings of the Symposium on Semiconductor Effects in
Amorphous Solids, New York, May 14-17, 1969

Symposium on Semiconductor Effects in Amorphous Solids, held in New York, May 14-17, 1969
J. Non-Cryst. Solids, 2 (Jan. 1970)

Symposium on Electrotechnical Glasses, held at
Imperial College, London, 30 Sept.-2 Oct., 1970
Society of Glass Technology, Thornton, 20 Hallam Gate Road,
Sheffield S10 5BT, England

Amorphous semiconductor switching devices
D. Adler
Tech. Engineering News, March 1969, pp. 7-9

A simple band model for amorphous semiconduct-
ing alloys
M. H. Cohen, H. Fritzsche, and S. R. Ovshinsky
Bull. Am. Phys. Soc., 14:311 (1969)

Simple band model for amorphous semiconduct-
ing alloys
Morrel H. Cohen, H. Fritzsche, and S. R. Ovshinsky
Physical Review Letters, 22:20 (1969)

The Ovshinsky effect
Kasturi L. Chopra
Ledgemont Laboratory, Letters to the Editor, Physics Today,
22:9 (March 1969)

Some aspects of electronics switching in amor-
phous semiconductors
L. William Doremus
Am. Ceram. Soc. Bull., 48:425 (1969)

Glassy bistable electrical switching and memory
device
C. F. Drake, I. F. Scanian, and J. H. Alexander
International Standard Electric Corp., U. S. Patent 3,440,588
(April 22, 1969)
46% $B_2O_3$, 15% CaO, 39% CuO

Switching and amplifying device components made
of semiconducting transition-metal glasses
A. Engel and O. Holzinger
Frequenz, 23:294-301 (1969)

Physics of instabilities in amorphous semicon-
ductors
H. Fritzsche
IBM J. Res. Develop., 13:515-521 (1969)

Narrow-band conduction in a quasi-amorphous lat-
tice
P. Gosar
Nuklearni Institut Josef Stefan, Ljubljana, Yugoslavia, IJS-
R-568 (Oct. 1969), 5 pp.
Presented at International Conference on Amorphous and
Liquid Semiconductors, Cambridge, Sept. 24-27, 1969

Short-range order in amorphous semiconductors
R. Grigorovici
J. Non-Cryst. Solids, 1:303 (1969)

Variational LCAO method for amorphous semicon-
ductors
A. I. Gubanov
Fiz. Tekhn. Poluprovod., 3(5):651-658 (1969)
Sov. Phys. — Semicond., 3(5):355-360 (1969)

Simplified OPW method for amorphous semicon-
ductors
A. I. Gubanov
Fiz. Tekhn. Poluprovod., 3(6):881-885 (1969)
Sov. Phys. — Semicond., 3(6):742-745 (1969)

Madelung constant for amorphous substances
A. I. Gubanov and Yu. N. Tsarev
Fiz. Tverd. Tela, 11:1077-1078 (1969)
Sov. Phys. — Solid State, 11:880-881 (1969)

Variational method for calculating the vibrations
of an amorphous solid
A. I. Gubanov
Fiz. Tverd. Tela, 11:928-933 (1969)
Sov. Phys. — Solid State, 11:758-761 (1969)

Negative resistance characteristics in amorphous
semiconductors
D. R. Haberland, R. Karmann, and P. Thoma
Z. Angew. Phys., 28(3):143-148 (1969)

Amorphous-semiconductor switching
H. K. Henisch
Sci. Am., pp. 30-41 (Nov. 1969)

The physical, chemical and biological aspects of
thin films. I. Thin films in electronics
R. W. Hoffman
Trans. New York Acad. Sci., 31(7):368-371 (1969)

Switching behavior of amorphous semiconductor
diodes
R. Holmstrom
Bull. Am. Phys. Soc., 14:310 (1969)

Switching and conduction behavior of amorphous
semiconductor diodes (Ovshinsky effect, ovonic
threshold switch diodes)
R. Holmstrom
Proc. IEEE, 57:1451-1453 (1969)

Electronic conduction in dielectric films
A. K. Jonscher
Thin Film Dielectrics, Montreal, Canada, Oct. 7-11 (1968),
pp. 3-42
Electrochem. Soc., New York (1969)

Electrical future for amorphous materials
A. K. Jonscher
New Sci., 42(649):18-20 (1969)

Electrical conduction in non-metallic amorphous
films
A. K. Jonscher and P. A. Walley
J. Vac. Sci. Tech., 6:662-669 (1969)

Contribution à l'étude de l'analyse thermique
J.-J. Kessis
Thèse Doct. Sci. Phys. Paris, 1970. Arch. Orig. Centre
Document. C. N. R. S., No. 3995 (Dec. 31, 1969), 84 pp.
Application to glass transitions

Activation energy spectra for relaxation in amor-
phous materials
R. M. Kimmel and D. R. Uhlmann
Bull. Am. Phys. Soc., 14:406 (1969)

Struktur und Bindungsverhältnisse in amorphen
Halbleitern
H. Krebs
Festkörper Probleme IX: Advances in Solid State Physics,
Plenary Lectures of the Professional Group "Semiconduc-
tor Physics" of the German Physical Society, Munich,
March 19-22, 1969, and Invited Papers of the European
Meeting of the IEEE "Semiconductor Device Research,"
Munich, March 24-27, 1969 (O. Madelung, ed.), Pergamon
Press, New York (1969), pp. 1-21

Electrical conductivity and chemical bonding in
crystalline, glassy and liquid phases
H. Krebs
J. Non—Cryst. Solids,1(6):455-473 (1969)

Glass research in review
N. J. Kreidl
Glass Ind., 50:15-17 (Jan. 1969)
26 refs.

Electrical conductivity of glass
B. Lengyel and Z. Boksay
Z. Phys. Chem. (Leipzig), 241:36-42 (1969)

Electronic conduction in non-crystalline solids
J. D. Mackenzie
Presented at the SEAS Symposium, New York, May 14-17,
1969, Contract N00014-67-A-0117, AD-690202, RPI-TR-4
(May 1969), 26 pp.

Preparation of properties of noncrystalline films
J. D. Mackenzie
RP-TR-3, Contract N00014-67-A-0117 (March 1969), 27 pp.
Presented at the International Conference on Thin Films,
Boston, Massachusetts (May 1, 1969)

Semiconductor glasses
Herbert F. Matare
Solid State Tech., 12:43 (1969)
A survey of the basic properties of semiconductor glasses is
given specifically with respect to electronic switches, and
the principal differences between crystalline semiconductors and those in glass form are stresses

Theory of electronic switching effect as a coop-
erative phenomenon
D. C. Mattis
Phys. Rev. Letters, 22:936-939 (1969)

Conduction and switching in non-crystalline ma-
terials
N. F. Mott
Contemp. Phys., 10:125-138 (1969)

Charge transport in noncrystalline semiconduc-
tors
N. F. Mott
Festkörper Probleme IX: Advances in Solid State Physics,
Plenary Lectures of the Professional Group "Semiconduc-
tor Physics" of the German Physical Society, Munich,
March 19-22, 1969, and Invited Papers of the European
Meeting of the IEEE "Semiconductor Device Research,"
Munich, March 24-27, 1969 (O. Madelung, ed.), Pergamon
Press, New York (1969) pp. 22-45

Conduction in non-crystalline materials.
III. Localized states in a pseudogap and near
extremities of conduction and valence bands
N. F. Mott
Phil. Mag., 19:835-852 (1969)

The equilibrium topography of sputtered amor-
phous solids
M. J. Nobes, J. S. Colligon, and G. Carter
J. Mater. Sci., 4:730-733 (1969)
Theory is for the sputtering of amorphous solid by an ion
beam and the changes in surface topography

Amorphous structures and the Ostwald rule
A. S. Nowick
Comments Solid State Phys., 2(5):155-160 (1969)

Amorphous semiconductors
S. R. Ovshinsky
Science, 5A:73-78 (1969)

Electronic conduction processes in glass
A. E. Owen
Sci. Tech. Aerospace Rep., 7:495 (1969)
Arsenide sulfide glasses; Ge-Se systems, transition-metal
oxide glasses

General characteristics of semiconducting glass
switching/memory diodes
A. D. Pearson
Bull. Am. Phys. Soc., 14:745 (1969)

Characteristics of semiconducting glass switch-
ing/memory diodes
A. D. Pearson
IBM J. Res. Develop., 13:510-514 (1969)

Thermal behavior of annealed organic glasses
E. B. Petrie
Bull. Am. Phys. Soc., 14:424 (1969)

Ovshinsky: Promoter or persecuted genius?
Physics Today (August 15, 1969), 673-677

Preparation of vitreous materials in the form of
well-defined hollow cylinders
N. S. Platakis, H. C. Gatos, and A. F. Witt
J. Electrochem. Soc., 116:510-511 (1969)

Effect of chemical structure on the mechanical
behavior of glassy polymers at cryogenic temperatures
J. M. Roe and Eric Baer
Bull. Am. Phys. Soc., 14:318 (1969)

The structure of amorphous solids
W. Ruland
Pure Appl. Chem., G. B., 18(4):489–515 (1969)

Experimental methods of infra-red spectroscopy to examine glass structures
P. Schleifer and B. Jarzmik
Szklo I Ceramika (Poland), 20(1):1–4 (1969)

Optische und elektrische Eigenschaften von amorphen Halbleitern
J. Stuke
Festkörper Probleme IX: Advances in Solid State Physics, Plenary Lectures of the Professional Group "Semiconductor Physics" of the German Physical Society, Munich, March 19–22, 1969, and Invited Papers of the European Meeting of the IEEE "Semiconductor Device Research," Munich, March 24–27, 1969 (O. Madelung, ed.), Pergamon Press, New York (1969), pp. 46–73

Progrès en recherche des materiaux dans les verres aux chalcogenures
M. Tanaka
J. Soc. Mater. Jap., 18(192):763–772 (1969)

Under what conditions can a glass be formed?
D. Turnbull
Contemp. Phys., 10:473–488 (1969)

Electronic properties of semiconducting solid solutions
Aleksei Borisovich Almazov
Consultants Bureau, New York (1968)
Glassy semiconductors, p. 63

Characterization of amorphous alloy films
B. G. Bagley, H. S. Chen, and D. Turnbull
Mat. Res. Bull., 3:159–168 (1968)

New nomenclature to describe the structure of vitreous systems and their related crystalline compounds
P. Beekenkamp and J. M. Stevels
Phys. Chem. Glasses, 9:64–68 (1968)

New type of energy band calculation in disordered systems
R. Grigorovici and R. Mansila
Mat. Res. Bull., 3:25–30 (1968)

Electrical properties of ionic and semiconducting glasses
R. M. Hakim
Ph. D. thesis, Dept. of Metallurgy and Materials Science, Massachusetts Institute of Technology, Sept. 1968

A. C. conduction in amorphous films
P. J. Harrop, G. C. Wood, and C. Pearson
Thin Solid Films, 2:457–466 (1968)

Energy conversion by means of amorphous insulating thin-films
M. Hartic and R. G. Muller
Bundesministerium für Wissenschaftliche Forschung, Germany, BMwF-FBK-68-40, September, 1968

Measurement of ionic transport in glass
K. Hughes, J. O. Isard, and G. C. Milines
Phys. Chem. Glass., 9:37–46 (1968)

Apparatus for chemical vapor deposition of oxide and glass films
Werner Kern
RCA Rev., 29:525–532 (Dec. 1968)

Semiconducting glasses — oxides and nonoxides
J. D. Mackenzie
Am. Ceram. Soc. Bull., 47(4):383 (1968)

Semiconductor glass
H. F. Matare
Intern. Elektronische Rundschau (Germany), 22(7):163–165 (1968)

Glass compositions and solid-state switching devices using these compositions
Minnesota Mining and Manfg. Co., British Patent 1,117,211, June 19, 1968
Sb 33–46.6, S 38.4–53, and I 8–25 at. %

Tunnel experiments on amorphous superconductors
G. V. Minnigerode and J. Rothenberg
Z. Phys. (Germany), 213(4):397–410 (1968)

Conduction in non-crystalline materials. I. Localized electronic states in disordered systems; II. The metal-insulator transition in a random array of centres
N. F. Mott
Phil. Mag., 17:1259–1268 (1968); 17:1269–1284 (1968)

Amorphous semiconductor switches
S. R. Ovshinsky
Am. Ceram. Soc. Bull., 47(4):383 (1968)

Solid state relay from tellurides
Stanford R. Ovshinsky
Energy Conversion Devices, Inc., U. S. Patent No. 3,395,445, August 6, 1968

Electronic conduction processes in glass 1963–1967
A. E. Owen
AD–676 451, Feb. 1968

Characterization of amorphous substances
J. A. Prins
Mat. Res. Bull., 3:217–218 (1968)

A new approach to the prediction of glass formation
P. T. Sarjeant and Rustum Roy
Mat. Res. Bull., 3:265–280 (1968)

New vitreous semiconductors
J. C. Schottmiller, D. L. Bowman, and C. Wood
J. Appl. Phys., 39:1663 (1968)

Physical phenomena in semiconductors with negative differential conductivity
A. F. Volkov and Sh. M. Kogan
Uspekhi Fiz. Nauk, 96(6):633–719 (1968)
Sov. Phys. — Usp., 11(6):881–903 (1969)

Effect of disorder on properties of solids
A. Weyl
Glass Indust., 49:433–437 (1968)

Range-energy data for keV ions in amorphous materials (computed from the theoretical analysis of Lindhard, Scharff, and Schiott (1963))
K. B. Winterbon
Atomic Energy of Canada, Ltd., Chalk River, Ontario, Canada, AECL-3194 (November, 1968) 37 pp.

# Amorphous Semiconductors

**184**

Measurement of the absorption coefficient of
glasses in the submillimeter range
  E. M. Dianov, N. A. Irisova, and V. N. Timofeev
  Sov. Phys. — Solid State, 8:2113 (1967)

Electronic properties of amorphous dielectric
films
  A. K. Jonscher
  Thin Solid Films, 1:213–234 (1967)

Photoconductivity and phototropy in noncrystal-
line solids
  J. D. Mackenzie
  Interaction radiat. solids, Proc. Conf., Cairo (1966) (Pub.
    1967), pp. 133–154
  Review of reversible phenomena

Semiconducting glasses
  J. D. Mackenzie
  Proc. Seventeenth Electron. Components Conf., Jackson,
    1967, Institute of Electrical and Electronics Engineers,
    Inc., New York, 1967, pp. 11–16

Determination of the distribution of amorphous
material in a mixed amorphous and crystalline
film by "amorphous" dark field electron microscopy
  J. Markali
  Select. Top. Struct. Chem. Oslo, Universitetsforlaget (1967)

Thin glass films
  W. A. Pliskin, D. R. Kerr, and J. A. Perri
  Physics of Thin Films (Georg Hass and Rudolf E. Thun, eds.),
    Academic Press, New York (1967), Vol. 4, pp. 257–324

Crystallization of amorphous films prepared by
vacuum-evaporation
  M. Shiojiri
  Mem. Fac. Industr. Arts Kyoto Tech. Univ. (Sci. Technol.)
    (Japan), No. 16, 1–18 (Dec. 1967)

Electronic properties of amorphous materials
  Jan Tauc
  Science, 158:1543–1548 (1967)

Optical properties and electronic structure of
amorphous germanium
  J. Tauc, R. Grigorovici, and A. Vancu

The Constitution of Glasses, a Dynamic Inter-
pretation
  Woldemar A. Weyl and Evelyn C. Marboe
  Interscience Publishers, New York (1967), Vol. I: Fundamen-
    tals of the structure of inorganic liquids and solids; Vol. II,
    Part 1: Constitution and properties of some representative
    glasses; Part 2: Electrical properties of glasses

Physical vapor deposition (sublimation, evaporation, sputtering or ion plating of metals, cadmium chalcogenides, gallium arsenide, glasses,
lithium or magnesium fluoride, iron, neodymium,
silicon or titanium oxides, or organics)
  C. M. Jackson, J. G. Kura, J. F. Shea, V. D. Barth, A. G.
    Ingram, C. E. Sims, and C. B. Voldrich
  Redstone Scientific Information Center, AD-803390, March
    31, 1966, 320 pp.

Incorporation of water into (glow discharge) vapor-deposited (glassy silicon, titanium, tin, boron and germanium) oxide films
  D. R. Secrist and J. D. Mackenzie
  Solid State Electron., 9(2):180–181 (1966)

Organic vapors decomposed in rf plasma yielded films
  seemingly isotropic and amorphous

A theoretical study of the glass crystallization
process by a differential thermal analysis sys-
tem
  A. I. Sherstyuk
  Soviet J. Opt. Tech., 33:77–81 (1966)
  Optiko–Mekhanicheskaya Promyshlennost (USSR), 33:28–32
    (1966)

Electrical and optical properties of elementary
amorphous semiconductors
  J. Stuke
  Conf. on electronic processes in low mobility solids, Shef-
    field, 1966 (University of Sheffield, Dept. of Glass Tech-
    nology, Sheffield, 1966), pp. 77–78

Recent trends in the development of conducting
glasses and related materials
  H. J. L. Trap
  Proc. of SIRA conf. on new materials and processes in in-
    strument manufacture, Eastbourne, 1965 (British Sci-
    entific Instrument Research Association, Chislehurst,
    1966), pp. 25–28
  Transition metal oxides or chalcogenides

Some peculiarities of the generation of a new
phase in melt and glasses
  V. N. Filipovich
  NTC Translation 69-12010-11B translated from Structural
    Transformations in Glasses at High Temperatures (N. A.
    Toropov and E. A. Porai-Koshits, eds.), Pub. Nauka,
    Moscow (1965), pp. 49–58

Hypothesis of orbital overlap shifting
  Lee Gildart
  J. Appl. Phys., 36:335 (1965)
  Switches

Quantum Electron Theory of Amorphous Conduc-
tors
  Aleksandr Ivanovich Gubanov
  Consultants Bureau, New York, 1965, 277 pp.
  (Translated from Russian by Albin Tybulewicz)

Preparation and properties of sulfide, selenide,
and telluride glasses (conclusion)
  A. David Pearson
  The Glass Industry (January 1965), pp. 18–21

Structure of Glass
  E. A. Porai-Koshits
  All-Union Conference on the Glassy State, Leningrad,
    (March 13–21, 1964), Nauka Press, Moscow (1965)
    Vols. 6 and 7

Physics of Non-crystalline Solids, Proc. Intern.
Conf., Delft, July, 1964
  J. A. Prins, ed.
  North-Holland Publishing Company, Amsterdam, and John
    Wiley and Sons, Inc., New York (1965)
  CONF-640708

Melting and Crystal Structure
  A. R. Ubbelohde
  Clarendon Press, Oxford (1965), pp. 294–306
  Melts and glasses

Some remarks on the development of the physical
constants of glasses
  A. Winter

Intern. Congr. on Glass (No. 7), Proc., Brussels (June-July, 1965), pp. 369.1-369.4
Can order from SLA as 68-11917-11B

Vitreous semiconductors (I)
B. T. Kolomiets
Phys. Stat. Sol., 7:359 (1964)

Vitreous semiconductors (II)
B. T. Kolomiets
Phys. Stat. Sol., 7:713 (1964)

Modern aspects of the vitreous state
J. D. Mackenzie
Butterworth, Inc., Washington, D. C. (1964)

Semiconducting oxide glasses: general principle for preparation
J. D. Mackenzie
J. Am. Ceram. Soc., 47:211 (1964)

Electrical properties of amorphous semiconductors
K. Moorjani and C. Feldman
Rev. Modern Phys., 36:1042 (1964)
This article constitutes a good review of existing theory concerning the electrical properties of amorphous semiconductors. Many of the properties of the highly regular crystalline compounds carry over into the glassy or amorphous state

Techniques for obtaining uniform thin glass films on substrates
W. A. Pliskin and E. E. Conrad
Electrochem. Tech., 2(7-8):196 (1964)

Some properties and structural elements of glasses and crystals of the same composition
E. I. Galant
Dokl. Akad. Nauk SSSR, 150(5):1100-1103 (1963)
Dokl. — Chem., 150(5):92-95 (1963)

Advances in Glass Technology
F. R. Matson and G. E. Rindone
Plenum Press, New York (1963)

Electronic conduction in glass
R. C. Nelson
J. Appl. Phys., 34:629-631 (1963)

Electrical Properties and Structure of Glass
O. V. Mazurin
Consultants Bureau, Inc., New York (1962)

Nature of the electrical conductivity of glasslike semiconductors
R. L. Myuller
Zh. Prik. Khim., 35(3):541-550 (1962)

New glass types with electronic conductance
H. J. L. Trap and J. M. Stevels
Verres et Refrac., 16:337 (1962)

[Title Not Given]
Lee Gildart and D. F. Clifton
U. S. Patent No. 2,968,014 (1961)
Switches

Electrical properties of glass, a bibliography
Robert Kepple
ANL-6426, July 1961
1940-1960

Non-crystalline Solids (Glass)
V. D. Frenchette
John Wiley and Sons, Inc., New York (1960)

Reversible low-voltage breakdown in stibnite
J. R. Davis and L. Gildart
Bull. Am. Phys. Soc., 3:218 (1959)

Physics of the glassy state. I. Constitution and structure. II. The transformation range. III. Strength of glass. IV. Radiation-sensitive glass
E. U. Condon
Am. J. Phys., 1954

Study of the amorphous state. XVIII. Electric conductivity of substances in the amorphous and crystalline states
P. P. Kobeko, E. V. Kuvshinskii, and N. I. Shishkin
J. Exptl. Theoret. Phys. USSR, 10:1071-1079 (1940)

Transition metal oxides, amorphous semiconductors, semiconducting glasses, Ovshinsky effect, and other switching (memory) materials—a literature review
John. T. Milek
Electronic Properties Information Center, Hughes Aircraft Co., Culver City, Calif., Interim Report No. 72 (including Update No. 1)

[Title Not Given]
Lee Gildart and D. F. Clifton
AD 203-361, July 31, 1958
Switches

## 17.b. Arsenides and Phosphides

Some electronic properties of glassy $CdGeP_2$
V. G. Fedotov, E. I. Leonov, V. N. Ivakhno, N. A. Goryunova, and I. I. Tychina
Fiz. Tekhn. Poluprovod., 3(11):1739-1741 (1969)
Sov. Phys. — Semicond., 3(11):1470-1472 (1970)

Seebeck coefficient in amorphous $CdGe_xAs_2$
P. Nagels, R. Callaerts, F. H. Hashmi, and M. Denayer
Phys. Stat. Sol., 41:K39-K42 (1970)

Semiconductive glasses in the system $ZnAs_2$-$CdAs_2$
Ya. A. Ugai, T. A. Zyubina, and K. B. Aleinikova
Izv. Akad. Nauk SSSR, Neorg. Mater., 6(2):266-270 (1970)
Inorg. Mater., 6(2):231-233 (1970)

The magnetic properties of $CdGeAs_2$ in the vitreous and crystalline states
G. A. Avetikyan, L. A. Baidakov, N. A. Goryunova, N. I. Kouzova, and E. O. Osmanov
Zh. Priklad. Khim., 42(10):2345-2346 (1969)

Semiconducting glasses based on $CdAs_2$
A. Hruby and L. Stourac
Mat. Res. Bull., 4:745-756 (1969)

Optical and thermal carrier activation energies in glasses of the $CdGe(As_xP_{1-x})_2$ system
Z. U. Borisova, N. A. Goryunova, N. I. Kouzova, E. O. Osmanov, and Yu. V. Rud'
Fiz. Tekhn. Poluprovod., 2(10):1548-1549 (1968)
Sov. Phys. — Semicond., 2(10):1292-1293 (1969)

Thermal conductivity of semiconducting amorphous $CdGeAs_2$
L. Stourac
Czech. J. Phys., B 19:681-684 (1969)

The Mössbauer effect in vitreous and crystalline specimens of the $Cd(Ge_xSn_{1-x})As_2$ system
  B. N. Veits, V. Ya. Grigalis, Yu. D. Lisin, G. V.
    Loshakova, E. O. Osmanov, and Yu. V. Rud'
  Latvijas PSR Zinatnu Akad. Vestis, Fiz. Teh. Zinatnu Ser.,
    5:26–30 (1969)

Reflectivity of amorphous and polycrystalline $CdGeAs_2$ in the 0.6—5.5 eV region
  A. Abraham, V. Vorlicek, and M. Zavetova
  Czech. J. Phys., B 18:958–959 (1968)

Propagation rate of ultrasonic waves and structure of glasses in the system Se-As, S-As and Se-Ge
  A. P. Chernov, S. A. Dembovskii, and S. F. Chistov
  Izv. Akad. Nauk SSSR, Neorg. Mater., 4(10):1658–1663
    (1968)
  Inorg. Mater., 4(10):1449–1452 (1968)

Investigations of some properties of vitreous and crystalline $CdGeP_2$
  N. A. Goryunova, S. M. Ryvkin, G. P. Shpenikov, I. I. Tichina,
    and V. G. Fedotov
  Phys. Stat. Sol., 28:489 (1968)

The structure of crystalline and amorphous $CdGeP_2$
  R. Grigorovici, R. Manaila, and A. A. Vaipolin
  Acta Cryst., B 24, Pt. 4:535–541 (April 1968)

High-temperature modifications of the semiconductor compounds $CdSnAs_2$ and $CdGeAs_2$
  E. O. Osmanov, Yu. V. Rud, and M. E. Stryalkovskii
  Phys. Stat. Sol., 26:85 (1968)

Diffusion of gold in crystalline and glassy samples of $CdGeAs_2$
  F. F. Kharakhorin and V. V. Aksenov
  Fiz. Tekhn. Poluprovod., 1(6):961 (1967)
  Sov. Phys. — Semicond., 1(6):805–806 (1967)

## 17.c. Boron

Observation of filament formation in amorphous films during switching
  C. Feldman and K. Moorjani
  Thin Solid Films, 5:R1–R4 (1970)
  Boron

Electrical conduction in amorphous boron and silicon
  K. Moorjani and C. Feldman
  J. Non-Cryst. Solids, 4:248–255 (1970)

Investigation of the electrical properties of boron
  Sh. Z. Dzhamagidze, Yu. A. Mal'tsev, R.R.Shvangiradze
  Fiz. Tekhn. Poluprovod., 2(3):387–392 (1968)
  Sov. Phys. – Semicond., 2(3):320–323 (1968)

Amorphous boron films
  Charles Feldman
  Mat. Res. Bull., 3:95–106 (1968)

Optical transitions in amorphous boron films
  K. Moorjani and C. Feldman
  Solid State Commun., 6:473–475 (1968)

Preparation and properties of thin film boron nitride
  Myron J. Rand and James F. Roberts
  J. Electrochem. Soc., 115:423 (1968)
  Clear, vitreous films up to 6000 Å thick

Radial distribution function of non-crystalline boron
  A. R. Badzian
  Mat. Res. Bull., 2:987–992 (1967)

Spectrographic analysis of high-purity amorphous boron
  O. F. Degtyareva and M. F. Ostrovskaya
  Zh. Anal. Khim., 22:1863–1869 (Dec. 1967)

Chemistry of Boron and Its Compounds
  Earl L. Muetterties, ed.
  John Wiley and Sons, New York (1967), 699 pp.

Dispersion of amorphous boron films in the near infrared
  A. M. Murphy
  J. Opt. Soc. Am., 57:845 (1967)

Infrared absorption in high-purity boron films
  W. Zimmerman, A. M. Murphy, and C. Feldman
  Appl. Phys. Letters, 10:71 (1967)
  Absence of absorption peak at 2 to 15 $\mu$

Ultrapure boron from halide intermediates
  A. F. Armington, J. T. Buford, and R. J. Starks
  Boron, Vol. 2: Preparation, Properties, and Applications.
    Based on papers presented at the 1964 Paris International
    Symposium on Boron (Gerhart K. Gaulé, ed.) Plenum
    Press, New York, (1965), pp. 21–33.

Conductivity, Hall effect, optical absorption, and band gap of very pure boron
  Wolfgang Dietz and Hans Hermann
  Ibid., pp. 107–118
  All samples examined were p-type, though they were in part
    largely doped with W, Ti, Mo, Zr, V, Si, and C. It was not
    possible to measure a Hall effect. Because of the sensi-
    tivity of the measurement, $\mu$ must be below 1 cm$^2$/V-sec.
    In the optical transmission spectrum, numerous bands ap-
    pear both in amorphous and crystalline B

Boron semiconductor devices
  Wolfgang Dietz and Hermann Helmberger
  Ibid., 301 pp.

Vacuum-deposited amorphous boron films
  Charles Feldman, Fred Ordway, William Zimmerman, III,
    and Kishin Moorjani
  Ibid., pp. 235–260

Electron paramagnetic resonance, electrical conductivity, and impurity diffusion in doped boron
  D. Geist
  Ibid., pp. 203–214

Purity of boron produced by the decomposition of diborane and subsequent zone-melting
  Ingeborg Hinz and Heinz Wirth
  Ibid., pp. 9–20

Magnetoresistance in elemental boron
  W. P. Lonc, S. J., and V. P. Jacobsmeyer, S. J.
  Ibid., pp. 215–223
  Crystalline, beta-rhombohedral boron and amorphous boron
    film; negative effect in the film

Semiconductor properties of boron
  W. Neft and K. Seiler
  Ibid., pp. 143-167

Carbon determination in hyperpure elemental boron using gas chromatography
  J. M. Walker and R. J. Starks
  Ibid., pp. 63-79

Activation energies of monocrystalline beta-rhombohedral boron
  R. A. Brungs and V. P. Jacobsmeyer
  J. Phys. Chem. Solids, 25:701 (1964)
  Band gap

Electrical conduction in amorphous films
  C. Feldman
  Nature, 203:964 (1964)

Proc. Intern. Conf. Phys. Semiconductors, Paris, 1964
  R. Grigorovici, N. Croitorie, A. Devenyi, and E. Teleman
  p. 423

Optical Properties of Semiconductors
  T. S. Moss
  Academic Press, New York (1959), p. 102
  Band gap

## 17.d. Chalcogenides

On the preswitching phenomena in semiconducting glasses
  W. W. Sheng and C. R. Westgate
  Solid State Commun., 9:387-391 (1971)
  Suggests combined effect of thermal runaway and phase transition; chalcogenides

Sur la formation de chalcogénures vitreux ternaires
  J. P. Suchet
  Mat. Res. Bull., 6:491-502 (1971)

Magnetoresistance and ESR in amorphous semiconductors
  S. C. Agarwal and H. Fritzsche
  Bull. Am. Phys. Soc., 15:244 (1970)

Conductivity and photopolarization properties of vitreous arsenic triselenide
  Rasum Andreichin
  J. Non-Cryst. Solids, 4:73-77 (1970)

Current-voltage characteristics of the glassy semiconductor $TlAsSe_2$ at high voltages
  A. M. Andriesh and N. Kroiton
  Fiz. Tekhn. Poluprovod., 4(3):563-565 (1970)
  Soviet Phys. – Semicond., 4(3):466-467 (1970)

Amorphous versus crystalline GeTe films. III. Electrical properties and band structure
  S. K. Bahl and K. L. Chopra
  J. Appl. Phys., 41:2196-2212 (1970)
  81 refs.

A simple test for double injection initiation of switching
  I. Balberg
  Appl. Phys. Letters, 16(12):(June 15, 1970)
  "Thermal microheating" model supported

Radial distribution studies of amorphous $Ge_x Te_{1-x}$ alloys
  F. Betts, A. Bienenstock, and S. R. Ovshinsky
  J. Non-Cryst. Solids, 4:554-563 (1970)

Structural studies of amorphous semiconductors
  A. Bienenstock, F. Betts, and S. R. Ovshinsky
  J. Non-Cryst. Solids, 2(1):347-357 (1970)
  $Ge_x Te_{1-x}$, with x = 0.11, 0.66, and 0.72

Mössbauer effect in the arsenic-selenium-tin semiconducting system
  Z. U. Borisova, L. N. Vasil'ev, P. P. Seregin, and V. T. Shipatov
  Fiz. Tekhn. Poluprovod., 4(3):533-536 (1970)
  Sov. Phys. — Semicond., 4(3):443-445 (1970)

Photoelectric effects in semiconducting amorphous chalcogenides
  T. Botila and A. Vancu
  Mat. Res. Bull., 5:925-932 (1970)

Properties of thin films of PbTe and SnTe deposited at temperatures between 4.2° and 300°K
  R. W. Brown, A. R. Millner, and R. S. Allgaier
  Thin Solid Films, 5:157-168 (1970)
  Electrical and optical properties; amorphous and crystalline forms compared

Conductivity changes associated with the crystallization of amorphous selenium
  Clifford H. Champness and R. H. Hoffman
  J. Non-Cryst. Solids, 4:138-148 (1970)

Phase diagrams of systems $As_2X_3-AsI_3$ (X = S or Se)
  A. P. Chernov, S. A. Dembovskii, and I. A. Kirilenko
  Izv. Akad. Nauk SSSR, Neorg. Mater., 6(2):262-265 (1970)
  Inorg. Mater., 6(2):228-230 (1970)

Temperature dependence of switching in $Tl_2SeAs_2Te_3$ glass
  N. Croitoru and L. Vescan
  Rev. Roumaine Phys., 15(1):107-108 (1970)

Nonohmic properties of some amorphous semiconductors
  N. Croitoru, L. Vescan, C. Popescu, and M. Lazarescu
  J. Non-Cryst. Solids, 4:493-503 (1970)

Investigation of the switching characteristics of the tellurium-arsenic-silicon glass films
  A. Csillag
  J. Non-Cryst. Solids, 4:518-522 (1970)

Vitrification in the Ge—Se—I and Si—Se—I systems
  S. A. Dembovskii and N. P. Popova
  Izv. Akad. Nauk SSSR, Neorg. Mater., 6(1):138-140 (1970)
  Inorg. Mater., 6(1):116-118 (1970)

Thermal stability of $AsSe_xI_y$ glasses
  L. I. Doinikov, V. N. Mikhailov, and G. M. Orlova
  Izv. Akad. Nauk SSSR, Neorg. Mater., 6(6):1177-1178 (1970)

Short range order in amorphous GeTe films
  D. B. Dove, M. B. Heritage, K. L. Chopra, and S. K. Bahl
  Appl. Phys. Letters, 16:138-140 (Feb. 1, 1970)
  Radial distribution analysis shows that the local order is not characteristic of the distorted rock-salt structure of bulk GeTe

Influence of copper and iodine impurities on the velocity and attenuation of ultrasound in amorphous arsenic triselenide
J. Durcek, L. Hrivnak, S. Kolnik, C. Musil, and F. Strba
J. Non-Cryst. Solids, 4:66-72 (1970)

Electrical conductivity of amorphous chalcogenide alloy films
E. A. Fagen and H. Fritzsche
J. Non-Cryst. Solids, 2(1):170-179 (1970)

Photoconductivity of amorphous chalcogenide alloy films
E. A. Fagen and Hellmut Fritzsche
J. Non-Cryst. Solids, 4:480-492 (1970)

Temperature and pressure dependence of interband optical absorption in a chalcogenide glass
E. A. Fagen, H. Fritzsche, S. Holmberg, and J. C. Thompson
Bull. Am. Phys. Soc., 15:245 (1970)

Laser-induced reversible optical changes in amorphous chalcogenide semiconductors
J. Feinleib, J. de Neufville, and S. C. Moss
Bull. Am. Phys. Soc., 15:245 (1970)

Reflectivity studies of the tellurium (germanium, arsenic)-based amorphous semiconductor in the conducting and insulating states
Julius Feinleib and S. R. Ovshinsky
J. Non-Cryst. Solids, 4:564-572 (1970)

Calorimetric studies on chalcogenide alloy glasses
H. Fritzsche and S. R. Ovshinsky
J. Non-Cryst. Solids, 2(1):148-154 (1970)

Electronic and vibrational reflection spectra of $CdGeAs_2$ in crystalline and vitreous states
N. A. Goryunova, E. F. Cross, L. B. Zlatkin, and E. K. Ivanov
J. Non-Cryst. Solids, 4:57-65 (1970)

The adsorption of oxygen on clean chalcogenide glass surfaces
Mino Green, M. J. Lee, and P. R. Simmons
Surface Sci., 23:409-410 (1970)

On electrical switching and memory effects in amorphous chalcogenides
Y. P. Gupta
J. Non-Cryst. Solids, 3:148-154 (1970)

Transient electrical measurements of amorphous chalcogenide thin films
J. E. Hall
J. Non-Cryst. Solids, 2:125-132 (1970)
Proceedings of the Symposium on Semiconductor Effects in Amorphous Solids, New York, May 14-17, 1969

Electrical conduction and switching in amorphous semiconductors
C. Hamaguchi, Y. Sasaki, and J. Nakai
Japan. J. Appl. Phys., 9(10):1195-1203 (1970)
As-Te, As-Te-Ga, and As-Te-Ga-Ge-Si

Heat capacities of As-S glasses
M. Hattori, K. Nagaya, S. Umebachi, and M. Tanaka
J. Non-Cryst. Solids, 3:195-204 (1970)

Optical properties of chalcogenide glasses
A. Ray Hilton
J. Non-Cryst. Solids, 2(1):28-39 (1970)

Photoconductivity and density of states for amorphous germanium telluride
W. E. Howard and Raphael Tsu
Phys. Rev., B 1(12):4709-4719 (1970)

Electrical and thermal properties of semiconducting glasses As-Te-Ge
S. Iizima, M. Sugi, M. Kikuchi, and K. Tanaka
Solid State Commun., 8:153-155 (1970)

Effect of stabilization on electrical conductivity in chalcogenide glass
S. Iizima, M. Sugi, M. Kikuchi, and K. Tanaka
Solid State Commun., 8:1621-1623 (1970)

High-frequency conductivity of arsenic selenide
E. B. Ivkin and B. T. Kolomiets
J. Non-Cryst. Solids, 3(1):41-45 (1970)
Vitreous

Electrical conductivity and surface layer formation in semiconducting $As_{50}Te_{45}I_5$
R. T. Johnson, Jr., and R. K. Quinn
Bull. Am. Phys. Soc., 15:1591 (1970)

Current-voltage characteristics of chalcogenide glass films
J. T. Kerr
J. Non-Cryst. Solids, 2(1):203-204 (1970)

Differential negative resistance and turn-off behavior of amorphous chalcogenide threshold switches
J. T. Kerr and Howard K. Rockstad
Bull. Am. Phys. Soc., 15:70 (1970)

"Lock-on" phenomenon in amorphous semiconductors
M. Kikuchi, S. Iizima, and M. Sugi
J. Japan Soc. Appl. Phys., 39 Suppl. 203-210 (1970)
Proceedings of the 1st Conf. on Solid State Devices, Tokyo, 1969
A high-temperature current path was formed at the switch-on, and then a lower-temperature thin filament appears and grows from the anode toward the cathode

Experiment on memory diode in amorphous semiconductor
Akira Kinoshita, Tomoyoshi Aono, and Tomoyasu Nakano
Japan. J. Appl. Phys., 9:411 (1970)
$As_2Se_3$-GeTe

Frequency dependent conductivity of vitreous $As_2Se_3$
M. Kitao, F. Araki, and S. Yamada
Phys. Stat. Sol., 37:K119-121 (1970)

Rectifying properties of junctions between vitreous $Tl_2Se As_2 Te_3$ and Ge, Si and InSb single crystals
B. T. Kolomiets, R. Grigorovici, N. Croitoru, and L. Vescan
Rev. Roumaine Phys., 15(2):129-131 (1970)

Radiative recombination in vitreous and single-crystal arsenic trisulfide and arsenic triselenide
B. T. Kolomiets, T. N. Mamontova, and A. A. Babaev
J. Non-Cryst. Solids, 4:289-294 (1970)

The thermally stimulated conductivity in vitreous and crystalline $As_2Se_3$
B. T. Kolomiets and T. F. Mazets
J. Non-Cryst. Solids, 3:46-53 (1970)

Effective mass of the current carriers in glassy arsenic chalcogenides
B. T. Kolomiets, T. F. Mazets, and Sh. M. Efendiev
Fiz. Tverd. Tela, 12:661–663 (1970)
Sov. Phys. — Solid State, 12(2):514–515 (1970)

Influence of pressure on the electrical and photoelectric properties of amorphous and single-crystal samples of $As_2Se_3$
B. T. Kolomiets and E. M. Raspopova
Fiz. Tekhn. Poluprovod., 4(1):157–161 (1970)
Sov. Phys. — Semicond., 4(1):124–127 (1970)

The absorption edge of amorphous $As_2S_3$
F. Kosek and J. Tauc
Czech. J. Phys., B 20:94–100 (1970)

Low acoustic loss chalcogenide glasses — a new category of materials for acoustic and acousto-optic applications
J. T. Krause, C. R. Kurkjian, D. A. Pinnow, and E. A. Sigety
Appl. Phys. Letters, 17(9):367–368 (1970)
IVA–VA–VIA monoxide glasses

A study of the anomalous Hall effect in the chalcogenide semiconductor glass systems
Dennis W. Kurtz
Thesis, Ohio School of Engineering, GE/EE/70–14; AD–710600 (Feb. 1970), 81 pp.

Conductivity of electrically switching chalcogenide glass
J. C. Male
Electronics Letters, 6:91 (1970)
Electrothermal mechanism; $As_{30}Te_{48}Si_{12}Ge_{10}$

Preparation and structure of thin films of germanium telluride and selenide
A. G. Mikolaichuk and A. N. Kogut
Kristallografiya, 15(2):353–357 (1970)
Sov. Phys. — Cryst., 15(2):294–298 (1970)

Thermal expansion and its related properties of As–S–Te glasses
T. Minami, M. Hattori, and M. Tanaka
Yogyo Kyokai Shi, 78(895):101–107 (1970)

Electrical properties of vitreous $Tl_2Te \cdot As_2Te_3$
P. Nagels, R. Callaerts, M. Denayer, and R. De Coninck
J. Non–Cryst. Solids, 4:295–303 (1970)

Superconducting amorphous semiconductors
Neuringer, H. H. Sample, J. A. Gerber, and J. de Neufville
Mass. Inst. of Technol., Francis Bitter National Magnet Lab., Quarterly Tech. Stat. Rept., July 1, 1970, to Sept. 30, 1970 (Sept. 30, 1970), pp. 10–11, QTSR 41 (to be published in Phys. Letters)

Structural changes related to electrical properties of bulk chalcogenide glasses
S. V. Phillips, R. E. Booth, and Peter W. McMillan
J. Non–Cryst. Solids, 4:510–517 (1970)

Electronic switching and memory device based on vitrified arsenic pentaselenide
Serge Renault, Richard Jansen, and Francois Gans
Compt. Rend., Ser. B, 270:1569–1572 (1970)

Internal space charge in amorphous chalcogenide switches
Howard K. Rockstad
Bull. Am. Phys. Soc., 15:332 (1970)

Hopping conduction and optical properties of amorphous chalcogenide films
H. K. Rockstad
J. Non–Cryst. Solids, 2(1):192–202 (1970)

Effects of valency on transport properties in vitreous binary alloys of selenium
J. Schottmiller, M. Tabak, G. Lucovsky, and Anthony T. Ward
J. Non–Cryst. Solids, 4:80–96 (1970)

Instabilities in semiconducting glass diodes
D. Shanefield and P. E. Lighty
Appl. Phys. Letters, 16(5):212–214 (1970)
Predominantly amorphous mixture of 81.5% Te, 18% As, and 0.5% Ge exhibits electrical memory effects, but continued cycling leads to unstable behavior; explanation based on the presence of varying amounts of crystalline material

Chemical and structural characterization of amorphous semiconducting materials in the system gallium-arsenic-tellurium-germanium
J. R. Shappirio, D. W. Eckart, and C. F. Cook, Jr.
J. Non–Cryst. Solids, 2(1):217–228 (1970)

Dielectric properties of amorphous semiconductor chalcogenide thin films
R. F. Shaw
Bull. Am. Phys. Soc., 15:245 (1970)

Optical properties, photoconductivity, and energy levels in crystalline and amorphous arsenic triselenide
R. F. Shaw, W. Y. Liang, and A. D. Yoffe
J. Non–Cryst. Solids, 4:29–42 (1970)

Electron microprobe analysis and radiometric microscopy of electric field induced filament formation on the surface of arsenic-tellurium-germanium glass
C. H. Sie
J. Non–Cryst. Solids, 4:548–553 (1970)

High-field photoconductivity of amorphous GeTe and GeSe films
P. J. Stiles, L. L. Chang, Leo Esaki, and Raphael Tsu
Bull. Am. Phys. Soc., 15:245 (1970)

Phenomenology of switching and memory effects in semiconducting chalcogenide glasses
H. J. Stocker
J. Non–Cryst. Solids, 2(1):371–381 (1970)

Mechanism of threshold switching in semiconducting glasses
Hans J. Stocker, C. A. Barlow, Jr., and D. F. Weitauch
J. Non–Cryst. Solids, 4:523–535 (1970)

Evidence for thermal pinching effect in chalcogenide amorphous semiconductors
Michio Sugi, Yasumasa Okada, Sigeru Iizima, Makoto Kikuchi, and Kazunobu Tanaka
Solid State Commun., 8:829–832 (1970)

Structure of memory-switching glasses I. Crystallization temperature and its control in Ge–Te glasses
T. Takamori, R. Roy, and G. J. McCarthy
Mat. Res. Bull., 5:529–540 (1970)

Electrical nature of the lock-on filament in amorphous semiconductors
K. Tanaka, S. Iizima, M. Sugi, and M. Kikuchi
Solid State Commun., 8:75–78 (1970)

Thermal effect on switching phenomenon in chalcogenide amorphous semiconductors
K. Tanaka, S. Iizima, M. Sugi, Y. Okada, and M. Kikuchi
Solid State Commun., 8:387-389 (1970)

Optical and magnetic investigations of the localized states in semiconducting glasses
J. Tauc, A. Menth, and D. L. Wood
Phys. Rev. Letters, 25:749-752 (1970)
$As_2S_3$

Far infrared reflectivity of $Tl_2Se-As_2(Se_xTe_{1-x})_3$ glasses
P. C. Taylor and S. G. Bishop
Bull. Am. Phys. Soc., 15:290 (1970)

Preswitching behaviour of amorphous chalcogenide semiconductor films
D. L. Thomas and A. C. Warren
Electron Letters, 6(3):62-63 (1970)

Switching properties and photoconduction in (single crystal) GaSe
R. H. Tredgold, R. H. Williams, and A. Clark
Phys. Stat. Sol., 3A, 407-410 (1970)
Analogous to those found in amorphous semiconductors

Photoconductivity and optical properties of amorphous GeTe
Raphael Tsu, Webster E. Howard, and Leo Esaki
J. Non-Cryst. Solids, 4:322-327 (1970)

Oxygen impurities and defects in chalcogenide glasses
A. Vasko, D. Lezal, and I. Srb
J. Non-Cryst. Solids, 4:311-321 (1970)

Thermal switching in semiconducting glasses
A. C. Warren
J. Non-Cryst. Solids, 4:613-616 (1970)

Field-enhanced conductivity effects in thin chalcogenide-glass switches
A. C. Warren and J. C. Male
Electron. Letters, Vol. 6, No. 18 (Sept. 3, 1970)

Threshold switching and thermal filaments in amorphous semiconductors
D. F. Weirauch
Appl. Phys. Letters, 16(2):72 (Jan. 15, 1970)
As-Ge-Se

dc Conductivity, optical absorption, and photoconductivity of amorphous arsenic telluride films
K. Weiser and M. H. Brodsky
Phys. Rev., B, 1(2):791-799 (1970)

Compositions and structures of the "lock-on" filaments in As-Te-Ge glasses
Yasumasa, Sigeru Iizima, Michio Sugi, Makoto Kikuchi, and Kazunobu Tanaka
J. Appl. Phys., 41:5341-5344 (1970)

Physicochemical properties of PbTe-SnTe-PbSe and PbTe-GeTe-PbSe alloys
G. T. Alekseeva, B. A. Efimova, D. I. Lainer, and L. M. Ostrovskaya
Izv. Akad. Nauk SSSR, Neorg. Mater., 5(12):2105-2109 (1969)
Inorg. Mater., 5(12):1793-1796

A study of the magnetic susceptibility of glasses of the Ge-Se system
G. B. Avetikyan, L. A. Baidakov, and L. P. Strakhov
Izv. Akad. Nauk SSSR, Neorg. Mater., 5(10):1667-1669 (1969)
Inorg. Mater., 5(10):1411-1413 (1969)

Electrical and optical properties of amorphous vs crystalline GeTe films
S. K. Bahl and K. L. Chopra
J. Vacuum Sci. Tech., 6(4):561-565 (1969)

Magnetic susceptibility of semiconducting arsenic chalcogenides in a vitreous state
L. A. Baidakov, L. N. Blinov, and L. P. Strakhov
Khim. Svyaz Krist. (1969), pp. 478-484

Comparative investigation of the optical properties of arsenic chalcogenides during transition from the crystalline to glassy state
M. L. Belle, B. T. Kolomiets, and B. V. Pavlov
Fiz. Tekhn. Poluprovod., 2(10):1448-1453 (1968)
Sov. Phys. — Semicond., 2(10):1210-1214 (1969)

Effect of the nature of chemical bonding on physicochemical properties of vitreous semiconductors
Z. U. Borisova
Khim. Svyaz Krist. (1969), pp. 462-470

Interaction of vitreous arsenic selenide with alkali metals
Z. U. Borisova and V. V. Bakulina
Zh. Prikl. Khim. (Leningrad), 42:1238-1243 (1969)

Space-charge-limited currents in vitreous antimony trisulfide films
T. Budinas, P. Mackus, A. Smilga, and J. Viscakas
Phys. Stat. Sol., 31:375 (1969)

Thermal breakdown and switching in chalcogenide glasses
P. Burton and R. W. Brander
Intern. J. Electron., 27:517-525 (1969)

Some experiments on switching-type amorphous semiconductors
P. Calella, S. DeFeo, J. Hall, P. Walsh, R. Nicolaides, and W. Doremus
Engineering Sciencws Lab., Feltman Research Labs., Picatinny Arsenal, Dover, N. J., ESL-IR-451 (May 1969)
Presented at American Physical Society, Feb. 3-6, 1969

Bistable electrical-switching behavior in glasses and other media: Effect due to production of a metallic bridge between the electrodes (Ovshinsky effect)
R. Chapman
Electron Letters, 5:246-247 (May 29, 1969)

Amorphous versus crystalline GeTe films. I. Growth and structural behavior
K. L. Chopra and S. K. Bahl
J. Appl. Phys., 40:4171-4178 (1969)

Non-ohmic properties of some amorphous semiconductors
N. Croitoru, L. Vescan, C. Popescu, and M. Lazarescu
Presented at International Conference on Amorphous Semiconductors, Cambridge, Sept (1969)
$As_2Te_3$ and As-Te-Ge-Si

Correlation of the glass point ($T_x$) and elastic constants with the structure of vitreous semiconductors
S. A. Dembovskii
Khim. Svyaz Krist. (1969), pp. 471-477

Comparison of the physicochemical properties of the chalcogens and basic chalcogenide compounds
S. A. Dembovskii
Izv. Akad. Nauk SSSR, Neorg. Mater., 5(3):463–471 (1969)
Inorg. Mater., 5(3):385–392 (1969)
Glass formers

Elastic constants, softening temperature, and structure of chalcogenide glasses
S. A. Dembovskii
Phys. Chem. Glass, 10:73–74 (1969)

Glass formation and some properties of glasses in a germanium-lead-arsenic-selenium system
A. M. Efimov and V. F. Kokorina
Opt. -Mekh. Prom., 36(10):43–48 (1969)

Silicon sulfide and selenide
Hans H. Emons, Siegfried Moehlhenrich, and Lothar Theisen
German (East) Patent 65,914 (March 1969), 2 pp.

Hopping conduction in an amorphous chalcogenide alloy film
Edward A. Fagen, S. R. Ovshinsky, and H. Fritzsche
Bull. Am. Phys. Soc., 14:311 (1969)

Rubidium metathio- and selenostibnites
Ya. G. Finkel'shtein, S. I. Berul', and N. P. Luzhnaya
Izv. Akad. Nauk SSSR, Neorg. Mater., 5(1):168–169 (1969)
Inorg. Mater., 5(1):140–141 (1968)

Electron microscope study of the diffusion of metals in amorphous arsenic triselenide films
L. A. Freeman, R. F. Shaw, and A. D. Yoffe
Thin Solid Films, 3:367–376 (May 1969)

Photostimulated conductivity in an amorphous chalcogenide alloy film
H. Fritzsche, Edward A. Fagen, and S. R. Ovshinsky
Bull. Am. Phys. Soc., 14:311 (1969)

Conductivity of As-Te-I films
R. B. Hilborn, Jr., and Kanti Prasad
J. Vac. Sci. Tech., 6:632–634 (1969)
Structural changes at high temperatures

Some properties of amorphous GeTe films
W. E. Howard and R. Tsu
Bull. Am. Phys. Soc., 14:428 (1969)

Crystallization of vitreous semiconductors $As_2(Se_{1-x}Te_x)_3$
Kenzo Igaki and Sumie Ito
Oyo Butsuri, 38(11):1032–1036 (1969)

Instability in amorphous arsenic sulfide films
S. W. Ing, Jr., Y. S. Chiang, and A. Ward
Am. Ceram. Soc. Bull., 48(4):442 (1969)

Glass-forming regions and structure of glasses in the system Ge-S
Y. Kawamoto and S. Tsuchihashi
J. Am. Ceram. Soc., 52:626–627 (1969)

"Memory exchange" in amorphous semiconductors
M. Kikuchi and S. Iizima
Appl. Phys. Letters, 15(10):323 (Nov. 1969)
As-Te-Ge-Si; switching + memory; thermal effects, filaments

Vitreous semiconductors
B. T. Kolomiets
Vestn. Akad. Nauk SSSR, 39(6):54–61 (1969)

Principal parameters of switches based on glassy chalcogenide semiconductors (Physics of switches based on glassy chalcogenide semiconductors and their parameters)
V. T. Kolomiets, E. A. Lebedev, and I. A. Taksami
Fiz. Tekhr. Poluprov., 3:312–314 (1969)
Sov. Phys. — Semicond., 3:621–624 (1969)

Mechanism of the breakdown in films of glassy chalcogenide semiconductors
B. T. Kolomiets, E. A. Lebedev, and I. A. Taksami
Fiz. Tekhn. Poluprov., 3:312–314 (1969)
Sov. Phys. — Semicond., 3:267 (1969)
Ge-Se-As-Te glasses

Refractive index of amorphous $As_2S_5$
F. Kosek and J. Cermak
Cesk. Casopis Fys., 19A:271–275 (1969)

Frequency dependence of the conductivity of a memory-type chalcogenide glass
B. P. Kraemer
Sci. B. thesis, Dept. of Chemistry, Massachusetts Institute of Technology, Sept. 1969

Density of glasses in the arsenic-selenium, arsenic-selenium-sulfur, arsenic-selenium-tellurium, and arsenic-selenium-thallium systems and density change of the arsenic-selenium glass due to heat treatment
Masanaga Kunugi, Rikuo, Ota, Takashi Yamagishi, and Seishiro Fukutani
Zairyo, 18:807–812 (1969)

Comments on the structure of chalcogenide glasses from infrared spectroscopy
G. Lucovsky
Mat. Res. Bull., 4:505–514 (1969)

Characteristic features of the flow of current and of the photoelectric processes in amorphous antimony triselenide in the presence of injecting and blocking contacts
V. M. Lyubin and V. S. Maidzinskii
Fiz. Tekhn. Poluprovod., 3(11):1675–1679 (1969)
Sov. Phys. — Semicond., 3(11):1408–1411 (1970)
Pt, Au, Al, and $SnO_2$ electrodes form blocking contacts; Sb or Bi form injecting contacts

Changes in the spectrum of local states in antimony triselenide during crystallization
V. M. Lyubin and V. S. Maidzinskii
Fiz. Tekhn. Poluprovod., 3(11):1702–1704 (1969)
Sov. Phys. — Semicond., 3(11):1430 (1970)

Composite system comprising As-S glass and aluminum
Shigeo Maruno
Japan. J. Appl. Phys., 8:530 (1969)

Transmission spectra near the absorption edge of glasses in the systems As-S, As-S-Te and Si-As-Te
T. Minami and M. Tanaka
Bull. Univ. Osaka Prefecture A (Japan), 18(1):165–174 (1969)

Thermal expansion and related properties of silicon-arsenic-tellurium glasses
T. Minami and M. Tanaka
Yogyo Kyokai Shi, 77(891):372–377 (1969)

Temperature dependence of the magnetic suscep-
tibility of As$_2$Se$_3$ at the crystal-melt and glass-
melt phase transitions
S. K. Novoselov, L. P. Strakhov, and L. A. Baidakov
Fiz. Tverd. Tela, 11(6):1564–1568 (1969)
Sov. Phys. – Solid State, 11(6):1266–1269 (1969)

Features of the volt-ampere characteristics of
"film-faced" threshold switches (ovonic) based
on chalcogenide glass
P. T. Oreshkin, A. S. Glebov, V. P. Oreshkin, et al.
Izv. Vyssh. Ucheb. Zaved., Fiz., 12(10):136–139 (1969)

Vitreous alloys containing a large amount of tellurium in an arsenic-germanium-tellurium system
V. R. Panus
Vestn. Leningrad. Univ., Fiz., Khim, No. 4:135–139 (1969)

Amorphous glass compositions
R. J. Patterson, and A. E. Tilton
Texas Instruments Inc., U. S. Patent 3,440,068, April 22,
1969; appl. Dec. 21, 1966, 4 pp.
Ge–Se–Te

Vitrification range and properties of glasses be-
longing to the system As$_2$O$_3$-S-Se
N. M. Pavlushkin, A. K. Zhuravlev
Izv. Akad. Nauk SSSR, Neorg. Mater., 5(3):595–597 (1969)
Inorg. Mater., 5(3):504–506 (1969)

Glass-formation region in a bismuth-germanium-
selenium system
A. V. Pazin and Z. U. Borisova
Vestn. Leningrad. Univ., Fiz., Khim., No. 4, 140–144 (1969)

Filamentary conduction in semiconducting glass
diodes
A. David Pearson and C. E. Miller
Appl. Phys. Letters, 14:280 (1969)

The chalcogenide semiconductor system: As$_2$Se$_3$-
Sb$_2$Se$_3$
N. S. Platakis, V. Sadagopan, and H. C. Gatos
J. Electrochem. Soc., 116:1436–1439 (1969)

Method for the production of amorphous cadmium
sulfide
Leonard E. Ravich
White Consolidated Industries, Inc., U. S. Patent 3,432,262,
March 11, 1969

Evidence for hopping conduction in amorphous
chalcogenide films
H. K. Rockstad
Solid State Commun., 7:1507–1509 (1969)

Di-phasic structure of switching and memory de-
vice "glasses"
R. Roy and Vera Caslavska
Solid State Commun., 7:1467–1473 (1969)
Even those which are really glasses consist of two non-
crystalline phases

Effect of manganese, iron, nickel, and cobalt on
the electrical conductivity and chemical stability of glassy arsenic selenide
Ya. Savan and Z. U. Borisova
Zh. Prikl. Khim. (Leningrad), 42(5):1017–1023 (1969)

Crystallization of glasses in the Cu-As-Se sys-
tem
Ya. Savan, I. I. Kozhina, G. M. Orlova, and Kh. Binder
Izv. Akad. Nauk SSSR, Neorg. Mater., 5(3):492–497 (1969)
Inorg. Mater., 5(3):410–414 (1969)

Heat capacities of glassy and liquid As$_2$S$_3$ and
As$_2$Se$_3$, technical report
U. E. Schnaus, C. T. Moynihan, R. W. Gammon, and P. B.
Macedo
Catholic Univ. of America, Washington, D. C., Vitreous
State Lab., Contract N00014-68-A-0506-0002, Rept. No.
TR-6; AD-698 842 (Dec. 1969), 18 pp.

Features of the volt-ampere characteristics of
diodes based on four-component chalcogenide
glass
V. A. Semenov
Izv. Vyssh. Ucheb. Zaved. Fiz., 12(12):30–33 (1969)

Nonrectifying solid state element
D. J. Shanefield, J. H. Battle, and E. W. Currier
International Telephone and Telegraph Corp., U. S.
3,453,583 (July 1, 1969)
Tl–Te–Se

Bulk and thin-film switching and memory effects
in semiconducting chalcogenide glasses
H. J. Stocker
Appl. Phys. Letters, 15(2):55 (July 15, 1969)

Switching characteristics of chalcogenide glass
M. Sugi, M. Kikuchi, S. Iizima, and K. Tanaka
Solid State Commun., 7:1805–1807 (1969)

Transient photoconductivity in chalcogenide glas-
ses
Mark D. Tabak
Xerox Corp, Abstract 2-11, Proc. Third International Conf.
on Photoconductivity, Stanford Univ. (August 12–25, 1969)
Solid State Commun., 7:i-xxiii (1969)

Aging of vitreous arsenic-selenium photoconduc-
tors
M. P. Trubisky and J. H. Neyhart
Appl. Opt., Suppl., 3:59–63 (1969)

On the structure of glasses in the system As-S.
III. The properties and structure of sulfide glas-
ses
S. Tsuchihashi and Y. Kawamoto
J. Ceram. Soc. Japan, 77:35–39 (1969)

Properties of TlBiX$_2$ thin films, where X is sul-
fur, selenium, and tellurium
A. S. Tsytko, S. A. Dembovskii, I. I. Ezhik, and V. A.
Bazakutsa
Izv. Vyssh. Ucheb. Zaved., Fiz., 12:154–157 (1969)

Mössbauer effect in tin-containing chalcogenide
glasses
B. N. Veits, V. Grigalis, Yu. D. Lisin, and Z. Konstants
Latv. PSR Zinat. Akad. Vestis, Kim. Ser., 6:744–745 (1969)

Negative capacitance in amorphous semiconduc-
tor chalcogenide thin films
R. Vogel and P. J. Walsh
Appl. Phys. Letters, 14:216–217 (1969)

Thermoelectric efficiency of thin layers of $A^{III}B^{V}C_2^{VI}$
L. G. Voinova, S. A. Dembovskii, and V. A. Bazakutsa
Izv. Vyssh. Ucheb. Zaved., Fiz., 12:152–154 (1969)
TlSbTe$_2$

Temperature dependence of Ovshinsky-type devices
O. E. Wagner
J. Appl. Phys., 40:4212–4213 (1969)

Conduction and electrical switching in amorphous chalcogenide semiconductor films
P. J. Walsh, Ruth Vogel, and E. J. Evans
Phys. Rev., 178:1274–1279 (1969)

Switching mechanism in chalcogenide glasses (Ovshinsky effect; Joule heating; nonohmic behavior)
A. C. Warren
Electron. Letters, 5:461–462 (1969)

On state of chalcogenide glass switches
A. C. Warren
Electronics Letters, 5(24):609 (1969)
The high-conductance state of chalcogenide glass switches is described in terms of Joule heating

Indirect transitions in glassy As$_2$S$_3$
E. L. Zorina
Opt. Spectrosc., 27:168–169 (1969)

Light absorption in amorphous semiconductive films of alkali-metal thioantimonites and selenoantimonites
E. L. Zorina, N. I. Gnidash, Ya. G. Finkel'shtein, S. I. Berul', and N. P. Luzhnaya
Izv. Akad. Nauk SSSR, Neorg. Mater., 5(12):2099–2104 (1969)
Inorg. Mater., 5(12):1788–1792 (1969)

Bismuth chalcogenide thin films
Robert M. Anderson, Hruska, Liedl, Meininger, and Pocker
Purdue University Progress Report, Oct. 1, 1967–Sept. 30, 1968, on Materials Sciences Research SD-102 (Dec. 1968), p. 21

Magnetic susceptibility of vitreous arsenic triselenide
L. M. Blinov, L. A. Baidakov, and L. P. Strakhov
Uch. Zap. Leningrad. Gos. Univ., Ser. Fiz. Nauk, 345:49–51 (1968)

Vitreous Bi-Se layers as near-infrared photodetectors
D. L. Bowman and J. C. Schottmiller
J. Appl. Phys., 39:1659 (1968)

Propagation rate of ultrasonic waves and the structure of glasses in the system Se-As, S-As and Se-Ge
A. P. Chernov, S. A. Dembovskii, and S. F. Chistov
Izv. Akad. Nauk SSSR, Neorg. Mater., 4(10):1658–1663 (1968)
Inorg. Mater., 4(10):1449–1452 (1968)

Study of the linear expansion of vitreous and polycrystalline selenium and As$_2$Se$_3$
S. F. Chistov, A. P. Chernov, and S. A. Dembovskii
Izv. Akad. Nauk SSSR, Neorg. Mater., 4(12):2085–2088 (1968)
Inorg. Mater., 4(12):1814–1816 (1968)

Selenium and Selenides
D. M. Chizhikov and V. P. Shchastlivyi
Collet's Publishers, Ltd., London and Wellingborough (1968)

Identification and properties of the chemical compounds TlAsS$_2$, TlAsSe$_2$, TlAsTe$_2$ in the glassy and crystalline states
S. A. Dembovskii
Izv. Akad. Nauk SSSR, Neorg. Mater., 4(11):1920–1926 (1968)
Inorg. Mater., 4(11):1671–1676 (1968)

Synthesis and certain properties of the compounds AsSI, AsSeI and As$_4$Te$_5$I$_2$
S. A. Dembovskii and A. P. Chernov
Izv. Akad. Nauk SSSR, Neorg. Mater., 4(8):1229–1232 (1968)
Inorg. Mater., 4(8):1079–1081 (1968)

Measurements of electrical conductivity and optical absorption in chalcogenide glasses
J. T. Edmond
J. Non-Cryst. Solids, 1:39–48 (1968)

Glass formation and physical-chemical properties of glasses of the antimony-germanium-selenium system
E. A. Egorova and V. F. Kokorina
Zh. Prikl. Khim. (Leningrad), 41(6):1200–1206 (1968)

Silicon monoselenide
H. H. Emons and L. Theisen
Z. Anorg. Allgem. Chem., 361:321–327 (1968)

Development of arsenic sulfide glasses since 1950
R. Frerichs
Am. Ceram. Soc. Bull., 47(4):398 (April 1968); 70th Annual Meeting of Am. Ceram. Soc., Chicago (April 20–25, (1968)

Electronic properties of chalcogenide glasses
T. J. Gray and L. D. Pye
Am. Ceram. Soc. Bull., 47(4):398 (April 1968); 70th Annual Meeting of Am. Ceram. Soc., Chicago (April 20–25, 1968)

Electrical conductivity and formation of chalcogenide glasses in germanium-arsenic-selenium and germanium-antimony-selenium melts
Robert W. Haisty and Heinz Krebs
Angew. Chem., Int. Ed. Engl., 7:947–948 (1968)

Chemical composition and glass formation in chalcogenide systems
A. Ray Hilton
Phys. Chem. Glasses, 9(5):148–152 (1968)

An electron diffraction study of the semiconductor CuAsSe$_2$
R. M. Imamov and I. I. Petrov
Kristallografiya, 13(3):412–416 (1968)
Sov. Phys. — Cryst., 13(3):335–339 (1968)

Chalcogenide glass
Minoru Imaoka
Univ. Tokyo, Japan, Kagaku To Kogyo (Tokyo), 21(3):326–336 (1968)
Review of range of glass-forming composition, structure, and properties:mechanical, thermal, optical, electrical; 50 refs.

Amorphous glass compositions
  Rowland Edward Johnson, Robert J. Patterson, and Andre
    E. Tilton
  Texas Instruments, Inc., Fr. 1,548,159 (Nov. 1968), 8 pp.
  $Ge_{35}Se_{60}I_5$

Mass spectrometric studies of laser-induced vaporization. III. The arsenic-selenium system
  Bruce E. Knox and Vladimir S. Ban
  Mat. Res. Bull., 3:885-894 (1968)

Drift mobility in amorphous $As_2Se_3$ films
  V. I. Kruglov, L. P. Strakhov, and N. A. Grishin
  Vestn. Leningrad. Univ., Fiz., Khim., 23(10):62-68 (1968)

Investigation of the band structure of amorphous $As_2Se_3$ by an ultrasoft x-ray spectroscopy method
  V. I. Kruglov and T. M. Zimkina
  Fiz. Tverd. Tela, 10(1):226-229 (1968)
  Sov. Phys. — Solid State, 10(1):170-172 (1968)

Preparation and electrical properties of mixed selenide-oxide glasses
  James E. Lang
  Thesis, Rensselaer Polytech. Inst., Troy, N.Y. (1968), 179 pp.
  University Microfilms, Ann Arbor, Mich., Order No. 69-6468

Mixed selenide-oxide semiconducting glasses
  J. E. Lang and J. D. Mackenzie
  Am. Ceram. Soc. Bull., 47(4):398 (1968)

Investigation of negative resistance in chalcogenide elements with fused-in electrodes
  O. V. Mitrofanov, V. F. Zolotaryev, V. A. Semenov, and
    A. P. Budennyi
  Izv. Vuz Fiz. (USSR), 12:119-121 (1968)

Thermal properties of vitreous Bi-Se alloys
  M. B. Myers and J. C. Schottmiller
  Am. Ceram. Soc. Bull., 47(4):402 (April 1968), 70th Annual
    Meeting of Am. Ceram. Soc., Chicago (April 20-25, 1968)

Photoconductivity of glassy $AsSe_x$
  G. M. Orlova, G. A. Nikandrova, and L. V. Octapenko
  Izv. Akad. Nauk SSSR, Neorg. Mater., 4(10):1646-1649 (1968)
  Inorg. Mater., 4(10):1438-1441 (1968)

Reversible electrical switching phenomena in disordered structures
  Stanford R. Ovshinsky
  Phys. Rev. Letters, 21:1450-1453 (1968)

Electrical conductivity of glasses in the arsenic-silicon-tellurium system
  V. R. Panus, Z. U. Borisova, and T. T. Alekseeva
  J. Appl. Chem. USSR, 41:2598-2600 (1968)

The Hall effect in vitreous semiconductors of the system As-Ge-Te
  V. R. Panus, Ya. M. Ksendzov, and Z. U. Borisova
  Izv. Akad. Nauk SSSR, Neorg. Mater., 4(6):885-888 (1968)
  Inorg. Mater., 4(6):778-781 (1968)

The thermo-emf in vitreous semiconductors of the system As-Ge-Te
  V. R. Panus, Ya. M. Ksendzov, and Z. U. Borisova
  Izv. Akad. Nauk SSSR, Neorg. Mater., 4(6):889-892 (1968)
  Inorg. Mater., 4(6):782-784 (1968)

Physical properties of glasses in the system arsenic-selenium-thallium-tellurium
  F. Pernot
  Verres Refract., 22(6):595-603 (1968)

New sulfide and selenide glasses: preparation, structure, and properties
  E. R. Plumat
  J. Am. Ceram. Soc., 51:499-507 (1968)

Semiconductivity in some chalcogenide glasses
  Lenwood D. Pye
  Ph. D. thesis, Coll. of Ceram. State Univ. of New York,
    Alfred, N. Y. (1968), Univ. Microfilms, Ann Arbor, Mich.,
    Order No. 68-16, 619, 127 pp.
  Phys. Rev., 178:1274 (1969)

Thermal conductivity of glassy semiconductors
  I. A. Rozov, A. F. Chudnovskii, and V. F. Kokorina
  Fiz. Tekhn. Poluprovod., 1(8):1159-1163 (1967)
  Sov. Phys. — Semicond., 1(8):969-972 (1968)

Electrical conductivity and microhardness of vitreous alloys of the arsenic-selenium-copper system
  Ya. Savan and Z. U. Borisova
  Izv. Akad. Nauk SSSR, Neorg. Mater., 4(12):2089-2093
    (1968)
  Inorg. Mater., 4(12):1817-1820 (1968)

Dielectric constants of the systems As-S and As-Se
  Akikazu Shibata
  Joint Meeting of Phys. Soc. Japan and Soc. Appl. Phys. 1968

Influence of Ge and Ag impurities on the thermal conductivity of semiconducting amorphous $As_2Se_3$
  L. Stourac, B. T. Kolomiec, and V. P. Silo
  Czech. J. Phys., B 18:92 (1968)

Examination of internal defects in chalcogenide glasses by infrared defectoscopy methods
  A. Vasko
  Mat. Res. Bull., 3:209-216 (1968)

Fields of crystallization of the system As-Se-Ge in the region of glass formation
  G. Z. Vinogradova, S. A. Dembovskii, and N. P. Luzhnaya
  Zh. Neorg. Khim., 13(5):1444-1450 (1968)

Capacitive effects in amorphous chalcogenide thin films
  R. Vogel and P. J. Walsh
  Picatinny Arsenal, Dover, New Jersey
  PA-TR-3797; AD-683467 (Dec. 1968), 13 pp.

Conduction and electrical switching in amorphous chalcogenide semiconductor films
  P. J. Walsh, Ruth Vogel, and E. J. Evans
  Picatinny Arsenal, Dover, New Jersey
  PA-TR-3765; AD-683465 (Nov. 1968), 25 pp.

The infrared absorption of vitreous $As_2Se_3$, $As_2Se_5$, and $AsSe_4$
  E. L. Zorina, S. A. Dembovskii, V. B. Velichkova and
    G. Z. Vinogradova
  Izv. Akad. Nauk SSSR, Neorg. Mater., 1(11):1889-1891 (1965)
  Inorg. Mater., 1(11):1708-1710 (1968)

Optical reflection spectra of the arsenic chalcogenides
  A. M. Andriesh and V. V. Sobolev
  Khimicheskaya Svyaz'v Poluprovodnikikh i Termodinamika
    (Chemical bonds in semiconductors and thermodynamics),
    Nauka i Tekhnika, Minsk, 1966 (Sept. 1967), pp. 212-216
  Both in the crystalline and vitreous form

Effect of the nature of the chemical bond on the electrical conductivity of glassy semiconductors
  Z. U. Borisova
  Zh. Fiz. Khim., 41(8):1942–1945 (Aug. 1967)

Electrical conductivity and microhardness of glasses in the arsenic–phosphorus–selenium system
  Z. U. Borisova and L. A. Krylova
  J. Appl. Chem. USSR, 40:52–55 (Jan. 1967)

Solid state switches
  A. S. Danfoss
  British Patent 1,083,154, Sept. 13, 1967; Ger. appl. May 5, 1964
  Te with the addition of elements from Groups IV and V

Antimony semiconductor materials
  A. S. Danfoss
  British Patent 1,083,154, Sept. 13, 1967; Ger. appl. May 5, 1964
  Sb sulfide, Sb selenide, switches

Effect of some elements on the optical absorption edge of vitreous $As_2S_3$
  G. Getov, P. Simidtchieva, M. Nikiforova, and R. Andreytchin
  Phys. Stat. Sol., 21:K 87 (1967)

Arsenic triselenide as photoconductor
  Hitachi, Ltd.
  French Patent 1,480,464, May 12, 1967

Drift mobility of carriers in glassy arsenic selenide
  B. T. Kolomiets and E. A. Ledbedev
  Fiz. Tekhn. Poluprovod., 1(2):300–301 (1967)
  Sov. Phys. — Semicond., 1(2):244–245 (1967)

Fluctuation levels in the glassy semiconductor $Ti_2Se \cdot As_2Te_3$
  B. T. Kolomiets, T. N. Mamontova, and G. I. Stepanov
  Fiz. Tverd. Tela, 9(1):27–30 (1967)
  Sov. Phys. — Solid State, 9(1):19–21 (1967)

Photoconductivity and phototropy in noncrystalline solids
  J. D. Mackenzie
  Interaction Radiat. Solids, Proc. Conf., Cairo, 1966 (publ. 1967), pp. 133–154
  Review of reversible phenomena

Hall effect measurement in semiconducting chalcogenide glasses and liquids
  J. C. Male
  Brit. J. Appl. Phys., 18:1543 (1967)

Dielectric properties of glass in the system As–S
  Shigeo Maruno
  Japan. J. Appl. Phys., 6:1474–1475 (1967)

GeSeTe — a new infrared-transmitting chalcogenide glass
  J. A. Muir and R. J. Cashman
  J. Opt. Soc. Am., 57:1–3 (1967)

Structural characterizations of vitreous inorganic polymers by thermal studies
  M. B. Myers and E. J. Felty
  Mat. Res. Bull., 2:715 (1967)

Effect of silver impurities on the fundamental absorption edge of glassy $As_2S_3$
  R. E. Andreichin, G. K. Getov, and P. A. Simidchieva
  Fiz. Tverd Tela, 8(6):1951–1952 (1966)
  Sov. Phys. — Solid State, 8(6):1546–1547 (1966)

Electronic conduction in $As_2Se_3$, $As_2Se_2Te$ and similar materials
  J. T. Edmond
  Brit. J. Appl. Phys., 17:979 (1966)

Optical transitions in crystalline and vitreous $As_2S_3$
  G. Getov, B. Kandilarov, P. Simidtchieva, and R. Andreytchin
  Phys. Stat. Sol., 13:K 97 (1966)

Thermal conductivity of arsenic–sulfur binary glasses
  Makoto Hattori, Tsutomu Minami, and Masami Tanaka
  Kogyo Kagaku Zasshi, 69(9):1737–1740 (1966)

Nonoxide chalcogenide glasses as infrared optical materials
  A. R. Hilton
  Appl. Opt., 5:1877–1882 (1966)

Non-oxide IV A–V A–VI A chalcogenide glasses:  I
  A. R. Hilton, C. E. Jones, and M. Brau
  Phys. Chem. Glasses, 7(4):105–112 (1966)

Non-oxide IV A–V A–VI A chalcogenide glasses: II
  A. R. Hilton and C. E. Jones
  Phys. Chem. Glasses, 7(4):112–116 (1966)

Non-oxide IV A–V A–VI A chalcogenide glasses: III
  A. R. Hilton, C. E. Jones, R. D. Dobrott, H. M. Klein, A. M. Bryant, and T. D. George
  Phys. Chem. Glasses, 7(4):116–126 (1966)

Bistable semiconductive glass composition
  ITT Industries, Inc.
  British Patent 1,141,229 (Cl. C 22c), Jan. 29, 1969, U. S. Pat. Appl. Dec. 20, 1966, 5 pp.
  As, Te, Ge

Semiconductive glass composition
  ITT Industries, Inc.
  British Patent 1,145,639, March 19, 1969; U. S. Pat. Appl. Dec. 21, 1966, 6 pp.
  As–Te–Se

Solid State Chemistry
  R. L. Myuller and others (Z. U. Borisova, ed.)
  Consultants Bureau Special Report, Plenum Press, New York (1966), 256 pp.

[Title Not Given]
  S. R. Ovshinsky
  No. 3,721,591 (Sept. 6, 1966)
  U. S. Patent No. 3,721,591 (Sept. 6, 1966)

The electrical properties of some chalcogenide glasses
  A. E. Owen, N. Clare, and S. Frank
  Conf. on electronic processes in low-mobility solids, Sheffield, 1966 (University of Sheffield, Dept. of Glass Technology, 1966), pp. 109–113

The infra-red transmission of telluride glasses
  J. A. Savage and S. Nielsen
  Phys. Chem. Glasses, 7:56–59 (1966)

Thermal expansion and its related properties of arsenic-sulfur glasses
Masami Tanaka, Tsutomu Minami, and Makoto Hattori
Japan. J. Appl. Phys., 5:185-186 (1966)

Infrared absorption spectra of arsenic-sulfur glasses
Shoji Tsuchihashi, Teruo Yano, Toshinori Komatsu, and Keiichiro Adachi
Yogyo Kyokai Shi, 74(855):353-361 (1966)

Compounds and crystal chemistry in polycomponent systems containing silver and tellurium— a review
C. R. Veale
J. Less-Common Metals, 11:50-63 (1966)
2- to 5-component systems

Infrared absorption of arsenic monoselenide
E. L. Zorina
Opt. i Spektr., 20(2):293-296 (1966)

Conductivity and some photoelectric properties of glass-like arsenic trisulfide (high resistivity dielectric)
R. Andreichin, P. Simidchieva, and M. Nikiforova
Compt. Rend. Acad. Bulgare Sci., 18(11):995-998 (1965)
Chem. Abstr. 64-11988

Dielectric loss and permittivity of the glassy semiconductor $As_2S_3$ containing silver as impurity
L. Bonchev and R. Andreichin
Compt. Rend. Bulgare Sci., 18(9):805-807 (1965)

Effect of some elements on the electrical conductivity and microhardness of vitreous arsenic selenide
Z. U. Borisova
Izv. Akad. Nauk SSSR, Fiz., 28:1293-1294 (1964)
Bull. Acad. Sci. USSR, Phys. Ser. (USA), 28(8):1195-1196 (Aug. 1964; publ. 1965)

Impurity photoconductivity in single-crystal and glassy arsenic selenide
B. T. Kolomiets and G. I. Stepanov
Fiz. Tverd. Tela, 7(9):2698-2700 (1965)
Sov. Phys. — Solid State, 7(9):2181-2183 (1966)

Chalcogenide glasses transmitting in the infrared between 1 and 20 $\mu$— a state of the art review
J. A. Savage and S. Nielsen
Infrared Phys., 5:195-204 (1965)

The effect of oxygen on the infra-red transmission of Ge-As-Se glasses
J. A. Savage and S. Nielsen
Phys. Chem. Glasses, 6:90 (1965)

Relationship of composition and density of arsenic-sulfur glasses
Masami Tanaka and Tsutomu Minami
Japan. J. Appl. Phys., 4:939 (1965)

Infrared absorption band near 12.51 $\mu$ of arsenic-sulfur glasses
Masami Tanaka and Tsutomu Minami
Japan. J. Appl. Phys., 4:1023-1024 (1965)

Direct formation in sheet form of arsenic-sulfur glasses by distillation under normal pressure
Masami Tanaka, Osamu Mukai, and Osamu Kamike
J. Ceram. Assoc. Japan, 73(3-1):51-55 (1965)

Formation and dielectric properties of glass in the system As-Te
Sachio Tsugane, Miyoshi Haradome, and Ryuichi Hioki
Japan. J. Appl. Phys., 4(2):77 (1965)

The electrical conductivity of vitreous compounds of general formula $AsSe_x B_y$ and $AsGeSe_x B_y$
L. I. Doinikov and Z. V. Borisova
Zh. Prikl. Khim., 37(7):1458-1462 (1964)
J. Appl. Chem. (USSR), 37(7):1453-1456 (1964)

Electrical conduction anomaly of semiconducting glasses in the system As-Te-I
D. L. Eaton
J. Am. Ceram. Soc., 47:554 (1964)

Vitreous semiconductors
N. A. Goryunova and B. T. Kolomiets
Voprosy Met. i Fiz. Poluprovod. Akad. Nauk, SSSR, Sbornik (1957), pp. 110-120

The phase diagram for the binary system indium — tellurium and electrical properties of $In_3Te_5$
E. G. Grochowski, D. R. Mason, G. A. Schmitt, and P. H. Smith

The Hall effect — Seebeck effect sign anomaly in semiconducting glasses
A. David Pearson
J. Electrochem. Soc., 111:753 (1964)

Preparation and properties of sulfide, selenide, and telluride glasses
A. David Pearson
The Glass Industry (Dec. 1964), pp. 666-669

The Hall effect in semiconducting glasses
W. F. Peck, Jr., and J. F. Dewald
J. Electrochem. Soc., 111:561 (1964)

Electrical conduction in rare-earth monoselenides and monotellurides and their alloys
F. J. Reid, L. K. Matson, J. F. Miller, and R. C. Himes
J. Phys. Chem. Solids, 25:969-976 (1964)

The Faraday rotation of diamagnetic glasses from 0.334 $\mu$ to 1.9 $\mu$
C. C. Robinson
Appl. Opt., 3:1163 (1964)
As-Se

Preparation of glasses transmitting in the infrared between 8 and 15 $\mu$
J. A. Savage and S. Nielsen
Phys. Chem. Glasses, 5:82 (1964)

Semiconducting materials. I. Semiconducting and associated properties of progressively crystallized infrared-transmitting glasses
Lyle H. Slack
AD-609107, 1964, 133 pp.
$As_2SeTe_2$

Infrared absorption spectra of arsenic-sulfur glasses prepared by evaporation under normal pressure
Masami Tanaka and Tsutomu Minami
J. Ceram. Assoc. Japan, 72(10):176 (1964)

Direct deposition in film form of arsenic-sulfur glasses by evaporation under normal pressure
Masami Tanaka, Shigeru Takimoto, and Osamu Kamike
J. Ceram. Assoc. Japan, 72(9):164 (1964)

The kinetics of dissolving $AsSe_{1.5}Ge_x$, $As_{1.5}Ge_x$, $AsS_{2.5}Ge_x$ glasses in solution of caustic soda
  V. N. Timofeyeva, G. M. Orlova, G. I. Ternovaya, and
  G. P. Tsayun
  Herals of Leningrad Univ.: Phys. and Chem. Ser. (Sept. 29,
    1964), pp. 148-157

Properties of glasses transmitting in the 8- to 14-micron region
  C. J. Billian
  Servo Corp. of America, Hicksville, N. Y., AD-297876 (1963)
  Compilations on optical, chemical, electrical, and physical
    properties are presented for 20 glasses

Current-voltage characteristics of point contact of glass-type semiconductors
  B. T. Kolomiets and E. A. Lebedev
  Radiotekhnika i electronika, 8:2097 (1963)
  Switching behavior in $TlAs(Se,Te)_2$

Space-charge-limited currents in amorphous arsenic trisulfide
  C. Bowlt
  Proc. Phys. Soc., 80:810 (1962)

Semiconducting glasses
  J. F. Dewald, A. D. Pearson, W. R. Northover, and W. F.
    Peck, Jr.
  J. Electrochem. Soc., 109:243C (Sept. 1962)
  As-Te-I; switching and memory

Chemical, physical, and electrical properties of some unusual inorganic glasses
  A. D. Pearson, W. R. Northover, J. F. Dewald, and W. F.
    Peck, Jr.
  Advances in Glass Technology, Plenum Press, New York
    (1962), pp. 357-365
  As-Te-I; switching and memory

Low-melting sulfide-halogen inorganic glasses
  S. S. Flaschen, A. D. Pearson, and W. R. Northover
  J. Appl. Phys., 31(1):219-220 (1960)

The role of impurity in the conductivity of vitreous $As_2SeTe_2$
  B. T. Kolomiets and T. F. Nazarova
  Fiz. Tverd. Tela, 2(1):174-175 (1960)
  Sov. Phys. — Solid State, 2(1):159-160 (1960)

Hall effect in vitreous materials of the $Tl_2Se \cdot As_2(Se,Te)_3$ system. II
  B. T. Kolomiets and T. F. Nazarova
  Fiz. Tverd. Tela, 2(3):395-396 (1960)
  Sov. Phys. — Solid State, 2(3):369-370 (1960)

Electron diffraction study of amorphous antimony sulfide
  L. I. Tatarinova
  Kristallografiya, 2(2):260-267 (1957)
  Sov. Phys. — Cryst., 2(2):251-258 (1958)

Electrical and optical properties of some $M_2^{V-B}N_3^{VI-B}$ semiconductors
  J. Black, E. M. Conwell, L. Seigle, and C. W. Spencer
  J. Phys. Chem. Solids, 2:240-251 (1957)

Preparation and some physical properties of $Bi_2Te_3$, $Sb_2Te_3$, and $As_2Te_3$
  T. C. Harman, B. Paris, S. E. Miller, and H. L. Goering
  J. Phys. Chem. Solids, 2:181-190 (1957)

Certain optical constants of oxide, sulfide, selenide, and telluride glasses
  Aniuta Winter
  Compt. Rend., 242(26):3057-3059 (1956)

X-ray crystallographic data on $As_2Te_3$
  Joseph Singer and Chester W. Spencer
  J. Metals, 7:144 (1955)

## 17.e. Germanium and Silicon

Electronic structure and optical spectra of amorphous semiconductors
  J. C. Phillips
  Phys. Stat. Sol., 44b:K1-4 (1971)

Structural, optical, and electrical properties of amorphous silicon films
  M. H. Brodsky, R. S. Title, K. Weiser, and G. D. Pettit
  Phys. Rev., B 1(6):2632-2641 (1970)

Electron paramagnetic resonance of ion-implanted donors in silicon
  K. L. Brower and J. A. Borders
  Appl. Phys. Letters, 16(4):169 (Feb. 15, 1970)

Détermination de la profondeur d'extraction des photoélectrons dans les couches minces de germanium
  G. Chabrier, J. Cornaz, J. P. Goudonnet, and P. Vernier
  Opt. Commun., 1:391-393 (1970)
  Extraction depth of photoelectrons in thin films of amorphous Ge

Properties of glow-discharge deposited amorphous germanium and silicon
  R. C. Chittick
  J. Non-Cryst. Solids, 3:255-270 (1970)

Structural, electrical, and optical properties of amorphous germanium films
  K. L. Chopra and S. K. Bahl
  Phys. Rev., B 1:2545-2556 (1970)

Piezoresistance in amorphous and polycrystalline germanium
  A. Devenyi, A. Belu, and G. Korony
  J. Non-Cryst. Solids (Netherlands), 4:380-390 (1970)

Electronic structure of amorphous of crystalline germanium: Photoemission and optical studies
  Terence M. Donovan
  Ph. D. thesis, Stanford University (1970), 153 pp.
  Available from University Microfilms, Inc., Ann Arbor,
    Mich., Order No. 71-12, 887

A high-density form of amorphous Ge
  T. M. Donovan, E. J. Ashley, and W. E. Spicer
  Phys. Letters, 32 A:85-86 (1970)

Optical properties of amorphous germanium films
  T. M. Donovan, W. E. Spicer, J. M. Bennett, and E. J. Ashley
  Phys. Rev., B2:397-413 (1970)

Neutron diffraction investigation of vitreous germania
  G. A. Ferguson and Marvin Hass
  J. Am. Ceram. Soc., 53:109-111 (1970)

Switching and temperature effects in lateral films of amorphous silicon
John E. Fulenwider and G. J. Herskowitz
Phys. Rev. Letters, 25:292–296 (1970)

Evaluation of the heat of crystallization of amorphous germanium
R. Grigorovici and R. Manaila
Nature, 226:143–144 (1970)

Conductibilité thermique de l'oxyde de germanium vitreux à basses températures
Kurt Guckelsberger and Jean–Claude Lasjaunias
Compt. Rend., Ser. B, 270:1427–1429 (1970)

Surface conductance of amorphous Ge induced by adsorbed gases
Marc Kastner and H. Fritzsche
Bull. Am. Phys. Soc., 15:244 (1970)

Electrical switching in amorphous layers of pyroactivated quartz containing impurities
V. F. Korzo
Fiz. Tverd. Tela, 11(7):1758–1762 (1969)
Sov. Phys. — Solid State, 11(7):1425–1428 (1970)

Electronic transport in amorphous silicon films
P. G. Le Comber and W. E. Spear
Phys. Rev. Letters, 31:509–511 (1970)

Energy gaps in amorphous covalent semiconductors
T. C. McGill and J. Klima
J. Phys. C, Solid State Phys., 3:L163–164 (1970)

The formation of amorphous Si by ion bombardment as a function of ion, temperature dose
F. F. Morehead, B. L. Crowder, and R. S. Title
Bull. Am. Phys. Soc., 15:396 (1970)

Electrical conduction in amorphous boron and silicon
K. Moorjani and C. Feldman
J. Non–Cryst. Solids (Netherlands), 4:248–255 (1970)

Electron tunneling into amorphous germanium
J. W. Osmun and H. Fritzsche
Appl. Phys. Letters, 16:87–89 (1970)

Photoemission from amorphous silicon
C. W. Peterson, J. H. Dinan, and T. E. Fischer
Phys. Rev. Letters, 25:861–864 (1970)

Structural model for amorphous silicon and germanium (technical report)
D. E. Polk
(Division of Eng. and Appl. Phys., Harvard Univ., Cambridge, Mass.), TR-29; AD-707 772 (May 1970), 24 pp.

Photoemission and optical studies of amorphous germanium
W. E. Spicer and T. M. Donovan
J. Non–Cryst. Solids, 2:66–80 (1970)

Electronic structure of amorphous Ge
W. E. Spicer and T. M. Donovan
Phys. Rev., 24:595–598 (1970)

Amorphous silicon thin films
R. G. Block
S. M. thesis, Dept. of Metallurgy and Materials Science, Massachusetts Institute of Technology, Sept. 1969

Solid amorphous Ge and As as examples of lattice–like amorphous substances
G. Breitling
J. Vac. Sci. Tech., 6:628–631 (1969)

Effect of thermal history on the properties of amorphous silicon films
M. H. Brodsky, K. Weisner, G. D. Pettit, and R. S. Title
Bull. Am. Phys. Soc., 14:311 (1969)

Electronic spectrum, k conservation, and photoemission in amorphous germanium
D. Brust
Phys. Rev. Letters, 23(21):1232–1234 (1969)

Electronic structure and optical absorption in noncrystalline semiconductors
D. Brust
Phys. Rev., 186:768 (1969)

Specific heat and heat of crystallization of amorphous germanium
H. S. Chen and D. Turnbull
J. Appl. Phys., 40:4214–4215 (1969)

Preparation and properties of amorphous silicon
R. C. Chittick, J. H. Alexander, and H. F. Sterling
J. Electrochem. Soc., 116:77–81 (1969)

Non–ohmic behaviour in amorphous germanium at high electric fields
N. Croitoru and L. Vescan
Inst. de Fizica, Academia Republicii Socialiste Romania, Bucuresti, Thin Solid Films, 3:269–276 (Apr. 1969)

Evidence for a sharp absorption edge in amorphous Ge (density 4.54, no states in gap)
T. M. Donovan, W. E. Spicer, and J. M. Bennett
Phys. Rev. Letters, 22:1058–1061 (1969)

Hole injection in the junction between amorphous Ge layers and n–type Ge single crystals
R. Grigorovici, N. Croitoru, L. Vescan, and M. Marina
Rev. Roumaine Phys., 14(2):199–206 (1969)

Short–range order in amorphous germanium
R. Grigorovici and R. Manaila
J. Non–Cryst. Solids (Netherlands), 1:371–387 (1969)

Density of "amorphous" Ge
T. B. Light
Phys. Rev. Letters, 22:999 (1969)

Interband optical properties of grain boundaries in germanium: an amorphous system
Jerrold L. McNatt and Paul Handler
Phys. Rev., 178:1328–1336 (1969)

Evidence of voids within the as–deposited structure of glassy silicon
S. C. Moss and J. F. Graczyk
Phys. Rev. Letters, 23:1167 (1969)

Effect of deposited metals on the crystallization temperature of amorphous germanium film
F. Oki, Y. Ogawa, and Y. Fujiki
Japan. J. Appl. Phys., 8:1056 (1969)

Optical properties of $GeO_2$ in the ultraviolet region
L. Pajasova
Czech. J. Phys., 19B:1265–1270 (1969)

Electroreflectance of disordered germanium films
H. Piller, B. O. Seraphin, K. Markel, and J. E. Fisher
Phys. Rev. Letters, 23:775 (1969)

Pressure derivatives of the elastic constants of vitreous germania at 25°, −78.5°, and −195.8°C
Naohiro Soga
J. Appl. Phys., 40:3382 (1969)

The fundamental absorption band of amorphous germanium
J. Tauc and A. Abraham
Czech. J. Phys., 19B:1246-1254 (1969)

Synthesis and properties of amorphous films of germanium nitride
L. L. Vasil'eva, T. I. Kovalevskaya, and S. F. Devyatova
Izv. Akad. Nauk SSSR, Neorg. Mater., 5(9):1537-1539 (1969)
Inorg. Mater., 5(9):1304-1305 (1969)

Characterization of amorphous alloy films
B. G. Bagley, H. S. Chen, and D. Turnbull
Mat. Res. Bull., 3:159-168 (1968)

The vibrational spectra of vitreous silica, germania, and beryllium fluoride
R. J. Bell, N. F. Bird, and P. Dean
Proc. Phys. Soc. Solid State Phys., Great Britain, 1(2):299-303 (1968)

Structure of germanium thin films
A. I. Bublik
Kristallografiya, 12(4):730-732 (1967)
Sov. Phys. — Cryst., 12(4):640-642 (1968)

Changes in the density of states of germanium on disordering as observed by photoemission
T. M. Donovan and W. E. Spicer
Phys. Rev. Letters, 21:1572-1575 (1968)

Electron-beam crystallization of silicon, germanium, and cadmium sulfide
John C. Evans, Jr.
NASA-TN-D-4522, April 1968
Amorphous films

Heterojunctions between amorphous Si and Si single crystals
R. Grigorovici, N. Croitoru, M. Marina, and L. Nastase
Rev. Roum. Phys., 13(4):317-325 (1968)
Surface preparation; ohmic contacts to amorphous and single-crystal Si

Optical constants of amorphous silicon films near the main absorption edge
R. Grigorovici and A. Vancu
Thin Solid Films, 2:105-110 (1968)

Electrical conduction through thin amorphous SiC films
T. E. Hartman, J. C. Blair, and C. A. Mead
Thin Solid Films, 2:79-93 (1968)

New interpretation of the electronic structure and optical spectrum of amorphous germanium
Frank Herman and John P. Van Dyke
Phys. Rev. Letters, 21:1575 (1968)

Infrared absorption of glassy silicon dioxide
M. Miller
Czech. J. Phys., 18(3):354-362 (1968)

Preparation and properties of noncrystalline silicon carbide films
C. J. Mogab and W. D. Kingery
J. Appl. Phys., 39:3640-3645 (1968)

The structure of amorphous germanium, obtained by argon ion bombardment of crystalline form
P. V. Pavlov and D. I. Tetel'baum
Dokl. Akad. Nauk SSSR, 175 (4):823-825 (1967)
Sov. Phys.—Dokl., 12(8):742-748 (1968)

Annealing of ion implantation damage in single crystal silicon
G. H. Schwuttke
IBM, AIME Conf. on preparation and properties of electronic materials; optical and nuclear radiation, Chicago (Aug. 12-14, 1968)
Session on ion implantation

Infrared spectra of vitreous germanium dioxide
Gouq-Jen Su and Benjamin Teh-Kung Chen
TR-a, AD-671194 (May 15, 1968) Contract Nonr-668(19)

Optical properties and electronic structure of amorphous Ge and Si
J. Tauc
Mat. Res. Bull., 3:37-46 (1968)

Electrical conduction in amorphous silicon and germanium
P. A. Walley
Thin Solid Films, 2:327-336 (1968)

Electrical and optical properties of amorphous germanium
A. H. Clark
Phys. Rev., 154:750 (1967)

The structure of amorphous silicon films
M. V. Coleman and D. J. D. Thomas
Phys. Stat. Sol., 24:K111 (1967)

Thermoelectric power in amorphous silicon
R. Grigorovici, N. Croitoru, and A. Devenyi
Phys. Stat. Sol., 23:621 (1967)

Thermoelectric power of amorphous germanium layers
R. Grigorovici, N. Croitoru, and A. Devenyi
Proceedings of the Colloquium on Thin Films, Budapest (April 20-23, 1965), pp. 213-218 (E. Hahn, P. B. Barna, and J. Peisner, eds.) (published 1967)

Hole injection in junctions between amorphous Ge layers and n-type Ge single crystals
R. Grigorovici, N. Croitoru, L. Vescan, and M. Marina
Phys. Stat. Sol., 24:K17 (1967)

Amorphous semiconductors (including Se, Ge, Si, and B)
Melpar, Inc., Falls Church, Va., Final rept., June 10, 1964-August 2, 1967, NASA-CR-89982 (August 1967), Contract NASw-934, 31 pp.

Amorphization of polycrystalline germanium films on irradiation with argon ions
P. V. Bavlev, D. I. Tetel'baum, E. I. Zorin, and R. V. Kudryavtseva
Fiz. Tverd. Tela, 12(1):155-157 (1967)
Sov. Phys. — Solid State, 12(1):134-136 (1967)

Electronic properties of amorphous materials
Jan Tauc
Science, 158:1543-1548 (1967)

Textural and electrical properties of vacuum-deposited germanium films
J. D. Williams and L. E. Terry
J. Electrochem. Soc., 114:158 (1967)
Film structure a function of substrate temperature and deposition rate

Properties and structure of thin silicon films sputtered on fused quartz substrates
H. Y. Kumagai, J. M. Thompson, and G. Krauss
Trans. AIME, 236:295 (1966)

Optical properties and electronic structure of amorphous germanium
J. Tauc, R. Grigorovici, and A. Vancu
Phys. Stat. Sol., 15:627 (1966)

Electrical properties of evaporated Si + Ge layers
R. Grigorovici, N. Croitoru, A. Devenyi, L. Vescan, and P. Barna
Rev. Roum. Phys., 10:649 (1965)

[Title Not Given]
R. Grigorovici, N. Croitoru, A. Devenyi, and E. Teleman
Proc. Intern. Conf. on Semicond. Phys., Paris 1964; Dunod, Paris (1965), p. 423

Chemical vapour deposition promoted by r.f. discharge
H. F. Sterling and R. C. G. Swann
Solid-State Electron., 8:653-654 (1965)

Electrical conductivity of evaporated layers of amorphous germanium
L. Reimer
Z. Naturforschung, 13A:536-542 (1958)

The structure of amorphous germanium and silicon
H. Richter and G. Breitling
ATS-37T91G, 1966, Z. Naturforschung, 13A:988-996 (1958)

## 17.f. Selenium and Tellurium

Constant difference in potentials in a layer of amorphous selenium adjacent to plasma
A. D. Andreev and L. V. Kuz'menko
Vestsi Akad. Navuk Belarusk. SSR, Ser. Fiz.-Mat. Navuk, No. 1, 129-132 (1970)

Piégeage des porteurs minoritaires dans les couches minces de selenium amorphe
D. Carles, C. Vautier, and A. Colombani
Thin Solid Films, 5:113-121 (1970)

Optical absorption, transport, and photoconductivity in amorphous selenium
E. A. Davis
J. Non-Cryst. Solids, 4(1):107-116 (1970)

Electronic transport properties of some low-mobility solids under high pressure
F. K. Dolezalek and W. E. Spear
J. Non-Cryst. Solids, 4(1):97-106 (1970)

Transformations of red amorphous and monoclinic selenium
H. Gobrecht, G. Willers, and D. Wobig
J. Phys. Chem. Solids, 31:2145-2148 (1970)

Photoemission from amorphous selenium
A. G. Leiga
J. Appl. Phys., 41:3227-3229 (1970)

Steady-state and transient photoemission into amorphous insulators
J. Mort and A. I. Lakatos
J. Non-Cryst. Solids, 4(1):117-131 (1970)

Current-voltage characteristics at high fields in amorphous selenium thin layers
L. Mueller and M. Mueller
J. Non-Cryst. Solids, 4(1):504-509 (1970)

Influence of sulfur on the behavior of the spherulitic crystallization in amorphous selenium
I. A. Paribok-Aleksandrovich
Fiz. Tverd. Tela, 11(7):2017-2018 (1969)
Sov. Phys. — Solid State, 11(7):1629-1630 (1970)

Photocrystallization of amorphous selenium
I. A. Paribok-Aleksandrovich
Fiz. Tverd. Tela, 11(7):2019-2020 (1969)
Sov. Phys. — Solid State, 11(7):1631-1632 (1970)

Piezo-optic properties of amorphous selenium at a wavelength of 1.15
W. C. Schneider and K. Vedam
J. Opt. Soc. Am., 60:800-804 (1970)

Capacitance characteristics of a selenium thin-film switching element
Z. V. Shapochanskaya, S. I. Konyaev, and Kh. I. Klyaus
Fiz. Tekhn. Poluprovod., 4(5):822-824 (1970)
Sov. Phys. — Semicond., 4(5):697-698 (1970)

The density of alpha-monoclinic (and amorphous) selenium
J. D. Taynai and M. A. Nicolet
J. Phys. Chem. Solids, 31:1651-1653 (1970)

Optical constants of amorphous selenium from x-rays to the far infrared
A. Vasko
J. Non-Cryst. Solids, 3:225-233 (1970)

Structure of amorphous arsenic and selenium according to the intensity curves
G. Breitling and H. Richter
Mat. Res. Bull., 4:19-32 (1969)

Thermally stimulated conductivity of glassy selenium films
Yu. A. Cherkasov and I. Yu. Yurkan
Fiz. Tekhn. Poluprovod., 2(7):1008-1010 (1968)
Sov. Phys. — Semicond., 2(7):835-837 (1969)

Transformationspunkt des glasigen Selens
G. Gattow and B. Buss
Naturwissenschaften, 56:35-36 (1969)

Electron Hall mobility in vitreous high-polymeric selenium
G. Juska, A. Matulionis, A. Skalas, and J. Viscakas
Phys. Stat. Sol., 36(2):K121-123 (1969)

Electron conductivity in thin films of amorphous selenium
Hiraku Kitagawa, Kitami Kogyo, and Tanki Daigaku
Kenkyu Hokoku, 2:457-463 (1969)

Relaxation time of anomalous photoconductivity in amorphous selenium treated with mercury vapor
M. I. Korsunskii, A. D. Volchek, K. S. Garger, and V. V. Klimenko
Izv. Akad. Nauk Kaz. SSR, Ser. Fiz.-Mat., 7(2):50-53 (1969)

The conduction of amorphous selenium in the glassy, high elastic, and viscous fluid states
G. G. Mamedalieva, S. I. Mekhtieva, D. Sh. Abdinov, and G. M. Aliev
Dokl. Akad. Nauk SSSR, 184(6):1354-1356 (1969)
Dokl. — Phys. Chem., 184(6):145-146 (1969)

Contributions to the study of optical and photoelectrical properties of hexagonal and amorphous selenium
L. Muller
Stud. Cercetari Fiz. (Rumania), 21(9):1001-1040 (1969)
95 refs.

Influence of an electric field on the refractive index of amorphous selenium
P. I. Perov, L. A. Avdeeva, M. I. Elinson, and G. V. Stepanov
Fiz. Tekhn. Poluprovodnikov, 3:183-187 (1969)
Sov. Phys. — Semiconductors, 3:153 (1969)

Bulk space charge and transient photoconductivity in amorphous selenium
M. E. Scharfe and M. D. Tabak
J. Appl. Phys., 40:3230 (1969)

Multiphonon processes in amorphous selenium
K. J. Siemsen and H. D. Riccius
J. Phys. Chem. Solids, 30:1897-1900 (1969)

Photogeneration effects in amorphous selenium
M. D. Tabak
Appl. Opt., Suppl., 3:4-7 (1969)

Volt-ampere characteristics of samples of amorphous selenium with anomalous conductivity
O. A. Trofimov and N. V. Sominskaya
Inst. Yad. Fiz., Alma-Ata, USSR, Vop. Obshch. Prikl. Fiz., Tr. Respub. Kong. (1st 1967), pp. 8-10 (M. I. Korsunskii, ed.), Izd. "Nauka" Kaz. SSR: Alma-Ata, USSR (1969)

Residual polarization in films of amorphous selenium treated with mercury
O. A. Trofimov and N. V. Sominskaya
Inst. Yad. Fiz., Alma-Ata, USSR, ibid., pp. 10-12

Conductibilité et photoconductibilité de couches de selenium amorphe
Claude Vautier, Daniel Carles, and Antoine Colombani
Thin Solid Films, 3:293-304 (1969)

Electroreflection of amorphous selenium
G. Weiser and J. Stuke
Phys. Stat. Sol., 35:747-753 (1969)

Ellipsometry of amorphous selenium on vacuum-evaporated gold
L. A. Weitzenkamp
Surface Sci., 16:353-364 (1969)

Screw and ring molecules in amorphous and crystalline chalcogens S, Se, Te
J. E. van Aken, J. A. Prins, R. Reijnhart, and F. Tuinstra
Mat. Res. Bull., 3:219-222 (1968)

Negative photoconductivity of amorphous selenium
I. P. Belyaev and B. A. Tazenkov
Uch. Zap., Leningrad. Gos. Pedagog. Inst. im. A. I. Gertsena, 384(1):111-122 (1968)

Photoelectromotive forces in amorphous selenium
I. P. Belyaev and B. A. Tazenkov
Uch. Zap., Leningrad. Gos. Pedagog. Inst. im A. I. Gertsena, 384:123-135 (1968)

Effect of thermal treatment on photoelectric and optical properties of amorphous selenium
V. G. Boitsov, I. P. Belyaev, and V. A. Popov
Gertsenovsk. Chteniya. Mezhvuz. Konf., Fiz. Poluprov Elektron, 21st, Leningrad (1968), pp. 130-133

An infrared spectrophotometric study of vitreous selenium-doped selenium dioxide
R. A. Burley
Phys. Stat. Sol., 29:551 (1968)

Courants limités par la charge d'espace dans les couches minces de selenium amorphe
Daniel Carles, Claude Vautier, and Antoine Colombani
Compt. Rend., 267:1101 (1968)

Selenium and Selenides
D. M. Chizhikov and V. P. Shchastlivyi
Collet's Publishers, Ltd., London and Wellingborough (1968)

Electronic processes in the photo-crystallization of vitreous selenium
J. Dresner and G. B. Stringfellow
J. Phys. Chem. Solids, 29(2):303-311 (1968)

Effect of metallic impurities on the electrical conductivity of selenium in the region of phase transition
Ya. I. Dutchak, V. Ya. Prokhorenko, and I. P. Klyus
Izv. Vyssh. Ucheb. Zavend., Fiz., 11(7):132-134 (1968)

Temperature dependence of the electrical conductivity and thermoelectric power of selenium near phase transitions
Ya. I. Dutchak, V. Ya. Prokhorenko, and I. P. Klyus
Fiz. Tekhn. Poluprovod., 2(5):752-753 (1968)
Sov. Phys. — Semicond., 2(5):625-626 (1968)

Effect of heat treatment and natrium admixtures on the temperature dependence of electrical conductivity of amorphous selenium
F. B. Gadzhiev, Ch. M. Askerov, and G. M. Aliev
Phys. Stat. Sol., 29:K47 (1968)

Faraday-Effekt an amorphem Selen
B. Garben and H. Seliger
Phys. Stat. Sol., 29:K27 (1968)

Thermochemistry of selenium. V. Thermochemical behavior of vitreous selenium
Gerhard Gattow and B. Buss
Z. Anorg. Allgem. Chem., 363:134-139 (1968)

Inelastic neutron scattering on solid and liquid tellurium and selenium
W. Gissler, A. Axmann, and T. Springer
IAEA Neutron Inelastic Scattering, Vol. 1 (1968), pp. 245-252

Quantitative phenomenology of optical absorption of fine Se powder compared with the permittivity of amorphous bulk Se
J. Gonella
Compt. Rend., 266(26):1611-1613 (1968)

Electron and hole drift mobilities in vitreous selenium
H. P. Grunwald and R. M. Blakney
Phys. Rev., 165:1006-1010 (1968)

Glass transition temperature and isothermal volume change of selenium
Shuichi Hamada, Tadao Sato, and Toshiaki Shirai
Bull. Chem. Soc. Japan, 41(1):135-139 (1968)

Optical properties of vacuum-evaporated selenium and tellurium
J. D. Hayes, E. T. Arakawa, and M. W. Williams
J. Appl. Phys., 39:5527 (1968)

Optical properties of vacuum-evaporated films of tellurium and amorphous selenium
J. D. Hayes, Jr., E. T. Arakawa, and M. W. Williams
ORNL-TM-2023 (Jan. 1968), 52 pp.

Atomic arrangement in vitreous selenium
Roy Kaplow, T. A. Rowe, and B. L. Averbach
Phys. Rev., 168:1068-1079 (1968)

Thermal expansion of vitreous selenium from $-190°$ to $+30°C$
R. K. Kirby and B. D. Rothrock
J. Am. Ceram. Soc., 51:535 (1968)

The role of mercury in the process by which anomalous photoconductivity appears in amorphous selenium
M. I. Korsunskii, A. D. Volchek, K. S. Garger, and V. V. Klimenko
Dokl. Akad. Nauk SSSR, 183:71 (1968)

Nature of the nonlinearity of volt-ampere characteristics during the anomalous photoconduction of amorphous selenium
M. I. Korsunskii, A. D. Volchek, K. S. Garger, and V. V. Klimenko
Izv. Akad. Nauk Kaz. SSR, Ser. Fiz.-Mat., 6(4):86-89 (1968)

Optical properties of amorphous selenium in the vacuum ultraviolet
A. G. Leiga
J. Opt. Soc. Am., 58(11):1441-1445 (1968)

Influence of crystallinity and oxygen impurities on the photosensitivity spectrum of selenium
S. I. Mekhtieva, D. Sh. Abdinov, and G. M Aliev
Fiz. Tekhn. Poluprovod., 1(12):1840-1844 (1968)
Sov. Phys. — Semicond., 1(12):1520-1523 (1968)

Photogeneration of carriers in vitreous selenium
D. M. Pai and S. W. Ing, Jr.
Phys. Rev., 173:729-734 (1968)

Epitaxial deposition of selenium
Yoshio Sakai and Hideki Fukuda
Japan. J. Appl. Phys., 7:303-304 (1968)
Effect of growth rate and substrate temperature on film structure

Low-temperature infrared photoconductors, semi-annual report, July 1-Dec. 31, 1967
Melvin L. Schultz
AD-670 009, May 29, 1968, Contract Nonr-2225(00)
Amorphous Se is unsuitable as the insulator layer if the photoconductor is to be a lead chalcogenide

Phonon thermal conductivity of amorphous selenium doped by germanium
L. Stourac, A. Vasko, I. Srb, C. Musil, and F. Strba
Czech. J. Phys., B18:1067-1073 (1968)

Field-controlled photogeneration and free-carrier transport in amorphous selenium films
Mark D. Tabak and Peter J. Warter Jr.
Phys. Rev., 173(3):899-907 (1968)

Photoemission studies on amorphous layers and single crystals of selenium
R. H. Tredgold, R. H. Williams, G. Kavalyauskene, and R. Keezer
Phys. Stat. Sol., 26:K 5 (1968)

Carrier recombination mechanism in amorphous selenium
J. Viscakas, V. Gaidelis, A. Matulionis, E. Montrimas, and A. Satas
Proceedings of the 9th International Conference on the Physics of Semiconductors, Moscow (July 23-29, 1968), Vol. 2, pp. 1290-1294, Publ. House "Nauka," Leningrad (1968)

Space-charge-limited currents and high electric field effect in vitreous selenium films
J. Viscakas, P. Mackus, and A. Smilga
Phys. Stat. Sol., 25:331-335 (1968)

Effect of dysprosium additions on some physical properties of high-purity selenium
G. B. Abdullaev, E. G. Akhundova, G. M. Aliev, and D. Sh. Abdinov
Dokl. Akad. Nauk Azerb. SSR, 23(2):14-16 (1967)

Effect of an electrostatic field on the crystallization of amorphous selenium
G. B. Abdullaev, K. P. Mamedov, A. I. Odobesku, and Z. D. Nurieva
Dokl. Akad. Nauk Azerbaidzh. SSR, 23(9):10-13 (1967)

Trapping processes in amorphous selenium
R. M. Blakney and H. P. Grunwald
Phys. Rev., 159:664-671 (1967)

Surface crystallization of vitreous selenium as induced by chemical vapors
Y. S. Chiang and J. K. Johnson
J. Appl. Phys., 38:1647 (1967)

The phenomenon of selenium vitrification
S. U. Dzhalilov and K. I. Rzaev
Phys. Stat. Sol., 20(1):261-266 (1967)

Investigation of the crystallization of vitreous $SSe_{20}$ and Se
M. El'mosli and Z. U. Borisova
Izv. Akad. Nauk SSSR, Neorg. Mater., 3(6):923-931 (1967)
Inorg. Mater., 3(6):827-834 (1967)

The Raman spectrum of trigonal, alpha-monoclinic and amorphous selenium
Aram Mooradian and George B. Wright
(Massachusetts Inst. of Tech., Lexington) Proceedings of the International Symposium on Physics of Selenium and Tellurium, Montreal, Canada (October 12-13, 1967), pp. 269-276; AD-694- 135

Characteristics of the voltage-current limit of the spatial load in thin layers of amorphous selenium
L. Muller
Studii si Cercetari de Fizica, 19(1):35-40 (1967)

Space-charge-limited currents and high electric field effect on the trapping process in amorphous selenium films

J. Viscakas, P. Mackus, and A. Smilga
Vilnyus. Gos. Univ. im Kapsukasa, Vilnius, USSR. Nauch.
Konf. Molodykh Uch. Litov. SSR, Rab. Obl. Fiz., Mat.
Kibern. (P. Brazdiunas, ed.) (1967), pp. 152-154
Akad. Nauk Kitov, SSR: Vilnius, USSR

Initial stages of the crystallization of selenium
spherulites and the mechanism of their forma-
tion
I. E. Bolotov and E. A. Murav'ev
Fiz. Tverd. Tela, 8(5):1585-1591 (1966)
Sov. Phys. — Solid State, 8(5):1259-1263 (1966)

Influence of impurities on the carrier mobility
in amorphous selenium
B. T. Kolomiets and É. A. Lebedev
Fiz. Tverd. Tela, 8(4):1136-1139 (1966)
Sov. Phys. — Solid State, 8(4):905-908 (1966)

Specific heat of amorphous selenium at low tem-
peratures
K. K. Mamedov, I. G. Kerimov, M. I. Mekhtiev, and M. I.
Veliev
Khimichieskaya Svyaz'v Poluprovodnikakh i Termodinamika
(Chemical bonds in semiconductors and thermodynamics),
1966, Nauka i Tekhnika, Minsk (Sept. 1967), pp. 174-178

On the phase transition in selenium
Z. D. Nurieva, K. P. Mamedov, and Yu. G. Asadov
Acta Cryst., 21, Pt. 7, Suppl., A 203 (Dec. 30, 1966)
Seventh Intern. Congr. Symp. Intern. Union of Crystallog-
raphy, Moscow (1966)

Anomalous shift of the fundamental absorption
edge of films and amorphous samples of selenium
under the influence of an electric field
L. N. Strel'tsov, N. M. Kiseleva, and P. S. Kireev
Fiz. Tverd. Tela, 8(3):788-789 (1966)
Sov. Phys. — Solid State, 8(3):980-981 (1966)

Infrared absorption spectra of amorphous sele-
nium
K. K. Tagiev and M. A. Talibi
Mater. Nauch. Konf. Molodykh Uch. Aspir. Akad. Nauk Azerb.
SSR (Ser. Fiz.-Tekh. Mat. Nauk) (1966), pp. 172-177
(L. Dement'eva, ed.), Izd. Akad. Nauk Azerb. SSR, Baku,
USSR

Cristallisation du Se vitreux
Y. Toma and T. Nakagawa
J. Chem. Soc. Japan, Pure Chem. Sect., 87(5):422-426 (1966)

Elastic constants of selenium in the hexagonal
and glassy phases
K. Vedam, D. L. Miller, and R. Roy
J. Appl. Phys., 37:3432 (1966)

Conductivity anomaly associated with the glass
transition in vitreous selenium
R. Chang
Appl. Phys. Letters, 6:231 (1965)

On the thermal conductivity of selenium
D. Sh. Abdinov and G. M. Aliyev
Izv. Akad. Nauk, AzSSR, Ser. Fiz.-Tekh. i Mat. Nauk, No.
2,109-114 (1964)

Thermal conductivity of selenium at low tempera-
tures
G. K. White, S. B. Woods, and M. T. Elford
Phys. Rev., 112(1):111-113 (1958)

Thermal conductivity of glassy selenium
H. J. Orthmann and K. Ueberreiter
Z. Kolloid, 147:129-131 (1956)

Some aspects of the crystallization and recrys-
tallization of vapor-deposited vitreous selenium
N. E. Brown and F. L. VerSnyder
J. Metals, 7(2):379-381 (Feb. 1955)

## 17.g. Transition Metal Oxides

Optical absorption properties of vanadate glasses
G. W. Anderson and W. D. Compton
J. Chem. Phys., 52:6166-6174 (1970)

Transport processes in amorphous $Cr_2O_3$ films
D. F. Barbe and S. S. Herman
J. Appl. Phys., 41:3116-3120 (1970)

Switching and negative resistance in thin films
of nickel oxide
J. C. Bruyere and B. K. Chakraverty
Appl. Phys. Letters, 16(1) (Jan. 1, 1970)

Electrical conductivity of glasses in the sys-
tems $P_4O_{10}-V_2O_5$ and $P_4O_{10}-WO_3$
R. H. Caley and M. Krishna Murthy
J. Am. Ceram. Soc., 53:254-257 (1970)

Glass formation and properties in the system
tellurium dioxide-vanadium pentoxide-cadmium
oxide
Ya. Dimitriev, M. R. Marinov, I. Ivanova, and M. Popov
Dokl. Bolg. Akad. Nauk, 23(5):507-510 (1970)

Nuclear magnetic resonance study of semicon-
ducting vanadium phosphate glass
P. W. France and H. O. Hooper
J. Phys. Chem. Solids, 31:1307-1315 (1970)

Switching effect in $VO_2$
M. Guntersdorfer
Solid-State Electron., 13(3):369-379 (1970)

Electrode effects and bistable switching of amor-
phous $Nb_2O_5$ diodes
T. W. Hickmott and W. R. Hiatt
Solid-State Electron., 13:1033-1047 (1970)

Magnetic resonance study of the $V_2O_5-P_2O_5$ semi-
conducting glass system
F. R. Landsberger and P. J. Bray
J. Chem. Phys., 53(7):2757-2768 (1970)

Conductivity and dielectric behaviour of molybdenum and vanadium phosphate semiconducting
glasses
M. Sayer, A. Mansingh, J. M. Reyes, and A. M. Smith
Bull. Am. Phys. Soc., 15(6):765 (1970)

Two switching devices utilizing $VO_2$
R. H. Walden
IEEE Trans. Electron Devices ED-7, 603-11 (1970)

Structural characterization and optical transmission studies of vanadate glasses
G. W. Anderson
Dept. of Physics, Univ. Illinois, Urbana, Contract AT(11-1)-1198, COO-1198-605 (Feb. 1969)

Infrared and fundamental absorption edge studies of $V_2O_5$-$P_2O_3$ glasses
Gordon Wood Anderson and W. Dale Compton
Bull. Am. Phys. Soc., 14:427 (1969)

Polarons in crystalline and non-crystalline materials (conduction in glasses containing transition metal ions, pp. 85 - 91)
I. G. Austin and N. F. Mott
Advances in Physics, 18:71 (1969)

Influence of $SiO_2$, $GeO_2$, $B_2O_3$ and $TeO_2$ on the electrical conductivity of glasses in the systems $P_4O_{10}$-$V_2O_5$ and $P_4O_{10}$-$WO_3$
R. H. Caley and M. K. Murthy
Am. Ceram. Soc. Bull., 48:442 (1969)

Electric switching phenomena in transition metal glasses under the influence of high electric fields
C. F. Drake, I. F. Scanlan, and A. Engel
Phys. Stat. Sol., 32(1):193-208 (1969)

Electronic conduction in transition-metal oxide films
Walter J. Frey, Thomas N. Kennedy, and John D. Mackenzie
Am. Ceram. Soc. Bull., 48:425 (1969)

Electroluminescence, bistable switching, and dielectric breakdown of $Nb_2O_5$ diodes
T. W. Hickmott
J. Vacuum Sci. Tech., 6(5):828-833 (1969)

A vanadium oxide film-switching element
T. N. Kennedy and F. M. Collins
AD-683368 (Feb. 1969), 27 pp.

Suppression of the semiconductor-metal transition in vanadium oxides
T. N. Kennedy and J. D. Mackenzie
J. Non-Cryst. Solids (Netherlands), 1:326-330 (1969)
$V_2O_3$ and $V_2O_4$ films

Bistable switching in $Zr$-$ZrO_2$-$Au$ junctions
K. C. Park and S. Basavaiah
Semiconductor Effects in Amorphous Solids (W. Doremus, ed.), North-Holland Publishing Co., Amsterdam (1970), pp. 284-291
Proceedings of symposium held at Holiday Inn, New York, May 14-17, 1969

Semiconductors produced by doping oxide-glasses with Ir, Pd, Rh, or Ru
C. C. Sartain, W. D. Ryden, and A. W. Lawson
Presented at the International Conference on Amorphous and Liquid Semiconductors (September 24-27, 1969), in the Cavendish Laboratory, Cambridge, England, RMIC Preprint

AC conduction in glasses containing transition-metal oxides
Anthony P. Schmid
Bull. Am. Phys. Soc., 14:427 (1969)

Polaronic effects in potassium vanadium phosphate glasses
H. F. Shaake and L. L. Hench

Semiconductor Effects in Amorphous Solids (W. Doremus, ed.), North-Holland Publishing Co., Amsterdam (1970), pp. 292-306
Proceedings of symposium held in New York, May 14-17, 1969

Photoconductivity in disordered nickel-oxide films
R. Tsu, L. Esaki, and R. Ludeke
(IBM Corp., Watson Research Center, Yorktown Heights, N. Y.), AD-704238; AROD-8511-1-P (Aug. 13, 1969), 6 pp.

Oxide glasses in light of the "ideal glass" concept: II. Interpretations by reference to simple ionic glass behavior
C. A. Angell
J. Am. Ceram. Soc., 51(3):125-134 (1968)
The relation between viscosity and electrical conductance

Switching phenomena in titanium oxide thin films
F. Argall
Solid State Electron., 11:535-541 (1968)

High-speed thermal switches based on vanadium dioxide
R. G. Cope and A. W. Penn
J. Phys. D (British J. Appl. Phys.), 1:161-168 (1968)

The amorphous to crystalline transition in barium titanate ($BaTiO_3$)
D. S. Gelles
S. M. thesis, Dept. of Metallurgy and Materials Science, Massachusetts Institute of Technology, January, 1968

Conduction in glasses containing transition metal ions
N. F. Mott
J. Non-Cryst. Solids, 1:1-17 (1968)

Electron transport phenomena in transition metal oxide glasses
H. F. Schaake and L. L. Hench
A center of competence in solid state materials and devices (Univ. of Florida, Gainesville), Contract No. F 19628-68-C-0058, AFCRL-68-0493, Scientific Report No. 2 (October 10, 1968), pp. 2-66

Evidence for the small polaron as the charge carrier in glasses containing transition metal oxides
A. P. Schmid
J. Appl. Phys., 39(7):3140-3149 (June 1968)

Spectral properties of tellurite glasses containing vanadium pentoxide
A. K. Yakhkind, N. V. Ovcharenko, and D. E. Semenov
Opt.-Mekh. Prom., 35(5):34-38 (1968)

Electron paramagnetic resonance study of the mechanism of electrical conductivity in the glasses of the ternary system $V_2O_5$-$P_2O_5$-$WO_3$
L. D. Bogomolova, V. N. Lazukin, and N. V. Petrovykh
Sov. Phys. Dokl., 12(11):1046-1049 (May 1968)
Dokl. Akad. Nauk SSSR, 177(2):310-313 (Nov. 1967)

Investigation of the properties of some ferroelectric crystallized glasses
V. V. Chkalova, V. F. Kalabukhova, V. S. Bondarenko, E. Zh. Freidenfel'd, and Z. P. Milberg
Izv. Akad. Nauk SSSR, Ser. Fiz., 31:1858-1860 (1967)

AC conductivity of a glass semiconductor
L. L. Hench and D. A. Jenkins
Phys. Stat. Sol., 20:327 (1967)
70 mol% $V_2O_5$–30 mol% $P_4O_{10}$

Thin-film switching elements of $VO_2$
K. van Steensel, F. van de Burg, and C. Kooy
Philips Res. Repts., 22:170–177 (1967)

Infrared spectra of $GeO_2$-$P_4O_{10}$-$V_2O_5$ glasses and their relation to structure and electronic conduction
Bh. V. Janakirama-Rao
J. Am. Ceram. Soc., 49:605 (1966)

Temperature-dependent Arrhenius factor for small-polaron conduction in glasses
H. R. Killias
Phys. Letters, 20:5 (1966)
DC conductivity, temperature dependence

Avalanche-induced negative resistance in thin oxide films
K. L. Chopra
J. Appl. Phys., 36(1):184–187 (1965)

Electrical properties of crystalline, glassy and liquid vanadium oxide
R. Hakim, T. Kennedy, and J. D. Mackenzie
Bull. Am. Ceram. Soc., 44(4):303 (1965)

Bistable switching in niobium oxide diodes
W. R. Hiatt and T. W. Hickmott
Appl. Phys. Letters, 6:108 (1965)

New glass compositions possessing electronic conductivities
Donald W. Roe
J. Electrochem. Soc., 112:1005 (1965)
This paper describes the use of oxides of the transition elements, V, Mo, W, Fe, Mn, and Ti, in various combinations with oxides of the alkali and alkaline earth metals and oxides of silicon, phosphorus, boron, aluminum, and zinc to produce glass compositions in which conductivity results from movement of electrons rather than ions

Final rept. on crystal chemistry studies for the period Oct. 1, 1961, to Jan. 31, 1965
Rustum Roy, S. Kachi, O. Muller, and W. B. White
Penn. State Univ., University Park, Penn. (Feb. 20, 1965),
U. S. Army Electronics Laboratories Contract No. DA-36-039-Sc89149, File No. 1067-PM-62-93-93
Glasses-transition metal oxides

Switching properties of thin NiO films
I. F. Gibbons and W. E. Beadle
Solid State Electron., 7:785–797 (1964)

Modern Aspects of the Vitreous State
J. D. Mackenzie
Butterworth, Inc., Washington, D. C. (1964)

Semiconducting oxide glasses
J. D. Mackenzie
NASA-CR-53199, 1964, NASA Grant NsG-100-60
$V_2O_5$-$P_2O_5$-RO

Electrical conduction in glasses containing vanadium pentoxide
H. Nester
Sc. D. thesis, Dept. of Metallurgy, M. I. T. (January 1964)

Preparation of ceramic semiconductors from high-vanadium glass
D. P. Hamblen, R. A. Weidel, and G. E. Blair
J. Am. Ceram. Soc., 46:499–504 (1963)

Low-frequency negative resistance in thin anodic oxide films
T. W. Hickmott
J. Appl. Phys., 33:2669–2682 (1962)

Electric and magnetic properties of $V_2O_3$ and related sesquioxides
A. J. MacMillan
(Massachusetts Inst. of Technology, Lab. for Insulation Research), Tech. Report No. 172; AD-291459 (Oct. 1962), 32 pp.

Preparation and properties of the glassy oxide semiconductors of the $V_2O_5$-$P_2O_5$-$RO_x$ system
L. A. Grechenich, N. V. Petrovykh, and V. P. Karpechenko
Fiz. Tverd. Tela, 2(9):2131–2139 (1960)
Sov. Phys. — Solid State, 2(9):1908–1915 (1961)

Properties of glasses of the system $V_2O_5$-$P_2O_5$
B. Nador
Steklo i Keram., 17:18 (1960)

Electrical properties of certain semiconducting oxide glasses
V. A. Ioffe, I. V. Patrina, and I. V. Poberovskaya
Fiz. Tverd. Tela, 2(4):656–662 (1960)
Sov. Phys. — Solid State, 2(4):609–614 (1960)

Electrical properties of some oxide semiconductor glasses
V. A. Ioffe, I. V. Patrina, and S. V. Poberovskaya
The Structure of Glass, Consultants Bureau, New York (1960), Vol. 2, p. 407

Synthesis and study of some vanadium glasses
V. P. Karpechenko and I. I. Kitaigorodskii
Steklo i Keram., 15:8 (1958)

Electrical conductivity of high vanadium phosphate glass
M. Munakata
Solid State Electron., 1:159 (1960)

The effect of monovalent positive ions on the electrical conductivity of high vanadium glasses
M. Munakata and M. Iwamoto
Yogyo Kyokai Shi, 68:59 (1960)

Vanadium pentoxide glass with a low electrical resistance
J. E. Stanworth, E. P. Denton, and H. Rawson
(British-Thomson-Houston Co. Ltd.)
German Patent No. 1,015,579 (Sept. 12, 1957)

The effect of low-valency vanadium phosphoric acid glasses
M. Munakata, S. Karvamura, J. Ashara, and M. Iwamoto
Yogyo Kyokai Shi, 67:344 (1959)

Semiconducting properties of some vanadate glasses
P. L. Baynton and others
J. Electrochem. Soc., 104:237–240 (1957)
Bibliography

Semiconductivity in glass systems $V_2O_3$-$P_2O_5$ and $V_2O_5$-$P_2O_5$
P. J. Roeder and A. L. Friedberg
COO-1198-220

# 18. Semiconductor Doping by Ion Implantation

## 18.a. General and Reviews

Performance of a sputter ion source and its application for implanted ion profile experiments
G. Brown and M. L. Renton
Nucl. Instr. Methods, 92:477–480 (1971)

Heavy ion-induced characteristic x-rays as a tool in solid state physics
J. A. Cairns
Nucl. Instr. Methods, 92:507–510 (1971)
A technique for elucidating, with high sensitivity, the concentration profile of an ion-implanted element

Effects produced by ion bombardment and implantation into thin films and surfaces
L. E. Collins, P. A. O'Connell, J. G. Perkins, F. R. Pontet, and P. T. Stroud
Nucl. Instr. Methods, 92:455–459 (1971)

Use of low-energy accelerators for ion implantation
P. J. Cracknell, M. Gettings, and K. G. Stephens
Nucl. Instr. Methods, 92:465–469 (1971)

Use of compound semiconductors in a sputtering ion source for ion implantation
R. M. Allen
Nucl. Instr. Methods, 84(2):325–326 (1970)

Ion implantation depth distributions: energy deposition into atomic processes and ion locations
D. K. Brice
Appl. Phys. Letters, 16:103–106 (1970)

Ion implantation target stage for an electromagnetic isotope separator
J. H. Freeman and G. A. Gard
AERE-R-6330 (Mar. 1970), 15 pp.

A simple ion source for implantation doping of semiconductors
P. S. Gwozdz and J. S. Koehler
Rev. Sci. Instr., 41:1677–1678 (1970)

Electron microprobe study of ion implantation
C. Legrand, C. Bahezre, and J. Le Duigou
Compt. Rend., B 271:88–92 (1970)

Method of simultaneous epitaxial growth and ion implantation
Ramzy G. Mankarious
(Hughes Aircraft Co.), U. S. Patent 3,520,741 (July 14, 1970)

Atomic Collision Phenomena in Solids
D. W. Palmer, M. W. Thompson, and P. D. Townsend, eds.
North-Holland Publishing Co., Amsterdam, London; American Elsevier Publishing Co., New York (1970)
Proceedings of an International Conference held at the University of Sussex, Brighton, England, Sept. 7-12, 1969

Colloquium on Ion Implantation
London, June 26, 1970; IEE 24, London (1970)

Investigation of ion-implanted crystals by means of directional effects in charged-particle reaction yields
E. Bogh
Proc. Roy. Soc. London, A 311:35–46 (1969)

Semiconductor doping by "ionic implantation"
R. B. Brocard
Toute Electronique, 36(334):112–114 (1969)
A review of developments discussed at the Solid State Devices Conf., Univ. Manchester, England (Sept. 1968)

Structure effects in low-energy electronic stopping (ions)
I. M. Cheshire, G. Dearnaley, and J. M. Poate
Proc. Roy. Soc. London, 311A:47–51 (1969)

The range and energy loss of implanted ions
G. Dearnaley
Proc. Roy. Soc., A 311:21–33 (1969)

Ion bombardment and implantation
G. Dearnaley
Rept. Prog. Phys., 32:405–491 (1969)
About 270 refs.

Doping solids with ions
Geoff Dearnaley and J. Harry Freeman
New Sci., 41:282–284 (1969)

Doping of semiconductors and semiconducting film. Vol. II
Defense Documentation Center, Alexandria, Va. (Cumulative volume, 271 refs.), AD-853000 (1969)

Channeling effect and its application to ion implantation phenomena
Lennart Eriksson
Thesis, Kungliga Tekniska Hogskolan, Stockholm (1969), 16 pp.

Use of isotope separators for ion implantation
J. H. Freeman
Proc. Roy. Soc. London, 311A:123–130 (1969)

Fabricating solid state devices by ion implantation
Alfred J. Gale
Ion Physics Corp., U. S. 3,434,894 (March 25, 1969)

Ion implantation by means of nuclear reactions
L. Grodzins
Proc. Roy. Soc. London, 311A:79–110 (1969)

Surface ionization source for ion implantation
D. M. Jamba
Rev. Sci. Instr., 40:1072–1074 (1969)

Ion implantation: a new method of doping semiconductors — I.
L. N. Large
Contemp. Phys., 10:277–298 (1969)

Ion implantation: a new method of doping semiconductors — II
L. N. Large
Contemp. Phys., 10:505–531 (1969)

The application of ion implantation to semiconductor devices
L. N. Large and K. G. Hambleton
Festkörper Probleme IX, Advances in Solid State Physics, (O. Madelung, ed.), Pergamon Press, New York (1969), pp. 316–337

Slowing-down of ions (review)
J. Lindhard
Proc. Roy. Soc. London, 311A:11–19 (1969)

Ion implantation in semiconductors
J. W. Mayer and O. J. Marsh
Applied Solid State Science. Advances in Materials and Device Research, Vol. 1 (Raymond Wolfe and C. J. Kriessman, eds.), Academic Press, New York (1969), pp. 239–342

Physical state of ion implanted solids
R. S. Nelson
Proc. Roy. Soc. London, 311A:53–61 (1969)

The equilibrium topography of sputtered amorphous solids
M. J. Nobes, J. S. Colligon, and G. Carter
J. Mater. Sci., 4:730–733 (1969)
Theory for the sputtering of amorphous solid by an ion beam and the changes in surface topography

Channeling in semiconductors and its application to the study of ion implantation
S. T. Picraux
California Institute of Technology, Pasadena, Calif., AD-689187 (April 1969), 123 pp.

Channelling studies in diamond-type lattices (diamond; silicon; germanium; gallium phosphide; gallium arsenide; gallium antimonide)
S. T. Picraux, J. A. Davies, L. Eriksson, et al.
Phys. Rev., 180:873–882 (1969)

Ion implantation. A method for doping semiconductors
A. Richardt
Vide 24(143):272–274 (1969)

A method of bombardment of metallic ions onto semiconductors by applying exploding-wire technique
Ryoji Takahashi, Takao Tuno, Mamoru Oshima, and Akihiko Kobayashi
Japan. J. Appl. Phys., 8:284–285 (1969)

Substitutional doping during ion implantation
W. W. Anderson
Solid State Electronics, 11:481–489 (April 1968)

Technology of ion implantation
D. E. Davies, T. C. Smith, and R. N. Cheever
Solid State Tech., 11:33–39 (1968)

Semiconductor processing by ion implantation
R. Dolan, B. Buchanan, and S. Roosild
AFCRL, Bedford, Mass., in ion implantation session at AIME Conf. on Preparation and Properties of Electronic Materials: Optical and Nuclear Radiation, Chicago (August 12–14, 1968)

Ion implantation in semiconductors — Part I. Range distribution theory and experiments
James F. Gibbons
Proc. IEEE, 56(3):295–315 (1968)

Ion implantation: a review of the process and its applications
J. F. Gibbons
Stanford Electronics Laboratories, Stanford, California, in ion implantation session at AIME Conf. on Preparation and Prop. of Electronic Materials: Optical and Nuclear Radiation, Chicago (Aug. 12–14, 1968)

Possibilities and limitations of ion implantation into semiconductors
Ph. Glotin
CEA-CONF-1256; CONF-681049-1; Tech. Note LETI/EI-490 (Dec. 10, 1968), 22 pp.
From Conference on Special Techniques for Semiconductors as Detectors, Ispra, Italy

An ion source for intense beams of negative heavy ions
G. Hortig, P. Mokler, and M. Mueller
NP-tr-1829
Z. Phys., 210:312–313 (1968)
Trans. C. R. Brightmore, AERE, Harwell, England

Development of ion implantation techniques for microelectronics
R. G. Hunsperger, H. L. Dunlap, and O. J. Marsh
(NASA, Electronics Research Center, Cambridge, Mass.), Contract NAS12-124, NASA-CR-86142 (Oct. 1968), 61 pp.

An ion implantation system which employs a velocity filter for mass separation
J. D. Macdougall, F. W. Anderson, K. E. Manchester, and P. E. Roughan
Electron and Ion Beam Science and Technology (Third International Conference held in Boston, Mass., 1968), Robert A. Bakish, ed.; Electrochemical Society, Inc., New York (1968), pp. 649–655

Method of making a semiconductor by ionic bombardment
  Kenneth E. Manchester
  Sprague Electric Company, U. S. Patent 3,390,019 (June 25, 1968)

Ion-beam interaction with semiconductors
  E. S. Mashkova and V. A. Molchanov
  Can. J. Phys., 46:713-717 (1968)

Ion implantation techniques to fabricate semiconductor nuclear particle detectors
  J. W. Mayer
  Nucl. Instrum. Meth., 63(2):141-151 (1968)

Ion implantation in semiconductors: Lattice disorder and electric effects
  James W. Mayer
  Trans. IEEE Nucl. Sci., NS-15(6):10-21 (1968)
  From Annual Conf. on Nuclear and Space Radiation Effects, Missoula, Mont. See CONF-680706

The observation of atomic collisions in crystalline solids
  R. S. Nelson
  Defects in Crystalline Solids, Vol. 1, North-Holland, Amsterdam; Interscience, Wiley, New York (1968), 282 pp.
  Review by J. W. Mayer in Science, 168:358 (1970)

The application of photolithographic techniques to the doping of semiconductors by ion implantation
  D. N. Osborne
  AERE-R-5631 (June 1968)

Requirements on beam techniques and systems for ion implantation doping of semiconductor materials and devices
  R. G. Wilson, R. R. Hart, D. M. Jamba, and S. A. Thompson
  Electron and Ion Beam Science and Technology (Third International Conference held in Boston, Mass., 1968), Robert A. Bakish, ed.; Electrochemical Society, Inc., New York (1968), pp. 640-648

Range-energy data for keV ions in amorphous materials (computed from the theoretical analysis of Lindhard, Scharff, and Schiott, 1963)
  K. B. Winterbon
  (Atomic Energy of Canada, Ltd., Chalk River, Ontario), AECL-3194, Nov. 1968, 37 pp.

Problem of obtaining a uniform volume concentration of implanted ions
  J. H. Worth
  AERE-R-5704 (March 1968) 37 pp.

A preliminary study of semiconductor structures produced by ion-implantation
  N. G. Blamires, D. N. Osborne, et al.
  July 1967, AERE-R-5519

A survey of channeling theories
  D. K. Brice
  Sandia Labs., Albuquerque, N. Mexico (Aug. 1967) 19 pp.
  SC-DC-67-1981

Application of ion implantation in the design and development of semiconductor devices
  J. F. Gibbons
  Presented at Conference on Application of Ionic Beams to Semiconductor Technology, Grenoble, France (May 1967)
  Editions Ophrys, Paris (1967)

Ion-implantation in semiconductors
  J. F. Gibbons
  Sixth annual report on materials research at Stanford University (July 1, 1966, to June 30, 1967), p. 40
  Report to ARPA (December 1967)

Problems and applications of ion implantation in semiconductors
  J. F. Gibbons and W. J. Kleinfelder
  Presented at Symposium on Test Methods and Measurements of Semiconductor Devices, Budapest, Hungary (April 1967)

Proc. Intern. Conf. Applic. Ion Beams Semicond. Tech.
  P. Glotin, ed.
  Editions Ophrys, Paris (1967)

High-energy ion implantation of materials, final report, December 28, 1964-December 28, 1966
  F. A. Leith, W. J. King, P. Mcnally, E. Davies, and C. M. Kellett
  AFCRL-67-0123, AD-651313 (January 1967) Contract AF 19(628)-4970

Apparatus for production and ion bombardment of thin films
  B. Navinsek and G. Carter
  Intern. J. Electron., 22(5):421-428 (1967)

Ion implantation sources
  R. G. Wilson
  Record IEEE Ninth Annual Symp. on Electron, Ion and Laser Beam Technology, Berkeley, 1967, San Francisco Press, Inc. (1967), pp. 22-23

Statistical range distribution of ions in single and multiple element substrates
  W. S. Johnson and J. F. Gibbons
  Appl. Phys. Letters, 9(9):321-322 (Nov. 1966)

The art of semiconductor doping by ion implantation
  Kenneth E. Manchester
  SCP and Solid State Technology, 9:48 (1966)

Doping of semiconductors, using ion implantation techniques
  G. D. Alton, K. E. Manchester, and C. B. Sibley
  Bull. Am. Phys. Soc., 11(4):530 (1966)

The doping of semiconductors by ion bombardment
  J. F. Gibbons, J. L. Moll, and N. I. Meyer
  Nucl. Instr. Methods, 38:165-168 (1965)
  Si, GaAs, CdS, ZnS

Electromagnetic isotope separators and their applications
  J. Koch and K. O. Nielsen
  Nucl. Instr. and Methods, Vol. 38 (1965)

Ion beams and solid state physics
  J. O. McCaldin
  Nucl. Instr. Methods, 38:153-164 (1965)

Progress in Solid State Chemistry, Vol. 2, Chap. 2
  J. O. McCaldin
  Pergamon Press, Oxford (1965)

Doping of crystal by ion bombardment to produce solid state detectors
T. Alvager and N. J. Hansen
Rev. Sci. Instr., 33:567 (1962)

A method of determining heavy ion ranges by analysis of alpha-line shapes
B. Domeij, I. Bergstrom, J. A. Davies, and J. Uhler
Arkiv Fysik, 24:399 (1963)

Range concepts and heavy ion ranges
J. Lindhard, M. Scharff, and H. Schiott
Mat. Fys. Medd. Dan. Vid. Selsk., 33:1-39 (1963)

Method for measuring impurity distributions in semiconductor crystals
N. I. Meyer and T. Guldbrandsen
Proc. IEEE, 51:1631-1637 (1963)

Solar cells produced by ion implantation doping
W. J. King and J. T. Burrill
Proc. 4th Photovoltaic Specialists Conf., Cleveland, Ohio (1962), Pt. 1-C, Sec. 82

Crystal "doping" by ion bombardment
F. M. Rourke, J. C. Sheffield, and F. A. White
Rev. Sci. Instr., 32:455 (1961)

Met. Soc. AIME Conf. on Prep. and Prop. of Electronic Materials
Aug. 12, 1968, Chicago
Session on ion implantation

Forming semiconductive devices by ionic bombardment
W. Shockley
U. S. Patent 2,787,564

## 18.b. Germanium and Silicon

Electron microscope observation of lattice disorder in ion-implanted silicon
M. Bertolotti, D. Sette, L. Stagni, and G. Vitali
Appl. Phys. Letters, 18:257-259 (1971)

Disorder in implanted semiconductors: Energy dependence and penetration depth
G. Della Mea, A. V. Drigo, P. Mazzoldi, G. Nardelli, and R. Zannoni
Phys. Stat. Sol., 4A:797-804 (1971)
Hg in Si

Carrier concentration profiles of ion-implanted silicon
R. Bader and S. Kalbitzer
Appl. Phys. Letters, 16(1):13 (Jan. 1, 1970)

Hall effect measurements on indium-implanted silicon
P. Bergamini, G. Fabri, and F. Pandarese
Appl. Phys. Letters, 17(2):18 (July 1, 1970)
Random and channeling directions; In and Sn; annealing effects; surface carrier concentration

Ion implantation depth distributions: energy deposition into atomic processes and ion locations
David K. Brice
Appl. Phys. Letters, 16(3):103 (Feb. 1, 1970)

Electron paramagnetic resonance of ion-implanted donors in silicon
K. L. Brower and J. A. Borders
Appl. Phys. Letters, 16:169-172 (1970)
After the anneal at ~900°C, the proportion of implanted Sb that is observed by EPR to be substitutional is significantly larger than for the implanted As

Depth distribution of EPR centers in 400 keV O$^+$ ion-implanted silicon
K. L. Brower, F. L. Vook, and J. A. Borders
Appl. Phys. Letters, 16:108-110 (1970)

Selective x-ray generation by heavy ions. Part I. The use of energetic heavy ions to generate characteristic x-rays from elements in a selective manner. Part II. Measurement of the concentration distribution of ion-implanted antimony in silicon by the use of selective heavy ion x-ray excitation
J. A. Cairns, R. S. Nelson, et al.
(Atomic Energy Research Establishment, Harwell, England), AERE-R-6408 (June 1970)

Effect of O$^+$ and Ne$^+$ implantation on the surface characteristics of thermally oxidized Si
N. J. Chou and B. L. Crowder
J. Appl. Phys., 41:1731-1738 (1970)

A study of diffused layers of arsenic and antimony in silicon using the ion-scattering technique
S. Chou, L. A. Davidson, and J. F. Gibbons
Appl. Phys. Letters, 17:23-26 (1970)

The role of damage in the annealing characteristics of ion-implanted Si
Billy L. Crowder
J. Electrochem. Soc., 117:671-674 (1970)

High-dose implantation of P, As, and Sb in silicon: A comparison of room-temperature implantations followed by a 550°C anneal and implantations conducted at 600°C
B. L. Crowder and J. M. Fairfield
J. Electrochem. Soc., 117:363-367 (1970)

ESR and optical adsorption studies of ion-implanted silicon
B. L. Crowder, R. S. Title, M. H. Brodsky, and G. D. Pettit
Appl. Phys. Letters, 16(5) (March 15, 1970)
Damaged-layer properties equivalent to those of amorphous Si; interferometry

Interaction between boron atoms and defects produced by Hg bombardment in silicon
G. Della Mea, A. V. Drigo, P. Mazzoldi, and G. Nardelli
Appl. Phys. Letters, 16(11):382 (May 1970)
He backscattering; annealing effect; B concentration effect

Investigation of ion implantation in diamond including the use of high-energy ion beam for crystal analysis
L. A. Davidson
Thesis, Center for Materials Research, Stanford Univ., June 1970

The implanted profiles of boron, phosphorus and arsenic in silicon from junction depth measurements
D. Eirug Davies
Solid-State Electron., 13:229-237 (1970)

Infrared absorption study of silicon implanted with high-energy nitrogen ions
Richard J. Dexter

Thesis, Virginia Polytechnic Institute, 1970, 71 pp.
Available from University Microfilms, Inc., Ann Arbor, Mich.,
Order No. 70-9911

Distribution of the concentration of phosphorus
in layered silicon structures prepared by ion im-
plantation
V. M. Evdokimov, A. A. Kukharskii, L. N. Strel'tsov, V. B.
Titov, and I. B. Khaibullin
Fiz. Tekhn. Poluprovod., 4(5):941-944 (1970)
Sov. Phys. — Semicond., 4(5):797-799 (1970)

The lattice location of boron ions implanted into
silicon
G. Fladda, K. Bjorkqvist, L. Eriksson, and D. Sigurd
Appl. Phys. Letters, 16:313-315 (April 15, 1970)
$^{11}$B(p, $\alpha$)$^8$Be reaction; effects of implantation temperature
and annealing

Depth profiles of the lattice disorder resulting
from ion bombardment of silicon single crystals
L. C. Feldman and J. W. Rodgers
J. Appl. Phys., 41(9):3776-3782 (1970)

Enhanced diffusion of ion-implanted Sb in silicon
Kenji Gamo, Atustoshi Doi, Kohzoh Masuda, Susumu Namba,
Shinji Ishihara, and Itsuro Kimura
Japan. J. Appl. Phys., 9:333-334 (1970)

ESR of nitrogen-ion-implanted silicon
K. Gamo, A. Doi, K. Masuda, and S. Namba
J. Japan. Soc. Appl. Phys., 39:Suppl. 78-81 (1970)
Proceedings of the 1st Conf. on Solid State Devices, Tokyo,
1969

Enhanced diffusion of high-temperature ion-im-
planted antimony into silicon
K. Gamo, K. Masuda, S. Namba, S. Ishihara, and I. Kimura
Appl. Phys. Letters, 17(9):391-393 (1970)

The amorphization of silicon crystals on irradia-
tion with fast ions
V. M. Gusev, Yu. V. Martynenko, and K. V. Starinin
Fiz. Tverd. Tela, 14(6):1050-1054 (1969)
Sov. Phys. — Solid State, 14(6):908-912 (1970)

Carrier concentration, mobility, resistivity and
impurity concentration of 400-KeV channeled
phosphorus ions in silicon
A. Johansen, J. S. Olsen, L. Sarholt-Kristensen, and
F. W. Martin
Radiation Effects, 3(1-2):65-72 (1970)

Hall effect measurements on Sb- and Ga- implanted
silicon; anneal behavior and comparison with
other species
N. G. E. Johansson and J. W. Mayer
Solid — State Electron., 13:123-130 (1970)

Technique used in Hall effect analysis of ion-im-
planted Si and Ge
N. G. E. Johansson, J. W. Mayer, and O. J. Marsh
Solid-State Electron., 13(3):317-335 (1970)

Ion Implantation in Semiconductors: Silicon and
Germanium
J. W. Mayer, L. Eriksson, and J. A. Davies
Academic Press, New York (1970)

Optical reflection studies of damage in ion-im-
planted silicon
T. C. McGill, S. L. Kurtin, and G. A. Shifrin
J. Appl. Phys., 41(1):246-251 (1970)

Enhanced diffusion and out-diffusion in ion-im-
planted silicon
O. Meyer and J. W. Mayer
J. Appl. Phys., 41(10):4166-4174 (1970)

Enhanced outdiffusion in ion-implanted silicon
O. Meyer and J. W. Mayer
Phys. Letters, 31 A:387-388 (1970)

Enhanced oxidation on ion-implanted silicon
O. Meyer and J. W. Mayer
Radiation Effects 3(1-2):139-140 (1970)

Analysis of Rb and Cs implantations in silicon by
channeling and Hall effect measurements
O. Meyer and J. W. Mayer
Solid-State Electron., 13:1357-1362 (1970)

Lattice location and dopant behavior of Group II
and VI elements implanted in silicon
O. Meyer, N. G. E. Johansson, S. T. Picraux, and J. W.
Mayer
Solid State Commun., 8:529-531 (1970)

Diffusion enhancement in germanium by light ion
implantation
R. L. Minear
Thesis, Center for Materials Research, Stanford Univ., TR
No. 4725-1; SU-SEL-70-003 (March 1970)

Characterization of junctions produced by medi-
um-energy ion implantation in silicon
Alain Monfret
(Commissariat à l'Energie Atomique, Grenoble, France),
CEA-R-3882 (June 1970), 109 pp.

The formation of amorphous Si by ion bombard-
ment as a function of ion, temperature and dose
F. F. Morehead, B. L. Crowder, and R. S. Title
Bull. Am. Phys. Soc., 15:396 (1970)

Enhanced diffusion in ion implanted silicon
S. Namba, K. Masuda, K. Gamo, A. Doi, S. Ishihara, and
I. Kimura
Radiation Effects, 6:115-120 (1970)
Sb

Radiation-enhanced diffusion of boron in silicon
D. G. Nelson
Thesis, Center for Materials Research, Stanford Univ.,
March 1970

Channeling study of boron-implanted silicon
J. C. North and W. M. Gibson
Appl. Phys. Letters, 16:126-129 (1970)

Concentration profile of boron ions implanted in-
to silicon with energies of 30 and 100 keV
V. M. Pistryak, A. K. Gnap, V. F. Kozlov, R. I. Garber,
A. I. Fedorenko, and Ya. M. Fogel'
Fiz. Tverd. Tela, 12(4):1281-1282 (1970)
Sov. Phys. — Solid State, 12(4):1005-1006 (1970)

Electrical characteristics of ion-implanted boron
layers in silicon
J. M. Shannon, R. Tree, and G. A. Gard
Can. J. Phys., 48:229-235 (1970)

Depth distribution of divacancies in 400-keV O$^+$
ion-implanted Si
H. J. Stein, F. L. Vook, and J. A. Borders
Appl. Phys. Letters, 16:106-108 (1970)

Infrared studies of the crystallinity of ion-implanted Si
H. J. Stein, F. L. Vook, D. K. Brice, J. A. Borders, and
S. T. Picraux
Radiation Effects, 6:19-26 (1970)
$^{11}$B, $^{64}$Zn, and $^{121}$Sb

Diffusion and defect annealing in silicon doped by phosphorus ion implantation
V. V. Titov
Phys. Stat. Sol., 2A:203-209 (1970)

Channeling studies of defects in silicon generated by H, He, B, N and P ion implantations
T. Tsuchimoto and I. V. Mitchell
J. Japan. Soc. Appl. Phys., 39 (Suppl.):82-87 (1970)
Proceedings of the 1st Conf. on Solid State Devices, Tokyo, 1969

Annealing of ion bombardment damage in Ge
E. Zwangobani and R. J. MacDonald
Phys. Letters, A32:308-309 (1970)

Electrical behavior of Group III and V implanted dopants in silicon
R. Baron, G. A. Shifrin, O. J. Marsh, and J. W. Mayer
J. Appl. Phys., 40:3702-3719 (1969)

Distribution of condensed defect structures formed in annealed boron-implanted silicon
R. W. Bicknell
Proc. Roy. Soc. London, 311A:75-78 (1969)

Investigation of ion-implanted crystals by means of directional effects in charged-particle reaction yields (iron; silicon)
E. Bogh
Proc. Roy. Soc. London, 311A:35-46 (1969)

X-ray investigation of lattice deformations in silicon induced through high-energy ion implantation
U. Bonse, M. Hart, and G. H. Schwuttke
Phys. Stat. Sol., 33:361-374 (1969)

Electron paramagnetic resonance of defects in ion-implanted silicon (relation to neutron damage; annealing effects)
K. L. Brower, F. L. Vook, and J. A. Borders
Appl. Phys. Letters, 15(7):208 (1969)

Annealing behavior of p-type layers formed by ion implantation of gallium in silicon
K. Bulthuis and R. Tree
Phys. Letters, 28A:558-560 (1969)

Annealing characteristics of n-type dopants in ion-implanted silicon
B. L. Crowder and F. F. Morehead, Jr.
Appl. Phys. Letters, 14:313-315 (1969)

Electron paramagnetic resonance in ion-implanted silicon
D. F. Daly and K. A. Pickar
Appl. Phys. Letters, 15(8):267 (Oct. 15, 1969)

Post annealing conductance behavior of implanted layers in silicon
D. Eirug Davies
Appl. Phys. Letters, 14(7):227 (April 1, 1969)
B and P; implantation temp. dependence

Range of implanted boron, phosphorus and arsenic in silicon
D. E. Davies
Can. J. Phys., 47:1750-1753 (1969)

Ion-implanted germanium particle detectors
G. Dearnaley, A. G. Hardacre, and B. D. Rogers
Nucl. Instr. Methods, 71:86-92 (1969)

Influence of n-type dopants on the lattice location of implanted p-type dopants in Si and Ge
L. Eriksson, G. Fladda, and K. Bjorkqvist
Appl. Phys. Letters, 14(6):195 (March 15, 1969)

Ion-implantation studies in silicon
L. Eriksson, J. A. Davies, and J. W. Mayer
Science, 163:627-633 (1969)

Implantation and annealing behavior of Group III and V dopants in silicon as studied by the channeling technique
L. Eriksson, J. A. Davies, N. G. E. Johansson, and J. W. Mayer
J. Appl. Phys., 40:842 (1969)

Ion-implantation doping of silicon for shallow junctions
John M. Fairfield and Billy L. Crowder
Trans. AIME, 245:469 (1969)

Evidence of a replacement reaction between ion-implanted substitutional Tl dopants and interstitial Si atoms
G. Fladda, P. Mazzoldi, E. Rimini, D. Sigurd, and L. Eriksson
Radiation Effects (GB), 1(4):249-256 (1969)
Tl in Si

Kinetics of thermal annealing of radiation defects in silicon doped by the ion-implantation method
V. M. Gusev and V. V. Titov
Fiz. Tekhn. Poluprovod., 3:3-10 (1969)
Sov. Phys. — Semicond., 3:1-6 (1969)

Changes of optical reflectivity (1.8 to 2.2 eV) induced by 40 keV antimony ion bombardment of silicon
R. R. Hart and O. J. Marsh
Appl. Phys. Letters, 14(7):225 (April 1, 1969)
Dose and implantation temp. dependence; lattice damage

Lattice disorder produced in Si by 40-keV boron and its effect on electrical behavior
R. R. Hart and O. J. Marsh
Appl. Phys. Letters, 15(7):206 (1969)

Defects in silicon during ion implantation
T. Ikeda, T. Tsuchimoto, and T. Tokuyama
(Central Research Lab., Hitachi, Tokyo, Japan), Record of the 10th Symposium on Electron, Ion and Laser Beam Technology, Gaithersburg, Md., May 21-23, 1969, San Francisco Press Inc., San Francisco, Calif. (1969), pp. 183-188

Crystal damage in ion-implanted silicon
Stephen M. Irving
Semiconductor Silicon (R. R. Haberecht, ed.), Electrochemical Society, Inc., New York (1969), pp. 433-444
From 1st Intern. Symp. on Silicon Materials Science and Technology, New York CONF-690506

Temperature dependence of RHS in aluminum-implanted layer in n-type single-crystal silicon
Tadatsugu Itoh, Taroh Inada, Masao Ishiki, and Kenshi Menabe
Appl. Phys. Letters, 14:255 (1969)

An autoradiographic technique for observing channeled penetration of low-energy ions
Cestmir Jech
Appl. Phys. Letters, 15(8):248 (Oct. 15, 1969)
Kr ions into Si

Electrical characteristics of silicon damage by low implantations
W. S. Johnson
Ph. D. thesis, Stanford University, April 1969

Electron microscopy of the growth of defects in Si under ion bombardment
K. Kleinhenz
Z. Naturforsch. 24A:912-917 (1969)

Fine structure of electron diffraction diagrams of Si after ion bombardment
K. Kleinhenz
Z. Naturforsch., 24A:918-921 (1969)

Ion implantation damage of silicon as observed by optical reflection spectroscopy in the 1 to 6 eV region
Stephen Kurtin, G. A. Shifrin, and T. C. McGill
Appl. Phys. Letters, 14(7):223 (April 1, 1969)
40-keV Ar and Sb; crystalline-to-amorphous transition; reflectivity decrease from 3.5 to 6 eV

The energy dependence of lattice disorder in ion-implanted silicon
D. A. Marsden, G. R. Bellavance, J. A. Davies, M. Martini, and P. Sigmund
Phys. Stat. Sol., 25:269-275 (1969)

Integrated E and dE/dx semiconductor particle detectors made by ion implantation
F. W. Martin
Nucl. Instr. Methods, 72:223-225 (1969)

Electrical and electron microscope observations on antimony-implanted silicon
M. D. Matthews
J. Mater. Sci., 4:997-1002 (1969)

The structure of ion-implanted gold layers in single-crystal silicon
M. D. Matthews and P. F. James
Phil. Mag., 19:1179-1188 (1969)

Radiation-enhanced diffusion of boron in silicon
D. G. Nelson, J. F. Gibbons, and W. S. Johnson
Appl. Phys. Letters, 15(8):246 (Oct. 15, 1969)
Effect of ion energy

Measurements and calculations of critical angles for planar channeling (silicon, germanium, gallium phosphide)
S. T. Picraux and J. U. Andersen
Phys. Rev., 186:267 (1969)

Temperature dependence of lattice disorder created in Si by 40 keV Sb ions
S. T. Picraux, J. E. Westmoreland, J. W. Mayer, R. R. Hart, and O. J. Marsh
Appl. Phys, Letters, 14(1):7 (January 1969)

Annealing studies of damage introduced by high-energy ion implantations of silicon
S. Roosild, R. Dolan, and B. Buchanan
IEEE Symposium on Nuclear and Space Radiation Effects, University Park, Pa. (July 8-11, 1969), 11 pp.
IEEE Trans. Nucl. Sci. NS-16, 33-36 (1969)

Annealing of high-energy ion-implantation damage in single-crystal silicon
G. H. Schwuttke and K. Brack
Trans. AIME, 245:475 (1969)

Some properties of ion-implanted boron in silicon
T. E. Seidel and A. U. MacRae
Trans. Met. Soc., AIME, 245:491 (1969)

Measurements of proton radiation damage and carrier diffusivities in silicon through observations of the drift of electron-beam-induced carriers
T. W. Sigmon
Thesis, Center for Materials Research, Stanford Univ., SU-SEL-69-077 (Dec. 1969)

Direct evidence of divacancy formation in silicon by ion implantation
H. J. Stein, F. L. Vook, and J. A. Borders
Appl. Phys. Letters, 14:328-330 (1969)

Low-energy irradiation of germanium single crystals with $He^+$, $Ne^+$, $Ar^+$, $Kr^+$, and $Xe^+$ inert gas ions
V. D. Tishchenko
Fiz. Tekhn. Poluprovod., 2(12):1825-1830 (1968)
Sov. Phys. — Semicond., 2(12):1518-1521 (1969)

The measurement of electrical activity and Hall mobility of boron and phosphorus ion-implanted layers in silicon
R. F. Webber, R. S. Thorn, and L. N. Large
Intern. J. Electron., 26:163-172 (1969)

Production and annealing of lattice disorder in silicon by 200-KeV boron ions
J. E. Westmoreland and J. W. Mayer
Appl. Phys. Letters, 15(9):308 (Nov. 1, 1969)

Bombardment of silicon monocrystals by positive intermediate-energy cesium ions
Sh. A. Ablyaev, V. M. Mikhaelyan, and V. P. Chirva
Izv. Akad. Nauk Uzb. SSR, Ser. Fiz.-Mat. Nauk, No. 2, 93-94 (1968)

Radiation damage and substitutional chemical impurity effects in single-crystal germanium bombarded with 40-keV $B^+$, $Al^+$, $Ga^+$, $Ge^+$, $P^+$, $As^+$, and $Sb^+$ ions
G. D. Alton and L. O. Love
Can. J. Phys., 46:695-704 (1968)
From Conf. on Atomic Collisions and Penetration Studies with Energetic (keV) Ion Beams, Chalk River, Ont. – CONF-670966

n-p junction obtained in silicon by means of ion implantation
S. Ascoli, F. Fioroni, D. Flori, R. Gislon, C. Papa, B. Rispoli, and U. Spoglia
(Comitato Nazionale per l'Energia Nucleare, Rome, Italy), RT/EL-(68) 11 (1968)

Lattice location of dopant elements implanted into Ge
K. Bjorkqvist, B. Domeij, L. Eriksson, G. Fladda,
A. Fontell, and J. W. Mayer
Appl. Phys. Letters, 13(11):379 (Dec. 1, 1968)

Improved profiles of electrical activity in boron implanted silicon
N. G. Blamires, M. D. Matthews, and R. S. Nelson
Phys. Letters, 28A:178 (1968)

Hyperabrupt junctions in Au-Si Schottky diodes by ion implantation
P. Brook and C. S. Whitehead
Electron. Letters, 4:335-337 (Aug. 9, 1968)

Anomalous penetration of Ga and In implanted in silicon
K. Bulthuis
Phys. Letters, 27A:193-194 (1968)

Nitrogen-ion implantation on p-silicon in the energy range between 20 and 215 keV
F. Cianfrone, U. Fasoli, and P. Mazzoldi
Nuovo Cimento, 57B:534-538 (1968)

Nitrogen donor level in silicon
A. H. Clark, J. D. MacDougall, K. E. Manchester, P. E. Roughan, and F. W. Anderson
Bull. Am. Phys. Soc., 13:376 (1968)

Implantation and detection of low-energy argon ions in silicon single crystals
James Comas and Carmine A. Carosella
J. Electrochem. Soc., 115:974-976 (1968)

Range and distribution of implanted boron in silicon
D. Eirung Davies
Appl. Phys. Letters, 13:243 (1968)

Experimental evidence for interstitial In and Tl in ion-implanted silicon
J. A. Davies, L. Eriksson, and J. W. Mayer
Appl. Phys. Letters, 12:255-256 (1968)

Implantation profiles of $^{32}$P channeled into silicon crystals
G. Dearnaley, J. H. Freeman, G. A. Gard, and M. A. Wilkins
Can. J. Phys., 46:587-595 (1968)

Channeling of medium-mass ions through silicon
F. H. Eisen
Can. J. Phys., 46:561-572 (1968)

Ion implantation doping of silicon for shallow junctions
J. M. Fairfield and B. L. Crowder
IBM, in ion implantation session at AIME conf. on preparation and properties of electronic materials: optical and nuclear radiation, Chicago (August 12-14, 1968)

Junctions (n-p) obtained in silicon by ion implantation
F. Fioroni, D. Flori, R. Gislon, C. Papa, and U. Spoglia
(Comitato Nazionale per l Energia Nucleare, Rome, Italy), RT/EL/68/11, 1968, 22 pp.

Conductivity and carrier density distributions in silicon doped by ion bombardment
I. A. Galaktionova, V. M. Gusev, V. G. Naumenko, and V. V. Titov

Fiz. Tekhn. Poluprovod., 2(6):787-791 (1968)
Sov. Phys.-Semicond., 2(6):656-659 (1968)

Electrical and physical measurements on silicon implanted with channeled and nonchanneled dopant ions
W. M. Gibson, F. W. Martin, R. Stensgaard, F. Palmgren Jensen, N. I. Meyer, G. Galster, A. Johansen, and J. S. Olsen
Can. J. Phys., 46:675-688 (1968)

Influence of temperature on phosphorus ion behavior during silicon bombardment
Phillippe M. Glotin
Can. J. Phys., 46:705-712 (1968)
Conf. on Atomic Collisions and Penetration Studies with Energetic (KeV) Ion Beams, Chalk River, Ont.

Influence of defects on channeling of phosphorus ions in silicon
Ph. Glotin
J. Phys., 29:926-936 (1968)

Doping effects of the 20-KeV P ions bombarding a Si target along the easy channeling (100) direction
Ph. Glotin
J. Phys. (Paris), 29(10):926-936 (1968)

Investigation of shallow p-n junctions in silicon prepared by bombardment with phosphorus
T. M. Golovner, V. V. Zaddé, A. K. Zaitseva, M. M. Koltun, and A. P. Landsman
Fiz. Tekhn. Poluprovod., 2(5):720-726 (1968)
Sov. Phys. – Semicond., 2(5):598-602 (1968)

Electrical properties of p-n junctions prepared by the bombardment of epitaxial silicon films with phosphorus ions
V. M. Gusev, M. I. Guseva, Yu. R. Nosov, G. É. Pines, Yu. S. Fedorovskii, and V. S. Tsyplenkov
Fiz. Tekhn. Poluprovod., 1(11):1728-1729 (1967)
Sov. Phys. – Semicond., 1(11):1432-1433 (1968)

Ion implantation of germanium
E. D. Hinkley and W. Matthews
ESD-TR-67-562, Lincoln Laboratory, Massachusetts Institute of Technology (January 1968), pp. 8-9

Doping of silicon by ion implantation
T. Itoh, T. Inada, and K. Kanekawa
Appl. Phys. Letters, 12:244-246 (1968)

Impurity distribution profiles in ion-implanted silicon
W. J. Kleinfelder, W. S. Johnson, and J. F. Gibbons
Can. J. Phys., 46:597-606 (1968)

Implantation of $^{57}$Fe into Si and Ge
G. L. Latshaw, G. D. Sprouse, P. B. Russell, G. M. Kalvius, and S. S. Hanna
Stanford University, Bull. Am. Phys. Soc., 13:1640 (1968)

Silicon solid-state detector produced by lithium ion implantation
Chul Chu Lee, Youn Kyu Ko, Do Kyong Kim, and Kwany Youl Sun
J. Korean Phys. Soc., 1:90-96 (1968)

Radiotracer studies of ion-implanted profile build-up in silicon substrates
K. E. Manchester
J. Electrochem. Soc., 115:656-660 (1968)

The electrical behavior of implanted bismuth in silicon
O. J. Marsh, R. Baron, G. A. Shifrin, and J. W. Mayer
Appl. Phys. Letters, 13:199 (1968)

Interstitial and substitutional components in ion-implanted silicon
J. W. Mayer, J. A. Davies, and L. Eriksson
Bull. Am. Phys. Soc., 13:376 (1968)

Ion implantation of silicon and germanium at room temperature. Analysis by means of 1.0-MeV helium ion scattering
J. W. Mayer, L. Eriksson, S. T. Picraux, and J. A. Davies
Can. J. Phys., 46:663 (1968)

The application of photolithographic techniques to the doping of semiconductors by ion implantation
D. N. Osborne
AERE-R-5631, June 1968

Semiconductor doping by high-energy 1-2.5 MeV ion implantation
S. Roosild, R. Dolan, and B. Buchanan
J. Electrochem. Soc., 115:307 (1968)

Donor behavior in thallium-implanted silicon
Gordon A. Shifrin, Ogden J. Marsh, and James W. Mayer
Bull. Am. Phys. Soc., 13:376 (1968)

Implantation of boron and phosphorus into silicon substrate and its application to range measurement
Takashi Tsuchimoto
Electron and Ion Beam Science and Technology. Third International Conf., Boston, Mass., 1968 (Robert A. Bakish, ed.), Electrochemical Society, Inc., New York (1968), p. 656

Investigation of the distribution of boron atoms in silicon doped by ion bombardment
E. I. Zorin, P. V. Pavlov, and D. I. Tetel'baum
Sov. Phys.—Solid State, 9:2874-2876 (1968)

Ion implantation of silicon. I. Atom location and lattice disorder by means of 1.0-MeV helium ion scattering
J. A. Davies, J. Denhartog, L. Eriksson, and J. W. Mayer
Can. J. Phys., 45:4053 (1967)

Analysis of Sb-implanted silicon by (p, p) scattering and Hall measurements
L. Eriksson, J. A. Davies, J. Denhartog, J. W. Mayer, O. J. Marsh, and R. Markarious
Appl. Phys. Letters, 10:323 (1967)

Electric effects in the ion bombardment of semiconductors
R. Gislon, B. Rispoli, and U. Spoglia
(Comitato Nazionale per l'Energia Nucleare, Rome, Italy), Activity coord. in the field of electron. and instr., Part 3 (1967), pp. 7-50

Ion implantation of phosphorus in germanium
E. A. Hinkley
Solid State Research Rept. No. 4, Mass. Institute of Technology Lincoln Laboratory, ESD-TR-66-566 (January 1967)

Properties of ion-implanted boron, nitrogen, and phosphorus in single-crystal silicon
W. J. Kleinfelder
SEL-67-015, Tech. Rept. No. K701-1, March 1967, ARPA Contract SD-87
Also AD-665965

Properties of ion-implanted boron, nitrogen, and phosphorus in single-crystal silicon
Walter J. Kleinfelder
Ph.D. thesis, Stanford Univ., Calif. (1967)
University Microfilms, Ann Arbor, Mich., Order No. 67-11043

Experimental distribution of ions implanted into single-crystal silicon
W. J. Kleinfelder, W. S. Johnson, and J. F. Gibbons
IEEE 9th Annual Symp. on Electron, Ion and Laser Beam Technology, Berkeley, Cal., May 1967

A comparison of the hot implantation behavior of several Group III and V elements in Si and Ge
J. W. Mayer, J. A. Davies, and L. Eriksson
Appl. Phys. Letters, 11:365-367 (1967)

Ion implantation of silicon. II. Electrical evaluation using Hall-effect measurements
J. W. Mayer, O. J. Marsh, G. A. Shifrin, and R. Baron
Can. J. Phys., 45:4073 (1967)

Boron ion injection doping of silicon
R. Ruth and F. H. Eisen
Proc. Conf. Applications of Ion Beams to Semiconductor Technology, Grenoble, France, May 24-26, 1967

Temperature dependence of the carrier density and mobility in silicon doped with boron and phosphorus by the ion implantation method
D. I. Tetel'baum
Fiz. Tekhn. Poluprovod., 1(5):712-717 (1967)
Sov. Phys. — Semicond., 1(5):593-597 (1967)

Deep (1-10μ) penetration of ion-implanted donors in silicon
R. W. Bower, R. Baron, J. W. Mayer, and O. J. Marsh
Appl. Phys. Letters, 9:203 (1966)

Implantation profiles for 40-KeV phosphorus ions in silicon single-crystal substrates
J. F. Gibbons, A. H. El-Hoshy, K. E. Manchester, and F. F. L. Vogel
Appl. Phys. Letters, 8:46-48 (1966)

Investigation of certain characteristics of p-n junction photovoltaic energy converters produced by ion bombardment
V. M. Gusev, V. V. Zadde, A. P. Landsman, and V. V. Titov
Fiz. Tverd. Tela, 8(6):1708-1712 (1966)
Sov. Phys.—Solid State, 8(6):1363-1366 (1966)
Si

Penetration of 2- to 11-keV lithium ions into silicon carbide single crystals
V. V. Markarov and N. N. Petrov
Izv. Akad. Nauk SSSR, Ser. Fiz., 30(5):925-926 (1966)

Junction counters produced by irradiation of silicon with dopant ions
F. W. Martin, S. Harrison, and W. J. King
IEEE Trans. Nucl. Sci., NS-13:22-29 (1966)

Formation of channeling patterns of the surface of ion-bombarded Si
  R. S. Nelson, D. J. Mazey, M. D. Matthews, and others
  Phys. Letters, 23:18-19 (1966)

Investigation of the diffusion of boron in silicon from a layer doped by ion bombardment
  P. V. Pavlov, V. A. Uskov, E. I. Zorin, D. I. Tetel'baum, and A. S. Baranova
  Fiz. Tverd. Tela, 8(9):2782-2783 (1966)
  Sov. Phys. — Solid State, 8(9):2221-2222 (1967)

Inversion layers formed on n-type germanium by bombardment with boron and aluminum ions
  P. V. Pavlov, E. I. Zorin, and D. I. Tetel'baum
  Fiz. Tverd. Tela, 8(6):1791-1795 (1966)
  Sov. Phys. — Solid State, 8(6):1425-1428 (1966)

Characteristics of photodiodes produced by bombarding silicon with boron ions
  P. V. Pavlov, E. I. Zorin, and D. I. Tetel'baum
  Fiz. Tverd. Tela, 8(9):750-752 (1966)
  Sov. Phys. — Solid State, 8(9):601-602 (1966)

Ion drift effect in silicon p-n junctions formed by sodium ion bombardment
  M. Waldner and P. E. McQuaid
  Proc. IEEE, 54:1187-1188 (1966)

Formation of $SiO_2$ films by oxygen-ion bombardment
  Masanori Watanabe and Atsutomo Tooi
  Japan. J. Appl. Phys., 5:737-738 (1966)

Doping of silicon by ion implantation
  K. E. Manchester, C. B. Sibley, and G. Alton
  Nucl. Instr. Methods, 38:169 (1965)

Implantation and channeling effects of alkali ion beams in semiconductors
  D. B. Medved, J. Perel, H. L. Daley, and G. P. Rolik
  Nucl. Instr. Methods, 38:175-177 (1965)
  Si

Alkali ion doping of silicon
  J. O. McCaldin and A. E. Widmer
  Proc. IEEE, 52:301-302 (1964)

Junction formation in silicon by positive ion bombardment
  R. R. Ferber
  IEEE Trans. Nucl. Sci., NS-10:15-20 (1963)

Silicon heavily doped by energetic cesium ions
  J. O. McCaldin and A. E. Widmer
  J. Phys. Chem. Solids, 24:1073-1080 (1963)

Doping of crystals by ion bombardment to produce solid state detectors
  T. Alvager and N. J. Hansen
  Rev. Sci. Instr., 33:567 (1962)

Detailed analysis of phosphorus-diffused layers in p-type silicon
  E. Tannenbaum
  Solid-State Electron., 2:123 (1961)

Properties of ionic bombarded silicon
  R. S. Ohl
  Bell Sys. Tech. J., 31:104-122 (1952)

Variations of physical and electrical concentrations of donors during thermal annealing
  Ph. Glotin and J. Grapa
  CONF-660414-5, AEC-CONF-66-119-9 (Second Intern. Cong. on Electron and Ion Beam Science and Technology, New York
  p-type Si

## 18.c. II—VI Compounds

Ion-implantation doping of zinc sulphide thin films
  M. R. Brown, A. F. J. Cox, W. A. Shand, and J. M. Williams
  Solid-State Commun., 9:37-40 (1971)
  With terbium, erbium, and manganese

New optical quenching bands in CdS crystals
  Mamoru Oshima and Ryoji Takahashi
  J. Electrochem. Soc., 118:489-494 (1971)
  A striking effect of bombardment with P, As, Sb, and Bi ion, each with an energy of 5 keV, is a considerable reduction in long wavelength, extrinsic response

Type conversion and lithium-ion-implanted ZnSe
  Y. S. Park and C. H. Chung
  Appl. Phys. Letters, 18:99-102 (1971)

Synthesis and characterization of thin ferroelectric and semiconducting films, final technical report, Oct. 1, 1966—Sept. 30, 1969
  Fred Chernow
  (Photoconductive Semiconductors and Devices Lab., Univ. of Colorado, Boulder), AFML-TR-70-9 (April 1970), 143 pp.
  The major portion of this report is concerned with ion implantation of single-crystal CdS
  Also AD 706 097

High-conductivity copper-ion-implanted CdS with room-temperature diffusion of copper
  P. S. Gwozdz and J. S. Koehler
  Bull. Am. Phys. Soc., 15:397 (1970)

Photoelectronic properties of ion-implanted CdS
  S. L. Hou and J. A. Marley, Jr.
  Appl. Phys. Letters, 16(11):467 (1970)
  Annealing effects; $P^+$-doped; electroluminescent

Some electrical and optical properties of cadmium sulfide containing phosphorus and arsenic
  B. Tell
  Bull. Am. Phys. Soc., 15:397 (1970)

Ion implantation of sodium, lithium, and neon in cadmium sulfide
  B. Tell, W. M. Gibson, and J. W. Rogers
  Applied Phys. Letters, 17(8):315-318 (1970)

Type conversion and p-n junction formation in ion-implanted ZnTe
  S. L. Hou, K. Beck, and J. A. Marley, Jr.
  Appl. Phys. Letters, 14:151 (1969)

Photoluminescence of oxygen in ZnTe introduced by ion implantation
  J. L. Merz and L. C. Feldman
  Appl. Phys. Letters, 15:129 (1969)

Ion-implanted junctions in II-VI compounds
  E. R. Pollard and J. L. Hartke
  Bull. Am. Phys. Soc., 14:115 (1969)
  CdS, ZnSe, ZnTe

# III—V Compounds

Selective doping of piezoelectric crystals by ion implantation
G. A. Shifrin, K. R. Zanio, D. M. Jamba, W. R. Jones, O. J. Marsh, and R. G. Wilson
(Hughes Res. Labs., Malibu, Calif., 90265), Semiannual report; 1 Jan. 1969 through 30 June 1969, Contract N00014-69-C-0171 (August 1969), 48 pp.
CdS and ZnO, and dopant ions of H, B, F, Al, Cl, and Ga

Phosphorus-ion-implanted CdS
W. W. Anderson and J. T. Mitchell
Appl. Phys. Letters, 12:334–336 (1968)

High-conductivity p-type CdS
Fred Chernow, Graeme Eldridge, Guy Ruse, and Lars Wahlin
Appl. Phys. Letters, 12:339–341 (1968)

Ion implantation of CdS
F. Chernow, G. Eldridge, G. Ruse, and L. Wahlin
(Univ. of Colorado, Boulder) Ion implantation session at AIME conf. on preparation and properties of electronic materials; optical and nuclear radiation, Chicago (August 12-14 1968)

Ion implantation of CdTe
J. P. Donnelly and A. G. Foyt
ESD-TR-67-562, Lincoln Laboratory, Massachusetts Institute of Technology (January 1968), p. 9

Type conversion and p-n junction in n-CdTe produced by ion implantation
J. P. Donnelly, A. G. Foyt, E. D. Hinkley, J. O. Dimmock, and W. T. Lindley
Bull. Am. Phys. Soc., 13:376 (1968)

Type conversion and p-n junctions in n-cadmium telluride produced by ion implantation
J. P. Donnelly, A. G. Foyt, E. D. Hinkley, W. T. Lindley, and J. O. Dimmock
Appl. Phys. Letters, 12(9):303–305 (1968)

Formation of $Hg_{1-x}Cd_xTe$ by Hg-ion bombardment of CdTe single crystals
Norman A. Foss
J. Appl. Phys., 39:6029 (1968)

Effect of ion bombardment on thin cadmium telluride films
G. A. Kachurin, A. E. Gorodetskii, V. M. Zelevinskaya, and L. S. Smirnov
Fiz. Tekhn. Poluprovod., 1(9):1427–1428 (1967)
Sov. Phys. — Semicond., 1(9):1187–1189 (1968)

Phosphorus ion implantation in cadmium sulfide
James Thomas Mitchell
Ph. D. thesis, Ohio State University, Columbus (1968), 152 pp.
University Microfilms, Ann Arbor, Mich., Order No. 69-4940

Bombardment of thin CdTe films with indium ions
G. A. Kachurin, A. E. Gorodetskii, Yu. V. Lobutets, and L. S. Smirnov
Sov. Phys.–Solid State, 9:375 (1967)

Doping cadmium telluride thin layers by ion bombardment
G. A. Kachurin, A. E. Gorodetskii, and L. S. Smirnov
Radiata. Fiz. Nemetal. Krist., Tr. Soveshch., Kiev, 1965 (Pub. 1967), Electron Paramagneti, 262-265 (I. D. Konozenko, ed.), Izd. "Naukova Dumka," Kiev, USSR
CONF-650995-1
Translation 7 pp. + Contents

Luminescence in hexagonal zinc selenide crystals
W. Y. Liany and A. D. Yoffe
Phil. Mag., 16:1153 (1967)
Heavy Mn ion bombardment

Irradiation of thin layers of CdTe by cadmium ions
L. S. Smirnov, G. A. Kachurin, and A. E. Gorodetskii
Fiz. Tverd. Tela, 9(3):719–722 (1967)
Sov. Phys. — Solid State, 9(3):564–566 (1967)

## 18.d. III—V Compounds

Ion-implantation induced optical absorption edge shifts in GaP
J. E. Davey, T. Pankey, P. R. Malmberg, and W. H. Lucke
Appl. Phys. Letters, 17(8):323–325 (1970)

Anneal behavior of defects in ion-implanted GaAs diodes
R. G. Hunsperger and O. J. Marsh
Met. Trans., 1:603–607 (1970)
AFCRL-67-0123, Final report, Contract AF 19(628)-4970, January, 1967
Zn and S in GaAs

Ion implantation in InAs and InSb (and the fabrication of mosaic i.r. detectors)
P. J. McNally
Radiation Effects, 6:149-153 (1970)

Ion Implantation of Bismuth into GaP. I. Photoluminescence
J. L. Merz, L. C. Feldman, and E. A. Sadowki
Radiation Effects, 6:285-291 (1970)

Properties of ion-implanted silicon, sulfur, and carbon in gallium arsenide
J. D. Sansbury
Thesis, Center for Materials Research, Stanford Univ., Aug. 1970

Effect of ion-implantation damage on the optical reflection spectrum of gallium arsenide
G. A. Shifrin and R. G. Hunsperger
Appl. Phys. Letters, 17(7):274–276 (1970)

Measurement of ion implantation lattice damage in (111) GaAs using the scanning electron microscope
E. D. Wolf and R. G. Hunsperger
Appl. Phys. Letters, 16(12):526 (June 15, 1970)
Coates–Kikuchi pattern degradation

Efficient doping of GaAs by $Se^+$ ion implantation
A. G. Foyt, J. P. Donnelly, and W. T. Lindley
Bull. Am. Phys. Soc., 14:372-374 (1969)

Conversion of GaAs to $Ga_{1-x}Al_xAs$ by implantation of $Al^+$ ions
R. G. Hunsperger and O. J. Marsh
Bull. Am. Phys. Soc., 14:273 (1969)

Electrical properties of zinc and cadmium-ion-implanted layers in gallium arsenide
R. G. Hunsperger and O. J. Marsh
J. Electrochem. Soc., 116:488-492 (1969)

Ion-implanted GaAs avalanche photodiodes
W. T. Lindley, J. P. Donnelly, A. G. Foyt, and R. J. Phelan, Jr.
Bull. Am. Phys. Soc., 14:369 (1969)

Ion-implanted mosaic infrared detectors
P. J. McNally and W. J. King
(Ion Physics Corp., Burlington, Mass.), 16th National IRIS
Meeting, Ft. Monmouth, N. J., May 1968, Proceedings of
the Infrared Information Symposia, NAVSO P-2315, Vol. 13,
No. 1 (April 1969)
InAs and InSb

Properties of ion-implanted GaAs diodes
P. E. Roughan and K. E. Manchester
J. Electrochem. Soc., 116:278-279 (1969)

Conductivity and Hall mobility of ion-implanted
silicon in semi-insulating gallium arsenide
J. D. Sansbury, J. F. Gibbons, et al.
Appl. Phys. Letters, 14:311-313 (1969)

The presence of deep levels in ion-implanted
junctions
R. G. Hunsperger and O. J. Marsh
Appl. Phys. Letters, 13:295 (1968)
Zn in GaAs

Zn and Te implantation into GaAs
J. W. Mayer, O. J. Marsh, R. Mankarious, and R. Bower
J. Appl. Phys., 38:1975-1976 (1967)

Implantation of zinc into gallium arsenide
J. B. Schroeder and H. D. Dieselman
Proc. IEEE, 55:125 (1967)

Mass analysis of ion beams from a low-voltage
spark ion source for ion-implantation doping of
semiconductors
R. G. Wilson and D. M. Jamba
J. Appl. Phys., 38:1976 (1967)
GaAs

The doping of semiconductors by ion bombard-
ment
J. F. Gibbons, J. L. Moll, and N. I. Meyer
Nucl. Instr. and Methods, 38:165 (1965)
Nd in GaAs

## 18.e. Miscellaneous Compounds

Penetration depths of ions in lithium fluoride,
zinc sulfide and cesium iodide
F. Fehsenfeld and A. Scharmann
Z. Phys., 230:435-442 (1970)

X-ray diffraction of ion-bombarded diamonds
M. I. Guseva, L. M. Ershova, K. V. Kiseleva, V. V.
Krasnopevtsev, and Yu. V. Milyutin
Kristallografiya, 15(3):523-527 (1970)
Sov. Phys. – Cryst., 15(3):441-444 (1970)

Ion implantation
K. Kandiah
Engineer., 209:21-24 (1970)
Doping of metal oxide semiconductors

Implantation of 10-80 keV lithium ions in dia-
mond
V. V. Krasnopevtsev and Yu. V. Milyutin
Fiz. Tekhn. Poluprovod., 4(5):837-846 (1970)
Sov. Phys. — Semicond., 4(5):709-716 (1970)

Resistive layers formed by ion implantation
into metal films
J. G. Perkins and L. E. Collins
Thin Solid Films, 5:R59-R62 (1970)

Ion implantation in diamonds
R. O. Carlson
Trans. Met. Soc. AIME, 245:483 (1969)

Diodes in silicon carbide by ion implantation
H. L. Dunlap and O. J. Marsh
Appl. Phys. Letters, 15(10):311 (Nov. 1969)

Semiconductors produced by doping oxide-glas-
ses with Ir, Pd, Rh, or Ru
C. C. Sartain, W. D. Ryden, and A. W. Lawson
The International Conference on Amorphous and Liquid
Semiconductors (24-27 Sept. 1969) Cavendish Laboratory,
Cambridge, England

Distribution and range of boron and phosphorus
ions used in bombardment of $SiO_2$
V. G. Volod'ko, E. I. Zorin, P. V. Pavlov, and D. I. Tetel'-
baum
Fiz. Tverd. Tela, 10(4):1048-1052 (1968)
Sov. Phys. – Solid State, 10(4):828-831 (1968)

Penetration of Li atoms with energies of 2-11
keV into SiC single crystals
V. V. Makarov and N. N. Petrov
Izv. Akad. Nauk SSSR, Ser. Fiz., 30(5):890-891 (1966)

Semiconducting diamonds produced by ion bom-
bardment
V. S. Vavilov, M. I. Guseva, E. A. Konorova, V. V.
Krasnopevtsev, V. F. Sergienko, and V. V. Tutov
Fiz. Tverd. Tela, 8(6):1964-1965 (1966)
Sov. Phys. — Solid State, 8(6):1560-1561 (1966)

Electronic Properties of Diamonds
F. C. Champion
Plenum Press, New York (1963), 132 pp.
Includes: The x-ray diffraction image topography and optical
interference pattern in diamond layers after ion bom-
bardment, by J. R. Parsons and C. W. Hoelke